CYNGOR
CEFN GWLAD
CYMRU

COUNTRYSIDE
COUNCIL
FOR WALES

K. Wöhrmann S.K. Jain (Eds.)

Population Biology

Ecological and Evolutionary Viewpoints

With 86 Figures and 69 Tables

Springer-Verlag Berlin Heidelberg New York
London Paris Tokyo Hong Kong

Professor Dr. K. Wöhrmann
Institut für Biologie II
Universität Tübingen
Auf der Morgenstelle 28
7400 Tübingen, FRG

Professor Dr. S.K. Jain
Department of Agronomy and Range Sciences
University of California
Davies, CA 95616, USA

ISBN 3-540-50802-3 Springer-Verlag Berlin Heidelberg New York
ISBN 0-387-50802-3 Springer-Verlag New York Berlin Heidelberg

Library of Congress Cataloging-in-Publication Data

Population biology: ecological and evolutionary viewpoints/K. Wöhrmann, S.K. Jain (eds.). p. cm. Includes bibliographical references.
ISBN 0-387-50802-3 (U.S.: alk. paper) 1. Population biology. 2. Ecology . 3. Evolution. I. Wöhrmann, K. (Klaus), 1928- . II. Jain, Subodh K., 1934- .
QH352.P574 1990 574.5′248--dc20

Typesetting by Thomson Press (India) Limited, New Delhi
Printing: Druckhaus Beltz, Hemsbach/Bergstrasse
Binding: Schäffer GmbH & CoKG, Grünstadt
2131/3145-543210 Printed on acid-free paper

A tribute to our late friend and colleague, Suresh Jayakar, in the words of William Wordsworth:

"Thou Shouldst have seemed a treasure house divine
O peaceful years; a chronicle of heaven-
Of all the sunbeams that did ever shine
The very sweetest had to thee been given."

Preface

This book, *Topics in Population Biology and Evolution*, was planned in a progression of volumes which originated from a recent series of symposia in this scientific area. Starting with a workshop in Berlin (1980), a group of European scientists met in Pavia, Italy (1982), Oberjoch, West Germany (1983), Sonderborg, Denmark (1985), Woudschoten, Netherlands (1986), and Barcelona, Spain (1988), to discuss with colleagues from overseas various problems of population biology and evolution. The meetings found expression in several symposia volumes: *Evolution and the Genetics of Populations*, eds. S.D. Jayakar and L. Zonta, Supl. Atti. Ass. Genet. Ital. Vol. XXiX (1982); *Population Biology and Evolution*, eds. K. Wöhrmann and V. Loeschcke, Springer (1984); and *Population Genetics and Evolution*, ed. G. de Jong, Springer (1988). The published contributions to the symposia reflected each time an update of recent investigations on an advanced level.

Many participants at the Tübingen meeting, in particular, raised interesting questions on the population genetic treatment of Darwinian fitness and genotypic evolution vis-à-vis the ecological interest in the evolution of phenotypes and whole organisms as units of evolutionary study. Many presentations included empirically rich and new advances in evolutionary ecology. Two of the subsequent volumes (Loeschcke 1988; DeJong 1988) covered the developmental, quantitative, and population genetic issues especially dealing with the genotype-phenotype correspondence and various constraints on evolution. This volume, planned at the invitation of Springer-Verlag, is aimed to cover some basic theoretical and experimental topics in populations genetics and ecology at a more general level so as to become useful in teaching graduate students as well as for the postdoctorals and researchers in other related fields. Several chapters emphasize theory (Chaps. 2, 3, 4, 12, 13), a few others present experimental results (Chaps. 5, 6, 7, 11), one chapter is specifically focused on the statistical problems of estimating mating system parameters (Chap. 3), one chapter reviews optimality in a wide variety of well-chosen contexts (Chap. 10), and one chapter on speciation problems is included to remind us of the ecological aspects besides the recent predominant molecular and population genetic treatments of speciation (Chap. 7). Several chapters cover recent advances in population biology theory extensively (and are therefore longer) (Chaps. 8, 9, 11, 12, 13), but in general, concepts rather than encyclopedic reviews are emphasized here.

We feel extremely fortunate in having found authors of various chapters so cordial and enthusiastic about this collaborative project; in addition to their very fine contributions, their patience is also gratefully acknowledged. We owe the late Suresh Jayakar a special debt of gratitude as he worked during his serious illness to complete his contribution. Helpful comments and discussions from Dr. Jürgen Tomiuk, and editorial assistance by Mrs. Rettenmeier are gratefully acknowledged.

In January 1989

Tübingen and Davis,
Klaus Wöhrmann,
Subodh Jain

Contents

Chapter 5

Chapter 6

Contributors

N. BARTON — Department of Genetics and Biometry
University College London
4 Stephenson Way
London NW1 2HE
UK

F.B. CHRISTIANSEN — Institute of Ecology and Genetics
University of Aarhus
Ny Munkegade 550
8000 Aarhus
Denmark

A. CLARK — Department of Biology
Pennsylvania State University
University Park
Pennsylvania 16802
USA

H. DINGLE — Department of Entomology
University of California
Davis CA 95616
USA

P. HEDRICK — Department of Biology
Pennsylvania State University
University Park
Pennsylvania 16802
USA

S.K. JAIN — Department of Agronomy and Range Sciences
University of California
Davis CA 95616
USA

V. LOESCHCKE
Institute of Ecology and Genetics
University of Aarhus
Ny Munkegade 550
8000 Aarhus
Denmark

A. VAN NOORDWIJK
Institute of Zoology
University of Basel
Rheinsprung 9
CH-4051 Basel
Switzerland

P. SCHMID-HEMPEL
Institute of Zoology
University of Basel
Rheinsprung 9
CH-4051 Basel
Switzerland

R.S. SINGH
Department of Biology
McMaster University
Hamilton
Ontario
L8S 4K1
Canada

R.C. VRIJENHOEK
Department of Biological Sciences
Rutgers University
New Brunswick NJ 08903
USA

K. WÖHRMANN
Biological Institute
University of Tübingen
Auf der Morgenstelle 28
D-7400 Tübingen
FRG

L. ZONTA
Department of Genetics and
Microbiology
University of Pavia
27100 Pavia
Italy

Introduction

K. WÖHRMANN AND S. JAIN

Perhaps fascinated by the diversity of living organisms, humans have always been curious about the origin of this diversity. Mayr (1982a) traced the history of evolutionary thinking to suggest that "one must make a distinction between the acceptance of evolution and the adoption of a particular theory explaining its mechanism". Whereas Lamarck's (1809) *Philosophie Zoologique* perhaps signified the first step toward accepting evolution, it was Darwin's (1859) *Origin of Species* which provided the scholarly and persuasive thesis for gradual evolution and speciation under natural selection. The gradual improvements in geological and biosystematic records on one hand, and numerous detailed studies of variation within species on the other, have made Darwinism a powerful scientific theory. Dobzhansky (1973) made the often-cited statement: "nothing in biology makes sense except in the light of evolution". However, although paleontologists described the evolution of species on the basis of fossil records, and although we have a lot of information on the factors of evolution, this does not at all mean that we understand evolution down to the smallest detail. Because of the complexity and multiplicity of interacting factors, particularly in the areas of population genetics and ecology as treated in this Volume, many open questions still remain to be answered.

In the past and still today, problems of evolution have been investigated from different viewpoints, which arose from both historical reasons and the highly varied subject matter of the investigations. Paleontologists collected a great amount of fossil records in order to document the lineages for a great number of groups. On the basis of these findings, the still disputed hypothesis of macroevolution was formulated, which proposes that **evolution in geological time happened by saltations**, and not by continuous changes, as postulated by most population geneticists. Taxonomists use karyotype, morphology and physiological characters to answer questions of descent and relationship between species. A major question is the degree of which classification hierarchy represents evolutionary events. Today, new methods to analyze DNA, the carrier of genetic information, or enzymes and other proteins, the primary gene products, provide a more reliable basis for systematics and the analyses of relationships (phylogenetic trees) between species and genera. Moreover, other kinds of characters connected with reproduction, life histories, and ecology are now used, based on the

Darwinian ideas of natural selection and adaptations as the primary source of diversity in life. More attention is now being paid to more precise descriptions of morphological and physiological characters, their mode of inheritance, and the developmental processes that result in different phenotypes.

By the rediscovery of Mendel's laws, **genetic aspects** were introduced in the theory of evolution by population geneticists. Based on the brilliant contributions by R.A. Fisher, B.S. Haldane, and Sewall Wright, population genetics was based on theoretical considerations. For simplicity, the complexity of genotypes was modeled in terms of one or two loci. Two main features of an evolutionary process through natural selection, e.g., the requirement of variation among genotypic fitness and population improvement due to allelic frequency changes, were established in Fisher's Fundamental Theorem of Natural Selection; in parallel, the rate of increase in mean population fitness provided many general theorems in Wright's treatments. These are necessary simplifications because at the end of each generation, with the exception of asexual organisms, genotypes/phenotypes become extinct and new ones are formed during gamete formation and fusion. The effect of selection, genetic drift, mating systems, mutations, and migrations on the genotypic/phenotypic composition of a population could be described and estimated. Changes in allele frequencies observed in laboratory experiments and in natural populations could be attributed to such factors.

R.A. Fisher was accustomed to thinking in terms of very large populations, so that random fluctuations were considered less important than the selective processes. On the other hand, Sewall Wright's shifting balance model with a three-phase process was based on the roles of local directional selection, drift, and interdeme selection, respectively, thus emphasizing interactive systems. As Ewens (1979) noted, this debate sometimes far outran the mathematical rigor of their views.

A second group of population geneticists was engaged in investigations on natural populations to study the distribution of genes in populations and their change over time and space. To distinguish this field from theoretical population genetics, it is named **ecological genetics.** The description of evolution considering genetical components is called evolutionary genetics. Population genetic theory, starting logically with one large panmictic population, is expanded step by step by introducing departures from random mating, by modeling subdivided populations with small demes, and selection under heterogeneous environments, in which interacting competitors and natural enemies (two or more species in a community) are also selective forces. Likewise, population ecology theory is developed from single population models to multispecies systems in which the number of species and their relative abundances are treated. Also, theoretical population genetics has recently become highly productive with advances in molecular genetics, mathematical and statistical, using computers for modeling, and in connection with the ecological developments at the population level. We now have a vast knowledge of population structures and of inter- and intraspecies variation.

The population genetical approach, however, does not often take into account **ecological viewpoints.** Many population genetics textbooks under-emphasize ecological parameters such as population size, population growth (and therefore, density factors), age structure, life history components, and the interaction with other species such as host-parasite and predator-prey interactions. On the other hand, ecologists handle populations as a unit and do not often distinguish genotypes or phenotypes. Hence, there is no explicit way to study change in the genetic composition of a population in many of the theoretical ecology writings (May 1976; Roughgarden 1979). Both approaches taken singly provide only restricted information on the population's biology and therefore on evolution.

Several Cold Spring Symposia during the 1950's and two notable Symposia (Baker and Stebbins 1965; Lewontin 1968) initiated a synthesis of population genetics, ecology, and systematics in a new evolutionary context. Likewise, Sammeta and Levins (1970) expressed the need for integration of the different approaches into **population biology**. This covers all biological processes within a population and biotic and abiotic influences on the population, genotype-phenotype relationship, and heterogeneous environments. This is of importance, since the population is the basis (unit) of evolution. Although selection, drift, etc., do act on the individuals, only the community of individuals may respond to changing environments in time and space, and is the starting point of new species. Therefore, it makes sense to correlate the biology of a population to evolutionary processes. Of course, the proposal of Sammeta and Levins has not been realized yet in full; however, to an increasing extent this idea is becoming more and more realized.

Thus, **the science of evolution has itself experienced major evolutionary turns.** Since the first publication of Darwin, numerous hypotheses have been established to explain evolution, and have been again refuted, or in combination with an antithesis the root for a synthesis was established. For example, for a long time competition was favored to be of decisive importance for the origin of species. Today, however, this hypothesis is questioned in its generality and explanatory powers. Already Darwin drew up a list of characteristics which are suspected to be of no service or disservice to the species, as they were nonadaptive. In 1930–40, there was a period in which random drift effects were attributed to small population size and polymorphic characters were regarded as examples of nonadaptive random drift. For Sewall Wright, random drift was an aid to selection, and in small subdivided populations one allele might occur, and in combination with migration, increase to a high enough frequency to spread through the entire population. Mather's work on polygenic traits led him to develop stabilizing selection as a mechanism for both maintenance and flow of potential genetic variation, frequency-dependency, competition, facilitation, and disruptive selection were also considered in some detail. Only later, in the 1950's, selection was more and more favored to explain the observed polymorphism in natural populations, and traits used as paradigms for neutrality were described now as being selected. In parallel, Kimura (1968a) postulated the "neutral theory of molecular evolution", which was renamed in 1983 after the great controversy between the "selectionist" and the "neutralist" as the "mutation-random drift theory", which suggests that most of the mutations are selectively neutral enough that their fate in the population and behavior is determined mainly by random drift.

Geographical patterns of gene frequencies and, in some examples, temporal trends, provided evidence for selection (e.g., melanism in *Biston*), or drift (flower color in *Linanthus*; blood groups in man). Over-dominance (heterozygote advantage) became the first theoretical explanation of many conspicious polymorphisms; genetic homeostasis (better buffering of heterozygotes) was the theme of an important book (Lerner 1954). Ecological interest soon provided many frequency-dependency examples (Clarke 1975). However, Endler (1986) found only a few examples at the enzyme level from among numerous published reports on selection. Endler concluded that "heterozygote advantage and stabilizing selection are neither universal nor important mechanisms that maintain genetic variation". Today, several other mechanisms are believed to be responsible alone or in interaction with others. One of these is sexual selection. Already Darwin (1882) realized that natural selection cannot explain many sexually dimorphic traits observed in nature, because such traits would decrease the survival ability of the carriers. Therefore, he postulated the theory of sexual selection, explaining that traits which might even lead to a decreased survivorship can be selected for if they increase the ability of an animal to mate.

How do populations of the same species adapt to contrasting environments, and how do different species adapt to the same environment? There are several difficulties in obtaining answers. First, a currency of adaptedness needs to be standardized; is it mean reproductive rate, or wave of advance of a gene combination, or the relative occupancy of territory, or stability of population numbers? Second, it is rarely clear how to separate the adaptive traits from the nonadaptive ones. Phylogenies, developmental correlations, hitchhiking of linked factors, and other reasons give spurious adaptive associations.

Just as evolutionary ideas have been evolving, there is also a change in the **characteristics of interest**. During the past decades enzyme polymorphism were the center of attention in population genetics. The introduction of electrophoreses techniques made it possible to investigate the genetic variation within populations and its change over time and space. Enzymes were investigated on the basis of the "one enzyme-one gene" hypothesis, mostly separated from the rest of the genome. Recent interest has switched to the variation on the DNA level on the one hand, and back again to polygenic traits on the other. At this point, population geneticists' and ecologists' interest meet, because fitness-relevant characters, like life-cycle components, are mostly determined by an unknown number of genes. Methods developed from quantitative genetics to solve plant breeding problems are also useful and are adapted to special questions in ecological genetics.

Defining ecology as the study of distribution and abundance of individual species, Andrewarth and Birch (1954) drew focus on the population numbers, their cycling, and regulation. In fact, population regulation with and without density-dependent factors and natural selection were the subject of numerous writings during the 1960's and 1970's. More recently, life histories and population structures have become central issues. Thus, population issues in genetics of adaptation and ecology of regulation both require evolutionary thinking.

The biology of populations depends not only on intrinsic factors but also on the **interactions with the living environment**. These interactions are manifold. There exists a competition for resources (food or nesting place), predators search for prey, parasites use hosts as nutrition, and others mimic the behavior or the morphological structure of related species. Again, such relationships were already observed and described by Darwin. Some of these have evolved in mutual response to each other, i.e., they have coevolved. The term **coevolution** was first coined by Ehrlich and Raven (1964) without being given a clear definition. For this reason, it is used to describe a wide range of interactions. Some authors use this term already in the case of a variety of species as a group changing in response to a variety of other taxa. Some authors use the term in a more restricted sense, if only one species experiences changes in response to a variety of others. This is called "diffuse coevolution." Sometimes a lack in response is accepted. The most restricted definition, however, as given by Janzen (1980) is an "evolutionary response in a trait of individuals in one population to the change in the first" (Abrahamson 1989). Of course, such changes in the trait must include changes in genotypic constitution of the population involved.

Since Darwin, **competition** between species is considered as a driving force in the evolution of species. Competition includes the negative influence of fitness components which is due to the use of the same limited resources (exploitative compeitition) or due to interferences (interference competition), e.g., fight for nesting places, etc. In many cases it could be proved that species became extinct by competition; however, only recently it was called into question whether competition is a necessary condition for evolutionary processes, whether each significant selection should be competitive selection, and moreover, it is doubted that in general, niches in one site are responsible for the stability of population (Den Boer 1985; Jacobs 1985). Certainly, there is no general answer to these questions. A case-by-case discussion has to be held, considering other important interactions like predator-prey, migratory processes, or life history of the populations. Investigations on models may help to solve these problems. Models are mostly based on simplified assumptions but can give information on the conditions which allow speciation by competition.

This introduction, largely a historical review, suggests that (a) different approaches taken together help us to understand evolution better, (b) in the history of evolutionary sciences, research has focused on kinds of phenotypic traits and responses to environments, and (c) last but not least, there is intensive discussion (too many debates and arguments?) on the relative importance of evolutionary mechanisms. It cannot be expected that only one strategy, e.g., drift, heterozygote advantage, or competition, was pursued during the evolution of species and ecosystems. These facts were kept in mind when the present book was planned. Evolution will be dealt with and investigated from several selected viewpoints of different fields of sciences. The importance of the biology of populations to evolutionary events will especially be considered. After a brief description of the genetic basis (Chap. 1), evolution is treated from the viewpoint of population genetics (Chaps. 2–4). Mendelian populations are considered mostly theoretically to focus the reader's attention on fitness components and selection within populations, the implication of mating systems to the genetic variability, and finally, to discuss population structure in the light of evolution. Chapters 5 and 6 are concerned with ecological genetics, with natural populations of asexual systems, and with the genetic variation and evidence for selection in plant populations. The evolution of species is discussed by means of a case study in *Drosophila* (Chap. 7). The evolution of life history traits receives considerable attention (Chaps. 8 and 9). It involves the study of the natural history, population biology, ecology, and physiology of organisms in order to understand the nature of parameters such as clutch size, life expectancy, and mortality. Such characters are of a quantitative nature and recent new analysis is provided by the methods of quantative genetics which were first developed for plant and animal breeding processes and are now adapted to the research on natural populations. In Chapter 10 also conceptional advances will be made by introducing such ideas as optimality theory in order to provide a new framework for analysis in behavior, ecology, and life history strategies. Finally, in Chapters 11–13, the interaction between populations from a community perspective is treated. To focus on the importance of genetics in evolution, examples of coevolution are given on the level of trophic interactions and of competition between species.

Investigations on the genetics and ecology of populations, their interrelationships, and their interaction with the physical and biotic environments, add to our understanding of the pattern and process of evolution. We envisage a pluralistic view of evolution, so that advances in molecular biology leading to a better description of genes and gene products on one hand, ranging to the problem of speciation based on ecological models of niche diversity on the other, all fit into a continuous spectrum of evolutionary research interests.

1 Genetic Variation: Prerequisite and Consequence of Evolution

K. WÖHRMANN

Introduction

Changes in the composition of a community of organisms are only possible if there are differences between the characteristics of its members. All traits have a genetical basis. Genetic variation within a species or a population is therefore a necessary prerequisite for a short-term or a long-term change. The information for trait development is stored in the DNA (deoxyribonucleic acid), and in the determination of such a character an unknown number of genes is always involved. Various factors are known to cause changes in this information: By mutation new information is created and by recombination numerous gene combinations are generated in each generation, and lead to a variety of different types within a population. Mutation rates and recombination rates themselves are a matter of evolution. Gene amplification, doubling of whole chromosomes or even genomes increase the total amount of DNA of the organisms and extend in this way the potential genetic variability. The genetic information is expressed in interaction with the environment. Some traits reveal a large phenotypic variation in dependence on the environment, others exhibit only a constant phenotype. Evolutionary forces act on the phenotype. Investigations on the evolution of organisms considering only phenotypic variation may describe evolutionary changes, do not, however, lead to an understanding of the processes involved.

Therefore, in this chapter an introduction into the genetic basis of evolutionary processes will be given, considering the concept of genetic code, kinds and origin of mutations, genetic recombination, genome size, gene activity in terms of structural and regulatory roles, and polygenic systems. This is to introduce the genetic basis of variation from molecular to organismic (phenotypic) level. Evolutionary genetics deals with origin and variation of all levels, including the adaptive and nonadaptive changes in the levels and organization of variation among populations and related species. Of course, in this chapter the genetic basis cannot be discussed down to the smallest detail, and therefore the reader is referred to the textbooks, e.g., Strickberger (1985) and Lewin (1987).

DNA the Basis of Inheritance

The development and function of organisms are determined and regulated by the deoxyribonucleic acid (DNA), with the exception of some viruses, where, instead of DNA, ribonucleic acid (RNA) is used. As early as in 1869, Miescher had extracted and described DNA. Its importance for the inheritance as the carrier of genetic information, however, was recognized only very much later by Avery et al. (1944).

The monomeric components of the nucleic acids are nucleotides, which consist of sugar (ribose in RNA, 2-deoxyribose in DNA), a phosphate group, and a heterocyclic base (pyrimidines or purines). The purines are adenine and guanine, the pyrimidines are cytosine and thymine. In RNA the last one is replaced by uracil. Investigations of DNA structure have revealed that DNA consists of two nucleotide chains, the sugar-phosphate axes, which are connected by hydrogen bridges between purines (adenine, guanine) and pyramidines (cytosine, thymine). Only thymine (T) can pair with adenine (A), and guanine (G) with cytosine (C) (Fig. 1). One amino acid is coded by three consecutive nucleotides in the DNA chain, called a codon. One gene consists of a series of codons which are read from one starting point to its termination point. Thus the structure of DNA not only allows storing information, but, in combination with the ability for replication of this code, the genetic information can be inherited from generation to generation.

The 20 known amino acids are each coded by the combination of three nucleotids (Table 1). The third base is frequently of no importance for the determination of the amino acid. For example, leucine is coded by combinations of CUU, CUC, CUA, and CUG. In fact, only 21 out of 64 possible combinations are realized. However, this is enough to produce an enormous variety within the DNA. Consider an informational segment for a structural gene of 600 nucleotides with 200 nonoverlapping groups of three bases. The calculation of possible different neighboring base sequences in this chain results in $21^{200} = 10^{264}$. This is an immense number!

The first knowledge of molecular genetics was gained for bacteria. Nowadays we know that the principle of the genetic code is realized in all organisms, in prokaryotes as well as in eukaryotes. However, there are fundamental differences in the structure of these two cells. Cells of eukaryotes have organelles containing

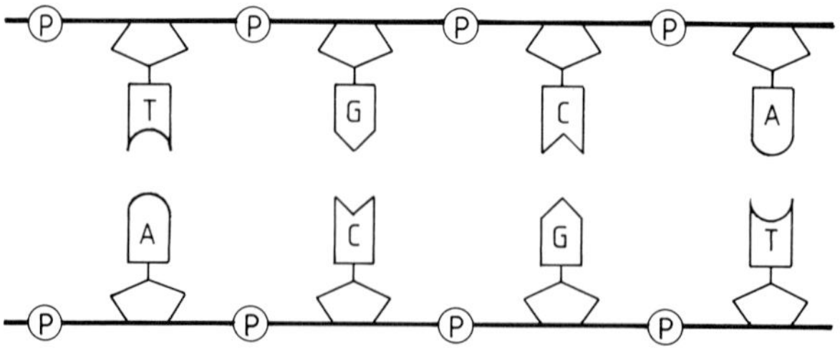

Fig. 1. Part of a DNA strand

Table 1. The genetic code, showing the amino acids or the termination signals specified by each of the 64 nucleotid triplets (codons) in messenger RNA (A: adenine, C: cytosine, G: guanine, U: uracil). The abbreviations for amino acids are *Ala*: alanine, *Arg*: arginine, *Asn*: asparagine, *Asp*: aspartic acid, *Cys*: cysteine, *Gly*: glycine, *Glu*: glutamic acid, *Gln*: glutamine, *His*: histidine, *Ile*: isoleucine, *Leu*: leucine, *Lys*: lysine, *Met*: methionine, *Phe*: phenylalanine, *Pro*: proline, *Ser*: serine, *Thr*: threonine, *Tyr*: tyrosine, *Trp*: trythophan, and *Val*: valine

First Letter	Second Letter: U	Second Letter: C	Second Letter: A	Second Letter: G	Third Letter
U	UUU Phe	UCU Ser	UAU Thr	UGU Cys	U
	UUC Phe	UCC Ser	UAC Thr	UGC Cys	C
	UUA Leu	UCA Ser	UAA Stop	UGA Stop	A
	UUG Leu	UCG Ser	UAG Stop	UGG Trp	G
C	CUU Leu	CCU Pro	CAU His	CGU Arg	U
	CUC Leu	CCC Pro	CAC His	CGC Arg	C
	CUA Leu	CCA Pro	CAA Gln	CGA Arg	A
	CUG Leu	CCG Pro	CAG Gln	CGG Arg	G
A	AUU Ile	ACU Thr	AAU Asn	AGU Ser	U
	AUC Ile	ACC Thr	AAC Asn	AGC Ser	C
	AUA Ile	ACA Thr	AAA Lys	AGA Arg	A
	AUG Met	ACG Thr	AAG Lys	AGG Arg	G
G	GUU Val	GCU Ala	GAU Asp	GGU Gly	U
	GUC Val	GCC Ala	GAC Asp	GGC Gly	C
	GUA Val	GCA Ala	GAA Glu	GGA Gly	A
	GUG Val	GCG Ala	GAG Glu	GGG Gly	G

DNA (nucleus, mitochondria, plastids, etc.), whereas these are not present in prokaryotes. At first, the most interesting organelle is the nucleus, in which the chromosomes are enclosed. The chromosome number is typically constant for each species. The sum of all genetic information on the chromosomes is called the genome. The content of DNA (genome size) varies considerably from species to species.

The **synthesis of proteins** takes place in the cell plasma (Fig. 2). For this, the DNA information is transcribed into the messenger RNA (mRNA) and transported to the place of protein synthesis in the plasma (transcription). The genetic code (or the message) is now translated into proteins (translation). The proteins catalyze, as enzymes, metabolic processes within cells, control the development of traits or form, as structural proteins, the protein-containing cell structures. Additionally, tRNA (transfer RNA) and rRNA (ribosomal RNA) are transcribed and, by this, ribonucleins are produced which are part of the structural machinery involved in the protein synthesis.

A particular sequence of nucleotides along a molecule of DNA (RNA in virus) which represents a functional unit is called a **gene**. In general, however, genes are very complex and exceedingly difficult to define. The idea that a "gene" codes for

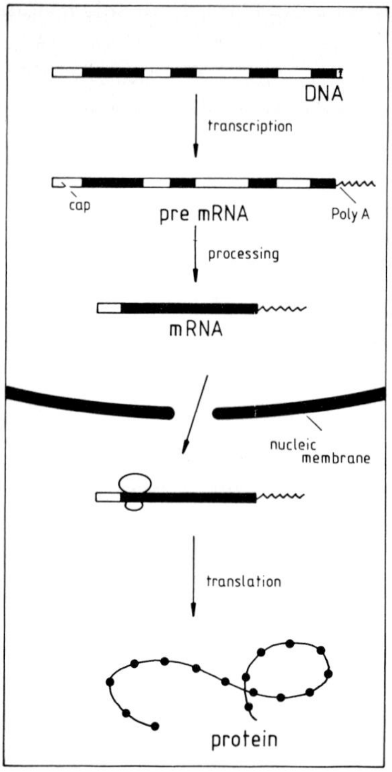

Fig. 2. Schematic representation of the processes of transcription and translation

an enzyme led to the "one-gene-one-enzyme" hypothesis which in its extension is called the "one-gene-one-polypeptide" hypothesis. Genes can be divided into those which code for polypeptides (structural genes) and those which are transcribed into tRNA, but are not translated into proteins (rRNA), and finally, those whose functional significance does not demand that they are transcribed at all.

Genes (coded by base sequences in the DNA) and primary gene products (amino acids and proteins) have provided many exciting research methods for understanding the way of inheritance and manipulation of genetic material.

Factors Causing Variation in DNA

Genetic variation is a prerequisite as well as a consequence of evolution. Several mechanisms are known which can cause changes in the base sequence and therefore in the genetic information of individuals and consequently within populations or species. In principle, mutations can occur in each cell of an organism, but they are of evolutionary relevance only if they are established in germ lines. Somatic mutations are passed on to the next generation only in asexually reproducing organisms. Mutations involving changes of single base pairs are called point mutations, which might be caused by endogenic or exogenic factors. Endogenic mutations are caused by mistakes during translation of DNA or during replication of DNA. Permanent changes in the genetic code become possible and are inherited from generation to generation.

A substitution of a purine for a purine (A–G) or pyrimidine for a pyrimidine (C–T) is called a transition, and a transversion is the replacement of a pyrimidine by a purine or vice versa (A or G to C or T). Transition agents are unable to revert transversion mutants and transversion agents do not revert transition mutants. There is experimental and theoretical evidence that transitions occur more frequently than transversions, as could be shown in hemoglobins (Vogel and Kopun 1977; Gojobori et al. 1982).

As already mentioned above, not all changes in base sequences become recognizable, e.g., a mutation in the third position of the codon for the amino acid valin (GUU, GUC, GUA GUG) does not affect the production of valin; the mutated codon still specifies valin. Even base substitutions resulting in an exchange of one amino acid by another must not result in changes in the function of the enzyme in question. This is particularly valid when the amino acids involved are of similar structure. Base substitutions with no effect on amino acids are called silent mutations.

Visible mutations are induced, for example, by a base substitution which causes an amino acid substitution within an active center of a protein (missense mutations) or results in a terminal signal (nonsense mutations). The latter leads to an interruption of the synthesis of the polypeptide chain and, therefore, to a loss of function. A loss of function is also induced by the insertion of an additional base pair (insertion) or the loss of base pairs (deletion). In such cases, the code will be read in a wrong frame, starting at the mutation point (frameshift).

Besides reading mistakes during replication, **recombination** between nonhomologous sequences also might cause heritable changes. The result of such a nonreciprocal recombination is the loss of nucleotide sequences, which is called back mutation or deletion. Such mutations cannot be reverted, in contrast to point mutations. Not all the mutations which are induced on the molecular level will be transmitted to the next generation. Mistakes in reading during replication may become corrected to a certain extent by specific repair enzymes.

Besides the natural instability of translation and transcribtion mechanisms, exogenic factors can lead to mutations. It is well known that changes in the base sequence may also be induced by **mutagenic substances** (e.g., UV-rays, X-rays or mutagenic chemicals). UV-rays cause the formation of thymine dimers within the base sequences and disrupt the reading of sequences in the correct manner, resulting in the above-mentioned frameshift. The chemical mutagen, bromuracil, being a base analog of thymine, cannot be distinguished by the cell. When bromuracil occurs in the very rare eno-constitution and matches with guanine, then during the next replication cytosine will be followed by guanine which results in the nucleotide pair G–C instead of A–T.

Table 2. Mutation rates of specific genes in various organisms. (After Dobzhansky 1970)

Organism and trait	Mutations per cell or gamete
Bacteriophage T2 (virus)	
Host range	3×10^{-9}
Lysis inhibition	1×10^{-8}
Escherichia coli (bacterium)	
Streptomycin resistance	4×10^{-10}
Lactose fermentation	2×10^{-7}
Zea mays (corn)	
Shrunken seeds	1×10^{-6}
Purple seeds	1×10^{-5}
Drosophila melanogaster (fruit fly)	
White eye	4×10^{-5}
Yellow body	1×10^{-4}
Homo sapiens (man)	
Aniridia (absence of iris)	5×10^{-6}
Hemophilia A	3×10^{-5}
Neurofibromatosis (tumor of nervous tissue)	2×10^{-4}

Ionizing rays (e.g., X-rays) supply energy to the molecules which can be destroyed. The results are serious changes: point mutations, deletions, chromosome breakage, etc. Such mutations might be repaired to a certain extent by specified repair systems. The mutation frequency varies between DNA segments (Table 2). There are sites on chromosomes which are characterized by a high mutation rate (hot spots).

Mutations in the widest sense lead to diversity between homologous sequences, and consequently to diversity between genes, individuals, and populations. An impressive example is given by the experiment of Kreitman (1983), who collected 11 individual genes on four continents from *Drosophila melanogaster*. He investigated the complete sequence at the ADH-locus, which is polymorphic with two alleles (ADH-f, and ADH-s) producing different activities. Five of the sampled individuals were from the ADH-f type and six from the ADH-s type. The molecular differences between both alleles are based on a single nucleotide substitution in the amino acid sequence. In addition, 42 other sites varied in the entire sequence, which was on average 2721 base pairs (bp) long. Several sites varied in their sequence length, indicating deletions and insertions. The results revealed that the ADH-f allele is less variable than the ADH-s allele. The kind of molecular similarities or dissimilarities, respectively, suggest that the f-allele arose recently from the s-allele and that there was only a short time to accumulate new mutations. This example illustrates impressively how genes become divergent. If such genes determine, e.g., life history traits like mating behavior, sexual maturity time, etc., mating barriers may be established between individuals or groups of individuals within the population, resulting in sexual isolation, and leading possibly to the origin of new taxa (see Chap. 7).

Population Genetic Consequences of Transposable Elements

One of the most important observations in genetics during the last decade is the existence of "transposable elements" in prokaryotes as well as in eukaryotes. These elements are able to self-replicate and are inserted in other chromosome segments after free replication. Several families of such transposable elements have been found in different species. All these elements are small, have terminal repeats and are found at different local sites in different genomes. They are not able to exist outside the genome. These elements permit the explanation of some phenomena which could not be understood by mechanisms of classical genetics. Such elements are mainly well described in *Zea mays, Drosophila* (Table 3), yeast and bacteria. The fact that they are able to reproduce themselves in the genome only led to the suggestion that transposons are "parasites" within the genome (host), and to the hypothesis of "selfish-DNA" (Doolittle and Sapienza 1980). The spreading within populations is guaranteed by transposition and recombination of the genetic material.

Little is known about the mechanisms of transposition within the genome, and its function is yet undiscovered in eukaryotes. The relevant gene products will still have to be identified. It is assumed that such elements have an important influence on gene expression. Recent investigations in *Drosophila*, however, have not supported this hypothesis, but indicate that insertion of elements into chromosomal sites is frequently accompanied by deleterious mutational effects, due to interferences with gene expression or the destruction of coding sequences of genes. This might often lead to a reduction in fitness of the host. An occasional benefit to the host may increase its fitness and may result in the spread and fixation of the elements. Examples are given by transposable elements involved in drug resistance and metal resistance in bacteria. Of population genetical interest are the rules which govern the frequencies of elements at chromosome sites, and the distribution of numbers of copies between individual sites (Charlesworth 1985).

In maize (*Zea mays*), results on unstable elements with relevance to population

Table 3. Transponsable elements in *Zea mays* and *Drosophila*. (After Lewin 1987)

Element	Copies genome	Length bp	Element	Copies genome	Length bp
	Zeamays		*Drosophilia*		
Ac	?	4563	Copia	20–60	5000
Mul	5	1567	FB	30	500–5000
Spm (En)	?	8287	P	50 or 0	500–2900

genetics are not yet available, but in *Drosophila* such investigations have been reported. Such elements are able to revert mutations to wild type or generate deletions of flanking material with an endpoint at the original site of mutations. The best-characterized type of transposon in *Drosophila* is **copia**. Its name reflects the presence of a large number of closely related sequences coding for abundant mRNA's. Copia is the paradigm for several other types of elements. The number of copia elements depends on the *D. melanogaster* strain, and varies between 20 and 60, and furthermore, copia elements show variability, although less in laboratory strains than in the natural populations. However, no estimates of the changes in the DNA of the elements exist. It is assumed that mobile elements in natural populations of *Drosophila* are of recent origin because in laboratory strains collected before 1950 no transposable elements were found.

Another family of transposons in *Drosphila* is called **FB-elements** (fold back). In spite of the variation in length of the inverted repeats, all members of the FB-family are homologous. This feature is explained by their structure, which consists of tandem copies of a simple-sequence DNA.

Smaller than FB-elements in size are **P-elements** (2.9 kb). They terminate in short inverted repeats and transpose only under certain conditions. The nature of the event might be an insertion, an excision or a chromosome break. In heterozygotes it may alter the expression of the alleles, leading to mosaics. The phenomenon of hybrid dysgenesis is correlated with the occurrence of P-elements. The reproduction of the organisms is disturbed by defects including chromosomal alterations, distorted segregation and sterility. P-elements affect only germ cells, and can occur in a large number in so-called P-males, but they are absent from chromosomes of M-flies. P-elements become active only in cytoplasm which is inherited from M-mothers (maternal contribution); but it is not simply a maternal effect, because it is inherited for more than one generation when there is a M-female in the line of descendants.

Investigations on P-elements were carried out mainly in natural populations of *D. melanogaster*. Wild strains collected from natural populations in the USA, Africa, and Western Europe are mainly P-strains. In Eastern Europe and Asia only M-strains have been found. It is assumed that P-elements have occur only a few years ago in North-American populations and since that time they have spread worldwide (Anxolabehere et al. 1986). With the exception of *D. melonogaster*, P-elements have not been found in other species of the *Melanogaster* group, but they were detected in the *Willistonii* group and in *D. subobscura*. On the basis of this information, a horizontal gene transfer between species has been assumed.

Evolution of Genome Size

The total amount of the genetic material (DNA) increases generally with the complexity of organisms (Fig. 3). The genome of a virus contains about 10^3 to 2×10^4 nucleotide pairs. On average, a bacterial genome has 4×10^6 base pairs.

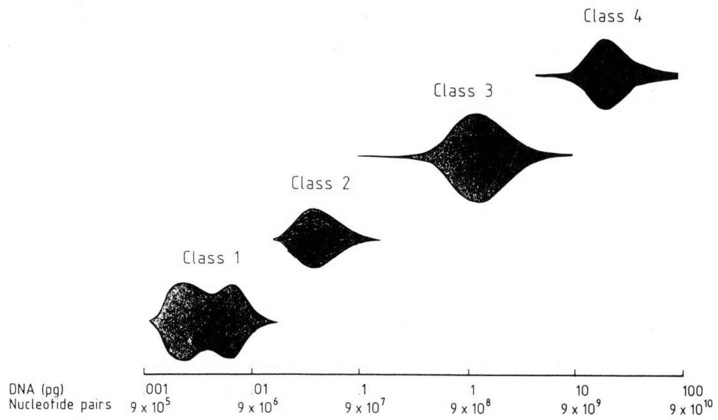

Fig. 3. Organisms classified according to their amount of DNA. *Class 1* bacteria; *class 2* fungi; *class 3* almost all animals, some plants; *class 4* many plants, salamanders, some non teleost fishes. (After Hinegardner 1976)

Among eukaryotes the fungi have already 10 to 22 millions, most of the animals and plants about 10^9, and the largest amount (10^{12}) is found in two multinucleated protozoa, *Chaos chaos* and *Urostyla caudata.* The DNA present in the genome is not totally organized in genes. The bacterial genome size is probably directly related to the number of genes, but in eukaryotes there is a vast amount of apparently nonexpressed DNA. The total amount of DNA sometimes is as high as 50,000-fold in excess of the protein coding requirements of the organism. Such estimations are based, for example, on the frequency of mutation sites and the number of bands on salivary gland chromosomes. The number of genes corresponds in this case to about 10% of the DNA measured. The same estimates have been also carried out in the other eukaryotes, and again, the amount of coding DNA is just a minor fraction of the total DNA.

Gene Families

The increase of the genome or the amount of DNA during evolution can be caused by several factors. One of these is unequal crossing-over or unequal gene conversion. If recombination does occur between nonallelic genes, nonreciprocal recombinants result of which one carries the duplicated gene and the other one a deletion. This event may happen several times, leading to "gene families". Its members may be clustered together in tandem genes or may be dispersed over the genome, ranging from extremes of two relatively adjacent genes to thousands of identical genes. A dispersal through the genome must have arisen by translocation.

An example of families with highly variable members is the immunoglobulin genes, and those with uniform members are the histone genes. In the human genome a sequence exists carrying the 18S rRNA and 28S rRNA and a nontransposable spacer from which 400 copies are dispersed over the whole human genome.

Partial gene duplication may result in a gene coding for a new amino acid sequence which has been described in human haptoglobin variants (Smithies 1964). Such a process might also have been important in the evolution of highly internally duplicated structures of immunoglobulin heavy chains (Ohno 1970,1984). The Bar-locus of *Drosophila* is an example for the duplication of a whole set of loci. Degeneration of duplicated genes leads to non-functional pseudogenes, which have been described in several species (see, for example, Zimmer et al. 1980).

Multiduplication of genetic information may explain at least partially the discrepancy between the DNA content and the number of genes in higher organisms.

Polyploidy

In polyploid organisms, the chromosome number exceeds that of the corresponding diploid individuals. Polyploid organisms may occur by failures during the reduction of the chromosome number in the first division of the meiosis, by double fertilization of eggs or by failures during mitotic divisions of cells. By such mistakes the chromosome number will be doubled and autopolyploids result. Further doubling of the chromosome numbers leads to organisms with a higher number of chromosome sets. Autopolyploid races are found in several plant species, e.g., in *Galium uliginosum* (2n = 22, and 44), *Valeriana officinalis* (2n = 14, and 28), *Pimpinella saxifraga* (2n = 18,and 36), *Rumex acetosa* (2n = 28, 42, and 56) (Schwanitz 1974). Autopolyploids developed in the course of time into different species by accumulation of point and chromosome mutations. Unequal segregation of chromosomes yields unbalanced gametes with even or uneven chromosome numbers. Gamates with less than normal numbers are mostly sterile, whereas those with higher numbers form aneuploids. Aneuploidy is mostly linked with polyploidy because the disturbance of the genetic balance is obviously less serious if at least two copies of each chromosome are still present.

Assuming the hybridization of species, the multiplication of the chromosome sets allows a stabilization of the genome, because in the resulting alloploids the pairing between chromosomes is regulated and doubling allows preferential pairing of chromosomes derived from the same species. This type of genome mutation clearly played a major role in the evolution of flowering plants. A number of examples are realized in our cultivated plants, e.g., *Triticum aestivum*, *Avena sativa*, *Brassica* species, and *Solanum tuberosum*. Series of allopolyploids exist (e.g., wheat 2n = 14, 28, and 42 or *Chrysanthemum* 2n = 18, 36, 54, 72, and 90), at least within 30% of the genera of flowering plants. It is assumed that flowering plants with basic numbers higher than 12 evolved by former multiplications.

In contrast, in the animal kingdom, polyploids are very rare, because polyploidy disturbs the sex chromosome mechanism. White (1973) has assumed that, caused by the breeding system, selfing in animals is not possible, and therefore, the organism are eliminated. However, there are a few examples like the fish family Salmonidae (trout, salmon) and Catostomidae (suckers) which descended from ancestors that became tetraploid about 50 m.y. ago (Allendorf and Thorgaard 1984). Polypoloid animals are more likely parthenogenetic (e.g. *Artemia*, *Solenobia* and *Curculionida*) or hermaproditic (*Dugesia*).

White (1973) has further assumed that polyploidy allows larger variability in a population. In comparison to three genotypes (AA, Aa, aa) in diploids, five are possible in 4n-autotetraploids (AAAA, AAAa, AAaa, Aaaa, aaaa). The relationship between the possible homozygotes and heterozygotes is on the diploid level 1:1 and on the tetraploid level 7:1. More mutations can be accumulated in polyploids at the same locus. In such a manner polyploids might be successful during evolution because of their potential for enlarged storage of genetic variation. This was indicated by investigations on artifical polyploids. Artifical polyploids became economically important only if hybrids of different varieties were used for multiplication. Introgression between two polyploids is more likely if both species have one genome in common (AABB and AACC). The genome A which belongs to both has a stabilizing effect in the case of newly arisen hybridization. However, the number of copies may very likely influence the fitness of an organism because of the variation in the quantity of the specific gene product (e.g., enzyme). If an overproduction reduces fitness, selection might result in dosage compensation and the activity of an enzyme can be lowered. This was observed in the tetraploid carp (*Cyprinus carpia*), in which the level of enzymes and RNA is regulated to the level of the related diploid species (Leipold and Schmidtke 1982).

Gene Activity

Gene Regulation

In principle two kinds of genes can be distinguished. Firstly, structural genes which code for enzymatic and structural functions, and also include genes for the rRNA and tRNA.The predominant number of genes in bacteria is of this category. The second group contains regulatory genes. A regulatory gene in the broadest sense is a gene which regulates the activity of structural genes or a gene which codes for a protein which functions alone or in combination with a corepressor in regulating the genetic transcription of the structural gene in an operon. In bacteria there are some very well-investigated examples for regulatory processes, whereas in eukaryotes the molecular basis of such a process is mostly unknown.

The regulation of an enzyme might be of importance for several reasons. In a varying environment it is important to an organism to react immediately on changing external factors by regulation of the enzyme production depending on, e.g., nutrition supply. This is of exceptional importance, e.g., to bacteria, which are exposed to a high extent to varying enviroments. A continuous production of enzymes, although the substrate in question is not at its disposal, means loss of energy and is probably of disadvantage to the organism. Enzymes which are produced in dependence on requirements are called **inducible enzymes**. By mutation at the regulatory gene they become **constitutive enzymes**, of which the product is not adjustable. Regulation happens during transcription and results from the interaction between a regulatory macromolecule and a sequence of nucleid acid. The macromolecule is most commonly a protein.

Gene complement and gene structure are generally the same in all cells of a multicellular organism. However, not all genes must be active at the same time and in each cell. Differences between different tissues occur in gene activity, and in mammals **tissue-specific funtions** of enzymes are well known. During ontogeny different gene groups become active at different stages, leading to differentiation of tissues. Which gene become active depends not only on the molecules within the cell but also on the condition of the neighboring cells, and on the interaction of the cell with the enviroment.

A model for the regulation of genes in eukaryotes is proposed by Britton and Davidson (1969, 1971) and Davidson and Britton (1973) (Fig. 4). In this model sensor genes (S) respond to the presence of stimulating substances by induction of the integrator genes (I) which transcribe to produce RNA. This RNA activates

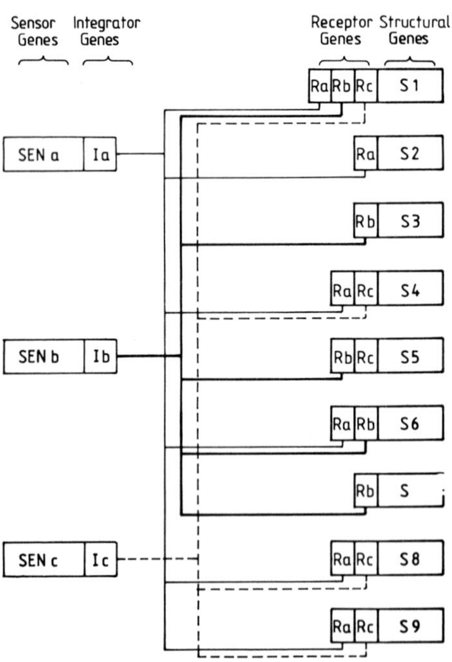

Fig. 4. Britten–Davidson model of gene regulation. *SEN*: sensor gene responds to the presence of particular inducing substances by activating adjoining integrators; *I*: integrator which transcribes to produce RNA; *R*: receptor, which is activated by RNA; *S*: structural gene which is caused by *R* to transcribe, eventually producing polypetide gene products. Several structural genes may be activated by a single sensor, and several sensors may be controlled by a master sensor. (After Britton and Davidson 1969, 1971)

receptors (R) which in turn cause structural genes to transcribe. Whole series of structural genes may become active by a single sensor. In such a system the gene functions may be changed by mutations within a gene or by changes in the sequences of genes, which occasionally result in altered patterns of gene expression. Just as a mutation in the structural gene influences the activity of the gene product, a mutation may occur in the regulatory gene. Such mutations can dramatically affect the production, though not the amino sequence, of specific groups. A mutation in the sensor genes or the integrator genes has an influence on several structural genes.

Neutrality

One of the most frequently discussed questions in the last decades is that of the importance of gene mutations (molecular substitutions) for the fitness of organisms (Kimura 1968a, see also Chaps. 2 and 4). Kimura has postulated that if the known rate of substitutions is extrapolated to the entire genome, Haldane's principle of the cost of natural selection could not be valid. He proposed that the evolution of most proteins, and therefore the genes coding for them, is for the most part due to chance. He created the "neutral theory", which assumes that for any gene a large proportion of mutations are harmful to the individual carrying the mutation, and because of their negative effect on the reproductive rate, these mutations will be rapidly eliminated from the population or kept in the population only in the order of magnitude of the mutation rates. The vast amount of changes in the proteins, however, is without importance to the fitness of the carrier and is called "selectively netural".

Since its first formulation, the neutral theory itself has evolved. Today, different alleles are no longer assumed to be strictly neutral and the theory has been renamed the "mutation, random drift theory" (Kimura 1983), which means that mutations are merely neutral enough with respect to natural selection for their fate to be determined merely by random drift. Evolution at the molecular level therefore consists for the most part of gradual replacement at one amino acid sequence by another functional equipment for the first. There might be favorable mutations, but they have only little effect on the total evolutionary rate of amino acid susbstitutions. This theory is hard to test experimentally, because the effect of one gene cannot be tested neglecting the other genes in the genome. Nevertheless, the theory is today widely accepted and makes it possible to explain most of the observed variation at the molecular level.

Gene Interaction

In diploid organisms each locus is present twice (homologous genes). New alleles originate by mutation at the loci which then have interact in a heterozygote condition. In the case of dominance, the phenotype of the heterozygote is identical with that of the homozygote. Dominance refers to the appearance of heterozygotes relative to the homozygotes. If the phenotype of the heterozygote is more similar to that of one homozygote than to the other, this is called partial dominance.

Most of the wild-type alleles are dominant over newly induced mutations. Despite the fact that alleles with a dominant effect sometimes occur, the question is raised as to the reason for these findings. Several hypotheses have been discussed to explain the evolution of dominance. (1) It was suggested that the recessive condition was caused by the absence of the normal allele. However, it was found in several examples that even deleterious alleles have dominant effects, and the presence-absence hypothesis was abandoned. (2) Fisher (1928) offered the hypothesis that a deleterious mutation would lead to intermediate fitness of the heterozygotes in relation to the wild homozygotes. In addition, modifier genes would change the expression of the heterozygotes toward that of the homozygotes. Continuous selection for such modifiers coding for high gene expression then would account for complete dominance. (3) The third hypothesis was proposed by Wright (1929), who proposed that the modifier genes would have major effects of their own, and their frequencies would be much more influenced by the selective advantage associated with their primary function than by their pleiotropic effects on the corresponding gene. Finally, the replacement of the allele by another in the population is considered. Assume a locus with two alleles which cause partial pigmentation in heterozygotes (A1A2). A third allele (A3) occurs, which leads to more intensive pigmentation than A1. A3A2 individuals are fully pigmented. In the case of strong selection for full pigmentation, the A1 allele will be replaced in the population and evolution of a higher degree of dominance will take place. However, today the evolution of dominance is still an unsolved problem and under discussion.

Some information on the change of dominance is given by investigations of Kettlewell (1973) on *carbonaria* forms of *Biston betularia*. Heterozygote melanic forms were phenotypically distinct from either homozygotes early in this period. After 80 years of selection, which has resulted in darker forms (industrial melanism), the heterozygotes became almost as dark as the homozygotes. This change in pigmentation may occur by allele substitution according to the last hypothesis. But matings between the British *carbonaria* forms and the more cryptic forms of closely related moth from Canada revealed that the British *carbonaria* was no longer dominant. This finding supports the second hypothesis.

Recombination

Despite strictly asexually propagating organisms, a specific genotype exists only for one generation. The unique gene combination will be eliminated by the death of the organism, and by forming gametes during meioses new gene combinations arise. Recombination of genes takes place firstly by crossing-over events between chromatids during pachytene in the first division, secondly by distribution of chromosomes during the first anaphase and thirdly during chromosome distribution in the second division. In the same direction mitotic crossing-over acts, which, however, occurs less frequently than meiotic crossing-over. All these events generate new gene combinations in each generation, and newly induced mutations will be combined with other genes in the genome.

The genes are organized on chromosomes (linkage groups) in a characteristic sequence which may be changed by chromosome mutations as duplications, inversions, and translocations. Such rearrangements may influence the gene expression, as mentioned already earlier.

On the other hand, the organization of genes in linkage groups allows inheritance of preferential combinations of genes. In some species, e.g., *Drosophila*, this tendency is increased by inversions as well as by the absence of crossing-over in the male sex. Asexuality and inbreeding systems common in different plants have the same effect. Summing up, antagonistic mechanisms exist keeping the formation of new genotypes and the breakdown of combination in balance. It is suggested that in constant environments constancy of gene combinations is of advantage, whereas in fluctuating environments variability is demanded (see Chap. 4).

Phenotype–Genotype Transforms and the Role of Polygenic Analyses of Variation

Genes individually associated with pronounced phenotypic effects are called **major genes** or oligogenes. Major genes control the production of discontinuous characters which are called qualitative characters. In contrast to this, **minor genes** or polygenes individually have a small effect. Major genes segregate clearly and are easily subject to Mendelian analysis, minor genes are not. The difference between major genes and minor genes is more or less arbitrary, because each character is determined always by a series of genes, and depends on allelic and nonallelic gene interactions. Polygenic traits are also called metric traits, because they are measurable on a continuous scale, like body weight. Meristic traits are also polygenic, like bristle numbers in *Drosophila*. The phenotypic expression of a polygenic character is subject to considerable modification by environmental influences. Biochemically there is no difference between major and minor genes. Both are involved in the production of enzymes or proteins. Special methods have been developed to investigate the inheritance of such quantitative traits (Chaps. 8 and 9).

The first indication that many genes might be involved in the determination of a trait was reported by Nilsson–Ehle (1909), who investigated several traits in wheat and crossed, for example, plants with white grains with those with red grains. The plants in the first generation produced red grain, too, but the color was much lighter. Selfing of these plants yielded in the F_2-generation different

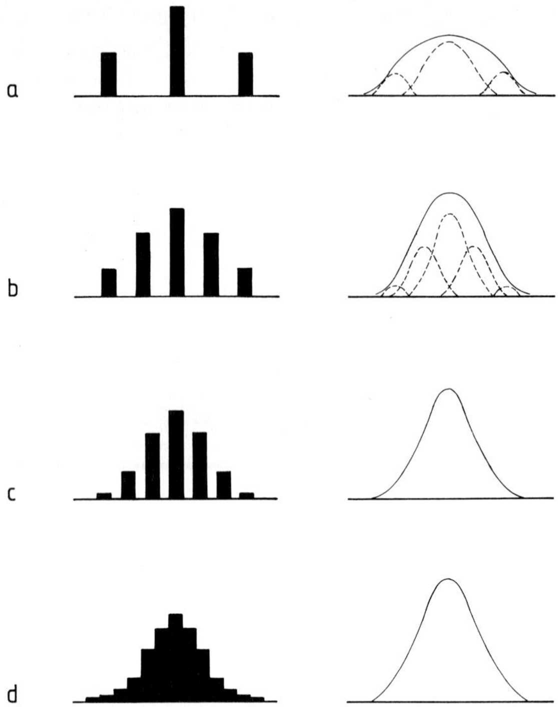

Fig. 5. Relative frequencies of genotypes in the F_2 of crosses between lines of plants differing in one (**a**), two (**b**) three (**c**) and six loci (**d**) (*left*). The distributions on the *right* side shows contributions of the environment to the phenotypic variation

grades of red. In order to explain this segregation pattern, Nilsson–Ehle proposed three loci responsible for the coloration. Only dominant alleles contribute a certain but equal amount to the trait. Under these assumption, the six color classes found in the second generation could be explained by the following genotypes: R1 R1 R2 R2 R3 R3, R1 R1 R2 R2 R3 r3, R1 R1 R2 R2 r3 r3, R1 R1 R2 r2 r3 r3, R1 R1 r2 r2 r3 r3, R1 r1 r2 r2 r3 r3, and r1 r1 r2 r2 r3 r3. The last genotype is colorless. In this example the contribution of each dominant allele can still be measured. Furthermore, the environment adds nothing or at least little to the expression of the trait. The influence of the environment may in fact mask the single gene action and incidentally lead to a continuous variation as can be seen from Fig. 5. This becomes of more importance with increasing number of genes and increasing influence of the environment.

Environmental influence on a metric trait was first demonstrated by Johannsen (1909), who selected different seeds of different size of a population of *Phaseolus vulgaris*. The progenies of the single seeds showed different distributions in seed size. However, in the following generations no further change in the distribution of each line could be found. Johannsen attributed the observed variation in each line to environmental factors only, and postulated the term "pure line".

The same individual can interact with the environment in different ways, producing quite different phenotypes. One most impressive example is the

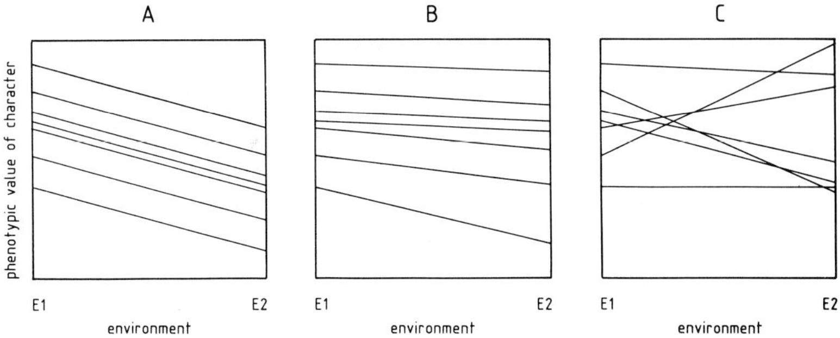

Fig. 6. Geographical representation of reaction norms. The mean value of a character state expressed by a given genotype in environment E1 is plotted on the *left* axis, the mean phenotype expressed in environment E2 is plotted on the *right* axis. **A** All genotypes respond similarly to a change in the environment. **B** The genotypes responds differently to the change in the environment, but there is no change in the rank order. **C** The genotypes respond differently to the environment and there is a change in the rank order. (After Via 1987)

variation of the leaves within the same plant of *Ranunculus aquaticus*: Leaves above the water surface are broad, flat, and multilobed. Leaves below the water are delicate and filamentously structured. The former are adapted to relative high sunlight intensity, the latter to use the diffuse underwater light in photosynthesis.

Genotypes may have different **reaction norms** in distinct environments. Firstly, consider a population which exhibits a normally distributed trait in a certain environment (E1 in Fig. 6A). A second environment (E2) may shift the distribution to a new average value, but with the same variance. In this case all genotypes respond equally to the change in the environment and thus no change in the rank order occurs. Something similar happens in the second case (Fig. 6B) where the genotypes respond to a change in the environment to a different extent. Although the rank order is still the same, the standard distribution is effected. In the third case the rank order is also changed by different responses of the different genotypes to the environment (Fig. 6c). The last two types of responses are called genotype-environment interactions.

Each genotype is able to respond within its reaction norm to changes in the environment. Woltereck (1909) defined a reaction norm as a set of all phenotypes that are displayed by a single genotype under different environmental conditions. The **phenotypic plasticity** is an environmentally produced change in the phenotypic expression of a genotype. Phenotypic plasticity has been shown for morphological, physiological, behavioral, and biochemical traits, and was evolved under natural selection. Via (1987) pointed out that phenotypic plasticity corresponds to populations in which there is natural selection toward different average phenotypic values in different environments, with gene flow between them. This in turn depends on the genetic variation in reaction norms available in the population. Therefore the study of evolution should consider adaptations to the environment. In Fig. 7 an informative example of the reaction norm in

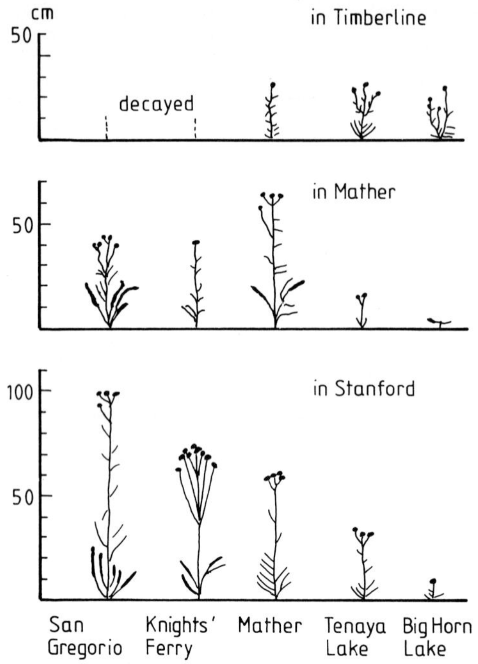

Fig. 7. Responses of *Achillea* clones originated from five different localities and grown at different altitudes. Stanford (sea level), Mather (4600 feet), and Timberline (10,000 feet). (After Clausen et al. 1948)

Achilea lanulosa is given. Several plants were collected at different sites, cloned and planted in locations with different environmental conditions. Different genotype-environment interactions become apparent.

Different traits exhibit a different amount of variation. It is generally suggested that traits related to the fitness of an organism are less variable than those which are not. Some traits are obviously identical in all individuals of a population. However, it was shown that, for example, the number of scutellar bristles of *Drosophila melanogaster*, which normally is four, could be decreased or increased, respectively, by selection under certain conditions. (Waddington 1960, 1957; Rendel 1967). This was an identification of existing genetic variability in the population with respect to this trait. The property of developmental pathways of obtaining a standard genotype in spite of genetic or environmental variation is called **canalization.** To expalin this observation it is assumed that a distribution of gene products (enzymes, proteins) corresponds to a distribution of phenotypes (Fig. 8). If each genotype corresponds to a certain phenotype, a linear gene-environmental factor/phenotype mapping function (GEPM), results (Scharloo 1987). The distribution may be shifted to higher or lower values by the action of major genes, the occurrence of mutations, by selection, or by environmental changes. The form of the mapping function is determined by developmental processes between gene product and phenotypic expression. If a threshold function is assumed for the GEPM, a bimodal distribution is transformed into a unimodal distribution. In the case of a sigmoid GEPM, the result of translation

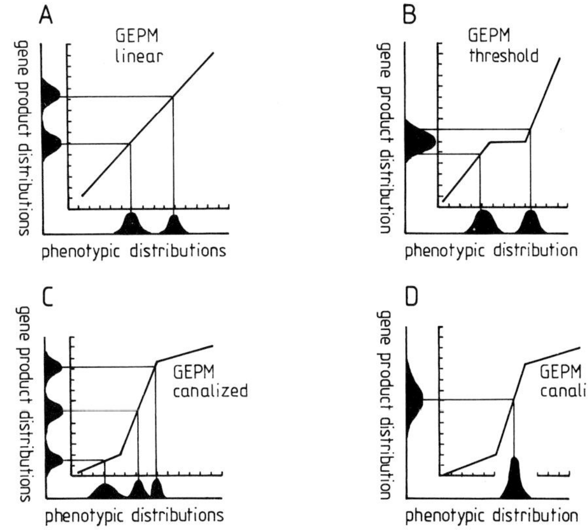

Fig. 8. Model for the effect of the shape of gene-environmental factor/phenotype mapping functions (GEPM's). On the ordinates normal distributions of gene products, e.g., bristle-forming or vein-forming substances, are considered. The variation is caused by genetic variation, variation of environmental factors and stochastic variation. The mean of the distribution can be shifted by a mutant, by artificial selection or by environmental factors. The normal distributions of the morphogenetic substance are translated by the mapping function in phenotypic distributions. The shape of the mapping function is determined by developmental processes between the gene product and phenotypic expression. When GEPM is linear (**A**) shifting the normal distribution will be reflected in a linear shift of the phenotypic distributions. When GEPM is a threshold function (**B**), unimodal frequency distributions of the gene products can result in bimodal phenotypic frequency distributions. When the GEPM is sigmoid (**C**), the distance between frequency distributions of the gene products is translated in relatively short distances in the steep part of the curve rather than beyond its range, and variability (**D**) is far smaller when the gene product distribution is translated in the steep part of the GEPM rather than in the other parts. (Scharloo 1987)

depends on the part of the curve. Gene products in the sigmoid part of the curve lead to phenotypic distributions with small variances; translations outside the steep slope result in distribution with larger variances. It becomes obvious from Fig. 8 (function C) that identical or similar phenotypes are determined by a great number of genotypes. The gene action in this system is buffered by properties of the developmental system. Besides bristle numbers, canalization of wing vein length in the *Drosophila* mutant ci^D was studied by Scharloo (1966, 1987). On the basis of this model, an explanation of the response to selection for higher or lower bristle numbers in *D. melanogaster scute* mutant is shown in Fig. 9. (see also Chap. 9).

Genotypic or genetic variability is a prerequisite, but also a consequence of the evolution of population and species. In 1958 Fisher stated the **fundamental theorem of natural selection** that the rate of increase in fitness of a population at any time is equal to its genetic variance in fitness at that time. (see also Chap. 2).

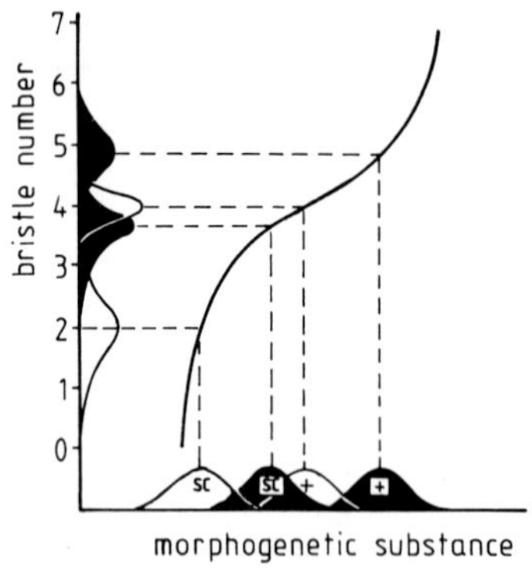

Fig. 9. Model for the formation of scutellar bristles in *Drosophila*. The number of the scutellars is supposed to be dependent on the amount of a morphogenetic substance. The sigmoid GEPM causes the canalization of bristle number in the class of four bristles. The morphogenetic substance is supposed to vary between individuals, both in flies carrying the mutant *scute* (*sc*) and in wild-type flies (±) according to a normal distribution (*horizontal axis*). Selection for a higher bristle number in the population shifts both distributions to higher values (*black*). Before selection the phenotypic variation in *scute* individuals (*vertical axis*) is larger than in wild-type flies; after the selection, when the *scute* flies approach the value 4, they have a small variability. In this selection line the variability of the non-*scute* flies increases as they pass out of the four-bristle class. (Scharloo 1987)

The strict formulation of Fisher's theorem applies to allelic variation in fitness at a single locus and assumes that the environment is constant. This is a very strong limitation. However, apart from the fact that the environment is not constant, the theorem is violated by a number of phenomena.

As already mentioned, many studies suggest that spontaneous mutation does not occur at random, and some of the bias present in mutations causes constraints on selection. DNA is organized in chromosomes by which the recombination rate between genes is reduced, and, moreover, even crossing-over events between homologous chromosomes are not randomly distributed. Gene activity is mostly not independent of other genes with the genome; epistatic effects are observed. The breeding system, especially in plants, can constrain adaptation because of the limitation of gene combinations, e.g., in the case of self-fertilization systems. Population structure acts in the same direction. Even in outbreeding plant populations it is difficult to avoid subdivided population structures if the species is animal-pollinated. Moreover, sexual selection, e.g., gametophytic competition, would restrict genotypic variability. Only recently has the role of developmental constraints in evolution been discussed. Scharloo (1987) defined developmental constraints as a bias in production of variant phenotypes or limitations on phenotypic variability caused by structure, character, composition, or dynamics of the developmental system. All these factors cause constraints at least on the speed of evolution of organisms, (for more detailed information see Loeschcke 1987b).

Acknowledgments. I thank S.K. Jain, M. Kittel, V. Loeschcke, R. Singh, and J. Tomiuk for helpful comments and suggestions on an earlier draft of this manuscript. My thanks also to Mrs. Stögerer for drawing the graphs.

2 Natural Selection: Measures and Modes

F.B. CHRISTIANSEN

Introduction

Natural selection is the driving force of Darwinian evolutionary change, and it occurs in any population whenever variation among individuals in the adaptation to their environment exists, where the environment of an individual is a broad concept covering both the physical environment and the biotic environment. The biotic environment is formed by the organisms which co-occur with the individual in question, e.g., its prey species, competitors, predators, and parasites, and it encompasses the population of conspecific individuals.

Variation among individuals in the adaptation to their environment is called **fitness variation**, and it is expressed as variation in the procreation of the individuals, in terms of both rate and numbers. This may occur as variation in the number of offspring per litter (Fig. 1), as varation in the total number of litters, as variation in the time of development to maturity, and as variation in the time elapsed between litters. In sexually reproducing organisms this may seem to refer to females only, but by assigning an offspring equally to the mother and the father individuals, the description of natural selection becomes quite general. The number of offspring depends critically on other aspects of the individual life. If an individual dies before sexual maturity is reached, then its total number of offspring is necessarily zero, and if a male does not find an opportunity to mate, then again its total number of offspring is zero.

The interesting evolutionary situations emerge when **phenotypic variation** in a population is correlated with the variation in some aspects of fitness. Then the rate of offspring production differs among the phenotypes, and if the phenotypic variation is heritable, then with time natural selection may result in changes in phenotypic characteristics of the population. Phenotypic variation that does not correlate with variation in fitness, however, is not expected to be subject to natural selection. When we consider natural selection with respect to the variation of a given phenotypic character, then we may compress our fitness description to the mean performance of the individuals in the various phenotypic classes. We may specify fitness in terms of characteristics of an individual of a given phenotype, like, for instance, probability of survival to maturity and mean number of litters. This collective description of individual fitness characteristics

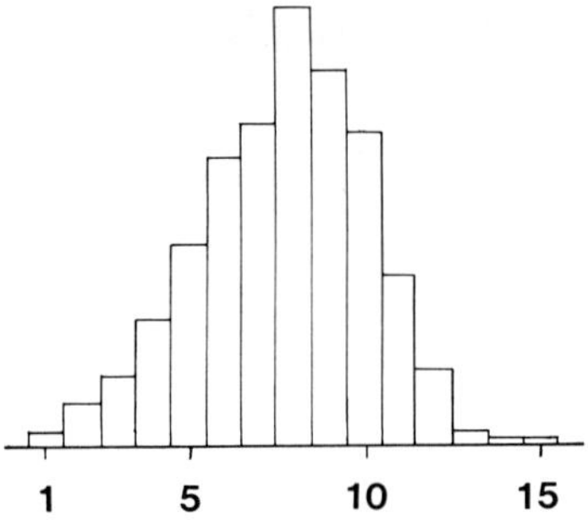

Fig. 1. Variation in the number of live young in the first litter of mice from a laboratory population, (After Falconer 1981)

will be used in the sequel, and it has the advantage that the description of adaptation is freed from purely accidental variation in number of offspring produced by the individuals. However, it is important to realize that this is a genuine collective description of fitness as the process of natural selection working on the phenotypic variation. The proper description of the average performance of individuals in a phenotypic class is not necessarily obtained in terms of the average individual. The average fecundity of a phenotype seems to be the average number of offspring produced by an individual, but characteristics of time such as, for instance, development time, usually have to be averaged on a logarithmic scale. Therefore, if a correlation exists between time of development and number of offspring, then simple averages are inadequate.

The above described definition of natural selection is straightforward and simple, but it is evident that statements made so far about the consequences of natural selection are somewhat vague. This is due to the nature of biological inheritance which, according to Mendel's rules, is indirect in the sense that the process of change is recorded only in the genetic composition of the population. The phenotypic composition of the population is an expression of the genotypic composition of the population in the sense that the phenotype of the individual is determined by its genotype and more or less random influences by its environment. Therefore, the relevant process which determines the evolutionary change in the phenotypic composition of the population is not natural selection as it occurs on the phenotypic variation, but the resulting selection induced on the genotypic variation behind the phenotypic variation. The process of evolution by natural selection may be illustrated as in Fig. 2, which shows a simplified scheme of Darwinian evolution by natural selection incorporating Mendelian indirect inheritance. Figure 2 addresses the effect of natural selection in an organism with nonoverlapping generations, and selection is assumed to occur only through mortality during development. Therefore, the effect of selection is to change the phenotypic composition of the population. The phenotype of an individual is determined partly by its genotype, and this induces a persisting relation between genotype and phenotype of the individual throughout life. The change in phenotypic composition of the population may therefore induce a change in the genotypic composition of the population. Only this induced change in genotypic

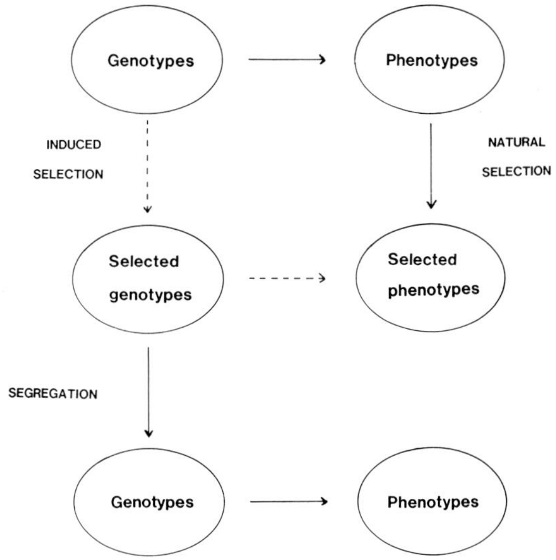

Fig. 2. A simplified scheme of Darwinian evolution by natural selection incorporating Mendelian indirect inheritance. For explanation see text

composition resulting from the influence of the genotype on the phenotype has a consequence for the genetic material which is passed on to the next generation. In sexually reproducing organisms the genetic material is passed to the gametes through segregation, and then the genotypic composition of the next generation is determined by the fertilization, or zygote formation. The change in genetic composition of the population is the only way in which natural selection can affect the phenotypic composition of the next generation.

Natural selection is a process going on throughout the whole life cycle of a species. Differences in survival among phenotypes from birth to the time of reproduction are probably the most widely known aspect of natural selection, and it is the basis of the popularized version of natural selection as **"the Survival of the Fittest"**. In the sense of Darwin (1872), fitness is understood as the character of the individual on which natural selection acts, and in this way it is defined as an abstraction used to quantify the properties of an individual in discussions of the consequences of natural selection. Fitness is thus an entity of evolutionary theory rather than an inherent property of the individual in a natural population.

The probability of survival from birth to maturity is only one aspect or component of natural selection. We will refer to it as **zygotic selection** to emphasize that this selection component modifies the population from the stage of zygotes to the stage of mature individuals. The process that is described under this heading actually may be rather complex, as, for instance, in insects, where it covers several developmental stages. However, it is clearly distinct from the component of **fecundity selection**, which occurs as a result of variations in the number of offspring of breeding individuals. Darwin included this selection component in his description of "the Survival of the Fittest", in that this phrase should not be read in only the material sense as survival of the individual, but rather in a metaphysical sense as the biological survival of the properties of the individual: "This preservation of favorable individual differences and variations, and the destruction of those which are injurious, I have called Natural Selection, or the Survival of the Fittest" (Darwin 1872).

To ascribe fitness values to individuals in fecundity selection is far simpler than the quantitative fitness description of zygotic selection. However, in genetic models of evolution in diploid sexually reproducing organisms, the technical handling of zygotic selection is considerably simpler than even rather simple models of fecundity selection. This difference is due to the assignment of fitness to the individual in zygotic selection as opposed to the assignment to the breeding pair in fecundity selection.

The third Darwinian fitness component is **sexual selection**, which describes the variation among sexually mature individuals in their probability of breeding, that is, the probability of realizing their potential for producing offspring. In a sense, sexual selection is the probability of "survival" from the stage of a mature adult to the stage of being a parent. Darwin separated this selection component from the two other selection components, in that these components of selection were envisioned as originating from variation in the individual adaptation to the environment, that is, they originate in nature. Sexual selection, on the other hand, originates from the processes involved in the sexual reproduction of the species, that is, it occurs because of variations in the reaction of individuals to each other. The evanescent nature of many aspects of sexual selection makes it a less appreciated biological phenomenon in many instances, although the existence of celebrated characters and behaviors connected with the individual procreation are widely known and treasured as contributions to the beauty of nature.

The cause for sexual selection is often the reaction of individuals to the phenotype of other individuals. This may bring about a choice of phenotype of the mate by one sex, competition or even combat within one sex to draw the attention of the other sex, or a combination of these phenomena. The action of sexual selection is therefore often intimately dependent on the phenotypic composition of the population.

Zygotic selection describes variation in the survival of the individual, fecundity selection describes variation among breeding pairs in their fecundity, and sexual selection describes the effect of sexual interactions among individuals. In sexually reproducing organisms yet another lifestage has to be separated, the lifestage of the gametes, or in general the haploid lifestage in the widespread alternation between haploid and diploid stages. Differential survival during the haploid stage, that is, from meiosis to gametic union in fertilization, is ascribed to **gametic selection**.

The evolutionary effect of natural or sexual selection on a given character depends on the nature of the inheritance of its traits and on the extent to which the phenotypic variation in the population is transmitted to the offspring population. Selection acts primarily on the phenotypic variation with the genetic material recording the changes in the population for all future time. The indirect nature of Mendelian inheritance, however, opens up for selection directly on variations related to the segregation and transmission of the genetic material. For instance, a gene responsible for meiotic drive in its favor may increase to quite high frequencies in the population even if it causes it carriers to have an injurious phenotype (e.g., the tailless trait in some natural populations of mice, Dunn 1956; Lewontin and Dunn 1960). In addition, the conservative transmission of genetic material makes it record the coancestry of individuals. The evolutionary spread of "socially valuable but individually disadvantageous characters" (Haldane 1932) seems a paradox unless this aspect of the nature and the transmission of the genetic material is taken into account. By recording the coancestry of the socially interacting individuals, the inclusive fitness of an individual with the socially valuable trait may be judged from the number of copies of its genes that will be transmitted to the next generation from itself and the other individuals in the group (Hamilton 1964a). This process of selection on groups of related individuals as a whole is termed kin selection (Maynard Smith 1964, 1982).

Heritable phenotypic variation, and even genotypic variation, which influences the performance of the individual, is the cause of selection. The distinction between natural selection as leading to "adaptation" of the individuals and sexual selection leading to seemingly mal-adapted traits is archaic, and reflects the belief that evolution is primarily a process leading to the adaptation of species to their environment. Even within the classical components of natural selection, zygotic selection and fecundity selection, the two components may pull in opposite directions (leading to an "adaptive" balance between the components), which makes the meaning of adaptive selection ambiguous. Sexual selection and genetic phenomena like meiotic drive exist and they are therefore participating in the process of shaping the species through evolution. The priggishness towards sexual selection should be abandoned by realizing that the kinds of supposedly mal-adaptive characters involved in sexual displays are no different from display characters involved in, for instance, antagonistic competitive interactions. Thus, selection in a natural population may be partitioned into the four fundamental components: zygotic selection, sexual selection, fecundity selection, and gametic selection, and their effect on the evolutionary change in the species is shaped by the genetics of the character in question.

The evolutionary consequences of a given aspect of natural selection are often studied in theoretical models where the induced selection on given genotypic variation is specified. The collective description of individual fitness characteristics is then carried further, in that fitness is described for the class of individuals belonging to a given genotypic class.

In the following, simple theoretical models of the action of natural selection are described and discussed. Other chapters of this book address the consequences of natural selection, so to avoid overlap, I will emphasize general aspects of the action of selection on genotypic variation to elucidate the problems and complications which arise due to the indirect nature of biological inheritance.

I will initially address the action of selection on the genotypic variation corresponding to allelic variation at an autosomal locus. The simplest genetic model assumes two alleles, A and a, with the corresponding genotypes AA, Aa and aa. The result of selection is monitored in terms of the change in gene frequencies p and q alleles A and a. The genotype frequencies of AA, Aa and aa are called g_{AA}, g_{Aa} and g_{aa}, so $p = g_{AA} + g_{Aa}/2$ and $q = g_{aa} + g_{Aa}/2$.

Differential Survival of Zygotes

The simplest models of the action of selection address situations where the various genotypes have different probabilities of survival to maturity, also called **zygotic selection**, and no other fitness variation is present. In addition we will assume that the population is large (i.e., we will consider fitness differences large enough to neglect the effect of random genetic drift). To simplify the discussion, we further assume nonoverlapping generations and a monoecious organism (each individual is both male and female). The genotype frequencies among zygotes are then determined from the gene frequencies among gametes and the rule of mating. The simplest mating rule is **random mating** or random union of gametes, where the zygotes are formed by uniting two randomly drawn gametes from those produced by the population of parents. With random mating, the genotypic frequencies among zygotes are in Hardy–Weinberg proportions, and the model is specified as in Table 1, where

$$W = p^2 v_{AA} + 2pq v_{Aa} + q^2 v_{aa} \qquad (1)$$

is the average probability of survival. The gene frequency among the adults, and therefore among the zygotes of the next generation, is

$$p' = \frac{pW_A}{W}, \qquad (2)$$

where $W_A = v_{AA}p + v_{Aa}q$ is the average probability of survival of allele A, and similarly $W_a = v_{Aa}p + v_{aa}q$ is the average probability of survival of allele a. The average probability of survival of a gene is $W_A p + W_a q$ which of course is W given by Eq. (1). Thus, when $W_A > W_a$ then $p' > p$ and the frequency of allele A increases, when $W_A < W_a$ then $p' < p$ and the frequency of allele A decreases, and when $W_A = W_a$ then $p' = p$ and the frequency of allele A is unchanged by selection. A state of the population where the studied process leaves the population unchanged is called an **equilibrium state**, and a gene frequency where

Table 1. Monoecious zygotic selection with random mating

Genotypes	AA	Aa	aa	Σ
Frequency in zygotes	p^2	$2pq$	q^2	1
Survival probability	v_{AA}	v_{Aa}	v_{aa}	
Frequency in adults	$\frac{p^2 v_{AA}}{W}$	$\frac{2pq v_{Aa}}{W}$	$\frac{q^2 v_{aa}}{W}$	1

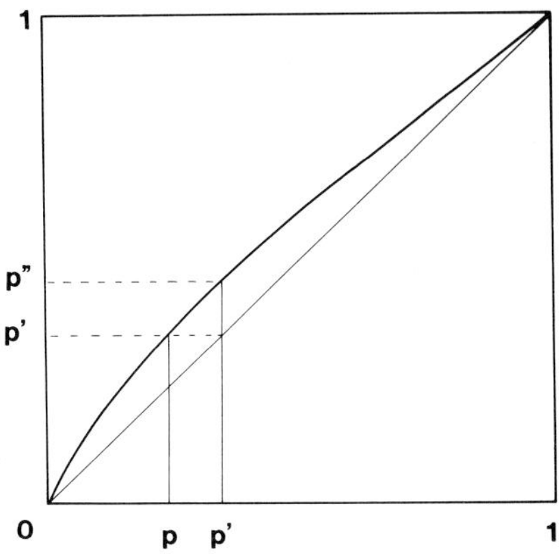

Fig. 3. A graphic representation of the recurrence Eq. (2) in the case of directional zygotic selection favoring allele A, $v_{AA} > v_{Aa} > v_{aa}$. The *curve* depicts the gene frequency in the offspring generation, p′, as a function of the gene frequency, p, among the parents. ($v_{AA}/v_{Aa} = 1.2$ and $v_{aa/Aa} = 0.5$)

$p' = p$ is called an equilibrium gene frequency. The two states $p = 0$ and $p = 1$ are always equilibria, and they are called the trivial equilibria.

If $v_{AA} \geqslant v_{Aa} > v_{aa}$, then $W_A > W_a$ for all gene frequencies p where $0 < p < 1$, so the frequency of allele A increases in a population where both alleles are present, and given enough time, the allele will be fixed (Fig. 3). Similarly, if $v_{AA} \leqslant v_{Aa} < v_{aa}$, then $W_A < W_a$ for all gene frequencies p where $0 < p < 1$, and given enough time, allele a will be fixed in the population. Thus, in those two cases, which are referred to as directional selection, only the trivial equilibria exist. One of these equilibria is an **unstable equilibrium**, in that a small disturbance, i.e., an introduction of the absent allele in a small frequency, will cause the population to move further and further away from the equilibrium. The other equilibrium is a **globally stable equilibrium**, in that a population, where both alleles are present initially, eventually will end up at that equilibrium.

If $v_{AA} > v_{Aa}$ and $v_{Aa} < v_{aa}$ or $v_{AA} < v_{Aa}$ and $v_{Aa} > v_{aa}$, then $W_A = W_a$ for a gene frequency $\hat{p}$ where $0 < \hat{p} < 1$. This equilibrium is characterized by the gene frequency

$$\hat{p} = \frac{v_{Aa} - v_{aa}}{2_{Aa} - v_{AA} - v_{aa}}. \tag{3}$$

The frequency of allele A increases or decreases dependent on whether $p > \hat{p}$ or $p < \hat{p}$ (Figs. 4 and 5). In the case $v_{AA} > v_{Aa}$ and $v_{Aa} < v_{aa}$, known as underdominant selection, $W_A < W_a$ when $p < \hat{p}$ and $W_A > W_a$ when $p > \hat{p}$, so $\hat{p}$ is an unstable equilibrium (Fig. 4). The trivial equilibria, on the other hand, are both **locally stable**: populations initially in the interval $0 \leqslant p < \hat{p}$ will end up at the equilibrium $p = 0$ and populations in the interval $\hat{p} < p \leqslant 1$ will end up at $p = 1$. For $v_{AA} < v_{Aa}$ and $v_{Aa} > v_{aa}$, known as overdominant selection, the equilibrium $\hat{p}$ is globally stable: a population with $0 < p < 1$ eventually will end up at $\hat{p}$, and both trivial equilibria are unstable (Fig. 5).

The extension of these results to the situation of dioecious organisms is not so simple. Random mating again provides the genotypic frequencies in terms of the

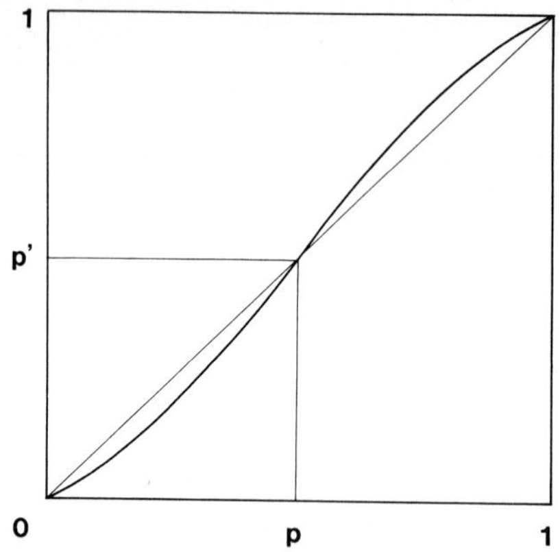

Fig. 4. As Fig. 3, but with under-dominant selection, $v_{AA} > v_{Aa}$ and $v_{aa} > v_{Aa}$. $(v_{AA}/v_{Aa} = v_{aa/Aa} = 2)$

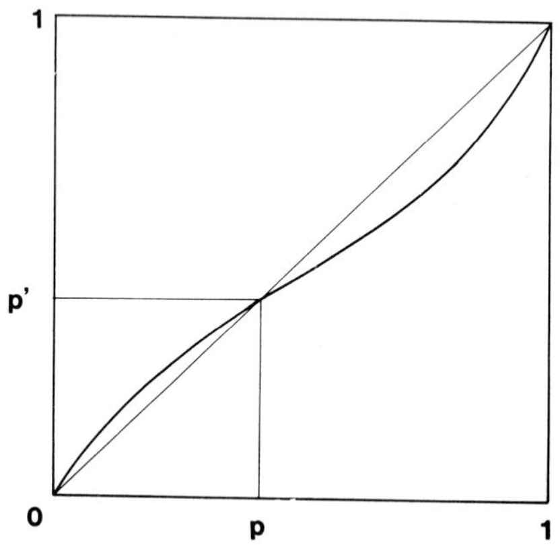

Fig. 5. As Fig. 3, but with over-dominant selection, $v_{AA} < v_{Aa}$ and $v_{aa} < v_{Aa}$. $(v_{AA}/v_{Aa} = 0.4$ and $v_{aa}/v_{Aa} = 0.6)$

gene frequency p_f ($q_f = 1 - p_f$) in female gametes and p_m ($q_m = 1 - p_m$) in male gametes, but we have to allow for the possibility that these gene frequencies are different. The model is outlined in Table 2, where

$$\begin{aligned} W_f &= u_{AA}p_fp_m + u_{Aa}(p_fq_m + q_fp_m) + u_{aa}q_fq_m, \\ W_m &= v_{AA}p_fp_m + v_{Aa}(p_fq_m + q_fp_m) + v_{aa}q_fq_m. \end{aligned} \tag{4}$$

The gene frequency among the adult females of the table is given by

$$p_f' = \frac{u_{AA}p_fp_m + (1/2)u_{Aa}(p_fq_m + q_fp_m)}{W_f}, \tag{5}$$

and for the males we obtain

$$p_m' = \frac{v_{AA}p_fp_m + (1/2)v_{Aa}(p_fq_m + q_fp_m)}{W_m}. \tag{6}$$

The gene frequency among the zygotes is $p = (p_f + p_m)/2$, but in view of Eqs. (5) and (6) it is evident that the gene frequency among adults, p', is not simply a

Table 2. Dioecious zygotic selection with random mating

Genotypes	AA	Aa	aa	Σ
Zygotes	$p_f p_m$	$p_f q_m + q_f p_m$	$q_f q_m$	1
Females				
Survival	u_{AA}	u_{Aa}	u_{aa}	
Adults	$\frac{u_{AA} p_f p_m}{W_f}$	$\frac{u_{Aa}(p_f q_m + q_f p_m)}{W_f}$	$\frac{u_{aa} q_f q_m}{W_f}$	1
Males				
Survival	v_{AA}	v_{Aa}	v_{aa}	
Adults	$\frac{v_{AA} p_f p_m}{W_m}$	$\frac{v_{Aa}(p_f q_m + q_f p_m)}{W_m}$	$\frac{v_{aa} q_f q_m}{W_m}$	1

function of p. This is because the genotypic proportions among the zygotes is determined by the gene frequencies in each sex (or equivalently by p and by $p_f - p_m$), and the result of selection depends on the genotypic proportions. Thus, we are stuck with a two-dimensional description of the state of our population, and the analysis is more complicated than in the monoecious case where the description is one-dimensional.

However, there are ways to obtain an impression of the evolutionary properties of the model. As before, we always have the two trivial fixation equilibria, and we may determine the stability properties of these equilibria by a local stability analysis. Let us therefore return to the monoecious model and consider the behavior of the population close to the equilibrium $p = 0$ where the allele a is fixed in the population. That is, we assume that allele A is rare, and that p is very small. Then Eq. (2) is approximately equivalent to

$$p' \approx p\frac{v_{Aa}}{v_{aa}} \tag{7}$$

where the error of approximation is of the order p^2. This approximation is obtained formally by a Taylor expansion of the right side of Eq. (2) where we neglect the terms of orders p^2, but in genetic terms what we do is to neglect the occurrence of the homozygote of the rare allele and assume that only the genotypes Aa and aa occur in the population. From Eq. (7) we deduce that the allele A increases in frequency when rare if $v_{Aa} > v_{aa}$ and decreases in frequency if $v_{Aa} < v_{aa}$. When $v_{Aa} = v_{aa}$ the increase or decrease of p is determined by the terms of order p^2, so the approximate equation leaves the stability undetermined.

The local analysis of the fixation equilibrium determines the condition for a rare allele to increase in the population. Biologically this situation is interesting in that it describes the condition for the **initial increase of a new allele** when introduced as rare into the population, either by mutation or by immigration. The fate of a single mutant or a single immigrant allele can only be determined by taking random genetic drift into account. However, this analysis provides the conditions that effectively prohibit increase of the allele, and that guarantee the eventual increase after repeated introductions, given that the fitness differences are sufficiently large (see Kimura 1983). An allele that increases initially in the population is called protected. The probability that a single gene of allele type A causes a successful invasion of this allele type into the population is of the order $\log(v_{Aa}/v_{aa})$ when A is protected. If both alleles are protected, then selection maintains both alleles in the population in a state termed a **protected polymorphism**, this occurs, e.g., with overdominant selection (Fig. 5).

In the dioecious case the assumption that allele A is rare provides the approximation

$$\begin{bmatrix} p_f' \\ p_m' \end{bmatrix} \approx \begin{bmatrix} \frac{u_{Aa}}{2u_{aa}} & \frac{u_{Aa}}{2u_{aa}} \\ \frac{v_{Aa}}{2v_{aa}} & \frac{v_{Aa}}{2v_{aa}} \end{bmatrix} \begin{bmatrix} p_f \\ p_m \end{bmatrix} \tag{8}$$

of Eqs. (5) and (6). This equation may be written as $\mathbf{p}' \approx \mathbf{W}\mathbf{p}$, so after t generations $\mathbf{p}^{(t)} \approx \mathbf{W}^t\mathbf{p}$ and the ultimate fate of the rare allele is determined by $\mathbf{W}^t$ which in turn is approximately proportional to λ^t, where λ is the eigenvalue of the matrix $\mathbf{W}$ with the largest numerical value (refer to any text in matrix calculus or in linear algebra, or to Gantmacher 1959). In Eq. (8) the largest eigenvalue is

$$\lambda_A = \frac{1}{2}\left[\frac{u_{Aa}}{u_{aa}} + \frac{v_{Aa}}{v_{aa}}\right]. \tag{9}$$

This eigenvalue determine the fate of allele A, and it is the average of the survival probability of the rare genotype relative to the common genotype in the two sexes (Parsons 1961).

The eigenvalue of initial increase in Eqs. (7) and (9) is given in terms of the survival probability of the rare genotype relative to that of the common genotype. In the same way, the recurrence Eqs. (2), (5), and (6) are unchanged if the probabilities of survival are multiplied with a factor independent of the genotype. This factor may be the reciprocal value of one of the genotypic survival values, and this transforms the survival probabilities into probabilities of survival of the genotypes relative to one of the genotypes. These values are often referred to as the **relative fitness values** or relative survival values of the genotypes. The description of theoretical results may often depend on which genotype is taken as the norm when the relative fitness values are calculated. For instance, the condition for allele A to be protected in the dioecious model is that $\lambda_A > 1$, where λ_A is given by Eq. (9). If we calculate the relative fitness values with the common genotype aa as the norm, then the result says that allele A is protected if the (arithmetic) mean of the fitness values of the heterozygote in the two sexes exceeds the fitness value of the common genotype. However, if the heterozygote is taken as the norm and the fitnesses are measured relative to the heterozygote in each sex, then allele A is protected if the harmonic mean of the fitness value of the common homozygote is less than the fitness value of the heterozygote (the harmonic mean is the reciprocal of the mean of the reciprocal values). This ambiguity in measures of relative fitness should be kept in mind when theoretical results are evaluated. The heterozygote is often used as a norm in one-locus two-allele models when the interest is the conditions for protected polymorphism, as this normalization produces the conditions for protection of allele A and allele a in the same form. One of the homozygotes is often used as the norm when the two alleles are viewed as inherently different, like the old and the new allele in the study of allele substitutions. In the following discussion of the dioecious model let $w_{fA} = u_{AA}/u_{Aa}$, $w_{fa} = u_{aa}/u_{Aa}$, $w_{mA} = v_{AA}/v_{Aa}$, $w_{ma} = v_{aa}/v_{Aa}$, and in terms of these relative fitness values the condition (9) for protection of an allele is illustrated in Fig. 6.

In the monoecious case the conditions for allele protection was a good predictor for the general dynamical behavior. This, however, is not true in the dioecious case, where more complicated dynamics may occur.

As a simple example showing the more complicated dynamics assume $w_{fA} = w_{fa} = w_f$ and $w_{mA} = w_{ma} = w_m$, so the condition for protected polymorphism (9) is $w_f + w_m > 2w_f w_m$. With this assumption the population always has an equilibrium solution to Eqs. (5) and (6) where $\hat{p}_f = \hat{p}_m = 1/2$. As with the trivial equilibria, we can analyze the local stability properties of this equilibrium. Assuming that p_f and p_m are close to this equilibrium, provide the approximation

$$\begin{bmatrix} p_f' - \dfrac{1}{2} \\ p_m' - \dfrac{1}{2} \end{bmatrix} \approx \begin{bmatrix} \dfrac{w_f}{1+w_f} & \dfrac{w_f}{1+w_f} \\ \dfrac{w_m}{1+w_m} & \dfrac{w_m}{1+w_m} \end{bmatrix} \begin{bmatrix} p_f - \dfrac{1}{2} \\ p_m - \dfrac{1}{2} \end{bmatrix} \tag{10}$$

of Eqs. (5) and (6). A small deviation from equilibrium will become smaller and eventually vanish when the leading eigenvalue, λ, of the matrix in Eq. (10), the stability matrix, is smaller than 1. Thus if $\lambda < 1$ the population will approach the

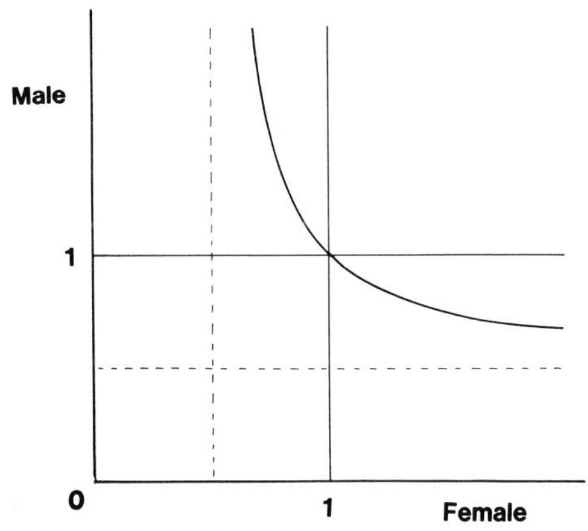

Fig. 6. The condition for initial increase of a rare allele A in a population monomophic aa in the dioecious model of zygotic selection. Fitnesses are measured relative to the heterozygote Aa, so on the *abscissa* is the relative fitness of the common type in females, u_{aa}/u_{Aa}, and on the *ordinate* is relative fitness in males, v_{aa}/v_{Aa}. Allele A is protected when in Eq. (9) $\lambda_A > 1$ which is fulfilled below the curve

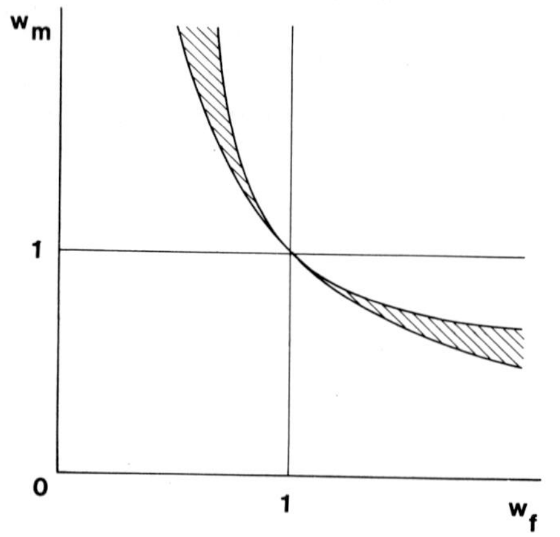

Fig. 7. The dynamics of the symmetric dioecious model where w_f is the relative fitness of the female homozygotes, AA and aa, relative to the female heterozygote, Aa, and where w_m is the relative fitness of the homozygotes in the males. *Below both curves* the only nontrivial equilibrium is $p_f = p_m = 1/2$, and this equilibrium is globally stable. *Above both curves* the only nontrivial equilibrium again is $p_f = p_m = 1/2$, and this equilibrium is unstable. *Between the curves* (*shaded region*) the central equilibrium, $p_f = p_m = 1/2$, is unstable and there is protected polymorphism (Fig. 6). In this case two additional equilibria exist and are locally stable [Eq. (13)]

equilibrium when initially close, so the equilibrium will be locally stable. If $\lambda > 1$ then an initial small deviation will eventually increase and the population will diverge from the equilibrium, which is then unstable. The leading eigenvalue of the stability matrix is

$$\lambda = \frac{w_f}{1 + w_f} + \frac{w_m}{1 + w_m}. \tag{11}$$

The equilibrium is stable when $\lambda < 1$ which (with some rearrangements) becomes the condition: $w_f w_m < 1$, and it is unstable ($\lambda > 1$) when $w_f w_m > 1$.

In general we have that if $w_f w_m < 1$ then $w_f + w_m > 2w_f w_m$. Therefore, when the equilibrium $\hat{p}_f = \hat{p}_m = 1/2$ is stable the polymorphism is protected. On the other hand, if $w_f + w_m < 2w_f w_m$ then $w_f w_m > 1$, so when neither allele A nor allele a is protected then the equilibrium $\hat{p}_f = \hat{p}_m = 1/2$ is unstable. Finally, when $(w_f + w_m)/2 > w_f w_m > 1$ then protected polymorphism prevails and the equilibrium $\hat{p}_f = \hat{p}_m = 1/2$ is unstable (shaded region in Fig. 7).

The general dynamics of the dioecious model in Table 2 can be found in Selgrade and Ziehe (1987). The population will always converge to a locally stable equilibrium, and apart from the trivial equilibria, at most two stable and at most one unstable equilibrium exist. When protected polymorphism prevails and the equilibrium $\hat{p}_f = \hat{p}_m = 1/2$ is locally stable, only three equilibria exist and the situation is similar to overdominance:

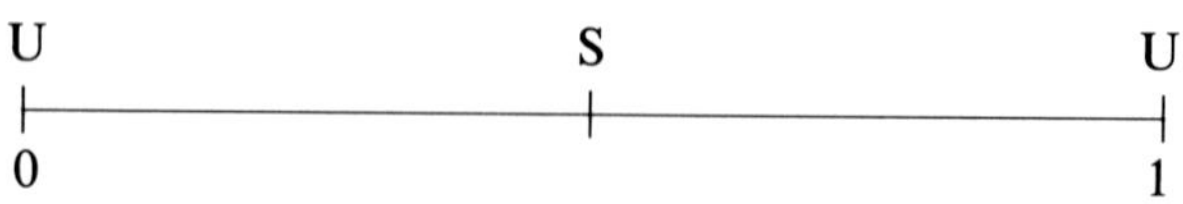

, (12)

where U and S designate unstable and stable equilibria, respectively. In the last case of the above example it can be shown that two equilibria placed symmetrically with respect to the equilibrium $\hat{p}_f = \hat{p}_m = 1/2$ exist and are locally stable:

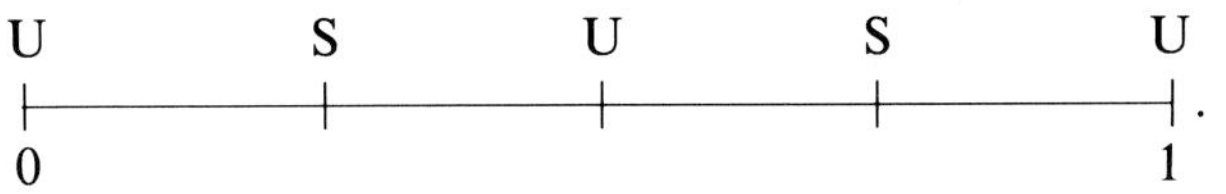

(13)

When neither allele is protected and the equilibrium $\hat{p}'_f = \hat{p}'_m = 1/2$ is locally unstable, then only three equilibria exist and the situation is similar to underdominance:

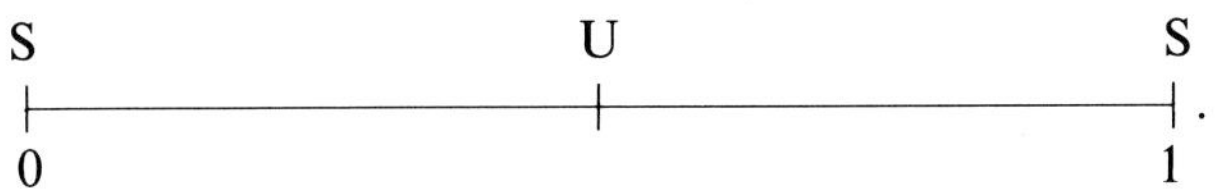

(14)

The symmetry of the example precludes the possibility of one protected allele and one unprotected allele. In this case we may either obtain the same situation as with directional selection in the monoecious case:

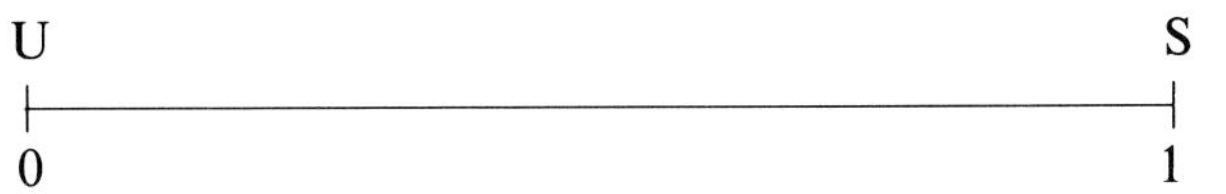

(15)

or a situation which resembles Eq. (13) pushed a bit to one side, i.e.,

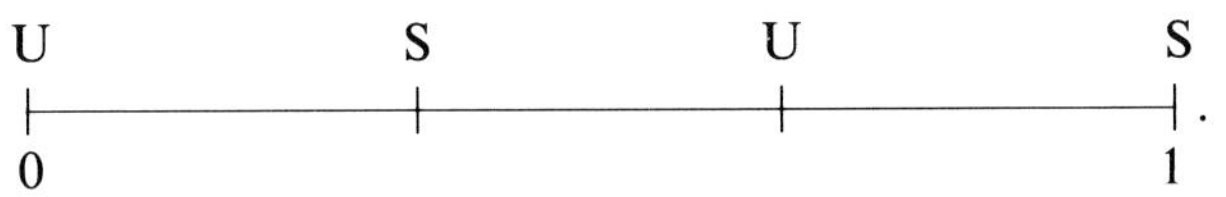

(16)

This last possibility is notable, since an analysis of the boundary equilibria concludes directional selection when in fact a stable polymorphism is possible. If the population was initially monomorphic aa and A became introduced as rare, then it would end up polymorphic, but if the population was initially monomorphic AA, then it would stay like that after an introduction of allele a as rare.

Figure 7 indicates that the deviant behavior shown in Eqs. (13) and (16) requires very different female and male relative fitness values. If the mode of selection with respect to the classification in the monoecious model is the same in females and in males, then the qualitative behavior of the dioecious is like a monoecious model. Thus, the far simpler monoecious model is a good guide in situations where at most slight sex differences in fitness exist, but as soon as the presence of genuine sex differences are of interest the dioecious model should be used.

The observational determination of the fitness values describing **zygotic selection** is in principle straightforward. If the genotypic proportions among zygotes and among adults are knwon, then the relative fitness values can be estimated simply as

$$w_{AA} = \frac{x_{AA} z_{Aa}}{z_{AA} x_{Aa}} \quad \text{and} \quad w_{aa} = \frac{x_{aa} z_{Aa}}{z_{aa} x_{Aa}}, \tag{17}$$

where x is the observed frequencies among adults and z is the observed frequencies among zygotes (Table 3). Assuming multinomial samples, these are

Table 3. Observation of zygotic selection

Genotypes	AA	Aa	aa	Σ
Frequency in zygotes	z_{AA}	z_{Aa}	z_{aa}	1
Relative survival	w_{AA}	1	w_{aa}	
Frequency in adults	x_{AA}	x_{Aa}	x_{aa}	1
Expected frequency in adults	$\frac{w_{AA}z_{AA}}{W}$	$\frac{z_{Aa}}{W}$	$\frac{w_{aa}z_{aa}}{W}$	1

Table 4. Zygotic selection in *Drosophila melanogaster*

Genetic system		Females		Males		Combined		Equilibrium gene frequency	
A	a	w_{AA}	w_{aa}	w_{AA}	w_{aa}	w_{AA}	w_{aa}	Female	Male
se	*h*	0.98	0.96	0.96	0.89	0.97	0.93	0.70	0.70
st	*th*	1.30	0.90	1.17	0.85	—	—	1.00	1.00
ri	*in*	0.99	0.69	0.89	0.84	—	—	0.80	0.78
ss	*ss*[a]	0.90	0.89	0.92	0.90	0.91	0.90	0.53	0.53

the maximum likelihood estimates of the relative fitness values. The difficult problem in this procedure is to acquire knowledge of the zygote population. However, in the laboratory the zygote population can be determined rather exactly by choosing appropriate parents. For instance, an F_1-cross, Aa × Aa, produces zygotes in the proportions $z_{AA} = 1/4$, $z_{Aa} = 1/2$ and $z_{aa} = 1/4$, and the backcross AA × Aa produces zygotes in the proportions $z_{AA} = 1/2$, $z_{Aa} = 1/2$ and $z_{aa} = 0$ (the other backcross, aa × Aa, gives similar proportions). Using this procedure, Bundgaard and Christiansen (1976) estimated zygotic fitness values for four genetic systems in *Dropsophila melanogaster* (Table 4). For two of the systems, the first and the last, the data showed no significant difference between the sexes, but for the others the zygotic selection in males and females was different. Three of the systems allowed for a stable polymorphic equilibrium under the assumption that zygotic selection is the only selective force working on the system. These observations are discussed further by Hedrick (1983).

Differential Fertility of Adults

In sexually reproducing organisms an individual has to be seen as an offspring of the mother individual as well as of the father individual. Therefore, the description of the dynamics of fecundity selection is more complicated, as its object is the mated pair. To simplify discussions suppose monogamy, i.e., that each breeding individual in the population participates in exactly one mating. Selection on the variation at an autosomal locus with the alleles A_1 and A_2 is described by the fertility $F_{ij,kl}$ of a mating between a female of genotype A_iA_j and a male of genotype A_kA_l, $i, j, k, l = 1, 2$. This description of selection is conceptually simpler than the description of zygotic selection, because fecundity is an observable quantity of any mating between two individuals, whereas probability of survival of a genotype is an abstraction from the frequency of survival among individuals of that genotype.

Now suppose that all genotypes are phenotypically equal except for variation in fertility. Then the genotypic frequencies among breeders equal the frequencies among zygotes and the sexes have equal genotypic frequencies. In a given generation let f_{11}, f_{12} and f_{22} be the frequencies among breeders of A_1A_1, A_1A_2 and A_2A_2, respectively. Finally, suppose random mating among breeding individuals. The equations governing fecundity selection are then given by

$$f'_{11} = (F_{11,11}f_{11}^2 + F_{11,12}f_{11}f_{12}/2 + F_{12,11}f_{12}f_{11}/2 + F_{12,12}f_{12}^2/4)/F, \tag{18a}$$

$$f'_{12} = (F_{11,12}f_{11}f_{12}/2 + F_{12,11}f_{12}f_{11}/2 + F_{22,12}f_{22}f_{12}/2 + F_{12,22}f_{12}f_{22}/2 + F_{12,12}f_{12}^2/2 + F_{11,22}f_{11}f_{22} + F_{22,11}f_{22}f_{11})/F. \tag{18b}$$

$$f'_{22} = (F_{22,22}f_{22}^2 + F_{22,12}f_{22}f_{12}/2 + F_{12,22}f_{12}f_{22}/2 + F_{12,12}f_{12}^2/4)/F, \tag{18c}$$

where F is the average fertility in the population:

$$F = \sum_{i=1}^{2} \sum_{j=i}^{2} \sum_{k=1}^{2} \sum_{l=k}^{2} F_{ij,kl}f_{ij}f_{kl}. \tag{19}$$

The gene frequencies $p_1 = f_{11} + f_{12}/2$ and $p_2 = f_{22} + f_{12}/2$ are not in general sufficient to describe the state of the population, because the genotypic frequencies will not be in Hardy–Weinberg proportions.

The dynamic difference between the zygotic selection models of the previous Section and this fecundity selection model originates from the fact that the offspring of a random individual in the population with rather high probability have the same genotype as the parent (probability 0.5 for a heterozygote), but the probability of the offspring to participate in the same type of mating as the parental mating may in some instances be very low. An extreme example is the situation that the fertility of the mating A_1A_1 with A_2A_2 is very high compared to the fertility of the A_1A_2 with A_1A_2 and that all other matings have a fecundity of zero; here the average fertility of the population may decrease between generations in contrast to the situation with zygotic selection (Pollak 1978).

Many special models of fecundity selection show a considerably simpler dynamic (see Feldman et al. 1983). If fertility is influenced by **one sex** only, usually the female sex, so that the fertility of any pair equals the fecundity of the female, i.e., $F_{ij,kl} = F_{ij}$ for all male genotypes A_kA_l, then fecundity selection changes the population in exactly the same way as zygotic selection working in females only.

Similarly, if fertility is the female fecundity modified by a male virility factor, i.e., $F_{ij,kl} = F_{ij} \cdot v_{kl}$ for all female genotypes A_iA_j and male genotypes A_kA_l, then fecundity selection changes the population in exactly the same way as zygotic selection in the dioecious model (Table 2).

The dynamic properties of any model of fecundity selection are preserved by the interchange of sex for some or all fecundity parameters, that is, the interchange of $F_{ij,kl}$ and $F_{kl,ij}$ for some of the genotypic combinations. In addition, the fecundity selection model given by the fertilities $F_{ij,kl}$, $i, j, k, l = 1, 2$ is indistinguishable in its dynamical properties from the fecundity selection model where the fecundity of female A_iA_j and male A_kA_l is $(F_{ij,kl} + F_{kl,ij})/2$. Thus any model of fecundity selection is equivalent to a model where the sexes are symmetric in their effect on the fertility of a mating.

Consider, for instance, the seemingly complicated fecundity scheme in Table 5. Here the male effect on the fecundity of A_1A_1 females is over-dominant, the male effect on the fecundity of A_2A_2 females is under-dominant, and the effect of the heterozygote male on the fecundity of A_1A_2 females is intermediate between the homozygotes. However, the fecundity scheme in Table 5 has the same dynamics as the scheme in Table 6, where fecundity is influenced by one sex only (here the female, but another procedure of interchange would give the male the effect on fecundity). This unisexual fecundity scheme also has the same dynamics as the scheme in Table 7 which is additive, in that the female genotypes contribute 7, 7.9 and 9.4, respectively, and the male genotypes contribute 4, 4.1 and 4.6, respectively, to the total fecundity of the pair.

This dynamical equivalence of unisexual fecundity selection and additive fecundity selection holds in general (Feldman et al. 1983), and the evolutionary dynamics of this kind of fecundity selection is similar to those of differential survial in one sex only. If fecundity is additive so that males add a fecundity contribution, i.e., $F_{ij,kl} = F_{ij} + V_{kl}$ for all female genotypes A_iA_j and male genotypes A_kA_l, then a model with the same dynamics is unisexual fecundity selection where $F_{ij,kl} = F_{ij} + V_{ij}$. If $F_{ij,kl} = F_{ij}$, then $F_{ij,kl} = (F_{ij} + F_{kl})/2$ provides a dynamically equivalent additive model of fecundity selection.

In animal populations the fecundity is often determined by the female alone. An example for a laboratory population of *Drosophila melanogaster* is shown in Table 8 (Bundgaard and Christiansen 1972). Table 8 is based on the number of eggs laid within 24 h of female imagines 4 days old. The table shows the mean

Table 5. Example of fecundity selection

Female Genotypes	Male genotypes A_1A_1	A_1A_2	A_2A_2
A_1A_1	31	32	31
$A_1A_2$1	31	32	34
A A_2	34	32	34

Table 6. Unisexual selection

Female Genotypes	Male genotypes A_1A_1	A_1A_2	A_2A_2
A_1A_1	31	31	31
A_1A_2	32	32	32
A_2A_2	34	34	34

Table 7. Additive fecundity selection

Female Genotypes	Male genotypes A_1A_1	A_1A_2	A_2A_2
A_1A_1	31.0	31.1	31.6
A_1A_2	31.9	32.0	32.5
A_2A_2	33.4	33.5	34.0

Table 8. Fecundity selection in a *Drosophila* population

Female Genotypes	Male genotypes A_1A_1	A_1A_2	A_2A_2	Mean
A_1A_1	46.9 ± 0.9	47.0 ± 0.9	44.7 ± 4.1	46.9 ± 0.6
A_1A_2	46.7 ± 1.0	47.0 ± 0.9	46.9 ± 0.9	46.9 ± 0.6
A_2A_2	41.9 ± 2.1	40.9 ± 1.1	41.3 ± 2.4	41.2 ± 0.8

number of eggs laid by females of the nine different mating types for an autosomal two-allele polymorphism, and next to each mean is the standard deviation. The table clearly shows that there are no indications of a male influence on fecundity, but the A_2A_2 females laid 5.7 eggs less than the A_1A_1 and A_1A_2 females. The analysis of data like these would often proceed with an analysis of variance specifying additivity of male and female effects. In terms of the dynamics of the resulting fecundity selection the above arguments show that this additive specification is equivalent to a hypothesis of a dynamic equivalent to unisexual effects. Similarly, an analysis of variance on logarithmic-transformed data specifies a hypothesis of dynamical equivalence to zygotic selection, perhaps with different viabilities in the two sexes.

The protection of the alleles is again determined from a local stability analysis of the trivial equilibria. The assumption that allele A_1 is rare provides the approximation $f'_{11} \approx 0$ and

$$f'_{12} \approx \frac{(F_{12,22} + F_{22,12})/2}{F_{22,22}} f_{12}. \tag{20}$$

Thus, allele A_1 is protected when the average fecundity of the two common matings in which the genotype A_1A_2 participates is larger than the fecundity of the individuals in the population that is monomorphic A_2A_2. (This analysis of local stability of the trivial equilibrium $p_1 = 0$ seems to be different from the analysis based on Eq. (8). However, the matrix in (8) has an eigenvalue of zero and the eigenvalue given by (9). The eigenvalue zero corresponds to the approximation $f'_{11} \approx 0$ and the eigenvalue in (9) corresponds to the factor in Eq. (20). So for multiplicative fecundities, $F_{ij,kl} = F_{ij} \cdot v_{kl}$, the two local stability analyses coincide.)

Table 9. Symmetric fecundity selection model

Genotypes	A_1A_1	A_1A_2	A_2A_2
A_1A_1	α	β	α
A_1A_2	β	δ	β
A_2A_2	α	β	α

To further our analysis of fecundity selection let us consider a simplified version (Table 9) of the symmetric model of fecundity selection suggested by Hadeler and Liberman (1975) and discussed by Feldman et al. (1983). In this model the two homozygotes are considered as symmetric in their effect on fecundity, and the sexes are symmetric. Thus, the simplification of this model is like the simplification in our symmetric dioecious zygotic selection model in Allele A_1 is protected when $\alpha < \beta$, and due to the symmetry this is also the condition for the protection of allele A_2 and for protected polymorphism. Again we find that an equilibrium with $\hat{p}_1 = \hat{p}_2 = 1/2$ and with $f_{11} = f_{22} = 1/4$ and $f_{12} = 1/2$ always exists, and we will refer to this equilibrium as the *symmetric equilibrium*. The local behavior around the symmetric equilibrium is governed by the equation

$$\begin{bmatrix} f'_{11} - \frac{1}{4} \\ f'_{22} - \frac{1}{4} \end{bmatrix} \begin{bmatrix} \frac{\alpha+\beta}{\alpha+2\beta+\delta} & -\frac{\alpha+\beta}{\alpha+2\beta+\delta} \\ -\frac{\alpha+\beta}{\alpha+2\beta+\delta} & \frac{\alpha+\beta}{\alpha+2\beta+\delta} \end{bmatrix} \begin{bmatrix} f_{11} - \frac{1}{4} \\ f_{22} - \frac{1}{4} \end{bmatrix}, \tag{21}$$

which is an approximation of Eqs (18) and (18) close to the symmetric equilibrium. The leading eigenvalue of the stability matrix is

$$\lambda = \frac{2(\alpha+\beta)}{\alpha+2\beta+\delta}, \tag{22}$$

so the equilibrium is stable ($\lambda < 1$) when $\alpha < \delta$ and unstable ($\lambda > 1$) when $\alpha > \delta$.

The symmetric model is equivalent to a unisexual fecundity model when it is additive, i.e., when $\alpha + \delta = 2\beta$. In this case the symmetric equilibrium is the only nontrivial equilibrium and it is stable if fecundity is overdominant ($\alpha < \delta$) and unstable if fecundity is underdominant ($\alpha > \beta$).

Next we will ask, when is the symmetric fecundity model equivalent to a **multiplicative fecundity model**, i.e., equivalent to a symmetric dioecious zygotic selection model? Thus the question is: when is it possible to write

$$\alpha = \delta w_f w_m \text{ and } \beta = \delta(w_f + w_m)/2? \tag{23}$$

The answer comes from the observation that w_f and w_m have to be the two roots of the equation

$$\delta x^2 - 2\beta x + \alpha = 0 \tag{24}$$

in order to fulfil Eqs. (23). Thus, we can find w_f and w_m only when Eq. (24) has two real roots, i.e., only when $\beta^2 > \alpha\delta$. When $\beta^2 < \alpha\delta$ the fecundity selection model of Table 9 is dynamically different from any of the zygotic selection models considered in the previous section, and in this class of models we may find patterns of evolutionary behavior characteristic of fecundity selection.

The dynamics of the model in Table 9 were described by Hadeler and Liberman (1975). Four different cases have to be distinguished (Feldman et al. 1983). For $\alpha > \delta$ and $\alpha > \beta$ three equilibria exist, the symmetric equilibrium, which is unstable, and the two trivial equilibria, which are stable [cf. Eq. 14]. Again for $\delta > \alpha$ and $\beta > \alpha$ three equilibria exist, the symmetric equilibrium, which is stable, and the two trivial equilibria, which are unstable [cf. to Eq. 12]. For $\beta > \alpha > \delta$ five equilibria exist, the symmetric equilibrium, which is unstable, the two trivial equilibria, which are unstable as well, the two additional equilibria, which are stable. This case always has $\beta^2 > \alpha\delta$, so it is equivalent to the dioecious zygotic selection model with the dynamics as illustrated in Eq. 13, Finally, for $\delta > \alpha > \beta$ five equilibria again exist, the symmetric equilibrium, which is stable, the two trivial equilibria, which are also stable and two additional equilibria, which are unstable:

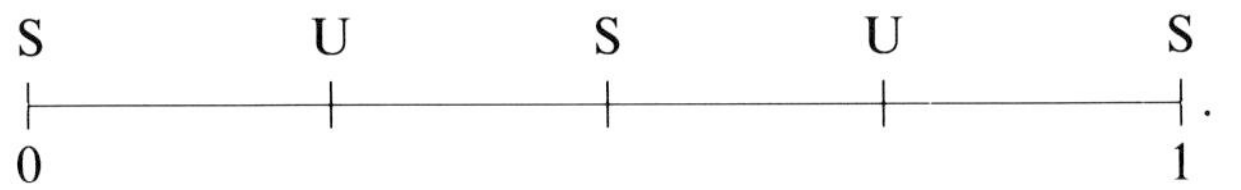

(25)

In this case we have $\beta^2 < \alpha\delta$, so it is never equivalent to a zygotic selection model. The dynamic behavior of this last case is found only for fecundity selection.

The symmetric dioecious model of zygotic selection only left two additional classes of dynamic behavior, namely, Eqs. (15) and (16), which both correspond to the situation in which one allele is protected and the other is not. This situation, of course, also has to be considered for fecundity selection. However, even for a symmetric situation more complicated dynamics may occur.The simple symmetric model in Table 9 may be generalized to the symmetric model shown in Table 10, and in this model Hadeler and Liberman (1975) have demonstrated the possibility of non-equilibrium dynamics, i.e., they have found cases (α and γ very different) where the population settles into a limit cycle where perpetual oscillations in the genotypic proportions occur.

The joint action of zygotic selection and fecundity selection can be modeled within the framework of the general fecundity selection model. Let z_{11}, z_{12} and

Table 10. General symmetric fecundity selection model

Genotypes	A_1A_1	A_1A_2	A_2A_2
A_1A_1	α	β	γ
A_1A_2	β	δ	β
A_2A_2	γ	β	α

z_{22} be the frequencies among zygotes, and describe zygotic selection as in Table 2, i.e., the survival in females is u_{11}, u_{12}, u_{22} and v_{11}, v_{12}, v_{22} in the males. Then it is easy to see that with random mating the equations governing the action of both selection components are again Eqs. (18) with z_{ij} substituted for f_{ij} and with the fecundities replaced by $u_{ij}v_{kl}F_{ij,kl}$. In this sense the fecundity selection model considered in this selection is the more general model, in that the dioecious zygotic selection model can be produced from the fecundity selection model.

Deviations from Random Mating

The genotypic proportions in a population of sexually reproducing organisms are determined by the pattern of gametic union which in turn is determined from the mating among individuals in the population and from the fecundities of these matings. So far we have looked at the effects of zygotic selection and of fecundity selection under the assumption that mating occurs at random (with the additional assumption of monogamy for fecundity selection). Deviations from this rule of random mating occur and are expected in many natural populations. In plants partial self-fertilization of ovules is a widespread phenomenon, and random mating among spatially proximate individuals may not produce random mating in an otherwise coherent and well-mixed population (see also Chap. 3).

Let us first consider the monoecious zygotic selection model as it applies to a plant that reproduces by mixed selfing and random mating (Kimura and Ohta 1971). Suppose that a fraction σ of the ovules are self-fertilized and that $1-\sigma$ are fertilized by a random pollen. In general, there is a differences between individuals formed by self-fertilization and individuals formed by outcrossing, in that the genotype at all the loci in an individual is influenced by the self-fertilization. An offspring formed by selfing has a higher chance of being a homozygote at any particular locus than an individual formed by outcrossing. This phenomenon is called **inbreeding** and usually inbred individuals show a lower general fitness than outbred individuals, a phenomenon known as inbreeding depression.

Without inbreeding depression the model can be described as in Table 11, where we use the mean probability of survival relative to the heterozygote which is given by

$$\begin{aligned} W = {} & [\sigma(f_{AA}+f_{Aa}/4)+(1-\sigma)p^2]w_{AA} \\ & +[\sigma f_{Aa}/2+2pq(1-\sigma)] \\ & +[\sigma(f_{aa}+f_{Aa}/4)+(1-\sigma)q^2]w_{aa}, \end{aligned} \tag{26}$$

and $p=f_{AA}+f_{Aa}/2$ is the gene frequency of allele A $(q=1-p)$.

With inbreeding depression the adult frequencies change to, for instance,

$$f_{AA}=\frac{\sigma(f_{AA}+f_{Aa}/4)vw_{AA}+(1-\sigma)p^2w_{AA}}{W^*}, \tag{27}$$

Table 11. Zygotic selection with partial selfing

Genotypes	AA	Aa	aa	Σ
Zygotes	$z_{AA} = \sigma(f_{AA} + f_{Aa}/4) + (1-\sigma)p^2$	$z_{Aa} = \sigma f_{Aa}/2 + 2pq(1-\sigma)$	$z_{aa} = \sigma(f_{aa} + f_{Aa}/4) + (1-\sigma)q^2$	1
Survival	w_{AA}	1	w_{aa}	
Adults	$f'_{AA} = \frac{w_{AA}z_{AA}}{W}$	$f'_{Aa} = \frac{z_{Aa}}{W}$	$f'_{aa} = \frac{w_{aa}z_{aa}}{W}$	1

where v describes the decrease in the fitness of inbred individuals due to the inbreeding depression ($v \leqslant 1$) and

$$W^* = [\sigma v(f_{AA} + f_{Aa}/4) + (1-\sigma)p^2]w_{AA} + [\sigma v f_{Aa}/2 + 2pq(1-\sigma)] + [\sigma v(f_{aa} + f_{Aa}/4) + (1-\sigma)q^2]w_{aa}. \tag{28}$$

However, this model with inbreeding depression is exactly the same as the model in Table 11 if we change the inbreeding fraction to

$$\sigma^* = \frac{\sigma v}{\sigma v + (1-\sigma)}, \tag{29}$$

which varies between zero and one. Thus, in the analysis of zygotic selection in a population with mixed selfing and random mating we may make the simplifying assumption that there is no inbreeding depression.

Without selection, i.e., if $w_{AA} = w_{aa} = 1$, the gene frequencies will stay constant, and the genotypic frequencies will converge to the proportions given in Table 12, where $F = \sigma/(2-\sigma)$. This convergence is rather fast since the distance to equilibrium is diminished by a factor $\sigma/2$ every generation, i.e., $z'_{Aa} - 2pqF = (\sigma/2)(z_{Aa} - 2pqF)$.

The condition for protection of allele A is found from the approximation

$$\begin{bmatrix} f'_{AA} \\ f'_{Aa} \end{bmatrix} \approx \begin{bmatrix} \sigma w_{AA}/w_{aa} & \sigma w_{AA}/(4w_{aa}) \\ 2(1-\sigma)/w_{aa} & (1-\sigma/2)/w_{aa} \end{bmatrix} \begin{bmatrix} f_{AA} \\ f_{Aa} \end{bmatrix}. \tag{30}$$

Table 12. Genotypic proportions with partial selfing and no selection

Genotypes	AA	Aa	aa	Σ
Parent	z_{AA}	z_{Aa}	z_{aa}	1
Offspring	$\sigma(z_{AA} + z_{Aa}/4) + (1-\sigma)p^2$	$\sigma z_{Aa}/2 + 2pq(1-\sigma)$	$\sigma(z_{aa} + z_{Aa}/4) + (1-\sigma)q^2$	1
Equilibrium	$p^2 + Fpq$	$2pq(1-F)$	$q^2 + Fpq$	

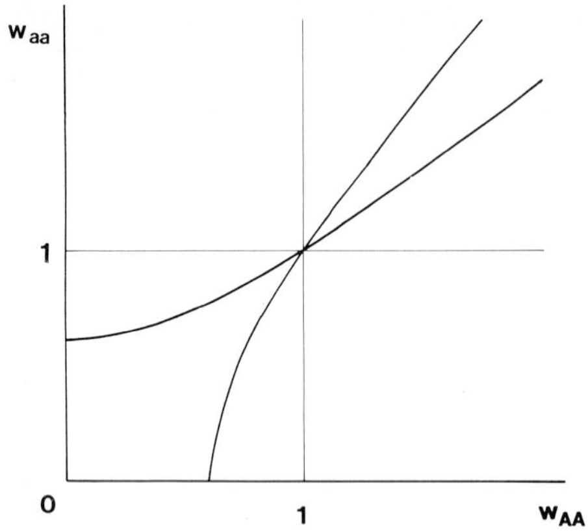

Fig. 8. The effect of partial selfing on the condition for protected polymorphism in the monoecious model for zygotic selection with fitnesses w_{AA}, 1 and w_{aa}. For random mating the condition for protected polymorphism is overdominance, $w_{AA} < 1$ and $w_{aa} < 1$ which is indicated by two *lines*. With partial selfing these lines bend, so the condition for protected polymorphism becomes more restrictive. For further explanation see text

The leading eigenvalue of the matrix in Eq. (30) is given by

$$\lambda = \frac{1 + \sigma(w_{AA} - 1/2) + [1 - \sigma + \sigma^2(w_{AA} - 1/2)^2]^{1/2}}{2w_{aa}}, \tag{31}$$

and A is protected when $\lambda > 1$. This condition for protection of an allele and the corresponding condition for protected polymorphism is shown in Fig. 8.

The condition for protection of allele A with random mating ($\sigma = 0$) is $w_{aa} < 1$ independent of the size of w_{AA}. However, as soon as selfing is allowed, then rare Aa individuals can produce AA offspring without the need for mating another rare Aa individual. Thus, the fate of the A allele when rare depends on the fitness of the homozygote AA, in that the increase is retarded when $w_{AA} < 1$ and it is enhanced when $w_{AA} > 1$ (Fig. 8).

The condition for protected polymorphism is more restrictive than for random mating. Protected polymorphism is guaranteed when w_{aa} and w_{AA} both are less than $1 - \sigma/2$, but when one of the fitnesses is larger than $1 - \sigma/2$ then the condition is that w_{aa} and w_{AA} cannot be too different (Fig. 8). When the condition for protected polymorphism is satisfied, then a globally stable equilibrium exists, and at this equilibrium we have zygotic frequencies as in Table 12 with $F < \sigma/(2 - \sigma)$. When both of the trivial equilibria are locally stable, then a nontrivial unstable equilibrium exists, and otherwise only the trivial equilibria exist. As with random mating, at most three equilibria exist (Kimura and Ohta 1971).

The model of partial selfing applies almost exclusively to monoecious plants capable of self-fertilization. However, similar results may be obtained from **assortative mating** where individuals with similar phenotypes have a higher probability of mating than expected from random mating. This phenomenon can give rise to a variety of mating patterns, many of which in themselves give rise to selection because the result is variation among phenotypes in their probability of producing offspring (see O'Donald 1980). However, some simple models of assortative mating do not give this side effect, and an example is given in Table 13. Here the females choose a mating partner of their own genotype with probability

Table 13. Genotypic assortative mating

Female genotypes	Male genotypes AA	Aa	aa	Σ
AA	$\alpha + (1-\alpha)f_{AA}$	$(1-\alpha)f_{Aa}$	$(1-\alpha)f_{aa}$	1
Aa	$(1-\alpha)f_{AA}$	$\alpha + (1-\alpha)f_{Aa}$	$(1-\alpha)f_{aa}$	1
aa	$(1-\alpha)f_{AA}$	$(1-\alpha)f_{Aa}$	$\alpha + (1-\alpha)f_{aa}$	1

α and choose a random male with probability $1-\alpha$, and Table 13 shows the genotypic proportions among the male partners of each type of female. The resulting rules for genetic change is just as with selfing in Tables 11 and 12 with α replacing σ.

The assumptions of the assortative mating model of Table 13 have the consequence that even a female of a rare genotype has a given, non-negligible probability of mating with a male of her own (rare) genotype. This may seem very unrealistic, but referring to the situation of selfing, it may be viewed as a model of spatially local mating, where the neighbors of a given individual have a higher than random chance of being of a similar genotype. This effect may also be studied by supposing a spatial structure of the population.

For simplicity assume that the neighborhood of a given individual is a collection of individuals distinct from the rest of the population, and we will term this collection a **subpopulation**. In every generation adult individuals may migrate between subpopulations, and in every generation the individuals in each population reproduce by random mating.

Consider two subpopulations, and let the frequency of allele A in population number $i (= 1 \text{ or } 2)$ be $p_i (q_i = 1 - p_i)$. Among the breeders of a population let a fraction m be immigrants and the rest, $1-m$, be born in the population where they breed. Then with zygotic selection the change in gene frequency is given by

$$p_1' = (1-m)p_1(w_{AA}p_1 + q_1)/W_1 + mp_2(w_{AA}p_2 + q_2)/W_2$$

$$p_2' = (1-m)p_2(w_{AA}p_2 + q_2)/W_2 + mp_1(w_{AA}p_1 + q_1)/W_1 , \quad (32)$$

where

$$W_i = p_i^2 w_{AA} + 2p_iq_i + q_i^2 w_{aa} \quad (33)$$

for $i = 1, 2$.

The model clearly leads to nonrandom mating in the whole population. In a sense this is assortative mating with respect to birth place. However, only when $p_1 \neq p_2$ does this kind of nonrandom mating have an effect, because if $p_1 = p_2$ then there will be random mating among the genotypes in the population. The effect of migration is to mix the two populations, so without selection the two populations eventually will become equal, i.e., $p_1' - p_2' = (1-2m)(p_1 - p_2)$. Zygotic selection will have the same effect when it pushes the population towards a globally stable equilibrium, as with directional selection (Fig. 3) or with overdominant selection (Fig. 5). However, underdominant selection (Fig. 4) may push initially different gene frequencies, $p_1 \neq p_2$, further apart, so the two subpopulations will maintain a gene frequency difference (Karlin and McGregor 1972a). If the two subpopulations were completely isolated from each other ($m = 0$), then four locally stable equilibria exist: $p_1 = p_2 = 0$, $p_1 = p_2 = 1$, $p_1 = 0$ and $p_2 = 1$, $p_1 = 1$ and $p_2 = 0$. The first two of these equilibria correspond to a homogeneous population, and the interesting equilibria in the present context are the two last equilibria. Suppose initially that $m = 0$, $p_1 = 1$ and $p_2 = 0$. Then assume that conditions in the populations change so m becomes positive, but small. Within each of the subpopulations this corresponds to a small perturbation of the gene frequencies away from the stable equilibrium, and selection will subsequently push the population back towards the equilibrium (Fig. 4). This selective push counteracts the perturbation due to recurrent immigration for sufficiently small immigration rates, and the population will settle on a selection-migration balance equilibrium where p_1 and p_2 are small.

Again we may obtain an overview of the properties of the model by considering the symmetric version where $w_{AA} = w_{aa} = w$. The model therefore always produces the equilibrium $p_1 = q_1 = p_2 = q_2 = 1/2$ which shows spatial homogeneity. At an equilibrium with $p_1 \neq p_2$ we must have $p_1 = q_2$ and $q_1 = p_2$ due to the symmetry. Thus, the equilibrium equation becomes

$$(p_1 - q_1)[wm(p_1^2 + q_1^2) + p_1 q_1(1 - w + 2wm)] = 0 \tag{34}$$

and this equation has roots with $p_1 \neq 1/2$ when $w > (1 - 4m)^{-1}$ or equivalently $m < (w - 1)/(4W)$ (Fig. 9). Thus, for low frequencies of migration in a homogeneous, but subdivided population, underdominant selection can stabilize a polymorphism showing spatial heterogeneity of the gene frequencies. This polymorphic state, however, does not correspond to a protected polymorphism: If one of the alleles has been fixed in the population, then the other allele cannot increase when introduced as rare.

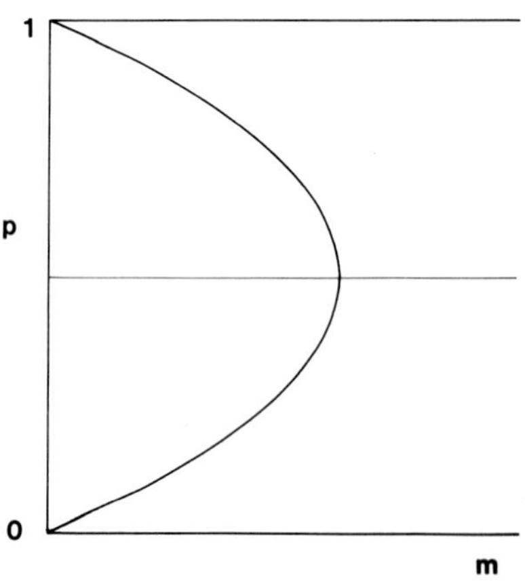

Fig. 9. The locally stable equilibria with symmetric underdominant zygotic selection ($w_{AA} = w_{aa} = w$) in two subpopulations as a function of the fraction *m* of immigrants in each population per generation. The pair of equilibria giving geographical variation exists for low values of m, and they fuse into the $p = 1/2$ equilibrium, giving geographical homogeneity when $m = (w - 1)/(4W)$

Inbreeding and spatial structure in the mating pattern are probably the most widespread and important sources of deviations from random mating in natural populations. However, they are both to a certain extent passive, in that the deviation from random mating is independent of the studied phenotype of the individual (unless the phenotype under study is the propensity to mate relatives including self-mating or the urge and ability to migrate). The assortative mating model (Table 13), on the other hand, is a model of mate choice or mate selection on the basis of phenotype, so it describes the reaction of individuals to other individuals in the population. Thus, we approach the first modeling of the reaction of an individual to the part of its biotic environment that consists of coexisting conspecific individuals.

Sexual Selection and Nonrandom Mating

The model of assortative mating in Table 13 assumed homogeneity among females in their propensity, α, to mate individuals with similar phenotypes. As soon as variation in this propensity is allowed, the assortative mating of the females induces selection on the genotypic variation among males. A very simple version of this is shown in Table 14. With the assumption of equal genotypic proportions among females and males, the proportion of genotype AA in the population of breeding males is given as

$$\begin{aligned} f^m_{AA} &= f_{AA}[\alpha_{AA} + (1-\alpha_{AA})f_{AA}] \\ &\quad + f_{Aa}[(1-\alpha_{Aa})f_{AA}] + f_{aa}[(1-\alpha_{aa})f_{AA}], \end{aligned} \tag{35}$$

so the "survival" of AA males from the population of adults to the population of breeding males is governed by the probability

$$v_{AA} = \alpha_{AA} + (1-\alpha_{AA})f_{AA} + (1-\alpha_{Aa})f_{Aa} + (1-\alpha_{aa})f_{aa}, \tag{36}$$

and we will address the effect as **sexual selection**. Thus, the sexual fitness value of AA males is dependent on the genotypic proportions in the population, in that it depends on the average propensity of females to show assortative mating:

$$\alpha_0 = \alpha_{AA}f_{AA} + \alpha_{Aa}f_{Aa} + \alpha_{aa}f_{aa}, \tag{37}$$

Table 14. Selective genotypic assortative mating

Female genotypes	Male genotypes: AA	Aa	aa	Σ
AA	$\alpha_{AA} + (1-\alpha_{AA})f_{AA}$	$(1-\alpha_{AA})f_{Aa}$	$(1-\alpha_{AA})f_{aa}$	1
Aa	$(1-\alpha_{Aa})f_{AA}$	$\alpha_{Aa} + (1-\alpha_{Aa})f_{Aa}$	$(1-\alpha_{Aa})f_{aa}$	1
aa	$(1-\alpha_{aa})f_{AA}$	$(1-\alpha_{aa})f_{Aa}$	$\alpha_{aa} + (1-\alpha_{aa})f_{aa}$	1

so the sexual fitnesses of the males are $v_{AA} = \alpha_{AA} + (1 - \alpha_0)$, $v_{Aa} = \alpha_{Aa} + (1 - \alpha_0)$ and $v_{aa} = \alpha_{aa} + (1 - \alpha_0)$.

The evaluation of these male sexual selection fitnesses, however, should be done with the recognition of the deviation from random mating. From Table 14 we obtain

$$p' = p/2 + (1/2)\{[\alpha_{AA} + (1 - \alpha_0)]f_{AA} + [\alpha_{Aa} + (1 - \alpha_0)]f_{Aa}/2\}, \tag{38}$$

which is the average of Eqs. (5) and (6) if we assume that the zygotes are not formed by random mating and if selection occurs in males only with the above calculated male sexual fitnesses. Thus formally the male sexual selection is equivalent to zygotic selection in the males with fitnesses given by the male sexual fitnesses. The model is different from the simple dioecious zygotic selection model, in that the male sexual selection fitnesses depend on the composition of the population. For this reason the selection is technically referred to as **frequency-dependent selection**, and it is the rule that selection is interractions between individuals or reactions to the biotic environment. In this model the frequency dependence of the sexual fitnesses of the males is particularly simple, because the composition of the population only enters as the average propensity toward assortment [Eq. 37], so it is mainly the strength of selection and not as much the pattern that changes with the genotypic frequencies.

To complete the description of selective assortative mating we need to describe the genotypic proportions in the offspring population. From Table 14 we obtain

$$f'_{Aa} = 2pq - \alpha_{AA} f_{AA} q - \alpha_{aa} f_{aa} p. \tag{39}$$

This frequency of heterozygotes is, as expected, always less than the frequency obtained from random union of the gametes produced by the females and the breeding males.

The model of Table 14 is a particularly simple version of models for sexual selection in the males exerted by **female choice** among the male phenotypes. The general characteristic of female choice models is that each female phenotype shows varying preferences of the different male phenotypes described by the preferences, c, which may in turn depend on the frequency of the various male phenotypes (Table 15). In this general model the male sexual fitnesses are given by the average frequency with which the male is chosen as a partner, e.g.,

$$v_{AA} = c(AA)_{AA} f_{AA}/C(AA) + c(Aa)_{AA} f_{Aa}/C(Aa) + c(aa)_{AA} f_{aa}/C(aa). \tag{40}$$

It is beyond the scope of the present text to go into a detailed discussion of these kinds of selection models. An authoritative discussion can be found in O'Donald (1980).

The model of Table 14 corresponds to the preferences $c(AA)_{AA} = (1 - \alpha_{AA}) + \alpha_{AA}/f_{AA}$, $c(AA)_{Aa} = c(AA)_{aa} = 1 - \alpha_{AA}$, etc., so it is a very simple, and not a particularly realistic model of the phenomenon of sexual selection by female choice. The simplest, and the classical model of female choice is that the choice coefficient, c, does not depend on the female type. In this case it is easy to see that the phenomenon is entirely equivalent to differences in the probability of

Table 15. Selection due to choice by AA females of male genotypes

	Male genotypes			
	AA	Aa	aa	Σ
Preference	$c(AA)_{AA}$ $c(AA)_{AA}f_{AA}$	$c(AA)_{Aa}$ $c(AA)_{Aa}f_{Aa}$	$c(AA)_{aa}$ $c(AA)_{aa}f_{aa}$	C(AA)
Frequency	$c(AA)_{AA}f_{AA}/C(AA)$	$c(AA)_{Aa}f_{Aa}/C(AA)$	$c(AA)_{aa}f_{aa}/C(AA)$	1

survival among male types with subsequent random mating. Therefore, the model becomes equivalent to the dioecious zygotic selection model with no selection in females and with the possibility of frequent-dependent fitnesses in the males.

A simple and widely discussed female choice model is the model of **rare male mating advantage** or minority effect (Petit 1954). Here the preferences may be equal among the females, and each male type is preferred less the more common it is. For instance, a rare genotype advantage in the males with a linear decline with male frequency in the female preference may be written as

$$c_{AA} = c_0 - c_1 f_{AA}, \tag{41}$$

where $c_0 > c_1 > 0$. With this model the male sexual fitnesses [Eq. (40)] are given by the simple (frequency-dependent) expression

$$v_{AA} = c_0 - c_1 f_{AA}/C, \tag{42}$$

where

$$C = c_0 - c_1(f_{AA}^2 + f_{Aa}^2 + f_{aa}^2). \tag{43}$$

By using the methods from the section on zygotic selection it is easy to obtain the obvious result that rare male genotype advantage always leads to a protected polymorphism, and with the present simple model the population always settles at a stable equilibrium with $p = 1/2$.

The characteristic of the female choice model is that all females are assumed to be mated independent of the types of males available. Therefore, the processes of mating inflicts no sexual selection on the females. A completely symmetric definition of **male choice** as a source of sexual selection in females may be given. The female choice model is a scenario where the individual female is presented with a choice among a number of males, and similarly the male choice model can be envisioned as a situation where the females attract males and the individual male chooses which female to approach. However, if we maintain the assumption that all females are mated, then male choice will never exert any selection on the females. Thus, a male choice model where all males are mated will only be a source of nonrandom mating, as for instance, with strict monogamy. An alternative to the male choice model is the **male competition model**, which like the choice model assumes that the females are passive, but which assumes that the males compete for the limited number of available females in the population. In a sense the females are considered as a limited resource for which the males compete.

The female choice model of Table 15 will in general lead to nonrandom mating between the breeding females and the breeding males. However, if the choice parameters are the same for each female, then the genotypic frequencies among the chosen males are the same for each female type, and the result is random mating between the females and the breeding males. The simple rare male mating advantage model considered above is an example where random mating prevails. Similarly, male mating competition may often be characterized by random mating.

In higher organisms the process of mating and breeding may contain many aspects important to the understanding of natural selection. A beautiful and well-worked example is the investigation of the polymorphism of plumage color in a Shetland population of the arctic skua by O'Donald (1983). A rough description of the polymorphism is that 0.25 of the population shows a pale plumage characterized by white underparts and 0.75 shows a varying degree of melanic plumage clearly distinct from the pale phenotype. The pale phenotype corresponds to the recessive homozygote aa at an autosomal locus, and the melanics are of the dominant phenotype, either Aa or AA. The skuas breed in early summer and breeding is initiated by pair formation. The breeding success varies with the date of breeding, in that early breeders on the average fledge 1.6 chicks and this number declines to less than one by the end of the breeding season. Pale males are slower than melanics in forming pairs, and on the average the pale males breed 4 days later than the melanics. O'Donald shows that this is due to a female preference for melanic males resulting in little, if any, deviations from random mating.

Sexual selection and nonrandom mating are intimately connected to fecundity selection. The pattern of mating is an important determinant of the effect of fecundity selection, and certain aspects of sexual selection may equivalently be viewed as aspects of fecundity selection. In a dioecious species a male produces offspring only when it participates in a mating, so in this sense male sexual fitness is an important aspect of the number of offspring that a given type of males produces.

The fecundity selection model previously considered was simplified by the assumption of monogamy, but to illustrate the range of possible interrelations between sexual selection and fecundity selection we may consider the opposite extreme, namely that repeated matings occur and the outcome of each mating is exactly one offspring. Here sexual selection and fecundity selection as seen from the male side fuse into one aspect of selection, where fitness is measured by the number of matings a male participates in. The male aspect of this best described as sexual selection, because the process is dominated by the mating participation. The female aspect of this process will often be dominated by the number of offspring an individual can produce, and in this case it will be an aspect of fecundity selection. However, if the limiting factor for the female fecundity is the number of matings, e.g., the number of males a female can attract, then the process comes closer to sexual selection.

The two simple models, female choice and male competition, presupposed no selection in females. Actually these two models represent extremes of a range of models. An intermediary model is, for instance, that the males compete for the opportunity to be chosen by the female. However, the two kinds of behavior described in the models are biologically fundamentally different. The male competition model is a model of sexual selection in males originating from the stride of individual males to secure their contribution to the offspring generation. This stride is no different from the stride to secure other necessary resources, such as food, for instance. In the female choice model, on the other hand, the agent of selection is the female, but the target is the male. This raises the question that has puzzled evolutionary biologists for more than a century: Why should a female choose among the males, what is the benefit to her?

The answer to this question is usually recognized to lie in the effect of the mating on the fecundity of a female or on the general fitness of her offspring. By choosing a good and able male she may maximize her fecundity and produce prime offspring. This may look like a sensible explanation until we start looking at the examples of ridiculous criteria for "good and able males" as given by various morphological or behavioral display characters, the peacock's tail being a well-known extreme. Although plausible explanations of this phenomenon may be produced, it is important to note that nothing in the Darwinian theory of evolution requires that female choice in relation to a particular character has to produce a benefit to the female that relates to that character. The only requirement for a plausible explanation is that the individual female benefits (or has benefited in the past) by making a choice, for instance, making choices to avoid deterimental hybridizations with other species. When a female choice has evolved for whatever reason, its occurrence influences the evolution of characters in the male. Only natural selection on variation in the female behavior will change the female choice, but as long as every female is mated there will be no selection on the variation of choice.

However, the female phenotype of mate choice may be founded on the same character that she chooses in the male. An example of this may be situations where mating occurs on characteristic substrates possibly in connection with male displays, tilts, and mating territory defence (lek behavior). Here variations in the choice of substrate may be expressed equally in females and males, so if again all females are mated, then random mating on each mating arena will produce female choice, resembling assortative mating. Parsons (1983) provides the example in Table 16 of lek behavior in two species of picture winged *Drosophila* on the underside of bracket fungi. Here an excess of males are always caught, whereas an excess of females are found on the soft fungi where the eggs are laid.

Sex pheromone production and perception is an example where the choice and the character chosen are different. A well-studied example in the European corn borer moth, *Ostrinia nubilatis*, has revealed three genes involved in the differentiation of two types of natural populations (Roelofs et al. 1987). The

Table 16. Sex ratios at mating sites and oviposition sites in *Drosophila*

	D. mycetophaga			*D. polypori*		
	Female	Male	Σ	Female	Male	Σ
Bracket fungi	27	131	158	53	97	150
Soft fungi	12	10	22	13	7	20

female moth attracts males by a sex pheromone, and two forms E and Z of that pheromone exist. The two naturally occurring types of females produce pheromone blends which are almost pure E and Z (fraction 0.99 and 0.03 of E, respectively), and these types correspond to the homozygotes, AA and aa say, at an autosomal locus. The natural populations are either monomorphic AA or monomorphic aa, and we will refer to these two types of populations as E and Z populations. The heterozygote Aa produce an intermediary blend (0.65 of E). Another autosomal gene with the alleles B and b determines the type of pheromone-responding olfactory cells in the male antennae. The olfactory cells of BB males respond much more strongly to the E pheromone than to the Z pheromone, and the cells of the bb males have the opposite characterization. The heterozygous Bb males have two types of olfactory cells corresponding to the cells of the two homozygotes. The E populations are monomorphic BB and the Z populations are monomorphic bb.

The response of males to the different kinds of females, however, is controlled by genes at a sex-linked locus with alleles S and s, say. The E populations are monomorphic for allele S and the Z populations are monomorphic for the allele s. The aabbss males are attracted strongly by pheromone mixtures which are almost pure Z (the pheromone mixture produced by aabbs females) and only few of these males react to more even mixtures of E and Z. The AABBSS males react strongly to almost pure E (the pheromone mixture of the AABBs females), but they are less choosy, in that they are mildly attracted to more intermediate mixtures between E and Z. The hybrid males AaBaSs are attracted moderately to almost pure Z and to intermediate mixtures, and only few are attracted to the almost pure E pheromone. The sex linkage of the gene for male response is disclosed by comparing paternal and material backcrosses of the hybride individuals. The hybrid males are expected to have the same genotype irrespective of the sex of the parents, but a hybrid female will carry only the allele transmitted from her father. Therefore, male offspring from the backcrosses of hybrid females to the paternal population will show the same phenotypes as the paternal population. The backcrosses of hybrid males to the maternal population will produce a mixture of pure and hybrid types.

The Estimation of Fitness Values

The different **selection components**, zygotic selection, fecundity selection and sexual selection, express themselves quite differently in the population. Zygotic selection occurs by the differential elimination of various types from the population: ideally we would see the genotypic proportions change through the development of the organism. Fecundity selection, at least the female part of fecundity selection, is seen from the number of offspring that the different individuals produce: it is evaluated as any other phenotypic trait. Sexual selection, at least the male part, may leave no directly observable trace in the population: after reproduction, breeding and nonbreeding males will often be indistinguishable. The male breeding population may in practice only be deduced from a study of the offspring, unless the breeding pair can be observed while breeding, as the arctic skua. The observational evaluation of the totality of natural selection working on some natural variation will therefore have to be shaped after the characteristics of the studied organism. However, there are still some general principles which may serve as a guide for the design of observations.

In the following we will consider the observation of natural selection that influences simple genotypic variation; to be exact, we will look at the observation of selection induced on some particular genetic variation. To simplify discussions, we will assume that we are studying an **autosomal locus** with two alleles, A and a.

In the theoretical development of the evolutionary consequences of natural selection, we studied each of the selection components in isolation assuming that other aspects of selection were absent. This assumption cannot be made in the context of observations unless corroborating observations exists. Therefore, it is important to free the design of the observational procedure from such assumptions, because otherwise serious biases are produced.

As an example let us return to the estimates of zygotic fitness values [Eq. (17)]. These estimates may be used whenever the genotypic proportions among zygotes are observed, or are otherwise known. Suppose that we observe a population generation after generation at the stage of adults (typical in insects like *Drosophila*) and assume, for simplicity, that we never see any difference between the genotypic proportions in the two sexes. In this population we do not know the genotypic proportions among zygotes, so we cannot estimate zygotic selection unless we get information about the zygote formation.

Let us see what will happen if we try to do as in the theoretical treatment in Table 1. We assume reproduction by random mating of the adults. Now we can calculate the zygotic frequencies from the gene frequencies among adults, $p = x_{AA} + x_{Aa}/2$ and $q = x_{aa} + x_{Aa}/2$, so the estimates [Eq. (17)] become

$$w_{AA} = \frac{x'_{AA}}{p}\frac{2q}{x'_{Aa}} \quad \text{and} \quad w_{aa} = \frac{x'_{aa}}{q}\frac{2p}{x'_{Aa}}, \tag{44}$$

where x'_{AA}, x'_{Aa} and x'_{aa} are the genotypic frequencies among adults in the next generation. If the zygotic frequencies are in fact the Hardy-Weinberg proportions corresponding to the gene frequencies p and q, then the estimates (Eq. (44)] are correct, but what happens if our assumption is wrong? This question was answered by Prout (1965), and the answer is that the estimates [Eq. (44)] may be very misleading.

Suppose we have been misguided by our preconception about the studied variation, in that zygotic selection is completely absent, and that selection only

Table 17. Selection among adults with random mating

Genotypes	AA	Aa	aa	Σ
Frequency in adults	p^2	$2pq$	q^2	1
Fitnesses	v_{AA}	1	v_{aa}	
Frequency in zygotes	p'^2	$2p'q'$	q'^2	1
Frequency in adults	p'^2	$2p'q'$	q'^2	1

occurs among adults after the time of observation (as, for instance, sexual selection). Assume for simplicity further, that there are no differences between the sexes. The model is summarized in Table 17 where

$$p' = p\frac{pv_{AA} + q}{p^2v_{AA} + 2pq + p^2v_{aa}} \tag{45}$$

and $q' = 1 - p'$. Now the estimates [Eq. (35)] become

$$w_{AA} = \frac{pv_{AA} + q}{qv_{aa} + p} \text{ and } w_{aa} = \frac{qv_{aa} + p}{pv_{AA} + q} \tag{46}$$

which are endoved with some odd properties. First, there seems to be two estimates, but they always have the property that $w_{AA}w_{aa} = 1$, so in reality there is only one. Second, the estimates are **frequency dependent**, and third, if the fitnesses in Table 17 are overdominant, so a stable polymorphism results, then the fitness estimates [Eq. (44)] produce the impression that the polymorphism is maintained by favoring the rare genotypes (Fig. 10).

Thus, the unjustified assumption of no sexual selection and no fecundity selection will produce estimates of zygotic fitnesses which do not bear any

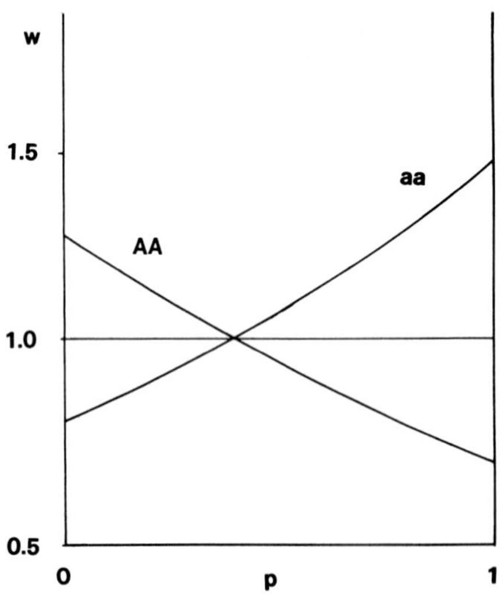

Fig. 10. The variation of the fitness estimates (Eq. (46)] under the model of Table 17 with $v_{AA} = 0.7$ and $v_{aa} = 0.8$. For explanation see text

relationship to the situation in the studied population. The bias is still there if zygotic selection is working in the population, the true zygotic fitnesses are just multiplied on the spurious values [Eq. (46)] (Prout 1965). Similarly, the effect is almost the same if differences between the sexes are allowed for the breeding components of selection (Christiansen et al. 1977a). For instance, with only sexual selection in males the estimates [Eq. (44)] has the property that $w_{AA} + w_{aa} = 1$.

The estimates [Eq. (17)] of fitnesses is founded on the assumption that the genotypic frequencies are known at the life stage before the selection begins and at the life stage after the selection ends. Therefore, to estimate zygotic fitness values we need to observe or to reliably estimate the genotypic proportions among zygotes in the population.

A general way to gain knowledge about the zygote population exists in organisms that protect and carry their offspring, so a sample of the adult population at the time of breeding may include combinations of parents and offspring, often in the way of motor-offspring combinations (Christiansen and Frydenberg 1973). As an example we may consider an esterase polymorphism in a population of the live-bearing marine fish *Zoarces viviparus*. The fish mate in late summer and the females go pregnant until mid winter, when they give birth. In autumn a sample of the adult population is procured, and the individuals are sorted into genotype and sex (Table 18). From each pregnant female a random offspring is chosen and genotyped to provide the mother-offspring combinations in Table 19. In addition, for each female, the total number of offspring is counted. The population of newborn offspring can then be reliably described from the mother-offspring combinations by weighting each mother-offspring combination by its average fecundity.

The information in the data extends beyond the estimation of genotypic proportions among zygotes. The selection components involved in breeding can be studied in the data. Primarily, the segregation of the alleles in the female can be observed in the mother-offspring combinations, as Mendelian segregation produces half of the offspring from heterozygous mothers as heterozygotes (1360 observed and 2696/2 = 1348 expected, and these two numbers do not differ more

Table 18. Adult data in *Zoarces viviparus*. (Christiansen et al. 1977b)

	Females		Adults		
Genotypes	Mothers	NPF[a]	Female	Male	Σ
AA	821	43	1008	693	1701
Aa	2696	161	3344	2332	5676
aa	2418	160	3029	2201	5230
Σ	5935	364	7381[b]	5226[b]	12607[b]

[a] Nonpregnant females.
[b] Includes samples not taken in the breeding season.

Table 19. Simple mother-offspring data in *Zoarces viviparus*. (Christiansen et al. 1977b)

Female Genotypes	Offspring genotypes AA	Aa	aa	Σ
AA	305	516		821
Aa	459	1360	877	2696
aa		877	1541	2418
Σ	764	2753	2418	5935
Gametes[a]	1641		2934	4575

[a]For explanation see text.

than expected in a random sample of 2696 individuals from a large population with equal numbers of heterozygotes and homozygotes). Genuine deviations from this expected ratio may be ascribed either to abnormalities in gamete production or to differential survival of the various types of gametes, **gametic selection**. The comparison between adult females and breeding females allows the estimation of female sexual fitnesses by the use of estimates similar to Eq. (17). The breeding males are represented as the male gametes that fertilized the females. For instances, the observation of an Aa mother with an AA offspring is an observation of an A-carrying male gamete. Similarly, every mother-offspring combination, except Aa with Aa, unambigously determines the allele supplied by the sperm. Thus, the 5935 mother-offspring combinations determine 5935 – 1360 = 4575 observed sperm types, as indicated in the last line of Table 19. Deviations from random mating reveal themselves as differences among females in the gene frequencies among the sperm that fertilized their offspring. The comparison between the population of adult males and the population of transmitted male sperm reveal the joint action of sexual selection and gametic selection in the males. Thus, the male sexual fitnesses cannot be estimated from data, only the average fitness of the two alleles can be estimated according to the joint action of sexual and gametic selection. This is a very incomplete description, because if male sexual selection is the only selection component present, then at a stable polymorphic equilibrium to gene frequency change is expected, and the two alleles have the same average sexual fitness in the males.

The statistical analysis of the data in Tables 18 and 19 is described by Christiansen et al (1977b). The breeding in the population can be satisfactorily described by random union of gametes produced by the adults, i.e., we need not invoke sexual, gametic or fecundity selection in the description of data. The genotypic proportions in adults and zygotes differ, and the fitness estimates [Eq. (17)] are given in Table 20, so the polymorphism is subject to underdominant zygotic selection.

The simple mother-offspring combinations determine a male gamete for almost every mother. Taking more than one offspring per mother would

Table 20. Zygotic selection in *Zoarces viviparus*. (Christiansen et al. 1977b)

Genotypes	AA	Aa	aa	Σ
Frequency in zygotes	0.129	0.461	0.410	1
Fitnesses	1.065	1.000	1.037	
Frequency in adults	0.135	0.450	0.415	1

detemine more male gametes per mother-offspring combination, so from sufficient offspring we can determine the genotype of the father. This would provide information on the genotype frequencies among breeding males and on the segregation of male gametes in terms of complete mother-offspring combinations (Christiansen 1980). However, this only works if the brood has a single father, i.e., only if the female mate only once for every brood. *Zoarces viviparus* females may mate more than once, so the scheme does not work. For truly monogamous species the sampling of complete mother-offspring combinations is possible (Siegismund and Christiansen 1985).

The description of selection components from mother-offspring data is founded on the same discreteness in the processes of selection as we assumed in our theoretical discussions. The estimation of fitnesses in Eq. (17) relies on a comparison between two discrete stages. Thus, in *Zoarces viviparus* when we refer to a "breeding population" this ought to refer to the collection of breeding individuals in nature which we have sampled. The breeding in *Zoarces viviparus* occurs in nature at a discrete time prior to our sampling, so the breeding population is well defined as a discrete unit. This turns out to be the important assumption in the use of mother-offspring combinations for the analysis of selection. *Zoarces viviparus* in fact has overlapping generations (it matures at an age of 2 years and breeds for many subsequent years), but although this complicates the analysis of data, age determination of the fish will allow an extended version of the analysis (Christiansen and Frydenberg 1976). In fact, the data that we referred to here covers a period of 4 years, and the year of birth of every age class can be determined, so the age structure of the samples can be interpreted in terms of the cohort structure of the population (Table 21). The zygotic selection in Table 20 is due to differential survival of zygotes to the first breeding at the age of 2 years.

If the breeding is not discrete, then a sample including mother-offspring combinations reflects only the genotypic composition of the females that happen to be breeding at the time of sample. As an example, we can return to the arctic skua. Here the phenotypic composition of the breeding pairs can be evaluated at any point in time by observing the nesting pairs. However, this may provide a very biased view of the process of breeding, because the male phenotypes differ in their average time of breeding. Early in the breeding season an excess of melanic males will be breeding, and this may change later in the season. In addition, the effective fecundity, measured in terms of fledgelings, varies through the season, so

Table 21. Year of birth for age classes in the data on *Zoarces viviparus.* (Christiansen et al. 1977b)

	Year of sample			
Age class (years)	1971	1972	1973	1974
0	1971	1972	1973	1974
1	1970	—	—	1973
2	1969	1970	1971	1972
3	1968	1969	1970	1971
4	1967	1968	1969	1970
5	1966	1967	1968	1969
6 +				

a proper description of the breeding process can be achieved only by following the breeding of every single pair through the season as O'Donald (1983) did. Mass samples of the population at discrete times will not provide proper information. In fact, the variation among the male morphs in the processes of breeding described by O'Donald would cause sexual selection even if all males were mated.

These complications are a general characteristic of organisms where the event of breeding cannot be separated into discrete coordinated stages like in *Zoarces viviparus*, and these complications can be overcome only by collecting information on the performance of the individual. In organisms like humans, without a defined breeding period, this requires demographic data for the evaluation of variation pertaining to natural selection (Bodmer 1968).

Selection with Overlapping Generations

In the theoretical treatment of the effect of selection in the first sections of this chapter we have studied the change in genetic constitution of the population from a parent population to an offspring population. These discrete generation models simplify the genetic considerations, because we can often separate the important phenotypic and ecological effects from the important genetic effects (Fig. 2). Nevertheless, we must realize that in many organisms generations overlap, as an individual may breed simultaneously with its parents and/or its offspring. This breaks the nice distinction between parents and offspring giving a discreteness in time, and replaces a layered structure, the **cohort structure**, which joins together individuals born at the same time into a cohort of individuals (Table 21). Suppose initially, as in *Zoarces viviparus*, that breeding occurs at discrete times, say once every year, and let $C(t,0)$, $t = \cdots, -2, -1, 0, 1, 2, \cdots$, be the number of offspring produced at breeding in the t'th year (we will for simplicity only consider the monoecious situation). Thus, $C(t,0)$ is the initial number of individuals in cohort t, and let $C(t,\tau)$, $\tau = 1, 2, \cdots m$, be the number of individuals of cohort t that survive τ years, where m is the maximal age possible. The number of individuals in the population before breeding in the t'th year is therefore given by

$$N(t) = \sum_{\tau=1}^{m} C(t-\tau,\tau). \tag{47}$$

At any point in time we must recognize the population as **age structured** with $N(t,\tau) = C(t-\tau,\tau)$ as the number of individuals of age τ at time t, and the fraction of individuals of age τ in the population at time t is $f(t,\tau) = N(t,\tau)/N(t) = C(t-\tau,\tau)/N(t)$. This quantity describes the age distribution in the population at time t and

$$1 = \sum_{\tau=1}^{m} f(t,\tau). \tag{48}$$

If the survival of an individual of age τ from one year to the next is $P(\tau)$ then $C(t+1,\tau+1) = P(\tau)C(t,\tau)$, so $P(m) = 0$. If the fecundity of an individual is $B(\tau)$ then

$$C(t,0) = \sum_{\tau=1}^{m} B(\tau)C(t-\tau,\tau) = \sum_{\tau=1}^{m} B(\tau)N(t,\tau). \tag{49}$$

Thus, if we write the age structure in a vector:

$$\mathbf{N}(t) = \begin{bmatrix} N(t,1) \\ N(t,2) \\ \vdots \\ N(t,m) \end{bmatrix}, \tag{50}$$

We can express the development of the population as

$$\mathbf{N}(t+1) = \mathbf{L}\cdot\mathbf{N}(t), \tag{51}$$

where **L** is the Leslie matrix:

$$\mathbf{L}(t) = \begin{bmatrix} P(0)B(1) & P(0)B(2) & \cdots & P(0)B(m-1) & P(0)B(m) \\ P(1) & 0 & \cdots & 0 & 0 \\ 0 & P(2) & \cdots & 0 & 0 \\ \vdots & \vdots & & \vdots & \vdots \\ 0 & 0 & \cdots & 0 & 0 \\ 0 & 0 & \cdots & P(m-1) & 0 \end{bmatrix}. \tag{52}$$

Similarly, we can write the age structure as the vector $\mathbf{f}(t) = \mathbf{N}(t)/N(t)$.

The Leslie matrix summarizes the life history of the organism. The age μ, where maturity is reached, is characterized by $B(\mu) > 0$ and $B(\tau) = 0$ for $\tau < \mu$, so the first line in the Leslie matrix has zero's until the μ'th position. The age-dependent mortality is characterized by the dependence of $P(\tau)$ on the age τ, in that the probability of survival to age τ is

$$l(\tau) = \prod_{\sigma=0}^{\tau-1} P(\sigma). \tag{53}$$

This function, $l(\tau)$, $\tau = 1,2,\ldots,m$, is know as the life-table of the organism. (The demographic description of an age-structured population will only be sketched here. For a more comprehensive treatment see Charlesworth 1980).

The growth of the population can be judged from the largest eigenvalue λ of the Leslie matrix, in that

$$N(t) = L^t N(0) \approx N(0)\lambda^t f(\infty), \tag{54}$$

where $\mathbf{f}(\infty)$ is the vector providing the **stable age distribution** (and an eigenvector to $\mathbf{L}$) to which the population age structure converges. Therefore, the population size grows exponentially with the factor λ: $N(t) \approx \lambda^t N(0)$, and $N(t,\tau) \approx \lambda^t N(0) f(\infty,\tau)$. Using this we obtain from Eq. (49) that $C(t,0) \approx \lambda^t N(0) B$, where B is the average fecundity in a population with a stable age structure, and therefore that $C(t,\tau) \approx \lambda^t N(0) B l(\tau)$. Introducing these two expressions in Eq. (49) provides Euler's equation:

$$\sum_{\tau=\mu}^{m} l(\tau)B(\tau)\lambda^{-\tau} = 1. \tag{55}$$

which has the growth factor λ as its largest root. At the same time we obtain the stable age distribution as

$$f(\infty,\tau) = \frac{l(\tau)\lambda^{-\tau}}{\sum_{\sigma=1}^{m} l(\sigma)\lambda^{-\sigma}} \tag{56}$$

for $\tau = 1,2,\ldots,m$. The population grows when $\lambda > 1$ and this happens when each individual on the average leaves more than one offspring, i.e. $R_0 > 1$, where

$$R_0 = \sum_{\tau=\mu}^{m} l(\tau)B(\tau). \tag{57}$$

The growth factor λ of the population summarizes the survival and the fecundity of the individuals in the population. If we form a mixed population of two types, A and B, of individuals that do not interbreed, then the population will eventually be dominated by the type with highest growth factor. This is seen from Eq. (54) which implies that the ratio between population size of type A and the population size of type B will change approximately with the factor λ_A/λ_B every time unit. However, if we introduce genetic variation in the survival probabilities P and the fecundities B, then it is clear that we cannot predict the development of the population from a summary measure like λ, and the general considerations of selection in an age-structured population are quite complicated.

We may nevertheless obtain some qualitative description of selection by considering the condition for protection of an allele. Let us assume that reproduction is by random mating, so in a population almost monomorphic aa we only have to consider the possibility that the rare allele A exists in Aa individuals. The growth in the monomorphic aa population is characterized by the growth factor λ_{aa} corresponding to the survival probabilities $P_{aa}(\tau)$,

$\tau = 1, 2, \ldots, m$ and the fecundities $B_{aa,aa}(\tau), \tau = 1, 2, \ldots, m$. The outcome of selection is specified in terms of survival probabilities and fecundities of aa and Aa individuals. This situation parallels the above situation of two types with no interbreeding, in that half of the offspring from Aa × aa matings are ascribed to the Aa individual and half to the aa individual, and it does not change anything in the process if we ascribe all Aa offspring to the Aa individual and all aa offspring to the aa individual. Thus, we may consider the Aa individuals as perpetuating themselves without intercrossing with the common aa individuals, which as usual are assumed not to be affected by the intruding Aa individuals. Thus, the relative proportion of Aa individuals will increase when they are rare when $\lambda_{Aa,a} > \lambda_{aa}$ and it will decrease when $\lambda_{Aa,a} < \lambda_{aa}$, where $\lambda_{Aa,a}$ is the growth factor in a hypothetical population with survival probabilities like the heterozygote, i.e., $P_{Aa}(\tau)$, $\tau = 1, 2, \cdots, m$, and fecundities like those of the matings between the heterozygote Aa and the homozygote aa, i.e., $B_{Aa,aa}(\tau)$, $\tau = 1, 2, \ldots, m$. The condition for protected polymorphism in an age-structured population is therefore $\lambda_{Aa,a} > \lambda_{aa}$ and $\lambda_{Aa,A} > \lambda_{AA}$.

When fecundity selection is absent, i.e., if all genotypes have the same **age-specific fecundities**, the conditions for allele protection simplifies, in that $\lambda_{Aa,a} = \lambda_{Aa,A} = \lambda_{Aa}$, say. The three growth factors λ_{AA}, λ_{Aa} and λ_{aa} act much like the zygotic fitnesses in the nonoverlapping generation model (Norton 1928; Charlesworth 1980). When the three factors are ordered there will be directional selection for the allele with the highest growth factor, and a trivial equilibrium will be globally stable. With underdominance both trivial equilibria will be locally stable, and they will be the only stable equilibria. Finally, with overdominance the population has a globally stable polymorphic equilibrium.

The general problem in handling a population with overlapping generations is that we do not obtain the same simplification from the assumption of random mating as we got in the case of non-overlapping generations. We cannot jump to consider a single variable, the gene frequency in the population. In essence what happens is that zygotic selection works in a simple way within each cohort, in that it begins with genotypic frequencies in Hardy–Weinberg proportions and selection gradually shapes these proportions. However, different cohorts do not begin with the same gene frequencies, unless the population is at a gene frequency equilibrium, so the process of selection is different in each cohort. Further, the individuals may act as parents at many different times during their life.

The same results hold for selection when the model is extended to a model where breeding may occur at any time, a continuous time model. This model is easier to handle in certain situations, so we will briefly introduce it. The survival probabilities $P(\tau)$ are replaced by the age-dependent death rate $d(\tau)$, so the probability of survival for an individual of age τ from time t to $t + \Delta t$ is $1 - d(\tau)\Delta t$ when Δt is small. Similarly, the probability that an individual of age τ reproduces in the interval from time t to $t + \Delta t$ is $\beta(\tau)\Delta t$ when Δt is small, so on the average an individual of age τ leaves $B(\tau)\beta(\tau)\Delta t$ offspring in the interval from time t to $t + \Delta t$. The description of reproduction is usually compounded into the age-dependent birth rate per individual given by $b(\tau) = B(\tau)\beta(\tau)$. The life-table then becomes

$$l(t) = \exp\left[-\int_0^t d(\tau)\,d\tau\right]. \tag{58}$$

The population will eventually grow exponentially, i.e., $N(t) = N(0)e^{rt}$, and the Euler equation to determine the growth rate r is

$$\int_\mu^m l(\tau)b(\tau)e^{-r\tau}dt = 1. \tag{59}$$

The conditions for allele protection are as before with $\lambda = e^{rt}$, so we may judge the fitness of the various genotypes from the parameter r.

This continuous time model and the model of discrete but overlapping generations both show exponential increase in population size. This is a general property of density-independent population growth models, i.e., models where the population growth is seen as a result of the characteristics of the individual independent of the other individuals in the population. These models are also called **Malthusian growth models** after Malthus, who in 1798 gave the first comprehensive discussion of the biological consequences of these models. In this tradition, the growth rate parameter, r, is called the Malthusian parameter.

A particularly simple model emerges when the birth and death rates are assumed to be independent of age. In this case the growth rate is simply $r = b - d$. With genetic variation in the death rate we can describe the evolution of the population as

$$\begin{aligned} \frac{dN_{AA}}{dt} &= p^2bN - d_{AA}N_{AA} \\ \frac{dN_{Aa}}{dt} &= 2pqbN - d_{Aa}N_{Aa} \\ \frac{dN_{aa}}{dt} &= q^2bN - d_{aa}N_{aa}, \end{aligned} \tag{60}$$

where $N(t) = N_{AA}(t) + N_{Aa}(t) + N_{aa}(t)$ and $p(t) = 1 - q(t) = N_{AA}(t) + N_{Aa}(t)/2$. With no selection the genotypic frequencies converge to the Hardy–Weinberg proportions with a rate equal to the birth, i.e.,

$$\frac{dF}{dt} = -bF, \tag{61}$$

where F is a measure of deviation from Hardy–Weinberg proportions, $F = 1 - N_{Aa}/(2pqN)$ see Table 12). The deviation from Hardy–Weinberg proportions is produced by selection, both by the direct distortion caused by selection, and by the mixing of cohorts with different gene frequencies. However, if the rate of return to the Hardy–Weinberg proportions, b, is large compared to the rate of distortion. Then we may approximate Eq. (60) by a system where the genotypic proportions at any time are in Hardy–Weinberg proportions. This approximation is called the **slow selection approximation**, and using this we can describe the evolution in terms of the change in gene frequencies:

$$\frac{dp}{dt} = pq[(d_{AA}p + d_{Aa}q) - (d_{Aa}p + d_{aa}q)], \tag{62}$$

i.e., the change in gene frequency is proportional to the difference in average death rate between the alleles. With this simplification the analysis of the model proceeds as the analysis of the monoecious zygotic selection model for nonoverlapping generations.

Fitness Maximization

The slow selection model with random mating [Eq. (62)] has a score of nice properties, one of which is the **fundamental theorem of natural selection** (Fisher 1930). This theorem stipulates that the mean fitness of the population will never decrease. To see this we differentiate the mean fitness in the population,

$$W = p^2 d_{AA} + 2pqd_{Aa} + q^2 d_{aa}, \tag{63}$$

and obtain

$$\frac{dW}{dt} = 2[(d_{AA}p + d_{Aa}q) - (d_{Aa}p + d_{aa}q)]\frac{dp}{dt}. \tag{64}$$

The bracketed expression is shared by dp/dt, [Eq. (62)], so by assuming $0 < p < 1$, Eq. (64) simplifies to

$$\frac{dW}{dt} = \frac{2}{pq}\left[\frac{dp}{dt}\right]^2, \tag{65}$$

which shows that the mean fitness W always increases except in the case where $dp/dt = 0$, i.e., except when the population is at a gene frequency equilibrium. Thus, by studying the form of W as a function of p we can depict the nontrival gene frequency equilibria as the extremum points. In addition, the locally stable equilibria are the local maxima of W as a function of p.

The time derivative of W is the rate of change in the mean fitness of the population, and Eq. (65) links it to the rate of change in the gene frequency. This is, however, a rather technical expression, and Fisher gave an intuitively more appealing expression for dW/dt. Taking dp/dt from Eq. (62) provides Eq. (64) in the form

$$\frac{dW}{dt} = 2pq[d_{AA}p + d_{Aa}q) - (d_{Aa}p + d_{aa}q)]^2, \tag{66}$$

and Fisher recognized the right side of this equation as a variance. Let $W_A = d_{AA}p + d_{Aa}q$ and $W_a = d_{Aa}p + d_{aa}q$ be the average fitnesses of the two alleles, then $W = pW_A + qW_a$ and a bit of calculation shows that Eq. (66) is equivalent to

$$\frac{dW}{dt} = 2[p(W_A - W)^2 + q(W_a - W)^2]. \tag{67}$$

Thus, the rate of change in mean fitness of the population equals twice the variance in mean fitness of the alleles. This variance $V_H = p(W_A - W)^2 + q(W_a - W)^2$ is called the gametic variance in fitness ("H" for haploid) and twice this variance, $V_A = 2V_H$, is the so called **additive variance** in fitness. The additive variance of a character can be understood as measuring the part of the genotypic variation with respect to the character that is transmissible to the offspring, in that the covariance between parents and offspring is proportional to V_A. The additive variance is an important tool in quantitative genetics (see, e.g., Falconer 1981).

We now have Fisher's fundamental theorem of natural selection as the rate of increase in mean fitness of the population equals the additive variance in fitness in the population, i.e.,

$$\frac{dW}{dt} = V_A. \tag{68}$$

The population is at gene frequency equilibrium when the additive variance vanishes (i.e., the correlation in fitness between mother and offspring is zero).

This property of the slow selection model was interpreted by Kimura (1958) as showing that the mean fitness of the population not only increases, but also that the gene frequencies in the population changes in such a way that the mean fitness changes most rapidly. (It turns out that such a statement is dependent on the measure of distance that we use between the different gene frequencies. Kimura's maximum principle relates to the Shahshahani metric, which is a natural metric for measuring distances between different gene frequencies, see Akin 1979).

Selection due to differential survival of zygotes in the monoecious non-overlapping generation model share some of the properties of the slow selection model. The recurrence Eq. (2) may be written as

$$p' - p = \frac{pq}{2W}\frac{dW}{dp} \tag{69}$$

(Wright 1935). Thus, the change in gene frequency is always in a direction where the mean fitness increases as a function of the gene frequency. The polymorphic equilibria, i.e., the nontrivial equilibria in the interval $0 < p < 1$, correspond to the extremum points of the mean fitness. In a discrete model this does not necessarily guarantee that the mean fitness increases between generations, i.e., that $W' \geqslant W$, but this result is true in general (see, e.g., Ewens 1979). However, Fisher's formula [Eq. (68)] is not valid in the discrete model: the only property which is carried over is that the additive variance in fitness is zero at a polymorphic equilibrium.

For **fecundity selection** the fitness maximization result fails. We already discussed the example of Pollak (1978) where the average fecundity of the population may show a steady decrease as the population converges to a stable equilibrium. Within fecundity selection we may consider the dioecious model of zygotic selection, and even though its simple dynamics guarantee that in principle it is possible to define a mean fitness which increases every generation, this may only rescue the principle that fitness is maximized for zygotic selection. The example of a fecundity selection model with a limit cycle in gene frequencies (Hadeler and Liberman 1975) precludes the existence of a maximization principle.

For **zygotic selection** the increase of the mean fitness between generations, i.e., $W' \geqslant W$, holds in general for one locus with multiple alleles in a population that reproduces by random mating (Kingman 1961). However, it is a genuine one-locus result, and it breaks down already for two loci. To see this, consider, in addition to the locus with alleles A and a, a linked locus with alleles B and b, and let r be the frequency of recombination between the two loci. With respect to these loci four types of gametes exist, namely AB, Ab, aB and ab. In a given generation these gametes are produced in the frequencies x_{AB}, x_{Ab}, x_{aB} and x_{ab} ($x_{AB} + x_{Ab} + x_{aB} + x_{ab} = 1$), so the gene frequencies at the two loci are given by $p_A = x_{AB} + x_{Ab}$, $q_a = x_{aB} + x_{ab}$, $p_B = x_{AB} + x_{aB}$ and $q_b = x_{Ab} + x_{ab}$. By random mating the ten different genotypes are produced in the Hardy-Weinberg proportions (e.g., AB/AB occur in the frequency x_{AB}^2 and AB/Ab occur in the frequency $2x_{AB}x_{Ab}$).

Without selection the gene frequencies at one locus are preserved from generation to generation due to the conservative transmission of the genes from parents to offspring. For **two loci**, however, the gametes are not transmitted unchanged to the offspring in sexually reproducing organisms, because meiosis allows for recombination between the genes at the two loci. The frequency of recombination is r, so in the population the fraction $1-r$ of the gametes is produced as unchanged, so among these gametes the frequencies are again x_{AB}, x_{Ab}, x_{aB} and x_{ab}. The remaining fraction r of the gametes is produced by recombination between the two gametes that formed the individual, so the type of a gamete is given by the allele at the A-locus in one of the gametes in the individual and the allele at the B-locus in the other gamete. However, by random mating the types of the two gametes that formed an individual are independent of one another, so among the recombinant gametes the frequency of the four gamete types are $p_A p_B$, $p_A q_b$, $q_a p_B$ and $q_a q_b$. Therefore, the gamete frequencies of the next generation are

$$\begin{aligned} x'_{AB} &= (1-r)x_{AB} + rp_A p_B \\ x'_{Ab} &= (1-r)x_{Ab} + rp_A q_b \\ x'_{aB} &= (1-r)x_{aB} + rq_a p_B \\ x'_{ab} &= (1-r)x_{ab} + rq_a q_b. \end{aligned} \tag{70}$$

These frequencies are different from the parental frequencies x_{AB}, x_{Ab}, x_{aB} and

x_{ab} if $r > 0$ and if the parental frequencies are different from the **Robbins' proportions**: $p_A p_B$, $p_A q_b$, $q_a p_B$ and $q_a q_b$. For $r = 0$ recombination does not occur, and the gametes are transmitted unchanged so their frequencies are preserved. With recombination, however, the gamete frequencies change every generation, and they converge to the Robbins' proportions, as we have, e.g.,

$$x'_{AB} - p_A p_B = (1 - r)(x_{AB} - p_A p_B). \tag{71}$$

Therefore, without selection recombination will cause the gametic proportions to converge to an equilibrium with Robbins' proportions, and this equilibrium is termed linkage equilibrium.

With this result we are able to argue that the fitness maximization principle cannot hold in general for zygotic selection at two loci, so imagine that the zygote to adult probability of survival is given for each of the ten genotypes. First consider the situation of no recombination which from the arguments above is equivalent to a four-allele situation. Here the population will evolve in such a way that the mean fitness increases in every generation, and the population will eventually end up at an equilibrium which is a local maximum of the mean fitness function. Now assume that this equilibrium maintains both loci polymorphic. Second, allow some recombination between the two loci, consider the same fitnesses, and place the population at a local maximum of the mean fitness where both loci are polymorphic. Now it is evident that unless the gametic frequencies at this local maximum are in Robbins' proportions recombination will move the population away, so (at least for low values of the recombination frequency, r) the mean fitness of the population must decrease initially. The final state of the population may be an equilibrium which is usually not a local maximum of the mean fitness [the Robbins' proportions in general are not releated to polymorphic mean fitness maxima (Bodmer and Felsenstein 1967)] or it may be a limit cycle (Akin 1979).

The mean fitness principle survives in one respect: The initial increase of a new allele. Suppose that the population is polymorphic at the A-locus and monomorphic at the B-locus, e.g., it segregates the gametes AB and aB. In this situation the allele b is protected if the average fitness of individuals carrying the rare b allele is higher than the average fitness of the population. The rare allele increases if the mean fitness of the population increases initially.

Table 22. Two locus symmetric viability selection model

	Genotypes in B-locus		
Genotypes in A-locus	BB	Bb	bb
AA	$1-\alpha$	$1-\beta$	$1-\alpha$
Aa	$1-\beta$	1	$1-\beta$
aa	$1-\alpha$	$1-\beta$	$1-\alpha$

Note: The class Aa and Bb covers the genotypes AB/ab and Ab/aB.

Some of the properties of selection on the variation at two loci may be judged from the **symmetric viability model** of Table 22 which is a simplification of the symmetric viability model suggested by Lewontin and Kojima (1960). The recurrence equations corresponding to this model are found from the gamete production of the various individuals. For instance, the AB gamete may be produced by four genotypes that carry the gamete, i.e., AB/AB, AB/Ab, AB/aB and AB/ab, and genotype Ab/aB that does not carry the gamete. The production of AB gametes by the first three genotypes is given by Mendelian proportions, whereas an AB/ab individual produces AB gametes in the frequency $(1-r)/2$ and the frequency $r/2$ describes the production of an Ab/aB individual. Thus, the recurrence equations become

$$\begin{aligned} Wx'_{AB} &= x_{AB}[(1-\alpha)x_{AB}+(1-\beta)x_{Ab}+(1-\beta)x_{aB}+(1-r)x_{ab}]+rx_{Ab}x_{aB} \\ Wx'_{Ab} &= x_{Ab}[(1-\beta)x_{AB}+(1-\alpha)x_{Ab}+(1-r)x_{aB}+(1-\beta)x_{ab}]+rx_{AB}x_{ab} \\ Wx'_{aB} &= x_{aB}[(1-\beta)x_{AB}+(1-r)x_{Ab}+(1-\alpha)x_{aB}+(1-\beta)x_{ab}]+rx_{AB}x_{ab} \\ Wx'_{ab} &= x_{ab}[(1-r)x_{AB}+(1-\beta)x_{Ab}+(1-\beta)x_{aB}+(1-\alpha)x_{ab}]+rx_{Ab}x_{aB}, \end{aligned} \tag{72}$$

where W is the mean fitness (independent of r and equal to the sum of the right sides of the equations).

If the population is polymorphic at the A-locus and monomorphic at the B-locus, then an equilibrium with $p_A = q_a = 0.5$ exists and this equilibrium is $W = 1-(\alpha+\beta)/2$. When allele b is introduced as rate in a population polymorphic at the A-locus, then the change in the frequency of allele b is from Eq. (72) approximately governed by

$$\begin{aligned} Wx'_{Ab} &\approx x_{Ab}[(1-\beta)/2+(1-r)/2]+rx_{ab}/2, \\ Wx'_{ab} &\approx x_{ab}[(1-r)/2+(1-\beta)/2]+rx_{Ab}/2, \end{aligned} \tag{73}$$

where $W = 1-(\alpha+\beta)/2$. We could proceed to find the leading eigenvalue of the stability matrix given by Eq. (73), but the pronounced symmetries of the model allow the equation for the gene frequency q_b as the sum of the two above equations:

$$q'_b = q_b\frac{1-\beta/2}{1-(\alpha+\beta)/2}. \tag{74}$$

Therefore, allele b is protected whenever $\alpha > 0$. (The equation displaying the other eigenvalue is the equation for $x_{Ab} - x_{ab}$). Even with the simplicity of the model it depicts the general result: the numerator of the eigenvalue in Eq. (74) is the average fitness of the new genotypes and the denominator is the mean fitness of the old genotypes before the introduction of the new gene.

The simultaneous introduction of alleles a and b as rare in a population monomorphic AB/AB may also be considered. This event is unlikely if the source of variation is mutation, but it is conceivable as a result of a rare event of immigration or a sudden change of environment. The "mean" fitness at the monomorphic equilibrium is $W = 1 - \alpha$, and the recurrence equations become approximately

$$(1-\alpha)x'_{Ab} \approx x_{Ab}(1-\beta) + rx_{ab}$$
$$(1-\alpha)x'_{aB} \approx x_{aB}(1-\beta) + rx_{ab} \tag{75}$$
$$(1-\alpha)x'_{ab} \approx x_{ab}(1-r).$$

If $\alpha > \beta$ then both x_{Ab} and x_{aB} increase when rare, no matter what the size of x_{ab}. This is because allele a is protected when the B-locus is monomorphic and allele b is protected when the A-locus is monomorphic. If the gamete ab is initially present, then it increases in frequency for sufficiently tight linkage, i.e., when $\alpha > r$. Otherwise, recombination erodes gamete ab away until the gametes Ab and aB become sufficiently common to allow for the genotype Ab/aB to produce this gamete. When $\alpha > 0$ the genotype AB/ab has a larger fitness than the common genotype AB/AB, so recombination can halt the increase of an initially superior gamete. This becomes especially relevant in the case $0 < \alpha < \beta$, where the stability of the monomorphic AB/AB equilibrium depends on the recombination frequency r, in that it is stable when $r > \alpha$ and unstable when $\alpha > r$.

The symmetric model of Table 22 allows for four different kinds of stable equilibria (Lewontin and Kojima 1960; Karlin and Feldman 1970): (1) the monomorphic equilibria, (2) the equilibria where one locus is monomorphic and the other polymorphic, (3) the central polymorphism where $x_{AB} = x_{Ab} = x_{aB} = x_{ab} = 1/4$ (which are in Robbins' proportions), and finally (4) the high complementarity equilibria $x_{AB} = x_{ab} = (1-y)/4$ and $x_{Ab} = x_{aB} = (1+y)/4$, where

$$y^2 = 1 - \frac{2r}{2\beta - \alpha}, \tag{76}$$

and this last type of equilibria exists when $r = 0$ or $r \leqslant (2\beta - \alpha)/4$ (which requires $2\beta > \alpha$). We have already seen that the monomorphic equilibria and the one-locus polymorphic equilibria may be stable. The central polymorphism is stable when $\alpha > 0$ and $r \geqslant (2\beta - \alpha)/4$, i.e., when the high complementarity equilibria do not exist. On the other hand, if we assume that $\alpha > 0$, then the high complementarity equilibria are stable when they exist. A full description of the various dynamical patterns of this model is beyond the scope of this discussion. However, it is interesting to note that the central equilibrium and the four monomorphic equilibria are stable simultaneously when $0 < \alpha < \beta$ and $\alpha < r$ (Christiansen and Feldman 1975a; Feldman and Liberman 1979).

Figure 11 shows the variation of the mean fitness in the symmetric viability model with $\alpha = 0.4$ and $\beta = 0.3$ (so $2\beta > \alpha$) and $r = 0.1$ (so $r \geqslant (2\beta - \alpha)/4$). The three curve correspond to different initial gamete frequencies. The middle curve starts close to a one-locus polymorphic equilibrium ($W = 0.65$), and here W increases initially and continues to increase until the central equilibrium is reached ($W = 0.75$). The upper curve starts at the point $x_{AB} = x_{ab} = 0.5$ and $x_{Ab} = x_{aB} = 0$ which is the high complementarity equilibrium for $r = 0$ ($W = 0.8$). This starting point is a local maximum for the mean fitness, so W decreases

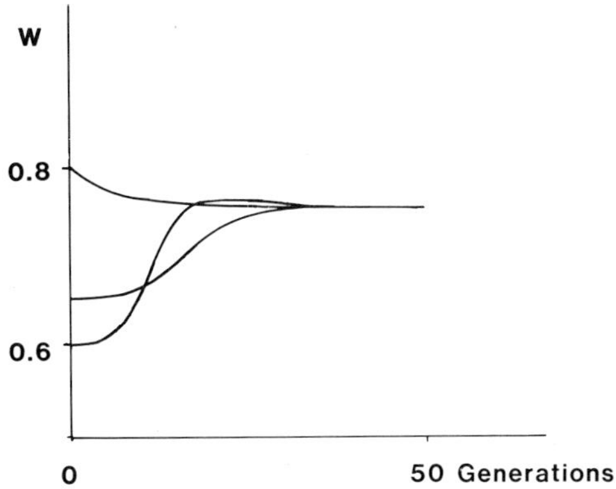

Fig. 11. The variation in mean fitness in the symmetric viability model with $\alpha = 0.4$ and $\beta = 0.3$ and $r = 0.1$. For explanation see text

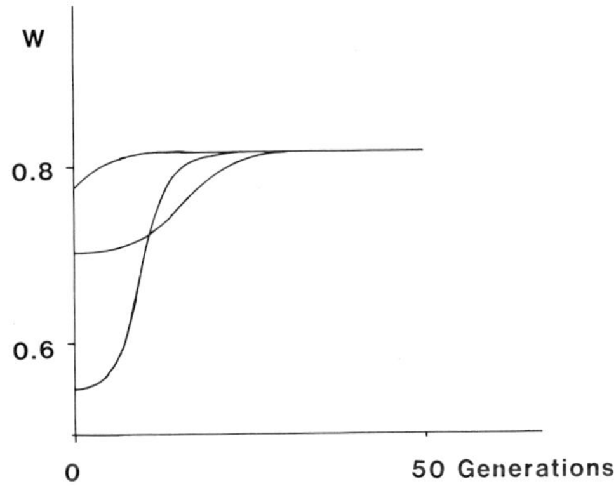

Fig. 12. The variation in mean fitness in the symmetric viability model with $\alpha = 0.45$ and $\beta = 0.15$ and $r = 0.1$. For explanation see text

initially, and here W displays a steady decrease until the central equilibrium is reached ($W = 0.75$). The lower curve starts close to a monomorphic equilibrium ($W = 0.6$), and after an initial rapid increase W overshoots the upper curve and then decreases until the central equilibrium is reached ($W = 0.75$). This pattern in essence is preserved when the high complementarity equilibria are stable ($r \leqslant (2\beta - \alpha)/4$).

Figure 12 shows the variation of the mean fitness in the symmetric viability model with $\alpha = 0.45$ and $\beta = 0.15$ (so $2\beta < \alpha$) and $r = 0.1$. The three curves show that W increases until the central equilibrium is reached ($W = 0.8125$). The middle curve starts close to a one-locus polymorphic equilibrium ($W = 0.7$), the upper curve starts at the point $x_{AB} = x_{ab} = 0.5$ and $x_{Ab} = x_{aB} = 0$ ($W = 0.775$ which is now a local minimum), and the lower curve starts close to a monomorphic equilibrium ($W = 0.55$).

The variation in mean fitness in Fig. 12 looks nice whereas the possibility of bumps on the "progressive" evolution in Fig. 11 may be rather disturbing. Is the model in Fig. 12 in some sense the nicer model? This question is hard to answer, because of few data on two locus selection. However, a nice model seems to be that the two loci are independent in their effect, e.g., fitness is given as a product of

Table 23. Estimated fitnesses for two inversion systems in *Moraba scurra*. (After Lewontin and White 1960)

	Karyotypes on chromosome CD		
Karyotypes on chromosome EF	BB	Bb	bb
AA	0.79	1.00	0.83
Aa	0.67	1.01	0.90
aa	0.66	0.66	1.07

two survival probabilities. In the symmetric model this corresponds to the specification: $1-\alpha=(1-\beta)^2$, and this implies that $2\beta>\alpha$, which is the bumpy situtation of Fig. 11. The difference between the two models is that the signs of the epistatic interactions (the fitness of a homozygote minus the fitnesses of the two adjoining single heterozygotes plus the fitness of the double heterozygote which equals 2β-α in the symmetric model) between the two loci are different, positive in Fig. 11 and negative in Fig. 12 (Christiansen 1988).

Table 23 shows data on two polymorphic inversion systems ($r=0.5$) in *Moraba scurra* which for convenience are renamed to our standard model. The epistatic interaction corresponding to the AB/AB homozygote is 0.13, to Ab/Ab −0.06, to aB/aB 0.34, and corresponding to ab/ab it is 0.52. Thus, the epistatic interactions are three positives to one small negative. Iterations of a model with the fitnesses as in Table 23 show that a two locus polymorphism is not possible (see Lewontin 1974).

The consideration of many loci does not resurrect the mean fitness principle. However, it survives in one respect; the initial increase of a new rare allele is conditioned on the mean fitness of carriers of the allele being higher than the population mean fitness, and this principle holds in a population polymorphic at any number of loci (Liberman 1988). Fisher's fundamental theorem of natural selection therefore has a limited applicability with the initial increase principle being the only aspect of the fitness maximization principle that survives general genetic considerations. However, when the population is at equilibrium then the additive variance in fitness is zero no matter what is the number of segregating loci (Ewens and Thomson 1977).

Selection and the Biotic Environment

Natural selection depends on the individual adaptation to the environment, so the description of natural selection in the previous sections of this chapter implicitly refers to the physical and the biotic environment of the considered organism. However, the environment changes and natural selection depends at any time on the particular environment in which the population of interest lives. Therefore, the process of natural selection varies with the characteristics of the biotic environment and becomes a function of the size and the phenotypic composition of the populations of the coexisting organisms. The most important component of the biotic environment of an individual is the population of conspecific individuals, so fitness depends on the size and composition of the population. The population genetic models that we have considered may be extended to this situation by making the fitness parameters density-dependent and frequency-dependent. The distinction between the two types of environmental influence on individual fitness is often of a purely technical nature, as selection dependence on the composition of a population usually varies with the density of the population.

Interspecific and **intraspecific interactions** are qualitatively different as causes for natural selection. The short-term effects of interspecific interactions on the evolution of a given species may be evaluated on the basis of the classical models of constant fitnesses. This simplification reflects that changes in a given species only amount to a change in the environment of other species. This environmental change may change selection in these other species and cause them to evolve and thereby change the environment of the original species. However, these future events are of no concern for the present action of natural selection and the present course of evolution.

Long-term results of selection due to interspecific interactions have to be produced in models that take the concurrent evolution in the interacting species into account (see Chaps. 12, 13). These models in essence describe the evolution of a particular species and the development of its biotic environment. Therefore, it is not surprising that the evolution is adequately predicted by simple models with constant fitnesses. A notable exception to this rule is the case when the interspecific interaction depends on the phenotypic composition of the populations. For instance, the forming of search images by predators may cause selection against common phenotypes in a prey species (Clarke 1979). This results in selection that resembles selection due to the rare male mating advantage.

The intraspecific interactions are an important cause for natural selection to occur, and variation in phenotypes related to intraspecific interactions produces frequency and density-dependent selection. The sexual selection models are examples of models of intraspecific individual interactions giving rise to frequency-dependent selection. However, even there it is reasonable to assume that the population density is important: female choice may be more important in a dense population than in a population where a female can hope to meet only one male. Other relevant types of intraspecific interactions are intraspecific competition and intraspecific mutualism, which share some evolutionary properties. However, populations have an inherent propensity to increase in numbers, so in any species competition for resources is the eventual cause for population size limitation. Therefore, intraspecific competition, in contrast to intraspecific mutualism, is an inherent aspect of natural selection.

The continuous time population genetic models of Fisher (1930) and Norton (1928) which we considered based their description of natural selection on Malthusian models of unlimited growth. These models may immediately be extended to density-dependent selection models by substituting the Malthusian models with the logistic model. This model is a modification of the Malthusian growth model, and it is the simplest model incorporating the effect of intraspecific competition, in that the birth rate per individual decreases and the death rate per individual increases linearly with population density to reflect competition for limiting resources. As in the Malthusian model all individuals of the logistic model population are considered to be identical, and all individuals are affected equally by the increase in population density, and the measure of population density is equally affected by each individual. Models introducing variation and thereby describing natural selection were introduced and analyzed by Kostitzin (see Scudo and Ziegler 1976, 1978) and by MacArthur (1962). A simple extension of the model in Eq. (60) allowing density dependence in the death rates is

$$\begin{aligned} \frac{dN_{AA}}{dt} &= p^2bN - (d_{AA} + \iota_{AA}N)N_{AA} \\ \frac{dN_{Aa}}{dt} &= 2pqbN - (d_{Aa} + \iota_{Aa}N)N_{Aa} \\ \frac{dN_{aa}}{dt} &= q^2bN - (d_{aa} + \iota_{aa}N)N_{aa}, \end{aligned} \tag{77}$$

where ι_{AA}, ι_{Aa} and ι_{aa} are coefficients describing the density-dependent death rate (Christiansen and Fenchel 1977). With no selection the population size becomes $N = (b - d)/\iota$ and the population again converges to the Hardy–Winberg proportions as described by Eq. (61). With selection the condition for protection of allele A (initial increase in a population monomorphic aa) is

$$\frac{b - d_{Aa}}{\iota_{Aa}} > \frac{b - d_{aa}}{\iota_{aa}}. \tag{78}$$

A further discussion of these models is given in Chapter 12.

The handling of the population genetic aspects of **density-dependent selection** is eased by the use of simple discrete time models (Anderson 1971; Roughgarden 1971, 1976; Clarke 1972). The model of Anderson and Roughgarden is a discrete time analog of the logistic model, but Clarke's model, also called the hyperbolic model, is more directly related to the continuous time logistic model. The Anderson–Roughgarden model specifies individual fitness values of the monoecious zygotic selection model of Table 1 as linearly dependent on the population size $W_{AA} = 1 + V_{AA}(K_{AA} - N)$, $W_{Aa} = 1 + V_{Aa}(K_{Aa} - N)$ and $W_{aa} = 1 + V_{aa}(K_{aa} - N)$. The population dynamic model corresponding to the Anderson–Roughgarden model is given by the recurrence equation

$$N' = N[1 + V(K - N)]. \tag{79}$$

In the difference equation formulation: $N' - N = VN(K - N)$, it refers more directly and simply to the differential equations of continuous time Lotka–

Volterra models.) As a model for population regulation it uses the heuristic parametrization of Gause (1934), in that it is parameterized in terms of the equilibrium population size, K, also called the carrying capacity, and a proportionality parameter V that provides the growth factor at low population densities as approximately $1 + VK$. The model is an approximation to the continuous time logistic model, so V should be assumed small (Christiansen and Fenchel 1977). The reason is that the factor $1 + V(K - N)$ becomes negative for N sufficiently large, and this always happens when V is sufficiently large. The model also involves a time lag in the density response, in that for moderately large values of V, cyclic or chaotic behavior is expected.

In general, model of density-dependent genotypic fitnesses leads to a maximization principle that parallels the fitness maximization principle for constant genotypic fitnesses (Sect. 8) in that selection on the variation at one locus in a panmictic population maximizes the equilibrium population size in the population (MacArthur 1962; Roughgarden 1976; Prout 1980).

The Anderson–Roughgarden model was extended to intraspecific competition by Matessi and Jayakar (1976). The genotypic fitnesses are specified as, e.g.,

$$W_{AA} = 1 + V[K_{AA} - N(p^2\gamma_{AA,AA} + 2pq\gamma_{AA,Aa} + q^2\gamma_{AA,aa})], \tag{80}$$

where the γ's are coefficients describing intraspecific competition. These fitness parameters are clearly density- and frequency-dependent (Matessi and Jayakar 1976, 1981; Christiansen and Fenchel 1977). The K's have a significance that parallels the K parameter in Eq. (79), but here the equilibrium population size in a monomorphic AA population is $K_{AA}/\gamma_{AA,AA}$. The fitness maximization principle for constant selection and the similar maximization principle for purely density-dependent selection both break down when frequency dependence is allowed. However, Matessi and Jayakar (1976, 1981) analyzed the situation of symmetric competition and found and extension of the maximization principle for density dependent selection (see Chap. 12).

The genetical simplifications of discrete population models requires discrete breeding and nonoverlapping generations, but the interaction among individuals of the population may occur in continuous time. This idea leads to another kind of discrete time logistic model (Poulsen 1979). Let N denote the population size immediately before breeding in a monoecious organism, and assume that these N individuals produce B offspring each. The initial number of offspring in the offspring population is then $C(0) = BN$, and the number of offspring at age t is denoted by $C(t)$. The individuals become mature after a fixed time of development, T, so immediately before breeding the population size is $N' = C(T)$. The offspring experience density-independent and density-dependent mortality, so the population size changes during development as

$$\frac{dC}{dt} = -(d + \iota C)C, \tag{81}$$

where the d is the density-independent death rate and ι is a coefficient describing the density dependent death rate. The probability of survival in a population at low density is approximately $D = e^{-dT}$, and the population then grows exponentially with the factor $\lambda = DB$ per generation. The recurrence equation in the population size is obtained by integration of Eq. (81), so

$$N' = \frac{DBN}{1 + \gamma(1 - D)BN}, \tag{82}$$

where $\gamma = \iota/d$. Therefore, the population size follows an integral curve of the continuous time logistic model at discrete time intervals. If $\lambda = DB > 1$, i.e., the population grows in size at low densities, then the population size converges to a globally stable equilibrium given by

$$N = \frac{DB - 1}{\gamma(1 - D)B}. \tag{83}$$

This population model specifies growth in terms of fecundity and death rates, so it may be used directly to formulate a model of natural selection. Let N_{AA}, N_{Aa}, and N_{aa} denote the number of the three genotypes in the population, the sum $N = N_{AA} + N_{Aa} + N_{aa}$ being the total population size. The parameters of the Poulsen model may be assigned for each genotype, so the density-independent death rates are d_{AA}, d_{Aa}, and d_{aa}, and the density-dependent death rate coefficients for genotype AA are $\iota(AA)_{AA}$, $\iota(AA)_{Aa}$ and $\iota(AA)_{aa}$, so $\iota(AA)_{aa}N_{aa}$ is the death rate inflicted on genotype AA by the presence of N_{aa} individuals of genotype aa in the population. For simplicity we will consider only zygotic selection and assume that all genotypes have the same fecundity B. With random mating the genotypic frequencies among offspring are the Hardy–Weinberg proportions, so the numbers of the various genotypes become

$$\begin{bmatrix} C_{AA}(0) \\ C_{Aa}(0) \\ C_{aa}(0) \end{bmatrix} = \begin{bmatrix} p^2BN \\ 2pqBN \\ q^2BN \end{bmatrix}, \tag{84}$$

where the gene frequencies are $p = (2N_{AA} + N_{Aa})/(2N)$ and $q = (2N_{aa} + N_{Aa})/(2N)$. These offspring show genotype-dependent mortality during development given by the equations

$$\begin{aligned} \frac{dC_{AA}}{dt} &= -(d_{AA} + \iota(AA)_{AA}C_{AA} + \iota(AA)_{Aa}C_{Aa} + \iota(AA)_{aa}C_{aa})C_{AA} \\ \frac{dC_{Aa}}{dt} &= -(d_{Aa} + \iota(Aa)_{AA}C_{AA} + \iota(Aa)_{Aa}C_{Aa} + \iota(Aa)_{aa}C_{aa})C_{Aa} \\ \frac{dC_{aa}}{dt} &= -(d_{aa} + \iota(aa)_{AA}C_{AA} + \iota(aa)_{Aa}C_{Aa} + \iota(aa)_{aa}C_{aa})C_{aa}. \end{aligned} \tag{85}$$

After development the three genotypes occur in the numbers $N'_{AA} = C_{AA}(T)$, $N'_{Aa} = C_{Aa}(T)$ and $N'_{aa} = C_{aa}(T)$.

The **density-dependent death rate** coefficients of Eq. (85) describe the effects of intraspecific competition, and the result may be selection on the genotypic variation that depends both on the density and on the composition of the population. Not every matrix of the ι parameters, however, produces density- and frequency-dependent selection. If all density-dependent death rate coefficients are equal, then selection becomes density-independent. The same happens if only the density-dependent death rate coefficients are independent of the genotype that experiences the death, i.e., if $\iota(AA)_{AA} = \iota(Aa)_{AA} = \iota(aa)_{AA} = \iota_{AA}$, $\iota(AA)_{Aa} = \iota(Aa)_{Aa} = \iota(aa)_{Aa} = \iota_{Aa}$ and $\iota(AA)_{aa} = \iota(Aa)_{aa} = \iota(aa)_{aa} = \iota_{aa}$. In this case Eq. (85) integrates to the (implicit) equations

$$\begin{aligned} N'_{AA} &= C_{AA}(0) D_{AA} G[C_{AA}(0), C_{Aa}(0), C_{aa}(0)], \\ N'_{Aa} &= C_{Aa}(0) D_{Aa} G[C_{AA}(0), C_{Aa}(0), C_{aa}(0)], \\ N'_{aa} &= C_{aa}(0) D_{aa} G[C_{AA}(0), C_{Aa}(0), C_{aa}(0)], \end{aligned} \tag{86}$$

where $D_{AA} = \exp(-d_{AA}t)$, $D_{Aa} = \exp(-d_{Aa}t)$ and $D_{aa} = \exp(-d_{aa}t)$ are the density-independent survival probabilities of the genotypes, and where the function G is given by

$$G[C_{AA}(0), C_{Aa}(0), C_{aa}(0)] = \exp[-\int_0^T (\iota_{AA}C_{AA} + \iota_{Aa}C_{Aa} + \iota_{aa}C_{aa})\,dt]. \tag{87}$$

This factor G is shared by the three equations (86) and it is independent of genotype. Therefore, it does not produce selection, and only the density-independent survival probability becomes relevant. Consequently, the model degenerates to the monoecious zygotic selection model of Table 1.

Intraspecific competition with variation in the impact of the presence of a given individual on other individuals does not produce either density dependence or frequency dependence in selection unless concurrent variation in the susceptibility of the individuals occurs. The result is frequency-dependent population regulation. The changes in the genetic composition of the population are governed by the density-independent death rates, and selection leads to an increase in mean survival of the population. The concurrent changes in the population size therefore need not bear any immediate relationship to the maximization of fitness (Poulsen 1979). A genotype which differs from other genotypes by being either more altruistic or more antagonistic will not be distinguished by selection unless it at the same time differs in its susceptibility to the behavior of the other individuals or in aspects of its density-independent fitness.

If the density-dependent death rate coefficients within each of the equations given by (85) are equal, that is, if $\iota(AA)_{AA} = \iota(AA)_{Aa} = \iota(AA)_{aa} = \iota(AA)$, $\iota(Aa)_{AA} = \iota(Aa)_{Aa} = \iota(Aa)_{aa} = \iota(Aa)$ and $\iota(aa)_{AA} = \iota(aa)_{Aa} = \iota(aa)_{aa} = \iota(aa)$, then selection becomes purely density-dependent. In this case Eq. (85) integrates to the (implicit) equations

$$\begin{aligned} N'_{AA} &= C_{AA}(0) D_{AA} F(C)^{\iota(AA)}, \\ N'_{Aa} &= C_{Aa}(0) D_{Aa} F(C)^{\iota(Aa)}, \\ N'_{aa} &= C_{aa}(0) D_{aa} F(C)^{\iota(aa)}, \end{aligned} \tag{88}$$

where $C(t) = C_{AA}(t) + C_{Aa}(t) + C_{aa}(t)$ is the total population size among offspring at age t, and where

$$F(C) = \exp\left[-\int_0^T C(t)\,dt\right]. \tag{89}$$

This factor depends only on the total density of the offspring population during development. In this sense selection is purely density-dependent. However, the variation during development in the total density, C, is determined by the variation in the density of each genotype as determined by the density-independent and the density-dependent death rates. Therefore, the variation in C depends on the initial composition of the population, and F is frequency-dependent. The frequency dependence, however, is weak, and the model has a property that parallels the purely density-dependent fitness models, namely that the average population size, F(C), after an initial period, increases during evolution (Iwasa and Teramoto 1980), a property which is not characteristic of other frequency-dependent selection models (Cockerham et al. 1972).

The detailed dynamic properties of the Poulsen model has not been described in general, but the condition for protection of allele A is

$$\frac{\log(BD_{Aa})}{\iota(aa)_{Aa}} > \frac{\log(BD_{aa})}{\iota(aa)_{aa}} \tag{90}$$

(Poulsen 1979). In this model the average rate of birth per time unit is $b = \log(B)/T$, and if this is inserted into (90) then we obtain the same condition as in the continuous time model, i.e., condition (78). Further discussions of this model are given by Christiansen and Fenchel (1977).

The Poulsen model is in some respects very detailed, and its analysis in more complicated situations is difficult (Christiansen 1984a, 1985). For the further analysis of intra-specific competition it has proven convenient to use models in the tradition of the Anderson–Roughgarden model (see Chap. 12). However, in other respects the model is simple, in that the impact of interactions is in terms of a mass action model where individuals meet in proportion to their densities. A comparison to the sexual selection models immediately suggests a deviation from these mass action models, where individuals first of all search for a mate and secondly may avoid further contacts after mating. These behaviors may in the context of competition result in functional relations different from that in Eq. (85). Some aspects of these interactions can be treated elegantly within the framework of game theoretical models (see Maynard Smith 1982; Hines 1987).

For the two relevant types of intraspecific eclogical interactions, intraspecific competition and intraspecific mutualism, we may conclude some common evolutionary properties. In both cases the survival and well-being of a given individual need not depend on the effect it conveys on the other individuals. The interaction of an individual with other individuals produces effects to both, but the aid or the harm done to other individuals at best may be of no concern for the fitness of the active individual. Indeed, such actions often may be inherently detrimental to the active individual. This poses, for instance, Haldane's (1932) problem of the evolution of socially valuable, but individually disadvantageous characters, also known as **altruistic** characters (see, e.g., Borman and Levitt 1980). In sedentary organisms the effect of interference competition between individuals is simpler, in that the detrimental effect of the interaction to one individual may be equivalent to the advantage to the other individual, as for instance in competition for space in barnacles. This opens up for neighbor effects giving deviations from random encounters just as preferential mating between neighbors could produce nonrandom mating. These effects are important in the evolution of both antagonistic and mutualistic traits or any other behavioral trait that confers a cost to the active individual. The most effective force influencing the evolution of both behavioral characteristics is **kin selection**, which under appropriate circumstances may confer the benefit of a certain behavior to relatives of the active individual, and thereby induces selection on the genes that influence the behavior (Hamilton 1964a; Maynard Smith 1964).

3 Mating Systems and Evolutionary Genetics

P.W. HEDRICK

... mating systems indeed hold a key place in the study of plant populations.... In this connection four points need stressing. First, reliable estimates of mating system parameters are a fundamental prerequisite of plant studies. Second, we need to go beyond the simple dichotomy of inbreeders vs. outbreeders by making allowance for the complexity of the diverse mating systems of different plant species.... Third, mating systems are flexible in space and in time, and such variation is likely to have profound effects on the genetic structure of populations.... Fourth, variation in mating systems is subject to genetic control, and mating systems themselves evolve.

Brown (1979)

Introduction

In population genetics random mating among individuals has traditionally been assumed. Obviously, however, many plant species (and some invertebrates) have mating systems with a high proportion of self-fertilization and possibly other mating types between close relatives. On the other hand, an extensive survey of natural populations of birds and mammals by Ralls et al. (1986) found little evidence of mating among relatives except for a few species, suggesting that random mating may be a reasonable approximation for most cases in these animals. Therefore, I will concentrate in this chapter on inbreeding mating systems in plants.

In the past decade, two developments have generated increased interest in mating systems in plants (Brown et al. 1985). First, although mating systems in plants have been inferred from flower morphology or estimated from visible markers in the past, the application of biochemical variants to directly measure the mating system has been an important advance. Second, theoretical developments concerning such topics as mating-system estimation procedures, the evolution of mating systems, and the impact of nonrandom mating on genetic variation have also stimulated research. As a result, it seems appropriate at this time to review a variety of evolutionary genetics topics of plant mating systems.

First, I will examine aspects of the procedures to directly estimate the mating system. Second, I will discuss the population genetic consequences of the mating system. In the final section, I will discuss how environmental and genetic factors affect the mating system. Related topics that I will not discuss are sex allocation (e.g., Charnov 1982), sexual selection (e.g., O'Donald 1980; Willson and Burley 1984), paternity analysis (e.g., Meagher 1986) and the evolution of dioecy (e.g., Charlesworth and Charlesworth 1978; Lloyd 1982).

Direct Estimation of the Mating System

The mating system in plants that have some proportion of self-fertilization has traditionally been described as a mixture of self-fertilization and random mating (e.g., Jones 1916; Fyfe and Bailey 1951). In this model, generally termed the **mixed-mating model**, each zygote is the result of an independent mating event that either occurs by self-fertilization, with probability *S*, or random mating (outcrossing), with probability T ($S + T = 1$).

If we assume that we are examining a codominant locus with alleles A_1 and A_2 having frequencies p_1 and p_2, respectively, then the proportions expected of the different progeny genotypes are as given in Table 1 (e.g., Brown and Allard 1970). For example, if the maternal genotype is A_1A_1, then whenever there is self-fertilization only A_1A_1 offspring genotypes are produced. When an A_1A_1, maternal genotype outcrosses, a proportion p_1 of the pollen will have the A_1 allele. Therefore, the total expected proportion of A_1A_1 progeny from an A_1A_1 maternal genotype is $S + Tp_1$.

There are a number of assumptions implicit in the mixed-mating model (e.g., Clegg 1980; Brown et al. 1985). We will discuss aspects of several of these assumptions in some detail below (see also Brown et al. 1985). The main assumptions are:

1. Each mating event is assumed to be the result of self-fertilization with probability S or random mating with probability T. For example, it is assumed that there is no assortative mating or mating between relatives (above that expected at random). Furthermore, S is assumed to be constant both spatially and temporally.
2. The expected frequency of the alleles in the pollen is identical for all maternal individuals. For example, it is assumed that there is no spatial or temporal variation in the allele frequencies in the pollen and there is no correlation in pollen allele frequencies caused by factors such as pollinator behavior or pollen morphology.
3. The rate of self-fertilization (or outcrossing) is independent of the maternal genotype. In other words, the mixed-mating model assumes that there are no genetically determined differences in selfing rates, a fact that would prevent any evolution of the mating system.
4. There is normal Mendelian segregation of the alleles in heterozygous maternal individuals for both pollen and ovule production and gametes containing these different alleles have equal success in producing zygotes. In other words, there is no segregation distortion or gametic selection.
5. There is no selection between fertilization and the determination of progeny

genotypes. This assumption implies that there is equal embryonic survival of selfed and outcrossed progeny.

Obviously these assumptions may not exactly hold in a number of species or populations. On the other hand, hopefully by understanding their importance in biasing estimation procedures, a better overall picture of the factors affecting the mating system may be obtained.

As an illustration of how the extent of self-fertilization can be estimated using the assumptions of the mixed-mating model, let us assume that the maternal parent has the genotype A_1A_1 (row 1 of Table 1). Symbolizing the observed number of heterozygous progeny as N_{12}, then

$$N_{12} = Tp_2N,$$

where there are N total progeny from the A_1A_1 maternal parents and p_2 is the frequency of allele A_2 in the pollen, making Tp_2N the expected number of heterozygous progeny. (If the maternal parent is A_2A_2, then p_1 can be substituted for p_2 and same expression used). The maximum likelihood estimate of the outcrossing rate is then

$$\hat{T} = \frac{N_{12}}{p_2N} \tag{1}$$

(Bailey 1951). If the frequency of the alleles in the pollen is known (p_2 is known), then the variance of the outcrossing estimate is

$$V(\hat{T}) = \frac{\hat{T}(1 - p_2\hat{T})}{Np_2} \tag{2a}$$

If an independent estimate of p_2 is obtained experimentally and it is assumed that there is no covariance between the estimates $\hat{T}$ and $\hat{p}_2$, and then the variance of the outcrossing estimate is approximately

$$V(\hat{T}) \cong \frac{\hat{T}(1 - \hat{p}_2\hat{T})}{\hat{N}p_2} + \frac{V(\hat{p}_2)\hat{T}^2}{\hat{p}_2^2} \tag{2b}$$

(Fyfe and Bailey 1951; Marshall and Brown 1974).

Table 1. The expected frequencies of progeny genotypes given different maternal genotypes when there is S self-fertilization and T outcrossing

Maternal genotype	Progeny genotype		
	A_1A_1	A_1A_2	A_2A_2
A_1A_1	$S + Tp_1$	Tp_2	—
A_1A_2	$\frac{S}{4} + \frac{Tp_1}{2}$	1/2	$\frac{S}{4} + \frac{Tp_2}{2}$
A_2A_2	—	Tp_1	$S + Tp_2$

Table 2. The progeny and their frequency with an A_1A_1 maternal parent when there is S_1 full-sib mating and T outcrossing.

	Male parent	Frequency	Progeny A_1A_1	Progeny A_1A_2
S_1	A_1A_1	$\frac{1}{4}(1+p_1)^2$	$\frac{1}{4}S_1(1+p_1)^2$	
	A_1A_2	$\frac{1}{2}(1+p_1)p_2$	$\frac{1}{4}S_1(1+p_1)p_2$	$\frac{1}{4}S_1(1+p_1)p_2$
	A_2A_2	$\frac{1}{4}p_2$	—	$\frac{1}{4}S_1p_2$
T	A_1	p_1	Tp_1	—
	A_2	p_2	—	Tp_2
			$1-\frac{1}{2}(1+T)p_2$	$\frac{1}{2}(1+T)p_2$

Assuming that the maternal parent can be heterozygous as well as homozygous, then the complete array of maternal and expected progeny genotypes is given in Table 1 for the mixed-mating model. Using either all maternal-progeny arrays, or only progeny arrays and inferring the maternal parent, then T and p_2 can be estimated using a likelihood approach (e.g., Clegg et al. 1978; Cheliak et al. 1983; Ritland 1986). The likelihood approach uses all the parent and progeny information available to estimate the parameters and is thereby generally preferable.

Inbreeding Other than Self-Fertilization

Unlike the conditions of assumption (1) above, let us assume that the outcrosses are not all matings with random individuals in the population but include a proportion of matings with related individuals (above that expected by chance). To examine this situation, Ritland (1984) has extended the likelihood approach of Clegg et al. (1978) to estimate the extent of "**effective selfing**". Effective selfing is defined as "the probability that an allele chosen at random from an individual's mate is identical by descent with either allele at the same locus in that individual". The implication is that other types of inbreeding besides selfing will increase the apparent extent of self-fertilization in the population.

Another method of quantifying the extent of inbreeding other than self-fertilization is to assume that there is a proportion of a given type of mating between relatives and the remainder of matings are random outcrosses (see Hedrick 1987a, for a more complete discussion). For example, let us examine a population in which a proportion S_1 of the progeny are produced by full-sib mating and the remainder T by random outcrossing ($T + S_1 = 1$). This type of partial inbreeding would be especially pertinent to organisms that do not self, such as dioecious plants or animals and self-incompatible plants.

Let us assume that the maternal parent is genotype A_1A_1, then there are three possible genotypes for the full-sib male parent. Because the maternal parent is

A_1A_1, then both her parents have given her an A_1 allele. Therefore, the probabilities that her full sib would get an A_1 or A_2 are $1/2(1 + p_1)$ and $1/2p_2$, respectively. Using these probabilities, the frequency of the three male full-sib genotypes are given in Table 2 along with the expected frequency of the progeny from the three mating types. Assuming that the observed number of heterozygous progeny is N_{12}, then

$$N_{12} = \tfrac{1}{2}(1 + T)p_2N$$

and the estimate of outcrossing is then

$$\hat{T} = \frac{2N_{12}}{p_2N} - 1. \tag{3}$$

For example, if $N = 100$, $N_{12} = 45$ and $p_2 = 0.5$, then the estimate of outcrossing is 0.8. In other words, 20% full-sib mating would be consistent with this progeny array.

Variation in Pollen Allele Frequencies

The mixed-mating model assumes that allelic frequencies in the pollen are the same for each maternal plant, assumption (2) above. However, if there is either microgeographical or temporal variation in allelic frequencies, then estimates of outcrossing may be biased downward (Ennos and Clegg 1982; see also Ellstrand and Foster 1983) when using the likelihood approach of Clegg et al. (1978).

One way to circumvent this problem is the method of Ritland and Ganders (1985), which uses maternal parents homozygous for two loci. In this case, both the allelic frequencies in the pollen and the selfing rate in a given progeny array can be simultaneously estimated. For example, if a maternal genotype is $A_1A_1B_1B_1$, then the frequency of the four possible progeny types (see Table 3) are

$$\begin{aligned} f(A_1A_1B_1B_1) &= S + Tp_1q_1 \\ f(A_1A_1B_1B_2) &= Tp_1q_2 \\ f(A_1A_2B_1B_1) &= Tp_2q_1 \\ f(A_1A_2B_1B_2) &= Tp_2q_2, \end{aligned} \tag{4}$$

Table 3. The expected frequency of different progeny genotypes given an A_1B_1/A_1B_1 maternal parent and self-fertilization of the amount S

Maternal genotype	Pollen gametes	Progenygenotypes			
		A_1B_1/A_1B_1	A_1B_1/A_1B_2	A_1B_1/A_2B_1	A_1B_1/A_2B_2
A_1B_1/A_1B_1	A_1B_1	$S + Tp_1q_1$	—	—	—
	A_1B_2	—	Tp_1q_2	—	—
	A_2B_1	—	—	Tp_2q_1	
	A_2B_2	—	—	—	Tp_2q_2

Table 4. The allelic frequencies for two loci in the pollen that fertilized five detasseled corn plants and that in a sample of possible pollen donors (p_i). (Bijlsma et al. 1986)

Locus	Allele	Plant 1	2	3	4	5	p_i
Prx 1	1	0.65	0.05	0.85	0.31	0.59	0.55
	2	0.29	0.53	0.06	0.35	0.39	0.25
	3	0.06	0.42	0.09	0.34	0.02	0.20
Est 4	1	0.70	0.17	0.68	0.38	0.91	0.57
	2	0.08	0.61	0.04	0.62	0.00	0.27
	3	0.22	0.22	0.28	0.00	0.09	0.16

where it is assumed there is no gametic disequilibrium, i.e., statistical association of alleles at different loci, in the pollen and the frequencies of alleles B_1 and B_2 and q_1 and q_2, respectively. Maximum likelihood estimates of the selfing rate and allelic frequencies are then

$$\begin{aligned}\hat{S} &= [f(A_1A_1B_1B_1)f(A_1A_2B_1B_2) - f(A_1A_2B_1B_1)f(A_1A_1B_1B_2)] \\ &\quad /f(A_1A_2B_1B_2) \\ \hat{p}_1 &= f(A_1A_1B_1B_2)/[f(A_1A_1B_1B_2) + f(A_1A_2B_1B_2)] \\ \hat{q}_1 &= f(A_1A_2B_1B_1)/[f(A_1A_2B_1B_1) + f(A_1A_2B_1B_2)].\end{aligned} \tag{5}$$

By looking at progeny arrays from different plants, estimates of selfing and allelic frequencies for each plant can be obtained and averaged to give an unbiased estimate of selfing (Ritland and Ganders 1985; see also Fig. 9 below).

An example of the importance of allele frequency variation in pollen was documented by Bijlsma et al. (1986) in an open-pollinated maize population. Although maize is generally an outcrossing species, selfing estimates for some loci have given rates significantly different from zero. They examined 100 seedlings from each of five detasseled plants and determined the allelic frequencies in the pollen fertilizing these plants. The two loci with the most variation over the five plants are given in Table 4. In fact, these are the two loci which also consistently had given outcrossing estimates less than unity (the average over six samples is 0.71). Bijlsma et al. (1986) suggested that a likely reason for the heterogeneity in the pollen pool and consequent low estimates of outcrossing was flowering time variation that "might cause both maternal plants and the pollen pool to be temporally heterogeneous in allelic composition for the marker loci".

One-Pollen Parent and Committed-Mating Models

Although the mixed-mating model may be appropriate for many wind-pollinated plants, for insect- or animal-pollinated plants with little pollen carryover, the

Table 5. The expected frequencies of different progeny genotypes given the one-pollen parent model

Parental genotypes	Progeny genotypes		
female × male	A_1A_1	A_1A_2	A_2A_2
$A_1A_1 \times A_1A_1$	1	—	—
$A_1A_1 \times A_1A_2$	S + T/2	T/2	—
$A_1A_1 \times A_2A_2$	S	T	—
$A_1A_2 \times A_1A_1$	S/4 + T/2	$\frac{1}{2}$	S/4
$A_1A_2 \times A_1A_2$	$\frac{1}{4}$	$\frac{1}{2}$	$\frac{1}{4}$
$A_1A_2 \times A_2A_2$	S/4	$\frac{1}{2}$	S/4 + T/2
$A_2A_2 \times A_1A_1$	—	T	S
$A_2A_2 \times A_1A_2$	—	T/2	S + T/2
$A_2A_2 \times A_2A_2$	—	—	1

pollen fertilizing different ovules within a flower or plant may be correlated in frequency, violating assumption (2) above. As an extreme, all pollen fertilizing ovules within a flower may be from one-pollen parent so that seeds will be full sibs instead of half sibs as in the mixed-mating model (e.g., Smyth and Hamrick 1984; Epperson and Clegg 1987). To describe this situation, Schoen and Clegg (1984, 1986) developed the one-pollen parent model in which the genotype of the male parent is specified (Table 5). For example, when there is selfing for the mating type $A_1A_1 \times A_1A_1$ (second row), only A_1A_1 progeny are produced. if outcrossing occurs for this mating type, half the progeny are A_1A_1 and half are A_1A_2.

Arnold (unpublished) has suggested another alternative model, the committed-mating model, in which a maternal parent either exclusively self-fertilizes or outcrosses by randomly mating with a single father. Such a model may be appropriate for some insect-pollinated plants or invertebrate hermaphrodites, such as some slugs or snails. Arnold (unpublished) also suggests that the sequential sampling techniques described by Sobel et al. (1986) may be appropriate to estimate the mating-system parameters of all three mating models.

Selection Before Determination of Progeny Genotypes

If selection occurs between fertilization and the determination of progeny genotypes, then assumption (5) does not hold. To examine this situation, let us assume the mean number of recessive lethals in a non-inbred zygote is x and the presence of each lethal is independent of every other. In this case, the probability of n lethals in a non-inbred individual is

$$P(n) = \frac{x^n e^{-x}}{n!}. \qquad (6)$$

Table 6. The expected numbers of progeny from an A_1A_1 maternal genotype when there is S self-fertilization and x is the average number of lethals per zygote

	Progeny genotype	
Maternal genotype	A_1A_1	A_1A_2
A_1A_1	$\dfrac{\left(S\left[\Sigma \frac{x^n - e^x}{n!}\left(\frac{3}{4}\right)^n\right] + Tp_1\right)N}{\bar{w}}$	$\dfrac{Tp_2N}{\bar{w}}$

The proportion of selfed progeny that are viable from an individual with n lethals, assuming the lethals assort independently, is $(3/4)^n$. Therefore, instead of a proportion S of the progeny from an A_1A_1 homozygote due to selfing, this proportion is reduced by the product of P(n) and $(3/4)^n$ for a given number of lethals (see Table 6). If we normalize by

$$\bar{w} = S\left[\sum \frac{x^n e^{-x}}{n!}(3/4)^n\right] + T, \tag{7}$$

then the estimate of outcrossing should be

$$\hat{T} = \frac{N_{12}\bar{w}}{qN}. \tag{8a}$$

However, because the size of $\bar{w}$ is generally not known, the estimate of outcrossing is generally

$$\hat{T}_L = \frac{\hat{T}}{\bar{w}}. \tag{8b}$$

Because $\bar{w}$ will be less than unity, the presence of recessive lethals (or other recessive fitness variants) will result in an overestimate of the outcrossing rate.

As an example, let us estimate outcrossing in a species with two lethals per zygote. In this case, the term in brackets in expression (7) becomes 0.605 so that $\bar{w} = 0.605S + T$. For example, if the actual outcrossing is 0.8, then the estimate of outcrossing, $\hat{T}_L$, would be 0.87, while if the actual outcrossing rate is 0.1, then the estimate would be 0.155. As we will discuss later, species with a high outcrossing generally have a higher inbreeding depression (genetic load), suggesting that the bias in T from selection is not as important with lower T values.

The presence of **polyembryony** and embryo competition appears to lessen this effect in conifers to some extent (e.g., Sorensen 1982; Griffin and Lindgren 1985). Even so, a number of conifer species have around 50% self-pollination but because of the high number of lethal equivalents, less than 10% of the sound seeds are from self-pollination (Sorensen 1982).

One approach to estimate the selfing rate when there are **deleterious alleles** present is based on the proportion of sound seed observed from artificial self-fertilization (p_S) and from artificial outcrossing (p_T). Given that the proportion of sound seed observed under natural pollination is p_N, then an estimate of selfing is

$$\hat{S} = \frac{p_T - p_N}{p_T - p_S}. \tag{9}$$

Charlesworth and Charlesworth (1987) have surveyed the gymnosperm literature with this technique and found that a number of species appear to have a high rate of self-fertilization.

In species that retain seeds on the tree for several years, there may be additional potential for selection. Several reports have documented higher outcrossing, as determined by electrophoresis, in older seeds (Cheliak et al. 1985; Moran and Brown 1980). A reasonable explanation for these observations is that the viability of selfed seeds is lower than that of outcrossed seeds, thereby resulting in an increase in the estimate of outcrossing over time.

Population Genetic Consequences of the Mating System

The major effect of inbreeding is to change the genotype proportions in a population. However, when there is selection in addition to partial inbreeding, the population genetic consequences depend upon the mode of selection. When two loci are considered simultaneously, partial inbreeding results in genotypic proportions different from that predicted from single loci. In addition, genetic hitchhiking may be important when there is high selfing. Overall, the effects of inbreeding on genetic variation is not straightforward and any predictions require careful analysis. In the following discussion, I will mention some of these possibilities to illustrate the potential complexity of the population genetics of partial-inbreeding organisms (see Allard et al. 1968; Brown 1979 for earlier reviews of the population genetics of inbreeding species).

Equilibrium Inbreeding

One effect of inbreeding is to reduce heterozygosity compared to Hardy–Weinberg proportions, i.e., increase the probability of homozygosity due to identity by descent. The equilibrium frequency of heterozygotes can be expressed generally as

$$H_e = (1 - \sum_i p_i^2)(1 - f_e), \tag{10}$$

where f_e is the equilibrium **inbreeding coefficient** and p_i is the frequency of the ith allele.

Given that there is a constant level of inbreeding of a particular type in each generation, then the heterozygosity will approach an equilibrium value over a few generations resulting from reduction of heterozygosity by inbreeding and the increase of heterozygosity due to random mating. To determine these equilibrium values for different mating systems, let us assume that, with probability S_j, the mating system will be repeated each generation (see Hedrick and Cockerham 1986 for a more complete exposition). First, let us assume S_0 is the proportion of selfing and f_t is the inbreeding coefficient in the tth generation. As a result,

$$f_{t+1} = S_0 \frac{(1+f_t)}{2} \tag{11a}$$

and assuming that $f_{t+1} = f_t = f_e$, then

$$f_e = \frac{S_0}{2 - S_0} \tag{11b}$$

(e.g., Wright 1969).

If we assume that the proportion of full-sib mating is S_1, then

$$f_{t+2} = \frac{S_1}{4}(1 + f_t + 2f_{t+1}) \tag{12a}$$

and

$$f_e = \frac{S_1}{4 - 3S_1}. \tag{12b}$$

In general, the equilibrium inbreeding coefficient is

$$f_e = \frac{S_j}{2^{j+i} - S_j(2^{j+1} - 1)} \tag{13}$$

where S_j is the proportion of mating resulting in an inbreeding coefficient of $(1/2)^{j+1}$, assuming no inbreeding in previous generations (see Hedrick and Cockerham 1986). The equilibrium heterozygosity is

$$H_e = \left(1 - \sum_i p_i^2\right)\left(\frac{1 - S_j}{1 - S_j[1 - (1/2)^{j+1}]}\right). \tag{14}$$

Let us compare the effects of different types of matings on the equilibrium heterozygosity. The easiest way is to compare H_e from expression (14) to the heterozygosity in a random-mating population. Figure 1 gives the ratio H_e to the heterozygosity with no inbreeding, termed "relative heterozygosity" here, for four types of mixed matings: partial selfing, partial full-sib mating, partial half-sib mating, and partial first-cousin mating, S_0, S_1, S_2, and S_3, respectively. Notice that the heterozygosity is not much reduced at low levels of inbreeding. For example, the heterozygosity is 0.858, 0.924, 0.96, and 0.98 the level in random-mating populations for populations with 25% S_0, S_1, S_2, or S_3, respectively. At increased levels of inbreeding, the reduction is greater for all types of inbreeding but lower for the more distant matings.

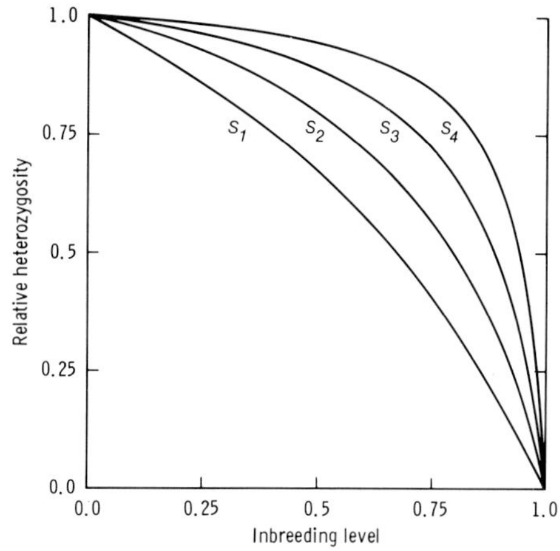

Fig. 1. The equilibrium heterozygosity relative to that found in a random-mating population for four different types of partial inbreeding (S_O: partial selfing; S_1: partial full-sib mating; S_2: partial half-sib mating; S_3: partial first-cousin mating). Hedrick and Cockerham 1986)

Let us assume that both selfing and full-sib mating are occurring in the same population, so the proportion of random mating is $1 - S_0 - S_1$. The inbreeding coefficient in generation $t + 2$ is

$$f_{t+2} = S_O\left(\frac{1 + f_{t+1}}{2}\right) + \frac{S_1}{4}(1 + f_t + 2f_{t+1}) \tag{15a}$$

The inbreeding equilibrium is approximately

$$f_e = \frac{2S_O + S_1}{4 - 2S - 3S_1} \tag{15b}$$

and the equilibrium heterozygosity is

$$H_e = \left(1 - \sum_i p_i^2\right)\left(\frac{1 - S_O - S_1}{1 - \frac{1}{2}S_O - \frac{3}{4}S_1}\right). \tag{16}$$

These expressions can be generalized to include more types of inbreeding as

$$f_e = \frac{\sum_j \frac{S_j}{2^{j+1}}}{1 - \sum_j S_j[1 - (1/2)^{j+1}]} \tag{17}$$

and

$$H_e = \left(1 - \sum_i p_i^2\right)\left\{\frac{1 - \sum_j S_j}{1 - \sum_j S_j[1 - (1/2)^{j+1}]}\right\}. \tag{18}$$

In human populations, when two or more types of consanguineous matings occur, the inbreeding coefficient is usually calculated as the inbreeding coefficient of the different types of matings weighted by the frequency of each mating type. The inbreeding coefficient calculated in this manner is usually called the average inbreeding coefficient (e.g., Cavalli–Sforza and Bodmer 1971) and is defined as

$$\alpha = \sum_{j=1}^{m} p_j f_j, \tag{19}$$

where p_j and f_j are the frequency and inbreeding coefficient, respectively, of the *jth* mating type. However, some of the individuals in the earlier generations may themselves be the offspring of consanguineous matings. In populations in which there are pedigrees extending over many generations, such matings may be known and incorporated into pedigrees. In many populations, such historical information is not known.

We can illustrate the difference between α and f_e for some data from human populations (see Hedrick 1986a). Table 7 gives the values of α, f_e, and α/f_e for the five countries with the highest α values given by Cavalli–Sforza and Bodmer (1971). As expected, the equilibrium inbreeding is always larger than the average inbreeding, the greatest difference occurring for Indian sample (a) where α is only 60.7% of f_e.

If it is assumed that a population is at inbreeding equilibrium, then expression (10) can be modified as

$$\hat{F}_I = 1 - \frac{\hat{H}}{\left(1 - \sum_i \hat{p}_i^2\right)}, \tag{20}$$

where $\hat{F}_I$, $\hat{H}$, and $\hat{p}_i$ indicate estimates of the fixation index (indirect), observed heterozygosity, and the frequency of allele A_i, respectively. Because a number of factors besides inbreeding can potentially influence the frequency of heterozygotes (e.g., Workman 1969; Hedrick 1983), the term fixation index is used instead of inbreeding coefficient.

Given the estimates of outcrossing $\hat{T}(\hat{T} = 1 - \hat{S}_0)$, then expression (11b) can be rearranged to give a direct estimate of the **fixation index**

$$F_D = \frac{1 - \hat{T}}{1 + \hat{T}}. \tag{21}$$

Table 7. The average inbreeding (α), equilibrium inbreeding (f_e) and the ratio α/f_e for the data from five human populations

	α	f_e	α/f_e
India			
a) Andra-Pradesh (1957)	0.032	0.053	0.607
b) Andra-Pradesh (1963)	0.020	0.025	0.781
Guinea	0.013	0.017	0.757
Japan	0.0046	0.0051	0.907
Brazil	0.0022	0.0023	0.956
Italy			
a) 1916–1920	0.0019	0.0019	0.954
b) 1921–1925	0.0016	0.0017	0.958

The difference between the indirect fixation index and the direct one should give an indication of the influence of other factors besides inbreeding on genotypic frequencies.

Brown (1979) describes what he terms the "**heterozygosity paradox**" in which species with low selfing, i.e., outcrossers, have a deficiency of heterozygotes and species with high selfing have an excess of heterozygotes compared to the inbreeding equilibrium level. In other words, he suggests that $\hat{F}_I > \hat{F}_D$ for outbreeders while $\hat{F}_I < \hat{F}_D$ for inbreeders. Some of the estimates he summarized are in Table 8 along with the means for all the outbreeders and inbreeders in his survey in which there were both indirect and direct estimates. Although on the average, $\hat{F}_I$ exceeds $\hat{F}_D$ for outbreeders (0.12) and the opposite occurs for inbreeders (−0.06), the effect does not appear to be present in every species.

Brown discusses a number of factors that can influence heterozygosity and suggests that a high genetic relatedness among mates due to small neighborhood size could be important in outcrossers (see also discussion in Kesseli and Jain 1985). As we discussed earlier, other types of inbreeding will lower the heterozygosity below that expected by selfing alone (e.g., Ritland 1984; Hedrick and Cockerham 1986). For example, if S_0 is estimated directly to be 0.1 and the observed heterozygosity is 0.889 that of Hardy–Weinberg proportions, then 0.167 full-sib mating in the population would explain this deficiency of heterozygosity (see Hedrick and Cockerham 1986).

Homosporous ferns have the potential for a mating system in which two gametes from the same haploid gametophyte can form a completely homozygous sporophyte by a process termed **intragametophytic selfing** (Klekowski 1979).

Table 8. Estimates of the fixation index by the indirect and direct approaches for some plant populations. (After Brown 1979). The right-hand column shows the difference between the two estimates of F

	Indirect	Direct		
Species	$\hat{F}_I$	$\hat{T}$	$\hat{F}_D$	$\hat{F}_I - \hat{F}_D$
Outbreeders				
Phlox drummondi	0.36	1.0	0.0	0.36
Zea mays	0.01	0.99	0.01	0.00
Liatris cylindracea	0.43	1.0	0.0	0.43
Eycalyptus pauciflora	0.14	0.68	0.19	−0.05
Mean[a]	0.19	0.92	0.07	0.12
Inbreeders				
Avena barbata	0.83	0.03	0.95	−0.12
Bromus mollis	0.73	0.10	0.82	−0.09
Hordeum spontaneum	0.98	0.01	0.98	0.00
Mean[a]	0.79	0.08	0.85	−0.06

[a]These are unweighted means for all species (13 inbreeders and 7 outbreeders) in which there were both indirect and direct estimates of outcrossing.

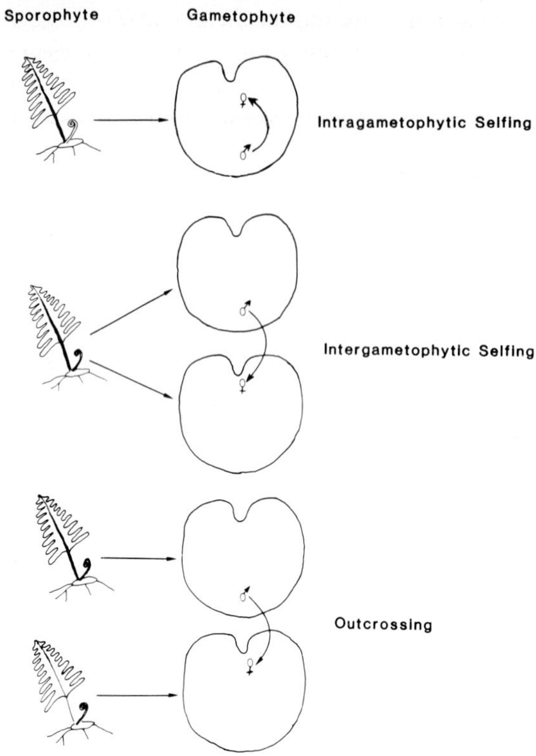

Fig. 2. Diagrammatic representation of the three possible types of mating in homosporous ferns. In nature, gametophytes are seldom larger than 5 mm. Intergametophytic selfing is equivalent genetically to selfing in seed plants

Figure 2 illustrates the genetic consequences of intragametophytic selfing as well as selfing and outcrossing (from Hedrick 1987b). If we assume that there is a proportion of selfing S, intragametophytic selfing S_I, and outcrossing $T (S + S_I + T = 1)$, then the equilibrium heterozygosity is

$$H = \left(1 - \sum_i p_i^2\right)\left(\frac{1 - S - S_I}{1 - \frac{1}{2}S}\right) \tag{22}$$

(Hedrick 1987b). If $S = 0$, then

$$f_e = S_I, \tag{23}$$

as shown by McCauley et al. (1985). Reaching the equilibrium in this case ($S_I > 0$, $S = 0$) takes only one generation as compared to approximately five generations when selfing is present.

Selection

Because inbreeding affects genotypic frequencies, selection in partially inbreeding populations is quite different than in random-mating populations. Both the dynamics of genetic change and the conditions for a polymorphic equilibrium differ. For example, the rate of incorporation of a new favourable recessive variant, when the selfing level is only 1/4 (or above) is much faster than for a random-mating population (e.g., Hedrick 1983). This effect occurs because inbreeding reduces the frequency of heterozygotes in which a new recessive variant is hidden by dominance and generates favorable homozygotes that can be selected. One implication of this is that if two populations differ in the rate of inbreeding due to genetic or environmental factors and a new variant is simultaneously introduced in both, then they may greatly differ in allelic frequency at some points in time.

When there is selection at a locus with two alleles and assuming the relative fitness of the heterozygote is unity, then depending upon the relative fitness of the homozygotes w_{11} and w_{22}, for genotypes A_1A_1 and A_2A_2, respectively, various selective outcomes are possible. For example, in a random-mating population when w_{11} and $w_{22} < 1$, then there is a stable equilibrium, as seen in the lower left quadrant of Fig. 3 (Hayman 1953; Hedrick 1983). However, when there is inbreeding, the sizes of the regions change and that for a stable equilibrium is reduced.

The conditions for a polymorphic equilibrium with partial selfing are

$$S < \frac{2w_{22}(1 - w_{22})}{w_{11} + w_{22} - 2w_{11}w_{22}} \tag{24a}$$

and

$$S < \frac{2w_{11}(1 - w_{11})}{w_{11} + w_{22} - 2w_{11}w_{22}} \tag{24b}$$

(Kimura and Ohta 1971). When there is high inbreeding and w_{11} and w_{22} are close to unity, i.e., little selection against the homozygotes, these conditions for

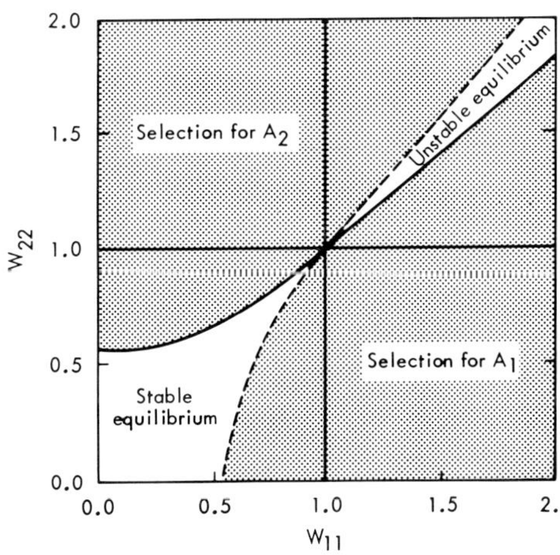

Fig. 3. The regions of equilibrium (*unshaded*) and directional selection when the relative fitnesses at a locus are w_{11}, 1, and w_{22} and there is 0.95 self-fertilization. (Hedrick 1983)

stability are quite restrictive so that there can be little asymmetry in the fitness values (see the unshaded region in Fig. 3). For example, if $S = 0.95$ and $w_{22} = 0.9$, then w_{11} must be between 0.888 and 0.91. One implication of this is that in a highly inbreeding species, slight differences in selection may lead to large differences in allelic frequencies. For example, if again $S = 0.95$ and $w_{22} = 0.9$, then when $w_{11} < 0.888$, A_2 would go to fixation while if $w_{11} > 0.91$, A_1 would go to fixation.

However, for symmetrical or nearly symmetrical selection models, the rate of allelic frequency change is quite slow. For example, where there is a heterozygous disadvantage, the homozygotes have equivalent fitnesses and there is high selfing, the allelic frequency only slowly moves away from the unstable equilibrium (Hedrick 1980a). This slow change occurs because with high inbreeding, there are few heterozygotes, making selection primarily between the two homozygotes. With complete symmetry, there is no difference in relative fitness between the homozygotes and the rate of response is mainly a function of the frequency of the few heterozygotes in the population.

When there is partial **intragametophytic selfing**, the conditions for a stable polymorphism are even more restrictive than for selfing (Hedrick 1987b). If we assume that s_1 and s_2 are the selection coefficients against A_1A_1 and A_2A_2, respectively, then the range of selection against A_1A_1 permissable to maintain a stable polymorphism is

$$s_2S_I < s_1 < \frac{s_2}{S_I}. \tag{25}$$

For example, if $S_I = 0.95$ and $s_2 = 0.1$, then s_1 must be between 0.095 and 0.105, i.e., w_{11} must be between 0.895 and 0.905.

Association of Fitness and Heterozygosity

A number of researchers have examined the association of various components of fitness (and other related measures) and individual heterozygosity as measured by electrophoresis (for a review see Mitton and Grant 1984). There are results suggesting that those individuals with higher heterozygosity have higher fitness although there are a number of studies showing no significant relation. One possible explanation for the positive association of individual heterozygosity and components of fitness is that as a group random-mated progeny may have both higher heterozygosity and fitness than inbred offspring, and that a given sample may composed of both types of progeny, thereby resulting in an overall positive correlation (e.g., Ledig et al. 1983).

To illustrate this possibility, let us draw a sample from a population with n equivalent loci having frequencies of p_{11}, p_{12}, and p_{22} for the three genotypes, say A_1A_1, A_1A_2, and A_2A_2, respectively, at locus A. If n_{11}, n_{12}, and n_{22} are the

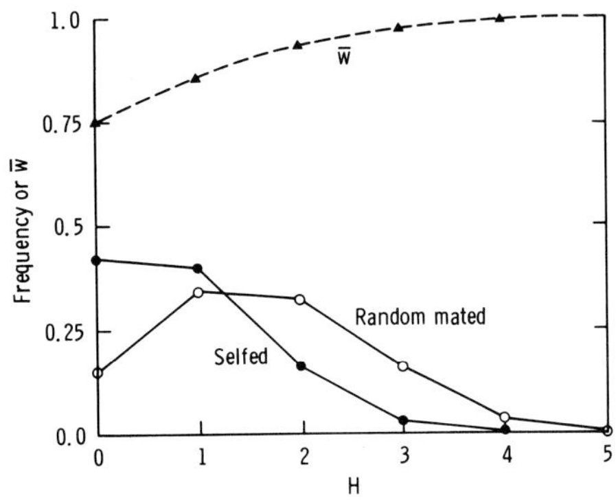

Fig. 4. The frequency of individuals with different numbers of heterozygous loci for random mated selfed progeny (*solid lines*) and the consequent fitness when it is assumed there are 25% self-progeny and the selfed progeny have half the fitness of the random-mated progeny

number of loci having 11, 12, or 22 genotypes, then the probability of this array in an individual is

$$P(n_{11}, n_{12}, n_{22}) = \frac{n!}{n_{11}!\,n_{12}!\,n_{22}!} P_{11} P_{12} P_{22}. \tag{26}$$

Therefore, the probability of i heterozygous loci out of n loci examined in an individual is

$$P(i) = \sum P(n_{11}, n_{12}, n_{22}) \tag{27}$$

when $n_{12} = i$.

As an example of the different distributions of heterozygosity for random-mated and inbred progeny, let us assume that for the random-mated sample $P_{11} = 0.04$, $P_{12} = 0.32$ and $P_{22} = 0.64$ and that for the inbred sample $P_{11} = 0.12$, $P_{12} = 0.16$, and $P_{22} = 0.72$. The latter may result from one generation of self-fertilization of the former array. Given that $n = 5$, the frequency of different numbers of heterozygous loci in individuals is given in Fig. 4. If it is assumed that the fitness of selfed progeny is 0.5 that of random-mated progeny and that 0.25 of the progeny are selfed, then the mean fitness is given by the broken line in Fig. 4. In other words, individuals with lower heterozygosity have lower fitness. However, this association is the result of the sample being composed of both random-mated and inbred progeny rather than random-mated progeny with higher heterozgosity having higher fitness.

Spatial Variation in Selection, Gene Flow and Inbreeding

One of the important types of selection models suggested for the maintenance of polymorphism is spatial variation in selection (e.g., Felsenstein 1976; Hedrick et al. 1976; Hedrick 1986b). When there is random mating over the area in which the different types of selection occur, then the conditions under which polymorphism may occur are generally quite restrictive (e.g., Maynard Smith and Hoekstra

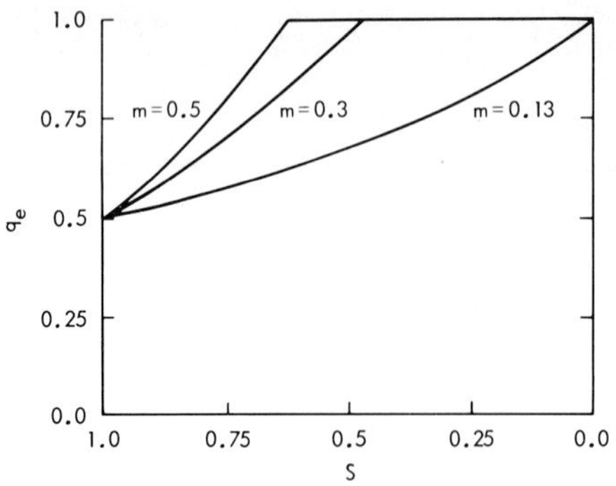

Fig. 5. The equilibrium allelic frequency for different levels of self-fertilization and three levels of gene flow when the relative fitnesses are 1.05, 1, and 0.95 in niche 1 and 0.8, 1, and 1.2 in niche 2. (Hedrick 1985a)

1980). However, limited gene flow between the environmental patches can allow a global polymorphism to be maintained that would be lost in a random-mating population (e.g., Christiansen 1975).

Let us include inbreeding in the form of partial selfing and specify gene flow as it occurs in plants to further examine this model (e.g., Hedrick 1985b). Assume that there is only pollen gene flow from one niche to the other and that gene flow and fertilization occur simultaneously and selection subsequently, say on seedling survival. We can imagine that a certain proportion of flowers produce selfed progeny while others outcross, some of which results in pollen gene flow.

How do different levels of selfing affect the potential for a polymorphism in this situation? Figure 5 gives the equilibrium values for three levels of pollen gene flow in two equally frequent niches when the relative fitnesses for genotypes A_1A_1, A_1A_2, and A_2A_2 are 1.05, 1, and 0.95 in niche 1 and 0.8, 1, and 1.2 in niche 2. Obviously, higher levels of selfing increase the likelihood of a global polymorphism, basically by causing a reduction in the extent of gene flow. For example, when $m = 0.5$ (the maximum gene flow in this case), there is no polymorphism possible for these fitness values with low selfing but when $S > 0.6$, a polymorphism does occur. This is in contrast to the heterozygote advantage model in which higher selfing makes a polymorphism less likely. In other words, balancing selection due to variation of selection in space seems a more likely mechanism of global polymorphism maintenance in predominantly selfed plants than heterozygote advantage. In addition, differential gene flow in male and female gametes (Gregorious and Namkoong 1983; Namkoong and Gregorious 1985) and maternal effects (Yokoyama and Schaal 1983) may result in broader conditions for a stable polymorphism in plant populations.

Two-Locus Inbreeding Equilibrium

When considering two loci in a mixed-mating model with no selection, the equilibrium genotypic frequencies are not simple extensions of the single-locus inbreeding equilibrium (Haldane 1949a; Bennett and Binet 1956; Weir and

Table 9. The genotypic frequencies at inbreeding equilibrium when $p_1 = q_1 = 0.5$, and $S = 0.95$. The values in parentheses are those if the two-locus genotypic frequencies were the product of one-locus genotypic frequencies

	A_1A_1	A_1A_2	A_2A_2
B_1B_1	0.2303 (0.2267)	0.0156 (0.0227)	0.2303 (0.2267)
B_1B_2	0.0156 (0.0227)	0.0164 (0.0023)	0.0156 (0.0227)
B_2B_2	0.2303 (0.2267)	0.0156 (0.0027)	0.2303 (0.2267)

Cockerham 1973). In fact, there is an excess of both double heterozygotes and of double homozygotes and a deficiency of single homozygotes when compared to single-locus inbreeding equilibria values (see Table 9 for an example when $S = 0.95$). Bennett and Binet (1956) showed that difference between the equilibrium frequency of double heterozygotes and the product of the single-locus equilibrium heterozygosities is

$$d = H_{AB} - H_A H_B = \frac{16CS(1-S)p_1p_2q_1q_2}{(2-S)^2}, \tag{28}$$

where

$$C = \frac{2 - S - 4c(1-c)(1-S)}{4 - 2S + 4Sc(1-c)}$$

and c is the **rate of recombination** between the loci. To illustrate the extent of this effect, Fig. 6 gives d for three levels of recombination. Obviously, the largest excess occurs for tight linkage and S values from 0.6 to 0.8.

The effect of inbreeding on two-locus genotypic frequencies is the result of identity disequilibrium, i.e., the frequency of gametes within individuals (e.g., Weir and Cockerham 1973). In fact, the asymptotic rate of decay of gametic

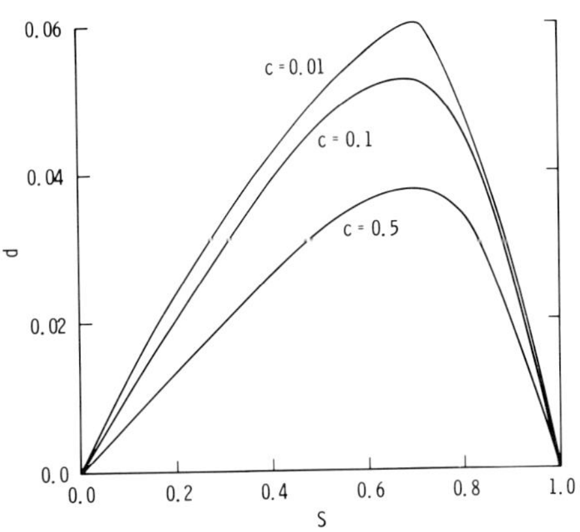

Fig. 6. The difference between the equilibrium frequency of double heterozygotes and the product of the single-locus equilibrium heterozygyosities for several recombination levels and a range of selfing

disequilibrium is affected in the same way by both selfing and recombination so that the gametic disequilibrium in generation t + 1 is

$$D_{t+1} = \frac{1}{2}\left(\frac{1+\lambda+S}{2} + \left[\left(\frac{1+\lambda+S}{2}\right)^2 + 2S\lambda\right]^{1/2}\right)D_t, \tag{29}$$

where $\lambda = 1 - 2c$ (Karlin 1969; Weir et al. 1972). This equivalence is due to a reduction in the production of new gametes from double heterozygotes caused either by lower recombination with tight linkage or by fewer double heterozygotes with inbreeding.

If there is intragametophytic selfing to the extent S_I, then the gametic disequilibrium in generation t + 1 is

$$D_{t+1} = D_t[1 - c(1 - S_I)] \tag{30}$$

(Hedrick 1987b). The rate of decay here is somewhat slower than that for the same degree of selfing, suggesting that populations with intragametophytic selfing might have more disequilibrium than populations with a similar amount of selfing.

When there is partial selfing and selection at one locus, then the effect on two-locus genotypic frequencies may be even more extreme than with no selection (e.g., Ohta and Cockerham 1974; Strobeck 1979). For example, Strobeck (1979) determined the two-locus equilibrium when one locus has a heterozygous advantage and the other locus is neutral. If we let the fitnesses at the selected locus be 1 – s, 1, and 1 – s for genotypes A_1A_1, A_1A_2, and A_2A_2, respectively, then Fig. 7 gives the apparent selection (s*) against homozygotes B_1B_1 and B_2B_2 at the neutral B locus. It is assumed here that the selected and neutral loci are unlinked. The maximum effect for S = 0.99 and no recombination occurs when $s \cong 0.5$. In this instance, the apparent fitness values at the B locus are 0.79, 1.0, and 0.79 for genotypes B_1B_1, B_1B_2, B_2B_2, a quite substantial apparent selective effect. When there is linkage between the selected and neutral loci, the apparent selection at the neutral locus is even larger. As a result, Strobeck concludes that "any estimates of fitness values based on one-locus theory such as the maximum likelihood estimates for four esterase loci by Allard et al. (1972b) are not

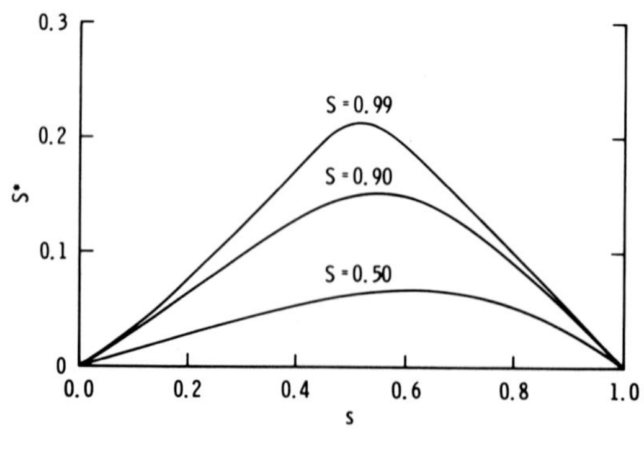

Fig. 7. The apparent selection s* at a neutral locus unlinked to a locus with selection s for three selfing rates. (Strobeck 1979)

estimating the fitness values at a single locus but the effects of all selected loci in the genome".

Golding and Strobeck (1980) investigated the effect of **finite population size** on gametic disequilibrium in a partial selfing population. They found that the disequilibrium generated was the same as in a random-mating population with the effective size

$$N_e = (1 - \tfrac{1}{2}S)N, \tag{31}$$

where N is the finite population size. The effective recombination value in this case is

$$c_e = \frac{c(1-S)}{(1-\frac{1}{2}S)}. \tag{32}$$

In other words, partial selfing reduces the effective population size by $(1-\frac{1}{2}S)$ and recombination by $(1-S)/(1-\frac{1}{2}S)$. Golding and Strobeck suggest that if $(1-S)c < 10^{-4}$, then gametic disequilibrium may be the result of genetic drift. For the three tightly linked esterase loci in barley (Allard et al. 1972b), $(1-S)c$ values are 0.000013, 0.000027, and 0.000034, suggesting that genetic drift may in fact be important in this instance.

Two-Locus Selection

Most of the initial two-locus selection studies in partially inbreeding organisms suggested that inbreeding and linkage have similar effects (e.g., Allard et al. 1968). This finding was supported by the fact that inbreeding and linkage retard the rate of decay of the gametic disequilibrium in the same way [see expression (29)]. However, when selection is present, this equivalence is no longer true. When two-locus genotypic frequencies are influenced by both inbreeding and selection, the combination of the Bennett–Binet effect and epistatic selection can result in a variety of outcomes. As an illustration, let us consider a symmetrical fitness array first investigated by Lewontin and Kojima (1960) (see Table 10). Without inbreeding, for there to be a stable two-locus polymorphism with gametic disequilibrium

$$c < \frac{a - 2b + 1}{4}, \tag{33}$$

where $a - 2b + 1$ is a measure of epistasis.

Table 10. A completely symmetrical two-locus fitness array

	A_1A_1	A_1A_2	A_2A_2
B_1B_1	a	b	a
B_1B_2	b	1	b
B_2B_2	a	b	a

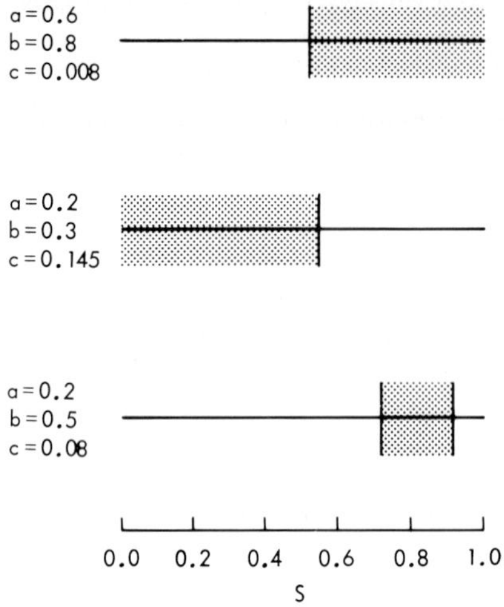

Fig. 8. The regions of two locus equilibria with gametic disequilibrium (*shaded*) for different levels of selfing and three fitness-recombination sets. (Holden 1979)

Three examples of particular fitness arrays were discussed by Holden (1979) to illustrate the **interaction of selection and inbreeding** (Fig. 8). In the top example, for no inbreeding ($S = 0$), only the polymorphism without disequilibrium is present. However, as S increases to 0.5 and above, a polymorphism with gametic disequilibrium results. In this case, more inbreeding and tighter recombination have analogous effects.

In the other two examples, this is not true. In the center example of Fig. 8 as inbreeding increases, the polymorphism with gametic disequilibrium that is present with no inbreeding disappears, the opposite of the top example. The bottom example presents an even more complex case in which the polymorphism with gametic disequilibrium exists only for an intermediate range of selfing. The basis for these results lies in the complex manner in which inbreeding and selection influence the frequencies of the single- and double-heterozygotes (see Holden 1979 for a discussion). In other words, gametic disequilibrium can increase or decrease as inbreeding is increased when selection is present. Furthermore, increased selfing, unlike tighter linkage, may result in a decreased fitness, a result shown by Holden (1979) and expanded by Charlesworth et al. (1979). Note that different populations, having the same selection but slightly different amounts of inbreeding, could have quite different gametic and genotypic frequencies due to these effects.

Recent investigations of two-locus **partial selfing models** have tried to find general implications for this complicated situation (e.g., Eshel 1978; Christiansen and Feldman 1983; Hastings 1984, 1985; Holzinger and Feldman 1985). For example, Hastings (1985) demonstrated that for high selfing levels (and free recombination), strong (twofold advantage for heterozygotes) and symmetrical selection is required to maintain gametic disequilibrium, in contrast to the situation for outcrossers and tight linkage. As a result of this and other considerations, he concludes that the "observed disequilibria in highly selfed plant populations are not likely to result from selection leading to an equilibrium".

Genetic Hitchhiking

If alleles at a neutral or nearly neutral locus are in gametic disequilibrium with alleles at a selected locus, then the neutral alleles may be carried along, or hitchhike, as selection changes allelic frequencies at the other locus (e.g., Thomson 1977). For hitchhiking to occur there must be initial gametic disequilibrium, a factor that slows the decay of the association, such as linkage or inbreeding, and selective change that occurs before the association disappears (Hedrick 1980b). In some cases, genetic hitchhiking may increase gametic disequilibrium over time between two neutral loci. Hedrick and Holden (1979) suggested that such a phenomenon is consistent with the increased disequilibrium observed in a composite cross in barley.

In two experimental populations of barley, *Hordeum vulgare*, started from crosses of collections from around the world, changes in allelic frequency and gametic disequilibrium have been observed for four esterase loci (Allard et al. 1972). I will concentrate on one of these populations, Composite Cross V (CCV), although a similar analysis could be carried out on the other population (see Muona 1982). In general, the allele frequency changes at the four loci were rather small (an average of 0.086 during 26 generations) for the two most common alleles at each locus but there were sizeable changes in the values of gametic disequilibrium between pairs of loci. The pairwise standardized gametic disequilibrium values (see Lewontin 1964 or Hedrick et al. 1978) for the three linked esterase loci (E_A, E_B, E_C) for all the generations examined are plotted in Fig. 9 (solid lines and open circles).

A genetic hitchhiking model can mimic these changes by assuming simple directional selection at a locus unlinked to the allozyme loci. The behavior of gametic disequilibrium is approximated quite well (as shown in Fig. 9) although other gametic arrays also satisfactorily mimic these changes. In these simulations, the reported recombination values were used, i.e., recombination rates between $E_A - E_B$, $E_A - E_C$ and $E_B - E_C$ were 0.023, 0.0048, and 0.0061, respectively, and selfing was assumed to be the estimated value of 0.9943. For the unlinked locus undergoing selection, there was a 20% selection difference between the homozygotes, and the heterozygote was assumed intermediate.

From the similarity of the observed and simulated values, it appears that change such as those observed in this population could be the result of hitchhiking of allozyme loci when there is directional selection at other loci in the genome. Because the experimental populations of barley represent the gene pools of a worldwide collection, it would be very unlikely that the original population was adapted to the Davis, California, environment, making directional selection in the population an obvious and simple expectation.

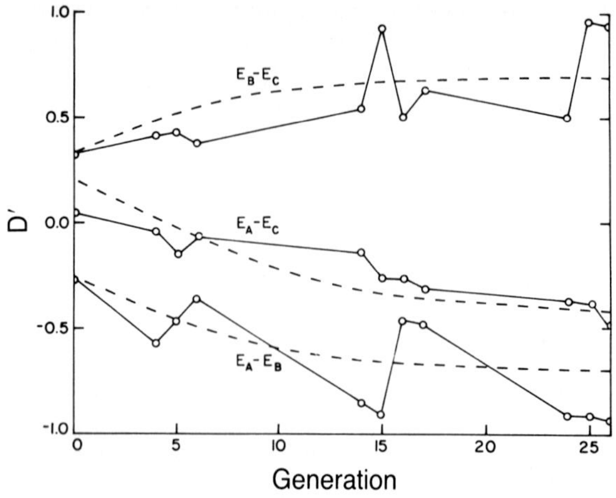

Fig. 9. The observed changes in gametic disequilibrium for three barley esterase loci E_A, E_B and E_C (*solid lines*) and simulations to mimic these changes (*broken lines*). (Hedrick and Holden 1979)

Furthermore, Clegg et al. (1978) have indicated their current belief which appears entirely consistent with a genetic hitchhiking model. They state that "What is measured at the marker loci is the selective flux transmitted throughout the genome by the correlational structure of the entire multilocus distribution. We can conclude that there is substantial selection, but we can isolate neither the phenotypic structures nor the genetic loci causally related to the selection". Recently, in an eight-generation study of these esterase loci in lines initiated from three different generations of CCV, Luckett and Edwards (1986) concluded that they "could not distinghish between the multi-loci selection and hitchhiking hypotheses, although we believe the latter to be the most plausible".

Factors Affecting the Mating System

Both environmental and genetic factors appear to affect the extent of self-fertilization. For example, biotic environmental factors such as low pollinator density or low plant density may result in an increase in self-fertilization in certain areas or at certain times. Abiotic environmental factors, such as wind, humidity, or elevation, may also be important in determining outcrossing rate (e.g., Clegg 1980).

Mating systems in plants may be affected genetically in two different ways. First, genetic variants may influence the rate of self-fertilization or outcrossing directly by changing attributes such as flower morphology or self-incompatibility systems. Genetic variants that enhance self-fertilization without diminishing pollen production should increase in frequency because they contribute twice as many alleles to their selfed progeny as to their outcrossed progeny (e.g., Fisher 1941). Second, self-fertilization increases the homozygosity of recessive deleterious alleles, thereby lowering the fitness of selfed progeny relative to that of outcrossed progeny. In other words, the expected increase of selfing alleles may be

Table 11. The whole plant fertility of three different genetic variants in five populations of *Primula vularis*. (Piper et al. 1984)

		Pin	Thrum	Homostyle
1982	1	213	238	255
	2	647	777	805
	3	316	143	513[a]
1983	1	144	144	364[a]
	4	112	40	316[a]
	5	163	19	374[a]

[a] $p < 0.01$.

counterbalanced by the lowered fitness of selfed progeny due to the presence of recessive deleterious alleles. For a detailed examination of various aspects of the evolution of selfing, see Jain (1976); Nagylaki (1976a); Lloyd (1979); Wells (1979); Jain (1984); Holzinger et al. (1984); Lande and Schemaske (1985), and Gregorius et al. (1987).

For **animal-pollinated species**, reduced pollinator activity in marginal populations may be sufficient to overcome the inbreeding depression of selfers and allow selfing to increase. In support of this notion, Piper et al. (1984) demonstrated that the self-fertile homostyle variant in *Primula vulgaris* has a higher number of seeds per plant than the self-incompatible pin and thrum morphs (see Table 11) in several populations in which homostyle variant was frequent. The effect was most striking in 1983, a wet and presumably inhospitable year for pollinators.

A number of studies have suggested there is both spatial and temporal variation in selfing rates (for a summary see Schoen 1982). However, there appears to be little interpopulational variation in some species when multilocus selfing estimates are used (e.g., Brown et al. 1985; Kesseli and Jain 1985). As with other phenotype characters, it is generally difficult from such observations to determine whether the variation is genetic, environmental, or some combination of both. In addition, as we discussed earlier, a number of factors such as variation in allele frequencies, other types of inbreeding, or selection can influence estimates of selfing.

Furthermore, when **measuring variation** in the mating system among plants, in addition to environmental and genetic factors, there is also statistical variation even if there is no true variation of selfing among plants. For example, Ritland and Ganders (1985) examined among-plant (family) variation of outcrossing rates in *Bidens menziesii*. Data from two of these populations are given in Fig. 10, where the histograms indicate the actual estimates and the closed circles are the distribution expected by chance.

First, note that the two populations differ in outcrossing rates with Puu Kooke having a lower rate of outcrossing. Second, the observed variance over plants is not significantly greater than that expected for Puu Kooke, while in the Akumoa

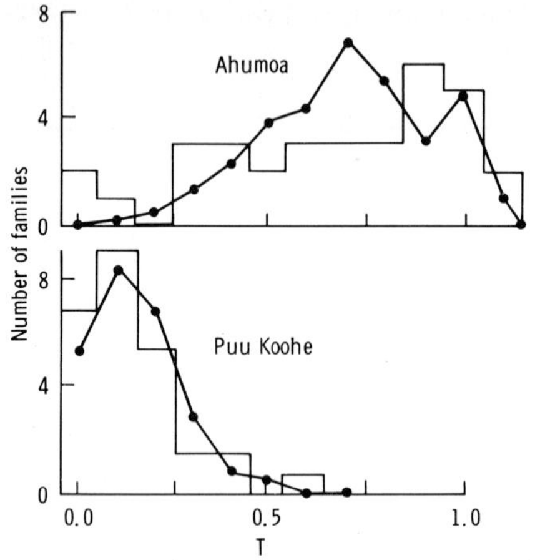

Fig. 10. The among-plant variation in outcrossing rate in two populations of *Bidens menziesii*. Histograms are the actual estimates while the lines are distributions estimated by computer simulation. (Ritland and Ganders 1985)

sample there is significant variation in the multilocus estimate. Although there is male sterilty in this species, thought to be environmentally influenced, Ritland and Ganders suggest that spatial variation in allelic frequencies and consangineous ratings may be the cause of the significant variation in outcrossing rates they observed in some populations.

Environmental and genetic factors affecting the mating system may have different evolutionary consequences. For example, if there is no selfing in one generation because of an environment fluctuation, the genotypes at all loci will return to Hardy–Weinberg proportions. On the other hand, an extreme bottleneck that would eliminate rare recessive lethals could allow the evolution of high selfing and permanently change the mating system. Let us examine some population genetic consequences of environmental variation in selfing and then discuss the evolution of the mating system.

Environmental Variation in the Mating System

As we just mentioned, whenever there is no inbreeding in one generation, heterozygosity returns to the Hardy–Weinberg proportion in the next generation. On the other hand, for complete selfing, heterozygosity is eliminated over several generations, and in general by five to ten generations, it is very low to zero, depending upon the type of matings. Nei (1975) and Brown (1979) have discussed this phenomenon and suggested because of the asymmetrical rate of return to inbreeding equilibrium that selfing estimates in a few generations may lead to the observed heterozygosity being higher than that expected from the inbreeding equilibrium.

When the selfing proportion varies over time and there is zero **autocorrelation between selfing levels**, the equilibrium genotypic proportions, actually the expected heterozygosity over a long period of time, are obtained by using the mean selfing level (Brown 1979; see also Allard et al. 1968). In other words, if $\bar{S}$ is the mean level of selfing over generations in a population, then the equilibrium heterozygosity is

$$H_e = 4pq\frac{1-\bar{S}}{2-\bar{S}}. \tag{34}$$

However, if there is a negative autocorrelation between inbreeding levels in different generations, the heterozygosity is somewhat higher than given by expression (34) while if there is a positive autocorrelation, it is lower.

As an extreme case, let us assume that selfing levels vary between two values. When there is an autocorrelation of 1 between levels, really infinitely long runs at each selfing level, then the equilibrium heterozygosity is

$$H_e = 2pq\left(\frac{1-S_1}{2-S_1} + \frac{1-S_2}{2-S_2}\right), \tag{35}$$

where S_1 and S_2 are the two different levels of selfing. In other words, the overall equilibrium heterozygosity is the arithmetic average of the equilibrium heterozygosity for each level. On the other hand, when the autocorrelation is -1, a switch in selfing level every generation, then the heterozygosity is

$$H_e = 2pq(1-\bar{S}) + \tfrac{1}{4}(S_1H_2 + S_2H_1). \tag{36}$$

If we assume $S_1 = 0$, then $H_1 = 2pq$ and Eq. (36) becomes

$$H_e = 2pq(1 - \tfrac{1}{4}S_2). \tag{37}$$

Obviously, the overall heterozygosity is closer in this case to that for the lower level of selfing.

Table 12 gives three numerical examples to illustrate these effects. The most extreme possible is given in the first two columns in which selfing varies between 0.0 and 1.0. In this case $\bar{S} = 0.5$ and the equilibrium heterozygosity for a zero autocorrelation is 0.667 that of Hardy–Weinberg proportions (and the same as if

Table 12. The relative heterozygosity H and fixation index F at equilibrium when the level of selfing varies between two values in different generations and there are different levels of autocorrelation between selfing in subsequent generations

	S_1, S_2					
	0.0, 1.0		0.0, 0.2		0.8, 1.0	
Autocorrelation	H	F	H	F	H	F
−1	0.75	0.250	0.950	0.050	0.188	0.812
−0.5	0.714	0.286	0.949	0.051	0.185	0.815
0	0.667	0.333	0.948	0.053	0.182	0.818
0.5	0.600	0.400	0.946	0.054	0.176	0.824
1	0.500	0.500	0.944	0.056	0.166	0.833

there is no variation in selfing). However, when the autocorrelation is -1, the heterozygosity is elevated to 0.75 of Hardy–Weinberg proportions and when it is 1, it is lowered to 0.5. Two other examples are given for a predominant outcrosser, such as most conifers, and a predominant inbreeder, such as some grasses. Notice that the autocorrelation has only a small effect on heterozygosity in these cases. It is possible that a negative autocorrelation between selfing rates due perhaps to weather may explain in part the aspect of the heterozygosity paradox of the excess of heterozygotes observed in predominantly selfing organisms. Schnee and Waller (1986) suggest that there is an alternation of selfing and outcrossing over generations in *Amphicarpaea bracteata*, presumably resulting in a negative autocorrelation of selfing rates.

Let us briefly examine how spatial **variation in selfing** may affect heterozygosity. If there is only pollen gene flow, then flow has no affect on genotypic frequencies, assuming equal allelic frequencies in different niches. Therefore, the equilibrium heterozygosity in niche i is

$$H_i = 4pq\frac{1 - S_i}{2 - S_i}, \tag{38}$$

where S_i is the level of self-fertilization in niche i.

On the other hand, when there is zygotic gene flow, then the heterozygosity in a given niche is a function of the extent of gene flow and self-fertilization in other niches. For example, if we assume that there are two niches, i and j, that exchange m migrants per generation, then at equilibrium

$$H_i = \frac{pq[(1 - S_i)(2 - S_j) + 2mS_i(1 - S_j)]}{1 - \bar{S} + \frac{1}{4}S_iS_j + m(\bar{S} - \frac{1}{2}S_iS_j)}. \tag{39a}$$

if $m = 0$, then

$$H_i = 4pq\frac{1 - S_i}{2 - S_i}, \tag{39b}$$

and when $m = 1/2$, then

$$H_i = 4pq\frac{1 - \bar{S}_i}{2 - \bar{S}_i}, \tag{39c}$$

In other words, zygotic gene flow approaches the effect of pollen gene flow $m \to 0$, i.e., each niche reflects the mating system present there, while as $m \to \frac{1}{2}$, the heterozygosity approaches that expected by the average self-fertilization over all niches.

When selection occurs in addition to variation in selfing, then the situation is more complicated. However, Brown and Albrecht (1980) showed that with heterozygote advantage and temporal variation in selfing, the average heterozygosity is lower than with constant selfing and the same selection. However, when there is selection, a negative autocorrelation over environments for selfing increases the heterozygosity as compared to no autocorrelation.

Genetic Factors Affecting the Mating System

As we stated above, the evolution of the mating system appears to be influenced by both genetic variants that influence the selfing rate and by fitness variants. For example, in outcrossing species, mutation to recessive fitness variants will result in the development of genetic load thereby inhibiting the increase of alleles that enhance the selfing rate. On the other hand, once a population has high selfing, then new recessive lethal and deleterious mutations will be quickly eliminated and no **genetic load** will accumulate. A species may go from an outbreeder to an inbreeder if genetic load is reduced by a bottleneck or polyploidy (Lande and Schemske 1985; Hedrick 1987c). Lande and Schemske (1985) suggested that the interaction of these genetic factors would result in either highly selfed species with low inbreeding depression (low genetic load) or in outcrossed species with high inbreeding depression. Note that these predictions may be influenced by dispersal patterns (Holzinger 1986), biparental inbreeding (Uyenoyama 1986), or other factors (e.g., Waller 1986).

The theoretical basis of the difference in the mutational component of genetic load of outbreeders and inbreeders for diploid species can be shown from an examination of the equilibrium allelic frequency of detrimentals maintained in a population by mutation-selection balance. When there is no breeding, the equilibrium frequency of a recessive detrimental allele is

$$q_e = \left(\frac{u}{s}\right)^{1/2}, \tag{40}$$

where u is the mutation rate to recessive allele a and the relative fitness of the genotypes AA, Aa and aa, are 1, 1, and $1 - s$, respectively.

Lande and Schemske (1985) discussed the extent of inbreeding depression when there is partial self-fertilization. Let us examine here the situation when there is a proportion of intragametophytic selfing S_I and the extent of outcrossing is $1 - S_I$. As a result, the frequencies of genotypes AA, Aa, aa, respectively, after selection are (after Hedrick 1987c)

$$\begin{aligned} P' &= [(1 - S_I)p^2 + S_I p]/\bar{w} \\ H' &= (1 - S_I)2pq/\bar{w} \\ Q' &= [(1 - S_I)q^2 + S_I q]/\bar{w}, \end{aligned} \tag{41}$$

where

$$\bar{w} = 1 - sq(q + S_I p)$$

and p and q are the frequencies of alleles A and a, respectively.

The expected change in allelic frequency due to selection is then

$$\Delta q_s = -spq(q + S_I p)/\bar{w}. \tag{42a}$$

The change in allelic frequency due to mutation is

$$\Delta q_m = up, \tag{42b}$$

so that the overall change in allelic frequency is

$$\Delta q = up - spq(q + S_I p)/\bar{w}. \tag{42c}$$

If we assume $\bar{w} \cong 1$, then

$$(1 - S_I)q_e^2 + S_I q_e - \frac{u}{s} = 0 \tag{43}$$

and

$$q_e = \frac{-S_I + \left[S_I^2 + 4(1 - S_I)\frac{u}{s} \right]^{1/2}}{2(1 - S_I)} \tag{44a}$$

when $S_I < 1.0$ [notice that because $S_I = f$, this is equivalent to the formulation by Haldane (1940)]. When the leftmost term in (43) is small (q small compared to S_I), then the equilibrium is approximately

$$q_e \cong \frac{u}{S_I s}. \tag{44b}$$

If $S_I = 1$ and $s = 1$, then

$$q_e \cong u. \tag{44c}$$

Let us now estimate the expected genetic load (inbreeding depression) by inbreeding a known amount f′ that causes the frequency of the recessive genotype to be

$$Q_f = q_e^2 + f' p_e q_e. \tag{45a}$$

When the inbreeding is intragametophytic selfing, then $f' = 1$ so that

$$Q_f = q_e. \tag{45b}$$

For a given locus i, the probability of survival is

$$v_i = 1 - s_i Q_{f(i)} \tag{46a}$$

and when $f' = 1$

$$v_i = 1 - s_i q_{e(i)}. \tag{46b}$$

If $S_I = 1$ and $s_i = 1$, this makes $q_e \cong u$, and then $v_i = 1 - u$. If we assume that there are n loci that can mutate to lethals and that they act independently, i.e., no epistasis on a multiplicative scale (e.g., Hedrick 1983), then the expected proportion of viable offspring assuming no gametic disequilibrium is

$$v = v_i^n \tag{47}$$

and the genetic load is

$$l = 1 - v. \tag{48}$$

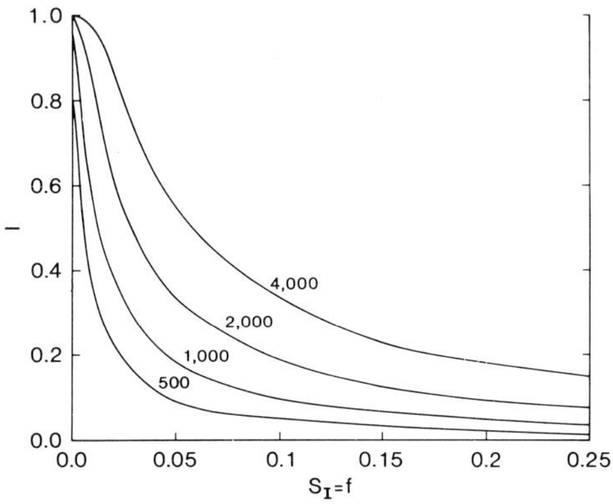

Fig. 11. The expected genetic load (l) resulting from intragametophytic selfing when the population is at mutation-selection-inbreeding equilibrium. There are either 500, 1000, 2000 or 4000 potentially lethal loci and the inbreeding coefficient in the population is $S_I = f$. (Hedrick 1987c)

The expected genetic load of progeny produced by intragametophytic selfing ($f' = 1$), taken from populations at equilibrium under mutation-selection-inbreeding is given in Fig. 11 when there are either 500, 1000, 2000, or 4000 potentially lethal loci given that all $s_i = 1$ and $u = 10^{-5}$. First, given an inbreeding equilibrium where $S_I = f > 0.25$, the progeny of intragametophytic selfing will have only a small genetic load, e.g., $l < 0.1$ for 2000 loci. In other words, for species that have experienced significant inbreeding in the past, the extent of genetic load will most likely be small. Second, a small amount of inbreeding will greatly reduce the extent of the genetic load. For example, if $S_I = f$ is only 0.05 and $n = 2000$, then the genetic load is only 0.33 compared to 1.0 when $S_I = f = 0.0$.

Schemske and Lande (1985), in a survey of natural plant populations, observed a bimodal distribution of outcrossing rate with few species having intermediate outcrossing rates, consistent with their theoretical predictions that an intermediate selfing rate is "unstable". These provocative findings have been questioned on several fronts. For example, among other comments, Waller (1986) points out that the sample of species is biased towards two groups: short-lived weeds and crop plants, both of which tend to be self-compatible, and gymnosperm trees which are known to be genetically variable and highly outcrossed. However, Aide (1986) suggests that if only wind-pollinated plants are examined, there is an even clearer bimodal distribution.

Although the selfing data in homosporous ferns are relatively few and recent, there appear to be few species with an intermediate outcrossing rate (Soltis and Soltis 1987). In addition, the extent of genetic load observed in diploid homosporous ferns suggests a bimodal distribution (e.g., Hedrick 1987c). In the 18 species in which estimates are available, 6 have a mean number of lethals less than or equal to 0.3 and 10 species have a lethal number greater than 1. Although there are only a few species in which both the genetic load and the mating system have been determined, the results are consistent with the prediction of Lande and Schemske (see Hedrick 1987c).

Charlesworth and Charlesworth (1987) in a comprehensive survey concluded that many organisms have a large amount of inbreeding depression. However, they feel that there is not sufficient evidence to state that there is necessarily low inbreeding depression in partially inbreeding populations. Furthermore, if the standard in a highly selfing organism to which inbred individuals are compared is a cross to an unrelated individual, then it appears that there is substantial genetic load.

Overview

Obviously, inbreeding adds significant complexity to understanding the evolution of many organisms. Proper estimation of the parameters of the mating system is a formidable task in itself. Understanding the population genetics of partially inbreeding organisms is challenging and there appear to be few simple predictions. In addition, the mating system itself may be affected by both environmental and genetic factors, resulting in a variety of population genetic consequences. Trying to understand the impact of the mating system in evolutionary genetics is an intriguing and humbling experience, but one which hopefully will enable us to further comprehend the evolutionary process.

4 Population Structure and Processes in Evolution

N. BARTON AND A. CLARK

Evolution and Population Structure

The strongest motivation for evolutionary biologists to study the effects of population structure is quite simply that most species of most organisms are subdivided, in the sense that not all pairs of individuals are equally likely to mate. Slight departures from panmixia will have no effect on evolution of the species if the subdivision has no effect on the distribution of genotypes. Many of the forces that affect genetic variation in panmictic populations have an influence on the degree of genetic differentiation, and many of the results will be surveyed in this chapter. Even in the "simplest" case of strictly neutral mutation, there are some intriguingly counter-intuitive results. The amount of migration among subpopulations necessary to make the population appear as one panmictic unit is surprisingly small.

In a single panmictic population, mutation to novel, neutral alleles tends to increase genetic variability, and random genetic drift tends to reduce this variability. In this strictly neutral setting, the forces of mutation and drift come to a steady state with a characteristic distribution of allele frequencies. If the mutation rate μ increases, the number of alleles present at steady state will increase, and the probability that a random pair of alleles will be identical (f) decreases according to $f = 1/(4N\mu + 1)$. If the population is subdivided, genetic differences can accumulate among individuals within groups as well as among groups. A natural way to quantify the hierarchy of genetic differentiation is with Wright's F statistics, and they have been very useful in both quantifying empirical observations of natural populations and in describing theoretical results.

Some of the deepest questions in population genetics concern the operation of natural selection in subdivided populations. A broad view of subdivision is taken in this chapter to include temporal environmental variation and familial aggregation. Selection in geographically subdivided populations can result in stable clines of gene frequency, and waves of gene frequency change, both of which have been well documented in nature. Natural selection can operate on phenotypes that are properties of the group, and the importance of such group selection remains controversial. There is still no hard answer to whether natural selection is more effective in subdivided populations (as Wright held) than in single panmictic populations, but many of the conditions that influence the answer to this question will be explored in this chapter.

Population Structure with Strictly Neutral Genetic Variation

Migration and Drift: The Island Model

The consequences of migration between a large mainland population and an incompletely isolated population were first considered by Wright (1931) in what is known as the island model. Wright formulated models of population structure using **inbreeding coefficients**, defined as expected correlations between uniting gametes. These were shown by Malécot (1948) to be equivalent to probabilities of drawing gametes that are identical-by-descent. First consider a finite population that is not subdivided. In such a population, the probability of drawing a pair of alleles that are identical-by-descent increases according to:

$$f' = 1/2N_e + [1 - (1/2N_e)]f, \tag{1}$$

where N_e is the effective population size, and f' is the inbreeding coefficient in the following generation. We will adhere to Wright's earlier nomenclature of using f for the inbreeding coefficient, which is equivalent to the probability of identity-by-descent of a pair of uniting gametes. In Wright's original formulation of the island model, if either allele is a migrant, it has a probability of zero of being identical-by-descent with other alleles on the island. This is equivalent to assuming that there are infinitely many islands, and that migration is equally likely between all pairs of island. It is also equivalent to assuming that there is one island that receives migrants at rate m from a continent that has an infinite population size. Hence both alleles must come from the island to be identical-by-descent: this possibility has probability $(1 - m)^2$. On the island then, the inbreeding coefficient increases according to:

$$f' = (1 - m)^2\{(1/2N_e) + [1 - (1/2N_e)]f\}. \tag{2}$$

At steady state, $f' = f$, and we obtain

$$f = \frac{(1 - m)^2}{2N_e - (1 - m)^2(2N_e - 1)}. \tag{3}$$

We use the word "steady state" here rather than equilibrium, because at steady state there is still a turnover of alleles, but the probability of identity of a pair of alleles settles to the steady state value given in Eq. (3). For values of m sufficiently

$$f = \frac{1}{4N_e m + 1}, \qquad \text{or equivalently,} \qquad H = \frac{4N_e m}{4N_e m + 1}. \tag{4}$$

small that we can ignore terms with m^2 (m less than 0.1 say), this is approximately This is of course identical in form to the steady state value of f under the infinite alleles model, where new alleles enter the population by mutation. In this instance, the new alleles that enter the island population by migration are equivalent to novel mutations.

The striking thing about Eq. (4) is that it suggests that very little migration is needed in order to prevent the build-up of identity-by-descent on the island, and hence very little migration is needed to prevent genetic divergence of the island (Fig. 1). In particular, if $4N_em \gg 1$, this implies that the absolute number of migrants is more than ten or so, migration will swamp the islands so that they behave as one panmictic unit. From Eq. (4) it also follows that the variance of allele frequency among islands in the infinite island model is $p(1-p)/(1+4Nm)$. As Nm increases, this variance approaches zero (the island become one panmictic unit); as Nm decreases, islands fix for one allele or the other, and the variance approaches the maximum possible value, $p(p-1)$. It should be noted that throughout this chapter, migration is assumed to be independent of genotype and and independent of degrees of relationship. If there were variation among genotypes in the probability of migrating, or if groups of kin migrated together (Rogers 1987), the genetic consequences of migration could be very different from the random migration models presented here.

A useful statistic derived from neutral models is the **effective number of alleles**, defined as 1/f (Kimura and Crow 1964). This is the number of alleles that the population would need to obtain the same f, if the alleles were equally frequent. For the infinite alleles model without population structure, the population arrives at a steady state between the influx of novel alleles at mutation rate μ per

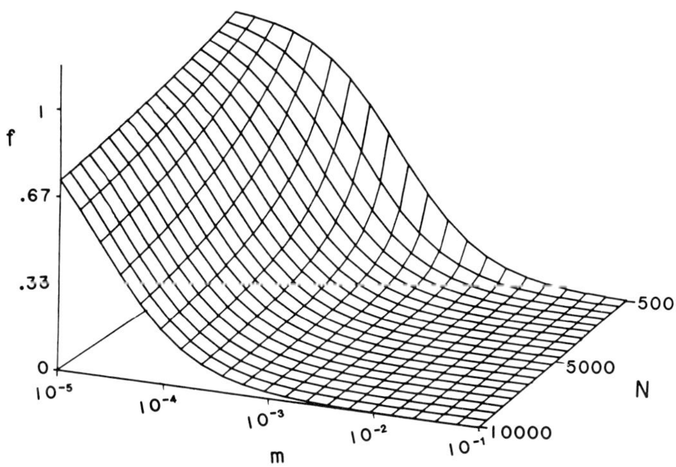

Fig. 1. The relation between the inbreeding coefficient (*f*), and the number of migrants per generation (*Nm*) at steady state in the island model. The function plotted is $f = 1/(4Nm + 1)$

generation and loss of alleles by drift. At this steady state, the effective number of alleles is

$$n_e = 4N\mu + 1. \tag{5}$$

Maruyama (1970) showed that for the n-island model (having n islands all equally accessible by migration), the steady state effective allele number is approximately

$$n_e = 4Nn\mu + 1 + n\mu/m. \tag{6}$$

The total population size is Nn, so by adding population structure, the effective number of alleles is increased by $n\mu/m$. It stands to reason that a more structured population (more islands or lower migration rate) will have a larger effective number of alleles, hence a lower f. A number of interesting extensions of the island model have been examined, including sex linkage, extranuclear genes and two loci. In the case of sex chromosomes, one needs to consider sex-specific migration (this will also complicate the case of autosomes, but this can be absorbed into the effective population size). Other properties of the island model are reviewed in Nagylaki (1983, 1985).

An intriguing extension of the island model has been developed for the non-Mendelian transmission of **mitochondrial** and **chloroplast DNA**. The infinite alleles model in this context was developed by Birky et al. (1983), and Takahata and Maruyama (1981). Takahata and Palumbi (1985) extend the model to consider the effects of population subdivision. In the case of mitochondria, there can be sampling at the stage of cell division as well as sampling in the formation of gametes. The multiple cell divisions within an individual lead to a balance between the generation of genetic diversity by mutation and the loss of this variation by sampling in the form of inter-cellular drift. With low mutation rates, we expect a "sorting out" of the intracellular variation, so that a heteroplasmic zygote will develop and mature and be likely to produce gametes with mitochondria of just one type. The dynamics will depend on both the number of mitochondria per cell and the number of cell divisions. We assume that the sampling of mitochondria is random, since no mechanism for segregation of mitochondria has been found. Under these conditions, Takahata and Maruyama (1981) found that the gene identity (sum of aquared allele frequencies) within an individual is approximately

$$I = 1/(1 + 2n\mu), \tag{7}$$

assuming no paternal contribution, n gene copies per cell and a rate of mutation per cell generation of μ. This is equivalent to the infinite alleles model, with $4N\mu$ replaced by $2n\mu$ because the infinite alleles model considers 2N gene copies. The gene identity within the population is

$$Q = \frac{1}{1 + (2N\lambda + 2n)\mu}, \tag{8}$$

where N is the number of females and λ is the number of cell divisions per generation. There is an implicit assumption that the number of cell divisions from the zygote to the cells that are scored is the same as the number of cell divisions in the germline. The results of the island model for mitochondrial variation can be summarized by saying that the effective population size is smaller, so higher f and lower genetic diversity is expected. It has been argued that the high levels of mitochondrial DNA diversity that are observed in some organisms are due to the elevated mutation rate and poor DNA repair function.

A more complete analysis of the island model requries consideration of several inbreeding statistics, discussed in Section 2.5.

The Diffusion Approximation

An important tool that has been used to discover many aspects of neutral genetic variation is the diffusion approximation. The application of diffusion theory to these problems was introduced by Wright (1945a), but diffusions have a long history in theoretical population genetics, beginning with Fisher (1922). The use of the diffusion approximation has made it possible to obtain results from otherwise intractable, but biologically very important models. In particular, Kimura (1955, 1964) made extensive use of diffusion equations in the study of fixation times, Maruyama (1971, 1972a, 1977) and Nagylaki (1978a) have used diffusions in deriving many important results for **neutral variation** in structured populations. Lucid derivations of diffusions can found in Roughgarden (1979) and in Crow and Kimura (1970). An important limitation of the diffusion approximation is that the means, variances and covariances of changes in allele frequencies per generation must all be small, and of order 1/N. For the present case, this implies that migration rates must be low.

Let $\phi(p, x)$ be the probability density of frequency x of allele A at time t, taking as an initial condition that at $t = 0$ the frequency was p. Wright (1945a) showed that for a strictly neutral gene, $\phi(p,x)$ follows the diffusion equation, formally known as the Kolmogorov forward equation:

$$d\phi/dt = \frac{1}{4N}\frac{\partial^2}{\partial x^2}[x(1-x)\phi_t(p,x)], \tag{9}$$

where t is time, which, despite being continuous, is often measured in generations. A simple geometric derivation of the forward equation is presented in Kimura (1955) and Crow and kimura (1970, pp. 373–376). The allele frequency x in a population of size N is not really a continuous variable; it has discrete steps of size 1/2N. This is called the forward equation because we start at a frequency p and predict the distribution of allele frequencies in a large set of populations over time. The solution to the forward equation can be represented as a sum of a series of eigenfunctions, each of which decays at a characteristic rate:

$$\phi_t(p,x) = p(1-p)\sum_{i=0}^{n} g_i(x,t), \text{ where} \tag{10}$$

$$g_i = i(i+1)(2i+1)F(1-i, i+2, 2, p)F(1-i, i+2, 2, x)e^{-(i+1)t/4N}$$

and F(,,,) represents the hypergeometric function. The solution to this equation is a distribution of allele frequencies that one would find in a collection of populations started under the same initial conditions. The populations would show a range of allele frequencies because of the stochastic nature of drift, but they are expected to follow the distribution $\phi_t(p,x)$. After a while, most populations will be fixed for one allele of the other, and there is an approximately uniform density of populations with intermediate frequencies. This "U-shaped" distribution ultimately leads to a distribution with all populations fixed for one or the other allele.

The above discussion is relevant to an unstructured, finite population with no mutation or selection. The diffusion equation for an island population that receives migrants at a rate m per generation is

$$d\phi/dt = \frac{1}{4N_e}\frac{\partial^2}{\partial x^2}[x(1-x)\phi] - m\frac{\partial}{\partial x}[x_m - x)\phi], \tag{11}$$

again assuming no mutation or selection, and that the frequency of allele A in the migrants is x_m. Note the second term, which reduces the rate of change of ϕ by amount that depends on the migration rate. The greater the migration rate, the more the island is swamped by the migration, and so the smaller the effect of drift on changes in allele frequency. The steady state solution to this equation is a beta distribution. This was derived by Wright (1973), and was discussed at length by Crow and Kimura (1956):

$$\phi(p,x) = \frac{\Gamma(4Nm)}{\Gamma(4Nmx)\,\Gamma[4Nm(1-x)]} x^{Nmx-1}(1-x)^{Nm(1-x)-1}. \tag{12}$$

The form of ϕ is presented in Fig. 2 for a range of values of Nm, where x_m, the allele frequency among the migrants, is 0.5. Note that for small Nm, drift

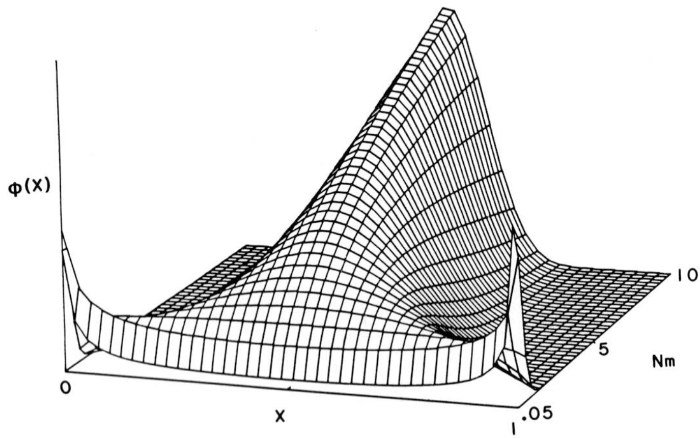

Fig. 2. The steady state probability density of populations undergoing drift and recurrent migration following Wright's island model. *N* is the population size on the island, *m* is the fraction of the population that arrives as migrants each generation, and *f* is the probability of drawing a pair of alleles identical-by-descent on the island

predominates, and most of the populations are fixed or nearly fixed. With large Nm, implying either a large population with minimal variance due to drift, or high migration rate, populations tend to have allele frequency closer to 0.5. Equation (12) is actually the solution for a whole class of models where allele frequency is affected by what Wright called "linear pressures". If the change in allele frequency is proportional to the allele frequency, as is the case for migration and mutation, then this is called a linear pressure. It is because of this property that the solutions to the island model bear such a strong resemblance to models with mutation.

The diffusion approximation has made it possible to treat more complex models as well, including two loci, multiple alleles, and models with natural selection operating in subdivided finite populations.

Isolation-by-Distance and Neighborhood Size

Wright (1943) realized that the island model made a restrictive assumption, in that no migrant allele could be identical by descent with any alleles already on the island. A more natural model might allow for **migration** among local subpopulations, with both the probability of migration, and the probability of identity-by-descent, decreasing with geographic distance. The classic study of *Linanthus parryae* (Epling and Dobzhansky 1942), showing marked geographic patterns of flower color polymorphism, further stimulated Wright's interest in these models.

Wright's treatment of isolation by distance was based on an ingenious argument, which introduces the concept of "neighborhood size", Nb. This is defined as the size of an ideal panmictic population which would give the same probability of identity-by-descent (f_1), as for two nearby genes in the actual structured population:

$$f_1 = 1/2Nb. \tag{13}$$

In a diploid population of density d, continuously distributed over a linear habitat, and with a normal distribution of dispersal distances, $Nb = d\sqrt{2\pi\sigma^2}$. In two dimensions, $Nb = 4\pi d\sigma^2$ (Wright 1943).

The neighborhood size is often thought of as an **effective population** size. However, this misleading, because although neighbourhood size is defined as that population size which would give the same probability of identity via the previous generation, the probability of identity via more distant ancestors will fall away more quickly in a structured population than in a single panmictic gene pool. More generally, we will see that the neighborhood size plays a quite different role from the effective size N_e in determining the relative importance of selection, migration, and drift.

Wright's argument uses the idea of "neighborhood size" to extend results on single populations to structured populations. In general, the homozygosity f can be calculated by summing over contributions to identity via ancestors $1, 2, 3 \cdots$ generations back. Let the probability of identity via an ancestor t ganerations back be f_t. If migration from a pool of unrelated genes, or mutation, produce novel alleles with probability m per generation, then the probability that two genes which descend from a common ancestor t generations back are still identical is $(1-m)^{2t}$. Thus:

$$f = \sum_{t=1}^{\infty} f_t(1-m)^{2t}. \tag{14}$$

In a single panmictic population of N diploid individuals, the probability of identity via an ancestor t generations back is:

$$f_t = (1/2N)(1-1/2N)^{t-1}. \tag{15}$$

Substitution of Eq. (14) gives the standard result Eq. (3). Wright argued that in a structured population, the ancestors of a pair of genes are likely to be further apart in space, the further back one traces the lineages. He assumed that the ancestral lineages trace a random walk, so that if the genes move in a normal distribution with variance σ^2 in one generation, the pool of potential ancestors t generations past can be thought of as occupying a normal distribution with variance $2\sigma^2 t$. Thus, the size of the pool increases in proportion to $\sqrt{t}$ in one dimension, and to t in two dimension. Thus:

$$f_t = \frac{1}{2Nb\sqrt{t}}\left(1-\frac{1}{2Nb}\right)\cdots\left(1-\frac{1}{2Nb\sqrt{t-1}}\right) \quad \text{in 1D} \tag{16a}$$

$$f_t = \frac{1}{2Nbt}\left(1-\frac{1}{2Nb}\right)\cdots\left(1-\frac{1}{2Nb(t-1)}\right) \quad \text{in 2 D.} \tag{16b}$$

It is simplest to follow the argument when neighborhood size is large, and m is small. Then, one can neglect the multiplying factors, and obtain:

$$f = (1/2Nb)\sum_{t=1}^{\infty}(1-m)^{2t}/\sqrt{t} \approx (1/2Nb)\sqrt{(\pi/2m)} \quad \text{in 1D} \tag{17a}$$

$$f = (1/2Nb)\sum_{t=1}^{\infty}(1-m)^{2t}/t \approx (1/2Nb)\ln(1/2m) \quad \text{in 2D.} \tag{17b}$$

The effects of **migration** and **drift** in one and two dimensions are qute different. In one dimension, the homozygosity includes significant contributions from distant generations ($t \sim 1/m$). It can be large even when the probability of identity in the previous generation is small (i.e., $Nb \gg 1$). In contrast, the homozygosity in a two-dimensional population is largely due to identity in the immediately preceding generations. It therefore depends strongly on neighborhood size, and only weakly on m. This has the important consequence that measurements of allele frequency variance in a two-dimensional population do not depend critically on the rate of long-distance migration or mutation. This was Wright's main application of his results on identity by descent.

Malécot (1948) developed a similar argument, which explicitly tracked the positions of the ancestors one, two, or three generations back, on the assumption that the genes move in a series of independent random steps of standard deviations. Although Malécot's argument differs slightly from Wright's, it gives the same results for large Nb. Malécot showed that the probability of identity-by-descent of two genes separated by a distance r decreases over a distance of order $1 = \sigma\sqrt{2M}$:

$$f(r) = (1/2Nb)\sqrt{\pi/2m}\, e^{-(r/l)} \qquad \text{in 1D} \tag{18a}$$

$$f(r) \approx (1/Nb)\, K_0(r/l) \qquad \text{in 2D,} \tag{18b}$$

where K_0 is the modified Bessel function.

Stepping Stone Models

Both Wright's and Malécot's arguments make the simplifying assumptions that the genes move through space in a series of independent random steps. This cannot be precisely correct in a continuously distributed population: Felsenstein (1976) has shown that their assumptions are incompatible with the density-dependent regulation which must act to maintain a more or less stable population structure. Because of the difficulty of formulating a rigorous model of a continuous population, most theoretical work has concentrated on "stepping stone" models, in which individuals are confined to discrete demes. It is often assumed that only migration to adjacent demes occurs.

In a series of papers, Maruyama (1969, 1970, 1971, 1972b, c) revealed several properties of stepping stone models. One such observation was that the average number of heterozygotes bearing a mutant alleles that appears in a population before the loss or fixation of the allele is 2N, where N is the population size. This is true regardless of the population structure. Similarly, the fixation probability of a mutation is invariant to population subdivision. Kimura and Maruyama (1971) demonstrate an important similarity between two-dimensional stepping stone models and the island model: that populations will be strongly differentiated if $Nm \ll 1$, while if $Nm > 4$ they behave as a panmictic unit (Fig. 3).

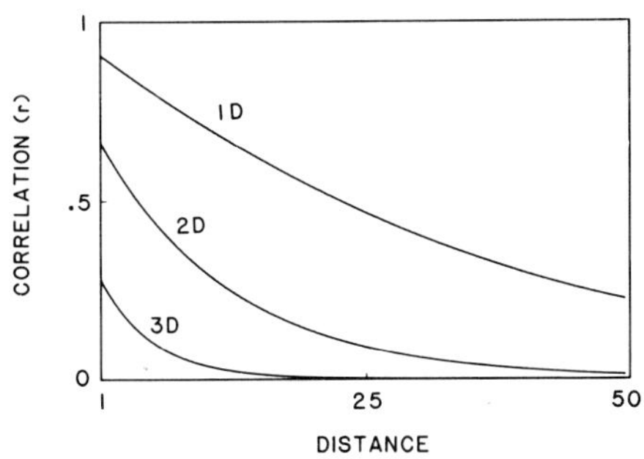

Fig. 3. The decay of correlation of uniting gametes with distance under the one, two- and three-dimensional stepping stone model

Statistical Measures of Population Structure

F Statistics

The most widely used statistical description of the degree of population subdivision makes use of the F statistics developed by Wright (1951). Despite the simple definition of these statistics, there is no universal agreement on the best approach to estimating them from population samples. There is some confusion in the literature about F statistics: they may refer to statistics calculated from a set of samples, to statistics describing the state of a partitioned set of populations, or to expectations over a random process. When defined as correlations of uniting gametes (see below), it is clear that they are true statistics, but they are also sometimes used in population genetic models as probabilities of identity. It is helpful to bear in mind that the aim of calculation F statistics is usually to make some inference about the underlying population structure, or in other words, to estimate the parameters of the population genetic models. They are the expected correlations of gametes, and are related to sample estimates only if the genetic variation is strictly neutral. Further discussion of this point is found in Wright (1978), Weir and Cockerham (1984), and Slatkin (1985).

The F statistics measure the departure from the expectation of panmictic proportions in the whole sample and within samples of local demes. The F statistics are equivalent to correlation coefficients with appropriate indexing of alleles. They measure the correlations between uniting gametes under various hypothetical and actual sampling schemes. F_{IT} is the correlation of uniting gametes relative to gametes drawn at random from the entire population. F_{IS} is the correlation of uniting gametes relative to gametes drawn at random from within a subpopulation (averaged over subpopulations). Finally F_{ST} is the correlation of gametes within subpopulations relative to gametes drawn at random from the entire population. The calculation of F statistics is most easily seen through the degrees of genetic diversity at the different levels. H_o is the observed heterozygosity averaged over the k subpopulations, and is:

$$H_o = \sum_{s=1}^{k} H_{o,s}/k. \tag{19}$$

H_S is the Hardy–Weinberg expectation of heterozygosity in subpopulation S, and the average of H_S taken over subpopulations is H_S.

$$H_S = 1 - \sum_i p_{i,s}^2, \tag{20}$$

where $p_{i,s}$ is the frequency of allele i in subpopulation s. Clearly, if $\bar{H}_S$ is greater than H_o then the average subpopulation is somewhat inbred, and the statistic Wright used to measure this is

$$F_{IS} = \frac{\bar{H}_S - H_o}{\bar{H}_S}. \tag{21}$$

Finally, the expected heterozygosity in the whole population, if it were unstructured, is calculated from the allele frequencies averaged across subpopulations, $\bar{p}_i$.

$$H_T = 1 - \sum_i \bar{p}_i^2. \tag{22}$$

The deviation of the total population from panmixia is then

$$F_{IT} = \frac{H_T - H_o}{H_T}, \tag{23}$$

and the measure of the degree of heterogeneity across subpopulations is

$$G_{ST} = \frac{H_T - \bar{H}_S}{H_T}. \tag{24}$$

The equations above are technically identical to Wright's definitions only in the case of two alleles. The above equations also fail to take into account the sampling bias introduced by **small samples**. With small samples, multiple alleles, and multiple loci there are several different methods that have been suggested for calculating F statistics. Weir and Cockerham (1984) analyzed five different approaches and found that of Reynolds et al. (1983) was superior in minimizing variance and sampling bias. Equation (24), derived by Nei (1973), is suitable for **multiallelic data**, and is usually distinguished by the symbol G_{ST}. Estimators of G_{ST} that are corrected for sampling bias are available (Nei and Chesser 1983). However, none of these various methods gives very different results: The most important source of error arises from uncertainty as to the appropriate model of population structure, rather than from sampling error.

Returning now to population genetic models, the F statistics correspond to parameters which can be defined in terms of identities-by-descent. We will use small letters for these parameters, reserving as before the capital letters for statistics. When there is no inbreeding within subpopulations, then $f_{IS} = 0$. This implies that $f_{IT} = f_{ST}$. f_{ST} is also written $\sigma^2/p(1-p)$, which is the ratio of the expected variance among subdivisions to the maximum possible variance among subdivisions (or "standardized variance"). The island model shows that at steady state $f_{ST} = 1/(1 + 4Nm)$. Combining these two expressions for f_{ST}, we obtain the variance among subpopulations as $\sigma^2 = p(1-p)/(1 + 4Nm)$, as discussed in Section 2.7. In rough terms, F_{ST} indicates the fraction of the genetic variation that is found among subpopulations, and is always positive. F_{IS} can be negative if there is systematic avoidance of mating with relatives. Random mating in a finite population will of course result in chance matings with relatives, leading to $F_{IS} > 0$. Finally, F_{IT} is generally positive, but it can have negative values if there is avoidance of consanguineous matings. The three F statistics developed by Wright are related by the equation

$$(1 - f_{IT}) = (1 - f_{IS})(1 - f_{ST}). \tag{25}$$

A convenient way to put the F statistics into a familiar context for hypothesis testing was developed by Cockerham (1969, 1973). Consider a two-allele polymorphism with alleles A and a with frequencies p and $q = 1 - p$ respectively. The allele are assigned a value $x = 1$ for allele A and $x = 0$ for allele a. The expectation of x is simply the allele frequency p and the total variance is $\sigma^2 = p(1 - p)$, which can be partitioned as follows:

$$x_{ijk} = p + a_k + b_{ik} + w_{ijk}, \tag{26}$$

where x_{ijk} is the value of the jth allele of the ith individual in the kth subpopulation. The effect a_k is for subpopulations, b_{ik} for the individuals within subpopulations, and w_{ijk} for alleles within individuals (inbreeding). In the terminology of analysis of variance, these are all random effects with associated variances σ_a^2, σ_b^2, and σ_w^2 respectively.

The F statistics are related to these components of variance as follows:

$$\begin{aligned} \sigma_w^2 &= (1 - F_{IT})\sigma^2 \\ \sigma_b^2 &= (F_{IT} - F_{ST})\sigma^2 \\ \sigma_a^2 &= F_{ST}\sigma^2. \end{aligned} \tag{27}$$

These are readily solved to give the F statistics:

$$\begin{aligned} F_{IT} &= \frac{\sigma_a^2 + \sigma_b^2}{\sigma^2} \\ F_{ST} &= \frac{\sigma_a^2}{\sigma^2} \\ F_{IS} &= \frac{\sigma_b^2}{\sigma_b^2 + \sigma_w^2}. \end{aligned} \tag{28}$$

Cockerham showed how to estimate these using standard variance components theory, as depicted in Table 1. By relating F statistics to analysis of variance, it is also possible to add additional levels of population subdivision, and to extract them as other components of variance.

The estimation of F statistics becomes more complicated with **multiple alleles** and **multiple loci**, and there is no single method that is universally the best. Long (1986) has presented a method that is a nice extension of Cockerham's approach. Rather than indexing the two alleles as 1 and 0, Long (1986) assigns a vector of

Table 1. Analysis of variance of allele frequencies in a subdivided population. (Cockerham 1969). Equal sample size within each subpopulation is assumed

Source	df	Mean squares	Expected mean squares
Among subpopulations	$M - 1$	MS_a	$\sigma_w^2 + 2\sigma_b^2 + 2N\sigma_a^2$.
Among individuals within subpopulations	$M(N - 1)$	MS_b	$\sigma_w^2 + 2\sigma_b^2$
Within individuals	N	MS_w	σ_w^2

length k − 1 for each allele at a locus having k segregating alleles. Rather than obtaining a simple variance, when we sample from a multiallelic locus the frequencies of alleles in the samples are no longer independent, and we have a sampling variance-covariance matrix. With multiple loci, Long (1986) simply appends the vectors for successive loci and ends up with a large covariance matrix. The F statistics now come from a partitioning of the covariance matrix using multivariate analysis of variance (MANOVA). Analyses of several sets of data (Long 1986) give reasonable results, but the sensitivity of the method to linkage disequilibrium and sampling error remain to be investigated.

Another approach to multiallelic and multilocus data, suggested by Weir and Cockerham (1984), is to use computer bootstrapping techniques, i.e., resampling from the data set over alleles and loci to obtain an empirical distribution of the F statistics using many subsamples of the complete data set. This method has much intuitive appeal: the confidence intervals are empirical, and problems such as linkage disequilibrium do not violate implicit assumptions of the method.

A practical problem that must be considered is that the models discussed so far have assumed that the population is in a steady state with respect to **mutation**, **migration**, and **drift**. If samples are taken from a population that is not at steady state, there may be biases in the F statistics that do not reflect the true population structure. Before considering the empirical problem, it is useful to examine theoretical attempts to characterize the structure of pre-steady state populations. Maruyama (1977) found that the leading eigenvalue for the recursion in f in an n-dimensional stepping stone model (starting with no genetic variation) was $1-(1/2N)$, which is identical to the case with no population structure. As will be seen in a subsequent section, the rate of approach to equilibrium of f in an infinite allele island model is of order $1/\mu$, whereas the rate of approach of f_{ST} is of order $1/m$. Of all the statistics that quantify population structure, it appears that G_{ST} should be least sensitive to small departures from steady state. When a population goes through a bottleneck, or a colony is founded by a small number of individuals, the population grows in size as new mutations accumulate and genetic diversity increases. In such a nonequilibrium population, the rate of increase of the number of alleles exceeds the rate of increase of heterozygosity (Nei and Li 1976; Nei et al. 1975; Watterson 1984b).

The use of molecular techniques to uncover genetic variation at the DNA level will have a profound impact on our ability to make inferences about population structure. In particular, **DNA sequence data** will afford the chance to detect identity-by-descent with reasonable confidence, and the details of the relationships between alleles will add a new depth to the analysis. Assuming that the technology will be developed for rapid and low cost cloning and sequencing of multiple isolates of the same gene, it is not premature to begin to consider how these data could be used. Genealogies of individual genes could be reconstructed, and the branching patterns within and between populations may provide the most powerful test for the effects of mutation, migration, recombination and drift (Griffiths 1980; Takahata and Nei 1985). By analysis of coalescence time, defined as the expected time since members of a sample had a single common ancestor

(Kingman 1982; Tajima 1983; Watterson 1984a; Nei 1987), using samples from within and among subpopulations, it should be possible to determine the degree of present-day structuring, and to some extent the past history of migration levels might be accessible. The inference of selective constraints from observations of excess diversity in third positions of codons (Kreitman 1983) could be extended by calculating F statistics separately for the three codon positions. By extending this type of analysis, it should be possible to infer how much of the diversity among subpopulations is selectivity constrained vs. neutral.

Recurrence Equations for F Statistics

One means of following the consequences of population subdivision is to formulate the recurrence equations for the F statistics, and see how they change over time as functions of mutation rates, migration rates, subdivision sizes and migration patterns. Much progress was made in this approach by Latter (1973) and Crow and Aoki (1982, 1984), and they discovered some important properties of subdivided populations.

Assume that m is the fraction of genes that are from migrants, N is the number of individuals in a subpopulation and there are n subpopulations. Further, let u be the rate of mutation from one allele to all of the other $K-1$ alleles. Latter (1973), Takahata and Nei (1984) and Crow and Aoki (1984) independently derived the equilibrium value of f_{ST}.

$$f_{ST} = \frac{1}{4Nma+1}, \tag{29}$$

where $a = [n/(n-1)]^2$. The remarkable thing about Eq. (29) is that the equilibrium value of f_{ST} is independent of the mutation rate and the number of alleles, provided $\mu \ll m$ and the population size is sufficiently large. In the limit as n increases, Eq. (29) approaches Wright's results from the island model, $f = 1/(1+4Nm)$, so the n-island model with mutation becomes equivalent to the infinite island model with no mutation, provided $\mu \ll m$. It was also shown that f_{ST} approaches equilibrium at a rate on the order of $1/m$. This is a remarkably rapid rate of approach to equilibrium, given that the rate in the nonstructured population with mutation and drift the rate is on the order of $1/\mu$. These results were obtained after several approximations and assumptions about the relative magnitudes of several parameters, but numerical simulations verify their robustness. Figure 4 shows trajectories of F_0 (the probability of identity of a pair of alleles from one subpopulation), f_1 (the porbability of identity of a pair of alleles from different populations) and f_{ST} for the infinite alleles n-island model (Crow 1986, p. 64):

$$\begin{aligned} f'_0 &= \{a[c+(1-c)f_0]+(1-a)f_1\}(1-\mu)^2 \\ f'_1 &= \{b[c+(1-c)f_0]+(1-b)f_1\}(1-\mu)^2 \\ f &= [f_0+(n-1)f_1]/n \\ f_{ST} &= (f_0-f)/(1-f), \end{aligned} \tag{30}$$

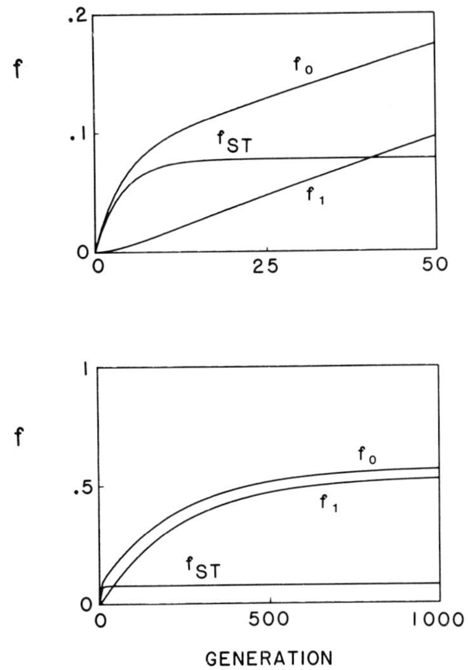

Fig. 4. Trajectories of f_0 (the probability of drawing a pair of alleles identical-by-descent within a subpopulation), f_1 (the probability of drawing a pair of alleles identical-by-descent from two different subpopulations, and f_{ST}. Equations (30) were iterated in both the *top* and the *bottom panel* with $N_e = 20$ (size of each island), $n = 10$ (number of islands), $\mu = 0.001$ (mutation rate), and $m = 0.1$ (migration rate)

where $a = (1-m)^2 + m^2/(n-1)$, $b = (1-a)/(n-1)$, and $c = 1/2N_e$. Figure 4 demonstrates that f_{ST} approaches equilibrium faster than the other inbreeding indices.

Population Structure of Linkage Disequilibrium

In a finite population with two segregating loci, sampling effects can generate linkage disequilibrium. Some of the sampling properties of two loci were examined by Hill and Robertson (1968) in a model of an unstructured population of size N, with no mutation. The mean value of the linkage disequilibrium (D) remains zero, but D^2 increases to some maximum value and then declines back to zero as genetic variation declines to eventual fixation for one gamete type. In the absence of recombination, Hill and Robertson (1968) showed that

$$E(D^2) = (1/15)p(1-p)q(1-q)[6(1-f) - 5(1-f^3) - (1-f)^6], \tag{31}$$

where p and q are allele frequencies at the two loci, and f is the inbreeding coefficient, which follows $f = 1 - e^{-t/2N}$. Since $E(D) = 0$, the variance in D is simply $E(D^2)$. With no recombination, D^2 reaches a peak at about $t = N$ generations, and then decays back to zero. When recombination is allowed, it was found that the rate of recombination only enters as a product with the population size. Similar results have been obtained by Ohta and Kimura (1969a) and Littler

(1973). The maintenance of linkage disequilibrium in steady state populations with recurrent neutral mutations was analyzed by Ohta and Kimura (1969, 1971) and Hill (1975).

Since linkage disequilibrium is generated in a finite nonstructured population, it is reasonable to assume that the structure of a subdivided population will be reflected in patterns of linkage disequilibrium. Variation among subpopulations in gamete frequencies leads to linkage disequilibrium due to a two-locus **Wahlund effect** (Feldman and Christiansen 1975), and migration among partially isolated subpopulations can generate linkage disequilibrium among **neutral loci** (Nei and Li 1973; Li and Nei 1974). There are two problems related to this that have been addressed. The first involves the theoretical prediction of patterns of linkage disequilibrium in structured populations, and the second involves the statistical assessment of population structuring inferred from multilocus samples. Ohta (1982) examined the infinite alleles, n-island model and a model where subpopulations undergo recurrent extinction and recolonization. She found that with limited migration, linkage disequilibrium can increase to high levels within a subpopulation, and that because of the population structure, linkage disequilibrium requires more dimensions to describe. Hence the analytical models immediately lead to the empirical problem of describing the structure of linkage disequilibrium.

By analogy with Wright's F statistics, Ohta (1982a, b) devised a series of five D statistics that partition the total variance of pairwise linkage disequilibrium. The total variance, D_{IT}^2, is the mean squared deviation of the observed frequency of gamete A_iB_j from the product of the average allele frequencies. D_{IT}^2, as well as its components, are calculated for all pairwise combinations of loci. D_{IS}^2 is the expected variance of linkage disequilibrium within a colony. D_{ST}^2 is the variance of correlation of genes of the two loci of a pair of gametes drawn from one colony relative to that of the total population. $D_{IS}'^2$ is the variance of the correlation of alleles on one gamete in a subpopulation relative to that in the total population, and $D_{St}'^2$ is the variance of disequilibrium of the total population. Ohta (1982) showed that the nonsystematic linkage disequilibrium expected as a result of random drift would result in the pattern $D_{IS}^2 > D_{ST}^2$ and $D_{IS}'^2 > D_{ST}'^2$. In the opposite case, where there are systematic linkage disequilibria, possibly generated by natural selection, one would observe $D_{IS}^2 > D_{ST}^2$, and $D_{IS}'^2 < D_{ST}'^2$. These statistics are computationally intensive, requiring calculation of up to the fourth moments of allele and gamete frequencies.

The sampling properties of Ohta's D statistics have not been thoroughly characterized, and unbiased estimators are not known. Simulation studies show that a commonly used estimator for mean squared linkage disequilibrium has undesirable sampling properties, and that in general, sample sizes required for statistical confidence in statistics of linkage disequilibrium are large (Hudson 1985). It would be desirable to have the diffusion theory giving the equilibrium distribution of D, analogous to the single locus $\phi(x)$. From this it should be possible to construct a sampling theory to test the distribution of pairwise linkage disequilibria, in a manner analogous to the Ewens–Watterson sampling theory for single locus variation.

Another approach to the analysis of multilocus structure of subdivided populations was developed by Brown and Feldman (1981). Rather than determining the structure of pairwise linkage disequilibria, this approach is based on properties of the **occupancy distribution**, f(K). K is defined as the number of heterozygous loci found in a random pair of gametes, and F(K) is the distribution of this number. It is clear that K will depend on the heterozygosities of the individual loci, and in fact E(K) is the sum of the single locus heterozygosities. The variance in K will depend on both the single locus heterozygosities and the pairwise linkage disequilibria, but not higher order disequilibria (Brown et al. 1980). The method describes population structuring through statistics relating the mean and variance across subpopulations in linkage disequilibria, and the covariance of allele frequencies over subpopulations. Other statistics quantify the degree of consistency of linkage disequilibria. Departures from panmixia are ignored, because the method is based on frequencies among random samples of gametes. Including **nonrandom association** of multilocus gametes would probably increase the number of degrees of freedom to the point that they would exceed the sample size. This method has the advantage of not requiring prohibitive sample sizes, but a disadvantage is that by collapsing multilocus data in thus way, information is lost.

Statistical associations among alleles of nuclear and mitochondrial genes afford another opportunity to infer population structure, particularly in hybrid zones. Asmussen et al. (1987) develop a hierarchical series of disequilibrium statistics and explore the consequences of mixed mating models. The statistical associations between mitochondria and nuclear genes can be used to infer the frequency of both reciprocal interspecific matings.

Inferences from Rare Alleles

Rare alleles can be particularly useful in inferences about population structure. We have seen that certain properties of neutral alleles are invariant under population subdivision, including the expected time to fixation. Deleterious alleles, however, have a shorter expected time to loss, and on average, rare alleles are more likely to be deleterious. Since these alleles are more likely to be younger, they are also more prone to show subdivision. This argument has been supported both by theory and by empirical observations of population samples (Kimura 1983). Furthermore, they will be less sensitive to variations in selection, relative to the effects of drift (Gillespie 1984).

In a series of papers, Slatkin (1980a), 1981a, 1985) capitalized on the inflated subdivision expected of rare alleles to devise means of quantifying population subdivision. His first approach was to examine "**conditional average frequencies**", which are average frequencies of alleles that are not present in all subpopulations. Inferences are made from the relation between the number of subpopulations bearing an allele and the average allele frequency of alleles present in a given number of subpopulations. The conditional average frequencies, $\bar{p}(i)$, are defined as the average frequency of alleles present in i of the sampled subpopulations. One expects that $\bar{p}(i)$ should increase as i increases, and this is the case. What is surprising about $\bar{p}(i)$ is that it is nearly independent of both the mutation rate and the level of selection. Using simulations of both the island model and the stepping stone model, Slatkin (1981) found that $\bar{p}(i)$ depends on the migration rate, and may be a useful predictor of gene flow (Nm). The lack of dependence of $\bar{p}(i)$ on mutation and selection levels means that results from different loci (having different rates of mutation) can be pooled. The conditional average frequencies were used to estimate Nm (Slatkin 1985a), and revealed a range from 9.9 in *Drosophila willistoni* to 0.1 in *Plethodon dorsalis.*

The results of Slatkin (1980a, 1981a, 1985) were largely inferred from computer simulations, and an analytical approach to the problem of population structure of rare alleles was taken by Slatkin and Takahata (1985). They consider a diffusion approximation of the models considered in the simulations, where the diffusions include a "**killing term**" to account for emigration. The predictions of $\bar{p}(1)$ from the analytical model (for a specified Nm) tend to be smaller than the simulation results, and this is explained by saying that private alleles in the simulations may have been present in more than one deme, hence inflating their frequency.

A more useful analytical approach to the study of population subdivision using rare alleles is to directly relate Wright's island model, and its predicted distribution $\phi(x)$ to the expected conditional average frequencies. Barton and Slatkin (1986) do this by showing that the expected frequency of an allele in a deme, given that it is present is:

$$p = \frac{\int_0^1 x\,\phi(x)\,dx}{(1 - a_0)}, \tag{32}$$

where a_0 is the probability that an allele is absent from a sample. In deriving Eq. (32) there is an implicit assumption that the population is in a steady state, or a "quasi-equilibrium" with respect to the gain and loss of alleles and there distribution across demes. Tests of this by numerical simulation show that the conditional average frequency attains steady state approximately as rapidly as F_{ST}, i.e., on the order of 1/m generations. Both Wright's F_{ST} and Slatkin's $\bar{p}(1)$ are statistical measures of the spread of the allele frequency distribution $\phi(x)$; in simulations, they give estimates of rates of gene flow (Nm) of similar accuracy (Barton and Slatkin 1986).

Empirical Studies of Population Structure

The literature is full of studies in which F statistics have been calculated, but we wanted to illustrate some of the important points raised in the preceding sections with studies that also test goodness-of-fit to neutral models. Since population subdivision results in a Wahlund effect, if one were to test an allele frequency distribution for goodness-of-fit to the infinite alleles model, using a pooled sample not known to contain subdivision, the observed distribution would have an excess genetic identity relative to that expected for the observed sample size and number of alleles. For a given total population size, population subdivision would result in more alleles, more singletons, and greater linkage disequilibrium. The literature contains a great diversity of empirical approaches to data that may reflect the effects of subdivision.

Drosophila pseudoobscura Est-5 and *Xdh*

Keith (1983) and Keith et al. (1985) survey two populations of Drosophila pseudoobscura and perform high resolution electrophoresis on *Esterase-5* and *Xanthine dehydrogenase* respectively. These loci are very allele-rich, and the populations showed a total of 41 *Est-5* alleles in a sample of 237 genes, and 20 *Xdh* alleles in a sample of 184 genes. In both cases it was observed that the most common allele was shared by the two populations, as were most of the more frequent alleles. In the *Xdh* data, the only cases when an allele is present in one population and absent from the other occur when the allele is observed only once. Given an observed sample size and count of distinct alleles, Evens–Watterson sampling theory (Ewens 1972, 1979; Watterson 1974, 1978) predicts an expected allele frequency distribution under assumptions of the infinite alleles model of neutral mutations and random drift. These goodness-of-fit tests were performed on both the *Xdh* and *Est-5* data, and in both cases there was a significant deviation in the direction of a deficiency of F, consistent with purifying selection.

The **Ewens–Watterson test** is based on the expected allele frequency distribution in a single finite population, assuming mutation to a possibly infinite number of alleles. Population subdivision clearly changes the expected allele frequency distribution within each subpopulation, and the pooled sample of n subpopulations each of size N is expected to have a distribution distinct from that of a sample from a population of size nN. Although an analog to Ewens–Watterson sampling methods that take account of population subdivision would be desirable, is seems unlikely that such a procedure would be very successful. Unless one had extrinsic estimates of rates of gene flow, then allowing another set of free parameters to be fitted by the allele frequency distributions would make it difficult to reject a good fit, even when the assumptions are violated. In the case of Keith (1983) and Keith et al. (1985), the deviations were in the direction of an excess in observed F over the expected value. The pooled sample showed even greater deviation in this same direction. They argue that because the departures are significant, either purifying selection is operating, or the populations are not yet at equilibrium [which would lead to deviations in the direction they observe (Nei and Li 1976)]. On the other hand, population subdivision can also lead to an apparent excess of homozygosity and number of rare alleles (Nagylaki 1985). Keith et al. (1985) argue that the migration rate must be high in order to obtain such similar allele frequency distribution in *D. pseudoobscura Xdh* and *Est-5*, an in fact a χ^2 test fails to reject the hypothesis that they represent samples from the same distribution. This makes it implausible that each sample is actually a pooling of heterogeneous subpopulations, and strengthens their claim for purifying selection. If they had obtained significant deviations between the James Reserve and Gundlach–Bundschu populations in the χ^2 test, the departures from the distribution predicted by Ewens–Watterson sampling theory could have been due to either subdivision or purifying selection.

Temporal and Spatial Variation in *Drosophila buzzatii*

A study of temporal and geographic patterns of allozyme variation in *Drosophila buzzatii* (Barker et al. 1986) illustrates a number of complications that real data introduce. Allozyme frequencies at six loci were collected from samples of *D. buzzatii* collected at ten different sites over a period of 4 years. The data include temporal as well as spatial variation, and the success of sampling varied with the weather. Within any given month the sample size varied from 9 to 1648 files. F-statistics were calculated treating the different samples across time as different subpopulations. Standard deviations of F statistics were obtained by jackknifing over samples. The F statistics showed significant deviations from zero when calculated across time, and F_{ST} was significantly heterogeneous across loci. There was much less spatial variation than temporal variation in F_{ST}, but individual cactus rots showed some heterogeneity. F_{IT}, a measure of inbreeding, was similar in the spatial and temporal tests, suggesting approximate equilibrium, but there were marked differences among loci. The principle conclusion from these analyses is that natural selection is reflected in the significant differences among loci in inbreeding coefficient, and the lack of spatial variation implies that the selective differences are exerted at the microenvironmental level. Inference of natural selection from heterogeneity across loci in spatial F statistics, such as the Lewontin–Krakauer test (Lewontin and Krakauer 1973) are flawed due to the fact that hierarchical subdivision would generate this heterogeneity among neutral loci (Robertson 1975; Ewens and Feldman 1976). Heterogeneity in temporal F statistics does not have this fault, and has been used previously to test neutrality (Gaines and Whittam 1980).

Pattern of Allozyme Variation in *Escherichia coli*

With the only exception that deviations from expected genotypic frequencies cannt be used to estimate inbreeding, the population structure of haploids can be examined in the same way as for diploids. There is now a large collection of papers concerned with the population structure of *E. coli* (Selander and Levin 1980; Whittam et al. 1983a, b, Ochman and Selander 1984). The paper of Whittam et al. (1983a) provides an excellent set of data for analysis of the population structure of linkage disequilibrium using Ohta's D statistics. Before reviewing this analysis, it is important to note that the average of F_{ST} taken over the 12 enzyme loci examined was 0.02, implying a remarkable degree of worldwide uniformity in average single-locus genetic variation in *E. coli*.

An important problem with D statistics is the lack of theory of their sampling characteristics, and Whittam et al. (1983a) deal with this is a simple but robust way. By calculating the statistics using Ohta's formulae directly, but repeating the process over computer-randomized allelic combinations, they obtained bootstrap confidence intervals in the D statistics. There were a total of 178 isolates that were scored, and a total of 66 pairwise linkage disequilibria could be calculated. This approach led to the conclusion that there was significantly greater total variance in linkage disequilibrium than would be obtained from an unstructured population. The partitioning of the total variance in D^2 showed that D^2_{IS} was greater than D^2_{ST} for most pairwise comparisons, implying that variation within subpopulations accounts for more of the total variance in disequilibrium than variation among subpopulations. This study was partly motivated by the desire to reveal the degree of which *E. coli.* genomes reflect naturally occurring recombination, and the failure to detect a relation between map position and D suggests that recombination is a rare phenomenon. Another approach, reconstructing genealogies from sequence data, detected recombinants with reasonable certainty, but this kind of study does not enable one to infer the population frequency of recombination.

Population Structure of Human mtDNA

One approach to the analysis of molecular sequence variation is to translate the variation back into allelic variation in functional genes. This was done by Whittam et al. (1986) on mtDNA restriction site data from 145 humans sampled from five geographic regions. By aligning the high resolution restriction map to the known sequence of human mtDNA, it was possible to identify alleles as unique combinations of presence and absence of restriction sites within each of 28 functional regions. One of the motivations for doing this was that the well-known tools of population genetic analysis, including F statistics and neutrality tests, could now be applied. The average diversity within regions ($\bar{H}_S$) was found to be 0.215, meaning that the expected probability of drawing a pair of alleles identical by descent from within a region was 0.785. The average G_{ST} was 0.063, with a range from 0 for some tRNA's that were depauperate in alleles, to 0.147 for cytochrome b. Generally the F statistics agreed with previous data on geographic structuring in human genetic variation in blood serum proteins.

Ewens–Watterson tests were also applied to the human mtDNA data by Whittam et al. (1986), and they found that 71% of the tests failed to reject neutrality. Of those that failed to fit, there was a consistent excess in observed gene identity (F). Since the Ewens–Watterson test is based on the expected allele distribution given an observed number of alleles and sample size, this implies that there was an excess allele richness relative to the genetic diversity. This is precisely what is predicted in growing populations that are not yet at steady state with respect to mutational gain of mutations and stochastic loss due to drift (Nei and Li 1976; Nei et al. 1975). Failure to fit the expected allele frequency distribution could also be due to hierarchical population subdivision within the identified geographic regions, or due to natural selection. It is clear that the added information gained by knowing the sequence differences among these alleles, including identification of synonymous and nonsynonymous substitutions, will be useful in weighing the relative merits of these hypotheses.

Natural Selection and Population Subdivision

In the preceding sections, we have dealt with the interaction between gene flow, mutation, and random sampling drift. We now consider how population structure interacts with selection. It is convenient here to divide the effects of population structure into those which arise within one generation of one random-mating population, and those which arise over longer times and longer distances. Individual movements within a generation may, of course, have many important biological functions and consequences. However, they can only affect the composition of the population by exposing the individuals to different physical, biological, or social environments, hence altering the numbers of offspring produced by different genotypes. In contrast, the flow of genes between partially isolated subpopulations acts as an additional and distinct process. We will therefore first discuss the effects of population structure on variation within a single random mating gene pool, and then turn to the interaction between selection, drift, and gene flow.

Within-Population Effects

In the simplest model of natural selection, each genotype has constant absolute fitness. The number of each type of gene will then either increase or decrease exponentially, independently of changes in the rest of the population. This process cannot continue for long: if the population is to remain within sustainable bounds, fitness must decrease with density. Darwin saw natural selection as arising not just from different reproductive rates, but rather from a Malthusian "struggle for existence" between individuals competing for a scarce resources (see also Chap. 20). The question is, how do the interactions which are essential for population regulation affect the evolutionary process?

A formulation that is central to the analysis of drift and selection in finite populations is the Wright–Fisher model. This assumes that the population is reduced to precisely N individuals in each generation, and that this regulation is independent of the genetic composition of the population. In this section, we investigate what happens when fitness depends on definite interactions between individuals, so that the population structure lies somewhere between the extremes of independent reproduction and absolute regulation. (These two extremes are known as **"hard" and "soft" selection** respectively, though there is some disagreement over their definition (Wallace 1975; Christiansen 1975; Walsh 1984)).

A major complication, which has hardly been dealt with in population genetics, is that the population may oscillate, or fluctuate chaotically, rather than setling to a constant equilibrium. Such instability is typical of the most obvious models of **density-dependent growth,** $N_{t+1} = N_t\bar{w}(N_t)$ (May 1973b). If all genotypes respond to density in the same way, so that relative fitnesses remain constant, then changes in population size will have no effect on gene frequencies. The effects of changing numbers on sampling drift can be approximated by using an effective number equal to the harmonic mean across generations. However, even with the restriction that relative fitnesses remain fixed, the mean fitness and hence the population size may depend on gene frequencies. This relation affects the genetic load associated with variation and substitution, and may also lead to group selection favoring genes which produce large and persistent populations (see next section).

If genotypes differ in their response to density changes, then selection coefficients will be density-dependent. This may simply be seen as another source of environmental variation, which may average out if the population fluctuates rapidly relative to changes in gene frequency. The most interesting effects occur when interactions depend on genotype (leading to frequency-dependent selection) or on relationship (leading to sib competition and kin selection). Such interactions my arise directly, or through subdivision of one random-mating population into separately regulated niches.

What are the effects of interactions which make individual fitnesses depend directly on the numbers and genotypes of others in the population? Are such interactions likely to maintain much genetic variation? How does competition between relatives affect the evolution of the breeding system? Can the effects of interactions between relatives be summarized by some measure of inclusive fitness? The latter question is bound up with the action of selection between groups, and will be deferred to the section on geographically subdivided populations (Sect. 3.2). Here, we consider the effects of population structure on genetic variation and the evolution of the breeding system (see also Chap. 3).

Does Environmental Heterogeneity Maintain Genetic Variation?

A pervasive idea in population genetics has been that environmental heterogeneity may in some way maintain genetic heterogeneity (for reviews, see Hedrick et al. 1976; Felsenstein 1976). This idea has been applied to the explanation of both genetic variation at individual loci, and the mechanisms of recombination and sexual reproduction which generate further diversity by producing new combinations of genes (Maynard Smith 1978a; Bell 1982). Here, we concentrate on the first (and simpler) issue.

The idea that environmental heterogeneity maintains genetic polymorphism was first quantified by Levene (1953). He imagined that, at the beginning of each generation, juveniles are scattered at random into a number of patches: there is no spatial subdivision of the population at reproduction, and no selection of particular patches by particular genotypes. Individuals spend their whole lives on a single patch: the environment is said to be "**coarse-grained**", in contrast to a "**fine-grained**" **environment**, in which individuals wander over many patches during their lives, and so all experience the same mix of selection pressures. Different genotypes may have different chances of surviving on different patches. However, if the numbers produced by each patch are directly proportional to the number of individuals which survive on the patch (that is, if selection is "hard"), this environmental heterogeneity in fitness will not by itself maintain any genetic heterogeneity. If, on the other hand, the numbers surviving on each patch are reduced by crowding of that patch (that is, if selection is "soft"), polymorphism may be maintained. The reason is that a rare genotype will find that the patch on which it survives best is less crowded, simply because that genotype is rare, and so it will gain an advantage. This negative frequency-dependent selection causes the rarer genes to increase, and so may maintain a balanced polymorphism.

To show how the conditions for polymorphism can be found, we model a haploid population, in which two alleles P, Q segregate at frequencies p, q. The results will also apply to a diploid model with complete dominance: we do not deal with the general diploid case, because then, polymorphism can be maintained simply by heterozygote advantage. [For example, in Gillespie's (1978) SAS-CFF model, polymorphism is maintained because the average fitness of the heterozygote is greater than that of either homozygote]. To keep notation consistent with the later treatment of between population variation, we label the patches by a continuous variable x; the sum over patches is then represented by an integral over the range of x (cf. Taylor 1976). The viabilities of the two genotypes in patch x are V_Q, V_P, and the proportion of individuals born in x is g(x). Then, with hard selection:

$$p^* = p \int g(x)[V_P(x)/\bar{V}]\,dx, \text{ where } \bar{V} = \int g(x)[pV_P(x) + qV_Q(x)]\,dx. \qquad (33)$$

The polymorphism would only be protected if allele P increases from low frequency [which happens if $\int g(x)V_P(x)dx > \int g(x)V_Q(x)dx$], and if Q increases when rare [which happens if $\int g(x)V_Q(x)dx > \int g(x)V_P(x)dx$]. A globally stable polymorphism is therefore impossible.

Now, suppose that selection is "soft", so that regardless of initial numbers and their subsequent survival, patch x produces a proportion d(x) of the adult population [$\int d(x)\,dx = 1$]. (The assumption here is that $g \gg d$, and that the viabilities are inversely proportional to local density. Then, the output from each patch is independent of g). The change in allele frequency is:

$$p^* = p \int d(x)[V_P(x)/\bar{V}(x)]dx, \text{ where } \bar{V}(x) = [pV_P(x) + qV_Q(x)]. \qquad (34)$$

The condition for protected polymorphism is now that $\int d(x)[V_P(x)/V_Q(x)]\,dx > 1$, and $\int d(x)[V_Q(x)/V_P(x)]\,dx > 1$. This condition can be satisfied if different alleles are favored in different patches. For example, if the relative viabilities are (0.1:0.9) in half the patches, and (0.9:0.1) in the other half, then the averages of $V_P(x)/V_Q(x)$ and $V_Q(x)/V_P(x)$ are both $\frac{1}{2}[9 + 1/9] = 41/9$, and there is a stable polymorphism.

Together with similar models of **temporal heterogeneity** (Haldane and Jayakar 1963), Levene's model has been greatly elaborated (Hedrick et al. 1976; Felsenstein 1976; Karlin 1982). It has inspired many attempts to correlate levels of variation—in particular, of electrophoretic variation—with measures of environmental "grain" (e.g., Selander and Kaufman 1973; Nevo 1978; Nevo et al. 1984). It has also formed the basis of models of sympatric speciation, where assortative mating evolves so as to match genotypes more closely to the patches in which they live (Maynard Smith 1966; Felsenstein 1981).

However, there are serious objections to the view that this form of environmental heterogeneity can account for a significant amount of genetic variation. Maynard Smith and Hoekstra (1980) and Hoekstra et al. (1985) point out that the conditions for polymorphism in the Levene model are extremely restrictive. Unless selection pressures are extremely strong, the fitnesses of the various genotypes must be nicely matched to the carrying capacities of the patches; slight changes destroy the polymorphism (Fig. 5a). We can see this by writing $V_P(x)/V_Q(x) = 1 + s(x)$, and expanding the conditions for a protected polymorphism in powers of s. This gives the conditions $\bar{s} > 0 > \bar{s} - \overline{s^2}$ (where $\bar{s}$ and $\overline{s^2}$ are the mean and mean square of s across patches). Temporal heterogeneity cannot maintain polymorphism in a haploid population, because

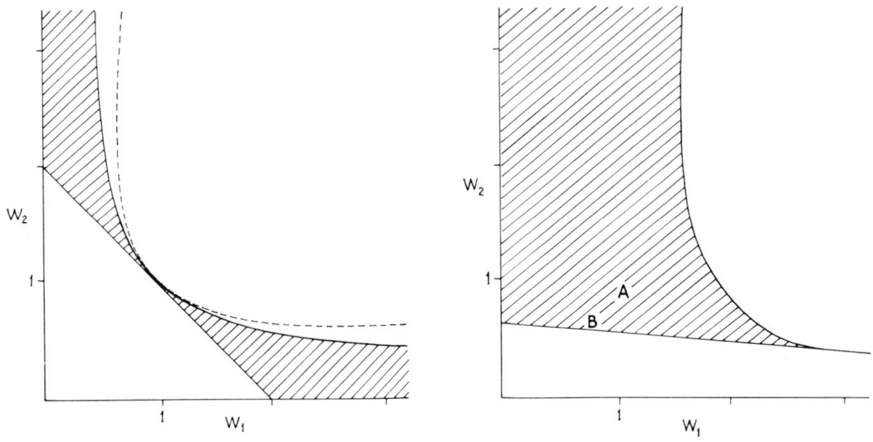

Fig. 5. a The conditions for stable polymorphism in a heterogeneous environment. Two patches of equal carrying capacity produce fixed and equal numbers of individuals. Two genotypes 7 (**A** and **B**) (in the haploid case), or two phenotypes (in the diploid case, with dominance), have relative viabilities $1{:}W_1$, $1{:}W_2$ in the two patches. The *shaded region* shows the range of viabilities for which a stable polymorphism is maintained $[(W_1 + W_2)/2 > 1,\ W_1 W_2 < 1]$. The *dotted line* shows the conditions when individuals return to the type of patch on which their mother was raised. **b** The conditions with habitat selection. Each type has a 75% probability of settling in its preferred patch

the gene with the highest geometric mean fitness will be fixed (Haldane and Jayakar 1963), but can ease the conditions for polymorphism in the Levene model (Ewing 1980). However, conditions are still restrictive when selection is weak [see Ewing 1980, Eq. (13)].

Such mechanisms cannot, therefore, maintain polymorphism at more than a few loci, and certainly cannot account for much electrophoretic variation. Even if the genetic load were reduced by epistatic interactions (for example, by truncation selection; Sved et al. 1967; Wills 1978), selection cannot act strongly at more than a handful of loci without causing an unacceptably high variance in fitness.

The simple Levene model can be strengthened slightly by supposing that individuals tend to return to the type of patch on which their mother was reared. However, such modifications do not make the conditions for polymorphism substantially less restrictive when selection is weak (Maynard Smith 1970; Fig. 5a). The model may also be strengthened by allowing some spatial subdivision of the mating population (Karlin 1982). Again, this does not help much: in general, although spatial subdivision can allow substantial variation between populations, it cannot maintain much variation within populations unless the scale of environmental heterogeneity is closely matched to the scale of spatial differentiation.

Polymorphism can be maintained more readily if different genotypes tend to begin in different patches, a situation termed "**habitat selection**". Suppose that the chance of beginning life in patch x depends on genotype $[g_P(x), g_Q(x)]$. Then:

$$p^* = p \int d(x)[g_P(x)V_P(x)/\overline{gV}(x)]dx, \text{ where } \overline{gV}(x) = \{pg_PV_P + qg_QV_Q\}. \qquad (35)$$

Polymorphism is protected if $\int d(x)[g_p(x)V_P(x)/g_Q(x)V_Q(x)]dx > 1$, and $\int d(x)[g_Q(x)V_Q(x)/g_P(x)V_P(x)]dx > 1$. Thus, variation in either viability or in habitat selection can promote stable polymorphism. Indeed, balanced polymorphism can be maintained even if all genotypes have the same chance of surviving in each patch (A in Fig. 5b), or even if individuals foolishly choose to live in the patches on which they have lower viability (B in Fig. 5b) (Templeton and Rothman 1981; Wilson and Turelli 1986). In fact, "habitat selection" has introduced a new and more powerful cause of polymorphism: if genotypes tend to live in patches whose productivities are separately regulated, then rarer types will necessarily increase, regardless of whether they survive well in these patches. Indeed, most models of negative frequency-dependent selection can be seen to depend on this process. For example, polymorphic patterns are maintained by Batesian mimicry because these patterns exploit different, and separately regulated, resources—in this case, the predators which have learned to avoid the various different patterns. Similarly, variations in female mating preferences can maintain polymorphism among males (O'Donald 1980): here, the separate limiting resources are the pools of females with different preferences. There is a close analogy with the maintenance of species diversity through specialized exploitation of different niches (May 1984; Futuyma and Slatkin 1983).

Wilson and Turelli (1986) point out that if polymorphism is first maintained through differential resource use, genotypes would not necessarily be fittest in the habitats which they happen to choose. Moreover, Wilson and Turelli show that heterozygotes will usually be less fit than either homozygote. This kind of polymorphism may therefore lead either to the perfection of polymorphic strategies within one species (through the evolution of dominance and other such modifications), or to sympatric speciation (through the evolution of assortative mating). However, recombination may still be a substantial genetic obstacle to these secondary adaptations (Felsenstein 1981).

It is important to realize, however, that unless habitat selection is strong, it is unlikely to maintain polymorphism. If different genotypes exploit similar mixes of resources, the conditions for polymorphism become just as restrictive as if they have similar viabilities in the different niches: only the product of habitat preferences and viabilities [$g(x)V(x)$] enters the equations. The main conclusions from these models are that first, stable polymorphism arises from the existence of a variety of limiting resources, rather than from heterogeneous selection per se; and second, whether polymorphism is maintained by differential viability or by habitat selection, polymorphism is only likely if there are substantial differences between genotypes.

All this suggests that environmental heterogeneity is a poor mechanism for the maintenance of variation. However, the models regard the environment as being either fixed, or varying with a stochastic distribution which is independent of gene frequencies. Coevolutionary interactions between species can maintain robust limit cycles (for example, between host and parasite, or predator and prey; Haldane 1949b; Futuyma and Slatkin 1983; Stenseth and Maynard Smith 1984; Hamilton and Zuk 1982). Behavioral interactions between members of the same species can maintain variation in the same way (Stenseth and Maynard Smith 1984). In a sense, variation is still being maintained by temporal heterogeneity of the environment, with the "environment" now consisting of other species or other genotypes; but this stretches the already strained notion of "population structure" too far.

Competition Between Relatives

Suppose that relatives tend to aggregate: for example, groups of siblings might grow together in the same patch, or might tend to exploit the same resources. (This aggregation arises solely through the breeding system, and not because relatives share heritable traits). If only a limited number of families exploit each patch, there will be an appreciable chance (F) that two genes in the same patch will be identical by descent, relative to the whole panmictic population. This can be seen instead in terms of gene frequencies: there will be a random variance (Fpq) in frequency $p(x)$ across patches, caused by sampling drift. Wilson (1977, 1982) has developed the point that subdivision into patches ("**trait groups**") has effects which depend only on the distribution of frequencies across patches; he emphasizes that variation in $p(x)$ due to random (i.e., binomial) allocation of individuals to patches is the least that can be expected.

We can first ask how this inbreeding and drift affects the maintenance of variation. It is easy to see that **competition between relatives** makes it less likely that soft selection can maintain a stable polymorphism. In the extreme case where the offspring in each path are genetically identical, there is no opportunity for selection within patches, and so the allele frequency in the next generation depends only on whether the parents select productive patches. No stable polymorphism is possible. Abbreviating the product $g(x)V(x)$ as $gV(x)$, we have:

$$p^* = E\left\{ \int d(x)[gV_P(x)p(x)/\overline{gV}]dx \right\}, \quad \text{where} \quad \overline{gV} = \{pgV_P + qgV_Q\}. \qquad (36)$$

Here, the expectation E is taken over the randomly varying allele frequencies $p(x)$. These are assumed to be independent of the nature of the patch ($g(x)$ or $V(x)$). Note that because this expression involves the ratio $p(x)/\overline{gV}(x) = p(x)/[pgV_P(x) + qgV_Q(x)]$, it cannot be written solely in terms of the mean and variance of allele frequency (p and Fpq): it involves the higher moments of p. This is true even if we restrict attention to the case of weak selection $[gV_P/gV_Q = 1 + s(x), s \ll 1]$, and to allele frequencies close to fixation. The condition for a protected polymorphism is then obtained by expanding in powers of s: $\bar{s}(1-F) - \overline{s^2}(F - F^*) > 0 > \bar{s}(1-F) + \overline{s^2}(1 - 2F + F^*)$. [$F^*$ is the limit $E(p^3)/E(p)$]. If a proportion F of patches are genetically homogeneous, and $(1-F)$ are formed at random, then $F = F^*$, whereas if a proportion 2F consist of full siblings, $F^* = F/2$. Whatever the exact distribution of allele frequencies, competition between relatives must narrow the range of parameters for which polymorphism can be maintained by environmental heterogeneity.

Competition between relatives has received most attention because of its possible importance in giving a short-term advantage to sex, **recombination**, **polyandry**, and other mechanisms that produce variable families (Williams and Mitton 1973; Maynard Smith 1978a; Taylor 1979; Bulmer 1980a; Bell 1982). In all the models, one or more families settle in each of many patches. Selection takes place within each patch, and the survivors then mate at random to found the next generation. There are three reasons why competition between relatives favors variable families. First, if there are several limiting resources in each patch, the output of each patch will increase if individuals use a greater variety of these resources. This mechanism will favor variability even if there is only one family per patch. Second, if selection favors extreme individuals, variable families will do better. In the two-allele model of soft selection discussed in the previous section, parents that produce a mixture of P and Q offspring will be at an advantage for this reason. Models of sib competition more often assume that there is some underlying "fitness potential", which varies additively and polygenically; the survivors from each patch are those with the highest fitness potential (Maynard Smith 1978a; Taylor 1979; Bulmer 1980a; Barton and Post 1986). This truncation selection, or indeed, any selection where fitness is a concave function of an additive variable, favors increased variance. Finally, even if the probability of survival from a patch is directly proportional to fitness potential, variability will still be favored if competitive ability within the patch is correlated with success after leaving the patch. This is because the weaker members of each family will have been eliminated without any net loss to the family. Striking examples are the selective abortion of chromosomally abnormal embryos in mammals, or the selective abortion of seeds (Stephenson and Winsor 1986).

Although these models show that sib competition can favor increased variability, it is hard to see that it can provide an advantage strong enough to outweigh the twofold disadvantage of sexual reproduction, or that it can be important in abundant groups with juvenile dispersal. It has, of course, proved difficult to find any convincingly strong and general mechanism favoring sex and recombination (Maynard Smith 1978a; Bell 1982).

Evidence

Does environmental heterogeneity (interpreted as the existence of diverse limiting resources) in fact maintain genetic variation? Are individuals more likely to compete with their relatives than with others? These two questions are entangled, because competition between relatives may cause direct selection for rare alleles, as well as indirect selection for genetic systems that promote variability.

There are many examples of polymorphisms which are maintained by **frequency-dependent selection**, arising through the exploitation of different resources by different genotypes (e.g., behavioral differences in Anopheles mosquitoes (Tabachnik et al. 1979) and the crustacean *Asellus* (Christensen 1977); variation in color and banding in *Cepaea* (Jones et al. 1977); and use of different pupation sites in laboratory cultures of *Drosophila* (De Souza et al. 1970)). In the most striking cases, the morphs differ by many characters, and have sometimes been misclassified as different species (e.g., different feeding morphs in some cichlid fishes (Kornfield et al. 1982) and *Lepomis* (Wilson 1988), and Batesian mimics in *Papilio dardanus* (Sheppard 1958)]. However, we do not know whether (as the theory suggests) polymorphism can be maintained only if there are large differences between genotypes, or whether these examples are instead only the easily detected representatives of a widespread phenomenon. Certainly, if a significant proportion of biochemical variation is maintained by balancing selection, that selection must be frequency-dependent. Heterozygosity in haploid and self-fertilizing species is similar to that in outcrossing diploids, and so heterozygote advantage cannot be of general importance (Milkman 1973; Hedrick and Holden 1979; Nevo 1983). Many highly heterozygous species show little large-scale geographic differentiation (e.g., *Drosophila*; Lewontin 1974; Slatkin 1985), and so spatially varying selection is unlikely to be a major source of polymorphism (see below). Minor molecular variants could cause substantial changes in the use of different chemicals or other resources. For example, the fast and slow alleles of alcohol dehydrogenase in *Drosophila melanogaster* have quite different activities; although it is not clear why, or how the polymorphism is maintained (Aquadro et al. 1986; Laurie–Ahlberg et al. 1980; Laurie–Ahlberg and Stam 1987; Oakeshott et al. 1982), these differences could give the strong selection pressures needed to give a robust polymorphism. On the other hand, it would be difficult to reconcile strong selection with the observed variation in fitness (Charlesworth 1987), and with the lack of strong linkage disequilibria between enzyme variants (Langley et al. 1978). There must also be some limit to the number of separate resources which could mediate frequency-dependent selection on large numbers of loci.

Better evidence may come from comparisons of **competition between homogeneous and heterogeneous groups.** These have usually been made in order to find the advantage to sex and recombination due to sib competition, but also give an idea of the net amount of frequency-dependent selection that might arise from genetic variation in resource use. For example, Perez–Tome and Toro (1982) have shown that competing *Drosophila* larvae survive about 5% better if they are half-siblings than if they are full siblings (though see Fowler and Partridge 1986; and Martin et al. 1988). Ellstrand and Antonovics (1984, 1985) and Kelley et al. (1988) directly compared sexual and asexual offspring in natural populations of the grass *Anthoxanthum*, and found similar (but stronger) effects. Analogous interactions are well known in agriculture, where mixtures of species or varieties often perform better when grown together than would be expected from their individual yields (Harper 1977; Barrett 1980).

What evidence do we have on the general relation between **enviromental heterogeneity** and polymorphism? Several experiments have compared levels of biochemical or morphological variation in *Drosophila* populations reared under different conditions (Beardmore 1961; Powell 1971; McDonald and Ayala 1974; Powell and Wistrand 1978; Mackay 1981). These show that both spatial and temporal variation consistently increase levels of variation; Mackay (1981) found that temporal variation was somewhat more effective than spatial. Given that the theory suggests that polymorphism can be maintained only under a limited range of conditions, and that spatial variability is much more effective than temporal, these experiments are surprising.

Because it is hard to measure (or even define) the degree of enviromental heterogeneity experienced by a species in nature, results from the many comparisons between species are equivocal. In an extensive survey of 1,111 species, Nevo et al. (1984) showed that levels of enzyme heterozygosity and polymorphism are better correlated with ecological characteristics (such as degree of specialization, and habitat complexity) than with demographic factors (such as population size, and species distribution). Nevo et al. argue that this supports the direct maintenance of the enzyme polymorphisms by environmental heterogeneity. However, such correlations are hard to interpret. If enzyme variation is largely neutral, then heterozygosity is determined by the product of the effective population size and the mutation rate to effectively **neutral alleles.** Unfortunately, this product bears no simple relation to any actual, observable, parameters of the present population. If most mutations are slightly deleterious, rather than being precisely neutral, the rate of production of effectively neutral mutations decreases with population size (Kimura 1983). This is because in a small population, a greater proportion of mutations will have effects so small as to be dominated by drift. Heterozygosity may be greatly reduced by past bottlenecks (Nei et al. 1975). The effective population size may depend on apparently "ecological" characteristics. For example, in Nevo et al.'s (1984) survey, geographic range is significantly and positively correlated with heterozygosity within the mammals. Nevo classes geographic range as an "ecological" characteristic, acting through its effect on habitat diversity; however, increased range would also be expected to increase the long-term effective population number, and hence to increase heterozygosity under the neutral theory.

These points also apply to laboratory populations, and may resolve the discrepancy with theory. Variation in an arbitrary set of morphological or biochemical markers is surely more likely to reflect the effects of sampling drift than the direct effects of selection. One can thus interpret the finding that variation in environmental conditions increases genetic variation as reflecting an increase in effective numbers. This may be because the actual numbers are greater in a variable environment, or because the variance in offspring number is smaller (Crow and Kimura 1970). More subtly, the genetic component of this variance may substantially alter the rate of **drift**, through hitchhiking with linked loci. Drift will generate random linkage disequilibria between the markers which we observe, and loci under direct selection; this will cause changes in marker frequency which parallel changes at the selected loci. Both substitutions of favorable alleles (Maynard Smith and Haigh 1974), and the elimination of deleterious mutations (Nei and Murata 1966) will cause random shifts elsewhere in the genome, and so will increase the rate of drift. In the extreme, overall heterozygosity in asexual bacterial populations is greatly reduced by the "periodic selection" of particular clones (Hartl and Dykhuizen 1981). Conversely, a balanced polymorphism at constant frequency will cause "associative overdominance" at other loci, which will reduce the rate of drift (Ohta and Kimura 1970). A likely example of this effect is described by Nadeau et al. (1982) and Nadeau and Collins (1983), who show that enzymes linked to the highly polymorphic H-2 locus in mice are themselves significantly more polymorphic. Clegg et al. (1980) performed an elegant perturbation experiment which suggests that, despite the fact that linkage disequilibrium between any particular pair of markers is weak, its cumulative effects may be significant: they showed that the rate of decay of artificially generated disequilibria was faster than expected by recombination alone, and argued that it reflected the stabilizing effects of selection at background loci.

Geographic Subdivision

Patterns of geographic variation have yielded important evidence for evolution, and for the action of natural selection; recently, much population genetics theory has focused on understanding the interaction between gene flow, selection, and drift (for reviews, see Felsenstein 1976; Slatkin 1985; Wijsman and Cavalli–Sforza

1984). A major issue has been whether gene flow binds a species into a single unit (Mayr 1963). If, on the other hand, populations can adapt and diverge in response to varying selection pressures, does this heterogeneity contribute to the maintenance of polymorphism? Does competition between different groups provide an additional selective process, which can favor adaptations that increase the absolute fitness of the whole group, rather than the relative fitness of individuals? This last question takes us back to selection within single structured populations; kin selection may best be seen as a special case of group selection.

Gene Flow: Definitions and Approximations

We have introduced the basic models in discussing the effects of gene flow and drift on strictly neutral variation. The simplest is where a single island receives immigrants from a large "mainland" source (Haldane 1931): a fraction m of genes in the island are replaced every generation. For convenience, it is usually assumed that the population size is reduced from a large number of gametes down to a fixed number of the adults, and that it is gametes which migrate. This extension of the basic Wright–Fisher model separates random sampling drift from deterministic migration. Because population size is not affected by selection, this model also ensures that the rate of migration (m) is independent of gene frequencies (see Sect. 2.1).

The **one-island model** extends to Wright's (1931) n-island model, in which each of n islands receives a fraction $m/(n-1)$ genes from each of the others. If the number of islands is very large, then this reduces to the one-island model, because the pool of migrants taken from all the demes can be treated as a fixed source. If $m = 1 - 1/n$, the model reduces to Levene's, since the population is now panmictic.

In the island model, there is no geographic structure: migration between any pair of islands is equally likely, and the rate of gene flow can be summarized by the single parameter, m. The proportion of genes which move between islands may instead vary with (say) distance. The pattern of gene flow can then be described either by the forward migration matrix g_{ij}, which gives the proportion of adults in i which will move to j, or by the backward migration matrix m_{ij}, which gives the proportion of adults in j that came from i. If population sizes are equal, these will be idential; otherwise, the proportion of migrants coming from larger subpopulations will be greater ($m_{ij} = g_{ij}/\Sigma_i c_i g_{ij}$) (Bulmer 1972).

Although numerical results, and some general theorems, can be obtained using these general matrix models (e.g., Bodmer and Cavalli–Sforza 1968; Karlin 1982), they are not easily understood, and the matrix parameters can be measured only with great effort. The generalization to a continuously distributed population, in which the probability that a gene at x will move to point y in the next generation is g(x, y), is still less tractable. Fortunately, a powerful diffusion approximation is available when gene flow is local (as, for example, in the stepping stone model), and changes are slow enough that one is concerned only with the accumulated effects of gene movements over several generations. Let the allele frequency at x be p(x). The effect of one generation's movement on p is:

$$p^*(x) = \int p(y)\,g(y,x)\,dy. \tag{37}$$

If the distribution of gene movements is isotropic, and the same everywhere [so that g depends only on $\delta = (x - y)$], and if allele frequencies vary smoothly, then p(x) can be expanded in a Taylor series:

$$p^*(x) = \int [p(x) + \delta(dp/dy) + (\delta^2/2)(d^2p/dy^2)..]g(\delta)\,d\delta. \tag{38}$$

Integrating by parts:

$$p^*(x) = p(x) - \tau(dp/dx) + (\sigma^2/2)(d^2p/dx^2). \tag{39}$$

Here, $\tau = \int \delta g(\delta)\,d\delta$ is the mean displacement; if there is no tendency for genes to move in any particular direction, $\tau = 0$. σ^2 is the mean square displacement, or (if $\tau = 0$) the variance in distance between parent and offspring. Variations in gene flow from place to place, and variations in density, can be included in a similar way, and add to the term in τ. For example, if there is a gradient in population density (d), $-\tau$ is replaced by $\sigma^2(dd/dx) - \tau$. These variations cause a directed flux of genes out of areas of high density and high dispersal (Nagylaki 1976b; Nagylaki and Moody 1980).

A similar derivation gives a continuous approximation to a population which is actually divided into **discrete demes**, provided that differences in allele frequency between neighboring demes are relatively small. Of course, no natural population is precisely described either by models of discrete demes of constant size, or by the continuous model of Eq. (37); density-dependent regulation of population size introduces serious complications both in discrete demes (May 1973b), and in spatial continua (Felsenstein 1975a; Barton 1988a).

The diffusion approximation is (as in so many other applications) surprisingly accurate. When neighboring demes exchange migrants in a linear series of stepping stones, the continuous approximation of Eq. (39) gives results which deviate by at most a few percent from the discrete values, even when there are large differences between demes. The approximation converges rapidly as the number of demes increases [$\approx \exp(-n^2)$] (Fig. 6; see Sect. 2.3). This accuracy is fortunate, because computer models of geographic variation almost always use a grid of demes with nearest-neighbor migration: the results should hold for any model of local dispersal. Agreement will be less good if dispersal is leptokurtic, so that there is a small fraction of long-distance migrants, as is often the case (Jones et al., 1981; Endler 1977). However, the distribution of gene movements will become more nearly normal over longer time periods. Extreme leptokurtosis can be included by allowing for a small fraction of genes which come from far away, in effect mixing the island model with the diffusion model (Malécot 1948; see Skellam 1973).

The effects of local gene flow in a variety of population structures depend on the single parameter σ. This should make field investigation of gene flow much

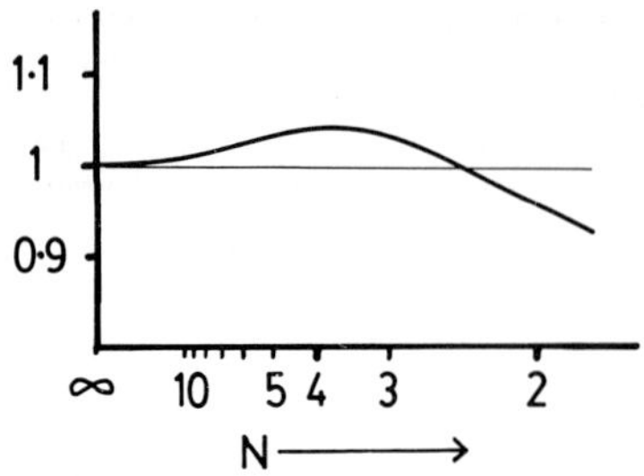

Fig. 6. Convergence to the diffusion approximation. The graph shows the width of a cline in a stepping stone model, relative to the continuous limit; this is plotted against the number of demes within one cline width, N Cline width is defined as 1/(maximum gradient), measured between the two central demes. The cline is maintained by selection against heterozygotes, giving a width of 41 in the continuous limit. Migration is fixed at $m = 0.4$ throughout

simpler, since one does not need accurate knowledge of the whole dispersal function $g(x, y)$ or m_{ij}. However, it is hard to measure σ even once, and harder still to measure the way σ varies in space and time. The problems in making direct measurements of gene flow by releasing marked animals or genes arise from the large sample sizes needed for accurate measurements (particularly of the small fraction of mobile individuals who contribute most to the variance); the disturbance caused by the measurement; and the difficulty of following through the whole life cycle (see Endler 1977, Chap. 2). The availability of polymorphic enzyme and DNA markers makes it possible to infer the relations between individuals in nature, and so may give better estimates of gene flow (albeit with rather small sample sizes; e.g., Meagher 1986).

The labor involved in directly measuring rates of gene flow has made indirect methods based on patterns of allele frequency variation attractive (Wright et al. 1942; Slatkin 1981a, 1985; see Sects. 2.3., 2.5). These depend on the fact that when geographic variation is caused by a balance between drift and gene flow, the spread of the distribution of allele frequencies across populations depends on the products Nm (in a set of demes) or $Nb = 4\pi d\sigma^2$ (in a continuum). Although estimates of these quantities are often referred to as "estimates of gene flow", it is important to remember that they give the number of genes that move, rather than the *proportions* that move. Here, we use the phrase "rate of gene flow" to refer to the latter measure (i.e., to m in a demic population, and σ in a continuous population).

Gene Flow and Selection

Will gene flow overwhelm local adaptations? Suppose that selection favors allele P in some deme, but that genes carrying allele Q migrate in at a rate m. For simplicity, assume a haploid population, and relative fitnesses $W_Q{:}W_P = 1{:}(1 + s)$. In this one-island model, migration is equivalent to irreversible mutation; $p^* = p(1 - m)(1 + s)/(1 + sp)$, where p is the allele frequency after migration. If m is low, there will be a stable balance at $= 1 - m(1 + s)/s$. However, if gene flow exceeds a certain critical rate $[= s/(1 + s)]$, the locally favored allele will be swamped: as m increases towards this value, p decreases smoothly to zero. A similar result applies when differentiation is promoted by selection towards alternative equilibria under a constant selective regime, rather than by different selection in different places. For example, suppose that (in a diploid population) heterozygotes have fitness $1 - s$, relative to homozygotes. Then, $p^* = p(1 - m)(1 - sq)/(1 - 2spq)$; selection can maintain the PP homozygote despite gene flow., provided that $s > 8m/(1 + m)^2$. As m increases towards the critical value, the frequency of the upper stable equilibrium agin decreases smoothly; however, as the critical point is crossed, p drops suddenly from $(1 + m)/4$ to zero.

Results from the one-island model are typical of more complicated situations: selection can maintain variation provided that gene flow does not exceed a certain threshold, where m is of the same order as s. This follows from the fact that if there is no gene flow, then selection can maintain variation if it favors different alleles in different places, or if several stable equilibria are available. A low level of gene exchange may introduce polymorphism, but will not immediately make the equilibrium unstable (Karlin and McGregor 1972b). From demic models such as that described above, one might think that selection has to be quite strong to preserve variation. However, we will see that very weak selection can preserve differentiation in spatially distributed populations, provided that genes only move short distances.

Clines

In any species which is distributed over a wide area, it is usual to find variation in the frequency of discrete alleles and continuous morphological characters. This variation may be organized into more-or-less smooth gradients, or clines; these may span distances from a few meters (as in adaptation to heavy metals; Antonovics et al. 1971) up to continental scales (as in the clines in alcohol dehydrogenase in *Drosophila*; Endler 1977; Barton and Hewitt 1985). How are these clines produced and maintained, and what do they tell us about the interaction between selection and gene flow?

If a cline is very broad, relative to the dispersal range of the organism, then gene flow can have no effect. For example, there is a smooth cline between Spain and Russia in the ability of clover to produce cyanide (Daday 1954). Exchange of seeds or pollen can hardly have an appreciable effect on such a shallow gradient: strong linkage disequilibrium between the two genes involved shows that selection is acting on the polymorphism, and would overwhelm any effect of gene flow (Ennos 1982). It seems that in this (and other) broad clines, selection maintains a balanced polymorphism with an equilibrium which directly reflects a gradient in some environmental factor.

Now, consider clines which are narrow enough to be influenced by gene flow. Suppose now that we can approximate gene flow by diffusion through a continuous habitatat rate σ. The simplest explanation of a smooth cline is that the differences involved are entirely **neutral**, and that the gradient is the transient relic of secondary contact between two different populations. In the extreme case, there would initially have been a sharp step ($p = 0$ for $x < 0$, $p = 1$ for $x > 0$); this would decay at a rate given by Eq. (39). The solution after t generations of mixing is a cumulative normal distribution, with variance $\sigma^2 t$ (Fig. 7a). The width of this cline can be defined as the inverse of the maximum gradient (other definitions are used in the literature—for example, the distance between 20% and 80% points—but give similar values for sigmoid clines), Then, the width is $\sqrt{2\pi\sigma^2 t}$. For example, with a dispersal range of $\sigma = 100$ m in a generation, neutral mixing would give mixing over a width of 2.5 km after 100 generations, 7.9 km after 1000 generations, 25 km after 10,000 generations, and so on. In some cases, it is known that dispersal rates are too high, and the cline too old, for a narrow cline to be explained by neutral mixing alone (see Endler 1977; Table 6, and Barton and Hewitt 1985). However, these calculations do demonstrate that diffusion is a slow process, which will produce very little mixing during the time for which a species' range is likely to remain stable. Models of isolation by distance (Sect. 2.3.2.4) give the same conclusion: over long timescales, neutral markers will be moved around by large-scale population movements, rather than by steady local diffusion.

Advantageous alleles will spread much more quickly. Suppose that haploid genotypes Q, P have fitnesses 1: 1 + s. Then, provided that selection is weak, and dispersal is localized, the spread of P will be approximately continuous in time and space, and the effects of gene flow and selection can be added together:

$$\partial p/\partial t = (\sigma^2/2)(\partial^2 p/\partial x^2) + spq. \qquad (40)$$

Fisher (1937) first investigated this equation, and showed that it admitted a range of solutions, each consisting of a wave advancing at a constant speed, and with constant shape. Wider clines would advance faster; however, there is a minimum cline width of $8\sigma/\sqrt{2s}$, and a corresponding minimum speed of $\sigma\sqrt{2s}$. The wave speed is indeterminate, and depends on the initial conditions; however, Fisher suggested (and later work confirmed) that irregular perturbations would slow the wave down to the minimum speed. Although there is no case in which all the variables are known, Fisher's model gives a reasonable description of the spread of (say) warfarin resistance in British rats. Separate resistance alleles arose in the Welsh borders and in Scotland shortly after warfarin was introduced. After a brief delay, the Welsh allele spread at a steady rate of 4km per year; given the strong initial selective advantage to the resistance allele (Greaves et al. 1977), this would imply a dispersal rate of the same order. There is a large mathematical literature which puts Fisher's equation into a rigorous framework, and further investigates its strange indeterminacy (e.g., Skellam 1951; Gazdag and Canosa 1974; Stokes 1976). Recently, an explicit solution to the limiting wave has been found (Ablowitz and Zeppetella 1979).

Suppose now that selection favors haploid genotype P on one side of an abrupt environmental transition (W_Q:$W_P = 1 + s$ for $x < 0$), and genotype Q on the left (W_Q:W_P = 1:1 − s for $x < 0$). Haldane (1948) first investigated this equation, and showed that a stable cline of width $\sigma\sqrt{6/s}$ would be maintained. Slatkin (1973), May et al. (1975); and Endler (1973) have explored similar models. Bazykin (1969) showed that even in a homogeneous environment, selection against heterozygotes could maintain a cline of width $\sigma\sqrt{8/s}$. Although the shape and width of clines maintained by a balance between selection and gene flow depend on the details, the general conclusion is that clines will be sigmoid, with width $\approx 2l = 2\sigma/\sqrt{2s}$ (Fig. 7). (Where the cline is due to neutral mixing for time t, s can be replaced by 1/t). This *characteristic scale*, denoted by the symbol l (Fisher 1937; Slatkin 1973) determines the distance over which the population can respond to selection, in the face of the homogenizing effects of gene flow.

Whether clines are caused by **neutral mixing**, the **advance of a favorable allele**, or a **stable balance between selection and dispersal**, they will have similar shape, and a width of order l. This regularity makes it hard to find out how clines are in fact maintained (Barton and Hewitt 1985). However, it suggests that one might be able to measure selection pressures from the cline width, relative to dispersal rate. For example, in the flightless grasshopper *Podisma pedestris*, there is a cline 800 m wide in the frequency of a chromosomal fusion. The movements of marked insects in the cline give a dispersal rate of $\sigma = 20$ m, and hence (assuming selection against heterozygotes) $s \approx 0.5‰$ (Barton and Hewitt 1981; Fig. 7). The cline in melanism in the peppered moth *Biston betularia* between Liverpool and North Wales has been compared with that expected from measurements of dispersal and selection (Mani 1981); on the whole, the fit is good.

Such calculations usualy show that quite weak selection could maintain apparently steep clines. However, they should be treated with caution, because direct estimates of dispersal may be unreliable, and because the estimates of selection can rarely be checked. In *Biston*, for example, there is an excess of melanics in North Wales, which may be due to frequency-dependent selection favoring rare alleles (Mani 1982), or to a few long-range dispersers. The estimates of selection on which the original models were based came from the rate of predation of moths that had been pinned to tree trunks; live moths are rarely found in such conspicuous positions, and so these estimates may be too large

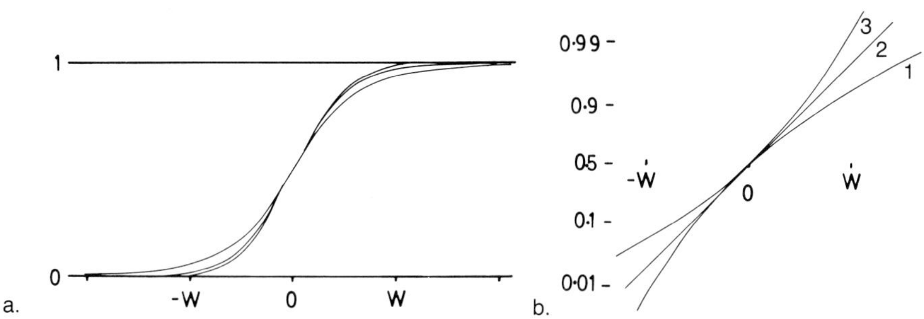

Fig. 7. a Clines maintained by a balance between dispersal and selection all have similar sigmoidal shapes. **b** Transformation to a logit scale [ln (p/q)] reveals differences in the shape of the tails; however, because these involve extreme gene frequencies, they can only be detected with large samples. Cline width is defined as w = 1/(maximum gradient); the graphs are scaled so that all clines have the same width. Three cases are shown: *1* Selection + s on one side of an ecotone, − s on the other ($w = \sqrt{6}l$). *2* Heterozygote disadvantage (w = 4 l.) *3* Introgression of neutral alleles for T generations ($w = \sqrt{4\pi}l$, where $l = \sigma/\sqrt{T/2}$). Looking at the *left-hand side* of the diagrams, these cases run from *1* at the top to *3* at the bottom; the order is reversed on the *right*

(Howlett and Majerus 1987). More generally, estimates of dispersal rate are often several orders of magnitude smaller than cline widths, implying either that gene flow is not involved, or (more likely) that the rate of gene flow is often underestimated (Barton and Hewitt 1985).

Regardless of the numerical details, it is clear that selection can and has produced local adaptations, despite free gene flow. Though it may be convenient to define biological species as "groups of actually or potentially interbreeding populations" (Mayr 1963), and although advantageous alleles can spread rapidly throughout a species' range, contemporary gene flow cannot be responsible for similarities over large areas (Ehrlich and Raven 1969; Grant 1980). Nagylaki (1975) has calculated the smallest region in which an allele can be maintained by selection. If the allele is favored by s inside a small pocket, but has disadvantage βs outside, then the pocket must be at least $2l\tan^{-1}(\sqrt{\beta})$ wide if the allele is to survive. Conditions for the survival of a recessive or underdominant allele are more stringent. If the pocket is wider than πl, then the allele can survive regardless of its disadvantage outside. These conditions extend in a straightforward way to give the probability that a single mutant will be established: if the pocket is wider than Nagylaki's (1975) threshold, then it has probability $< 2s$ of being established if it arises inside the pocket (Barton 1987). If selection is at least moderate, gene flow will have a negligible effect on alleles that are favored in regions more than $\approx 10\sigma$ across.

It is less clear whether spatially varying selection can account for much polymorphism within local populations. In contrast to the Levene model (Sect. 3.1), variation in selection can maintain a variety of alleles across a subdivided population. However, the scale of spatial variation must be close to the characteristic scale l if much polymorphism is to be maintained at any one place. If selection varies over a smaller scale, the allele which is fittest on average will be fixed everywhere, whilst if it varies over a larger scale, local populations will be fixed for one or the other allele. For example, suppose that allele P has

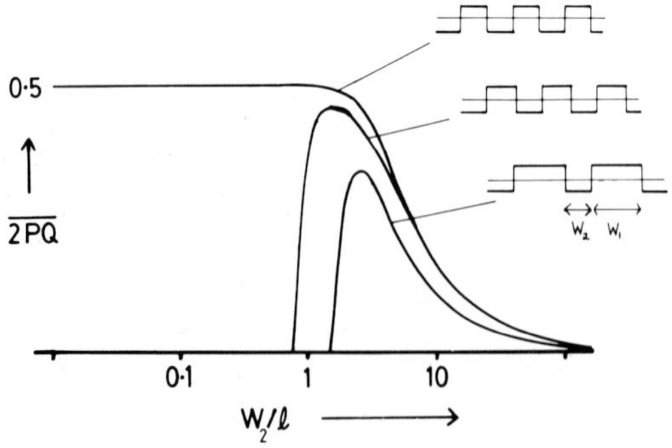

Fig. 8. The average heterozygosity (2PQ) maintained by spatial variation in selection. A codominant allele is favored by selection $+s$ in patches of size W_1, which alternate with patches of size W_2 in which the allele is at a disadvantage $-s$. In the *upper curve*, $W_1 = W_2$; polymorphism can be maintained even by fine-scaled variation ($W \to 0$). However, fine-scaled variation cannot maintain variation if one allele is at a net advantage. *Middle curve*: $W_1 = 1.1\,W_2$. *Lower curve*: $W_1 = 2W_2$. With asymmetry, significant polymorphism can only be maintained if the patch size is between $\approx l$ and $10l$.

advantage s in pockets of width w_1; these alternate with pockets of width w_2 in which the allele has disadvantage s ($s \ll 1$). If w_1, $w_2 \ll 1$, then polymorphism is stable only if $w_1 \approx w_2$; this sensitivity is analogous to that seen in the Levene model (Fig. 5.8). A stronger argument against the general importance of spatially varying selection is that most biochemical polymorphisms in most species show rather little geographic variation over the relevant scales (typically, $F_{ST} < 0.1$; Slatkin 1985); yet significant heterozygosity can only be maintained when accompanied by $F_{ST} > 0.1$.

Linkage Disequilibria

If two populations with two loci at different allele frequencies are mixed, then linkage disequilibrium will be generated (Li and Nei 1974; Feldman and Christiansen 1975). In the extreme case, a mixture of two population fixed for different gene combinations (*PU* and *QV*, say) will contain only those combinations, and so will be in complete disequilibrium. Recombination (or segregation if the loci are not linked) will break down these disequilibria, and so (provided that the differences in gene frequency remain constant) a balance will be reached. For example, consider two steady, linear clines across a chain of demes: at one locus, the frequency of the P allele increases by δp between each deme, whilst at the other, the frequency of the U allele increases by δu. Gene flow and recombination do not alter allele frequencies. The linkage disequilibrium immediately after migration will be $D^* = [(1-r)D + m\delta p\delta u]$; at equilibrium, $D = m\delta p\delta u/r$ (Li and Nei 1974). A similar result applies to a continuous population: then, $D = \sigma^2(dp/dx)(du/dx)/r$.

The term "**linkage disequilibrium**" is especially confusing here, because strong, steady associations can be maintained even between unlinked genes ($r = 1/2$). Such associations are often seen when populations which have differentiated at many loci meet and hybridize (Barton and Hewitt 1985). For example, the fire-bellied toad (*Bombina bombina*) hybridizes with the yellow-bellied toad (*B. variegata*) in a set of concordant clines about 1000 km long. The two taxa are fixed for different alleles at six unlinked enzyme loci, and differ in belly pattern (amongst many other differences). Hybrid populations from southern Poland show strong linkage disequilibrium between evey pair of enzyme markers. There are also correlations between enzyme markers and belly pattern, and the covariance between different pattern characters is greater in the *hybrid zone* (Fig. 9). The similarity across different pairs of loci, and different kinds of character suggests that these associations are due to the mixing of the divergent populations, rather than selection favoring particular gene combinations. Since the gene frequency gradients, the recombination rate (here, 1/2), and the strength of disequilibrium are known, the dispersal rate can be estimated as $\sigma \approx 890$ meters (Szymura and Barton 1986, and ms. in preparation). This estimate is about twice as large as direct measurements ($\sigma \approx 400$ m); the discrepancy may be because some disequilibrium is generated by selection, because σ varies from place to place and from time to time, or because long-distance migrants or juveniles were missed from the observations on toad movements.

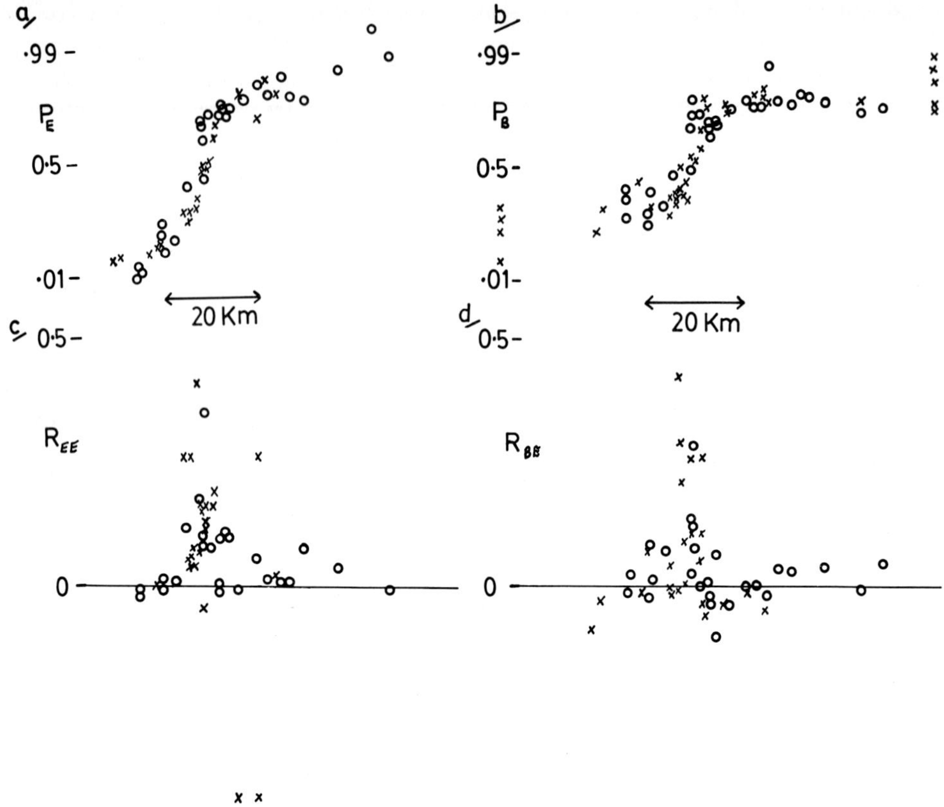

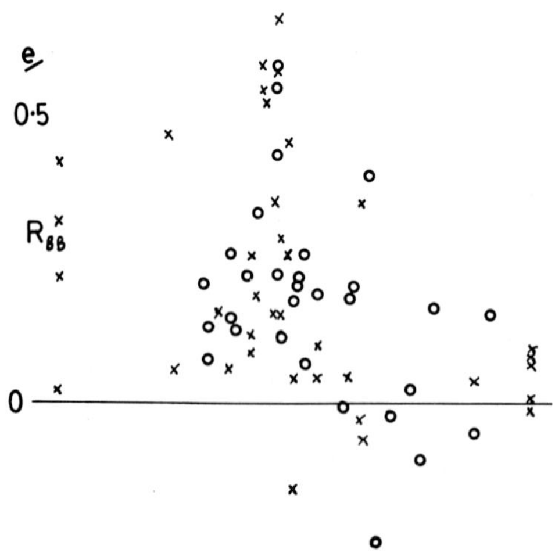

Fig. 9. Variation in diagnostic belly pattern elements, and electrophoretic markers, across the hybrid zone between *Bombina bombina* and *B. variegata*, at two transects in southern Poland, (Data from J Szymura and N Barton, ms in prep; see Szymura and Barton 1986). *Crosses* Krakow transect; *Open circles* Przemysl transect. **a** Average frequency of six electrophoretic markers, on a logit scale. **b** Average frequency of seven diagnostic pattern elements, on a logit scale. **c** Standardized linkage disequilibrium amongst enzyme markers. **d** Standardized covariance between pattern and enzymes. **e** Standardized covariance amongst pattern elements

What effect will such disequilibria have on the **clines**? When sets of alleles are associated in linkage disequilibrium, selection on one gene spills over onto all the others; the effective selection coefficient on gene i is $\Sigma_j s_j D_{ij}/p_i q_j$. Thus, all the clines involved will steepen. Clines will also be pulled together, although the effect is only significant if they already overlap (Slatkin 1975). When only two loci are involved, and selection acts in the same way on both, the effective selection pressure can at most be doubled. However, when many clines coincide, there is a strong feedback effect. Linkage disequilibrium strengthens the effective selection pressure, which steepens the clines, and so further strengthens the disequilibrium (Barton 1983). The final effect depends on the ratio between the selection pressure on each locus, and the harmonic mean recombination rate (s/r): when selection and recombination work at similar rates, strong disequilibrium produces steep clines.

Outside the hybrid zone, any alleles which have introgressed across from the other side will have had time to recombine onto the new genetic background. They will only experience the weak selection arising from their direct effects on fitness. Thus, one expects clines which show strong disequilibrium to change steeply at the center, but be flanked by long, shallow tails of introgression. This is so for *Bombina* (Szymura and Barton 1986), the frog *Rana pipiens* (Kocher and Sage 1986), the bat *Uroderma bilobatum* (Baker 1981; Greenbaum 1981; Barton 1982), and the grasshopper *Caledia captiva* (Shaw et al. 1985; Moran 1979). In these cases, it seems likely that linkage disequilibrium is producing a substantial barrier to gene flow. However, there are exceptions: in the hybrid zone between the mica *Mus musculus musculus* and *M. m. domesticus* in Jutland and Bulgaria, there is a stepped pattern, but no evidence of linkage disequilibrium (Hunt and Selander 1973; Vanderberghe et al. 1986), whilst there are zones where selection seems to be strong relative to recombination, and yet there is no evidence of a stepped pattern (e.g., *Podisma pedestris*, Barton and Hewitt 1981, and pattern races of the butterflies *Heliconius erato* and *H. melpomene*, Mallet and Barton in prep.).

In most of these examples, there is no evidence that the observed markers are themselves under selection: the biochemical differences might well be **entirely neutral**. However, this does not affect the argument: because the bulk of the selection acting at a locus in the center of the zone is induced by linkage disequilibrium, the shape of the central step is independent of the selection on any one locus, provided that selection is relatively weak ($s \ll \Sigma s$). The effect on a neutral locus is particularly simple: because the linkage disequilibria are proportional to the gradient at that locus (du/dx), the gradient at the center is proportional to the gradient at the edge (Barton 1986):

$$(du/dx)_{center} = (du/dx)_{edge}(\bar{w}_{edge}/\bar{w}_{center})^{(1/r)} \quad (41)$$

(This result applies in the limit where s/r is small: it is the first-order effect of linkage. Selection must not be frequency-dependent, but may involve arbitrary epistasis). The sharp step induced in a neutral cline by selection on linked loci represents a barrier to gene exchange: any localized barrier (for example, a river or mountain range) would give the same pattern (Nagylaki 1976b). Here, the barrier arises because introgressing genes must not only cross from one side of the cline to the other, but must also recombine into the new genetic background. It is a spatial analog of the hitchhiking effect (Maynard Smith and Haigh 1974).

Gene Flow, Drift and Selection

The combination of these three forces in any but the simplest cases raises considerable (and largely unsolved) mathematical difficulties. These stem from complications we have already encountered with gene flow and random drift. Nevertheless, the results that are known can be understood in a fairly straightforward way, by adding random noise to the results on gene flow and selection.

The simplest case, analyzed first by Wright (1931), is a single island under directional selection. With an infinite population, selection and (moderate) gene flow balance to give a stable equilibrium. The addition of sampling drift produces a distribution of allele frequencies, of the same form as that with mutation, drift, and selection. The variance of allele frequency is inversely proportional to $[1 + 4N(m + s)]$; thus, if more than ≈ 1 migrant is exchanged per generation, and if more than ≈ 1 individual is eliminated by sewlection per generation, migration and selection balance to maintain the population within a narrow range. Otherwise, the locally advantageous allele is likely to be lost. Wright's treatment of a variety of other models emphasized the importance of the dimensionless parameters 4Nm, 4Ns, and $4N\mu$ in determining the importance of drift relative to other evolutionary forces.

Once allele frequencies in more than one deme are involved, the mathematics becomes intractable. Wright was able to derive formulae for the stationary distribution of allele frequencies (and hence other quantities of interest) because the expected rate of allele frequency change can be written in the form $(pq/2)\,(dH/dp)$. (For example, with selection alone, $H = \ln(\bar{W})$]. Because the variance of fluctuations produced by drift is $pq/2N$, the (stochastic) diffusion approximation gives a stationary distribution $\exp(2NH)$. This method always works with one variable. However, it usually fails with more; in particular, neither migration nor recombination can be represented in this way, and so the effect of drift on subdivided populations, and on linked loci, is hard to analyze. One can find functions which give equations for gene frequency change of the form $dp_i/dt = \Sigma_j g_{ij}(dH/dp_j)$ (Barton 1979; Nagylaki 1989). If g_{ij} is positive, H will always increase, and one can think of the population as moving up an "adaptive landscape" However, this does not lead to simple conclusions about the effects of drift.

Most analytic results apply when **fluctuations in allele frequency** are small enough that the effects of selection are approximately linear. For example, weak selection against homozygotes (with fitnesses $1 - s$: 1: $1 - t$, $s, t \ll 1$) gives $\delta p = -(s + t)pq(p - \hat{p})$, where $\hat{p}$ is the equilibrium frequency. Provided that $p \approx \hat{p}$, $\delta p \approx -k(p - \hat{p})$. Because mutation and migration are also linear forces, Malécot (1948) and Wright's (1943) results (Sect. 2.) carry over to this case. Thus, drift in a continuous two-dimensional population with neighborhood size $Nb = 4\pi d\sigma^2$ produces fluctuations in a stable polymorphism with standardized variance $F_{ST} = \ln(1/\sqrt{2k})/Nb$. The fluctuations will be correlated over a characteristic distance $1 = \sigma/\sqrt{2k}$ (Malécot 1948). This linearization will be accurate provided that F_{ST} is small.

This method can be extended to deal with fluctuations in clines (Felsenstein 1975b; Slatkin and Maruyama 1975; Luskin and Nagylaki 1979). Now, both the stability of the equilibrium (k), and the variance generated by drift ($\approx pq$) vary across the cline. However, both numerical and analytic results are similar to what would be expected from the effects of drift on a uniform polymorphism. In one dimension, fluctuations are correlated over distances of the same order as the cline width, so that it remains smooth. Although it varies somewhat across the cline, the standardized variance F_{ST} is inversely proportional to the dimensionless product of neighborhood size and $\sqrt{k}$ (Nagylaki's β). On average, the cline becomes rather broader, although this effect is small (Slatkin and Maruyama 1975; Nagylaki and Lucier 1980).

One should be cautious in applying these results to nature, thought: the effects of drift in two dimensions are qualitatively different. By analogy with results on uniform polymorphisms, one would expect significant differences over short distances ($\ll 1$), and a standardized variance inversely proportional to neighborhood size and only weakly dependent on selection pressure. Although many samples are needed before one can say much about fluctuations in cline shape, these are broadly the patterns seen in the grasshopper *Podisma pedestris*. Chromosome frequency varies clinally, whilst there is no systematic change in allozyme polymorphisms. For both, there are significant differences between sites less than 30m apart. The magnitude of fluctuations is similar for all loci (Fig. 10; Barton and Hewitt 1981; Halliday et al. 1983).

Large fluctuations are harder to analyze: selection cannot then be treated as a linear force. This is unfortunate, because the most interesting questions concern these nonlinear effects, rather than minor fluctuations around stable polymorphisms. For example, if selection allows alternative stable equilibria (fixation of different chromosomal homozygotes, say), is drift likely to knock a population from one peak to another?

Lande (1979b, 1985) has made considerable progress with this problem, in the case where demes are almost completely isolated from each other. In the limit of no gene flow, the rate at which a new underdominant chromosome rearrangement will be fixed can be calculated from the single population result (Wright 1941; Crow and Kimura 1970). Lande then supposes that the new arrangement spreads through the colonization of gaps left by extinct demes. He shows that, if **extinction** and **recolonization** are random, the rate of establishment of new arrangements in the whole species is the same as the rate of change of single demes. This surprising result holds for the same reason that the rate of neutral molecular evolution equals the mutation rate at a single gene: all demes must trace back to a single ancestral deme. Low levels of gene flow can be included as perturbations to this scheme. Migration greatly reduces the chance of fixation of

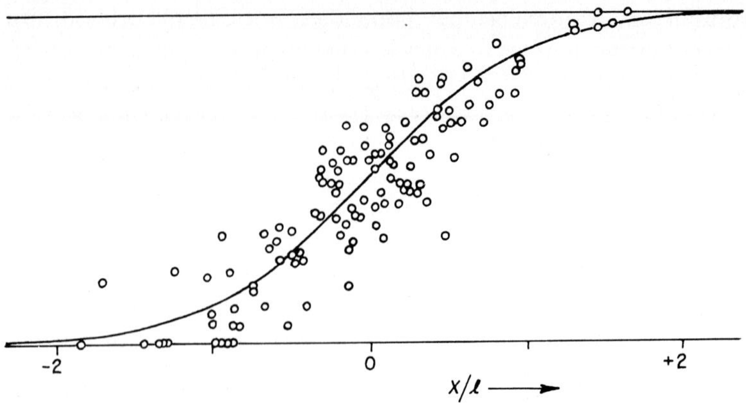
-2
0
+2
x/l

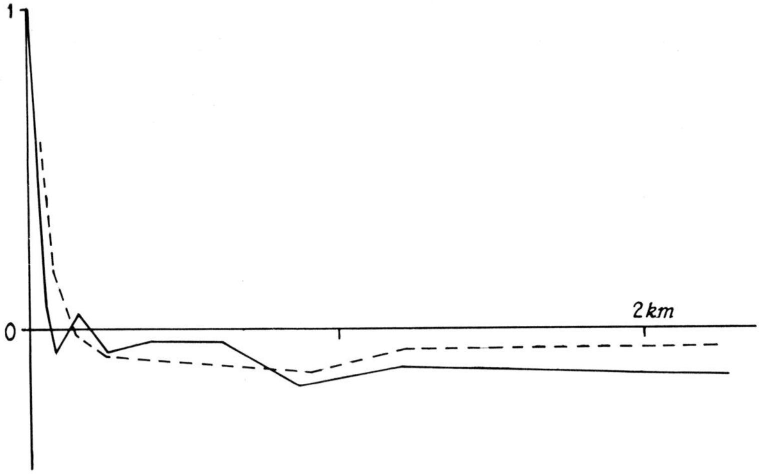
1
0
2 km

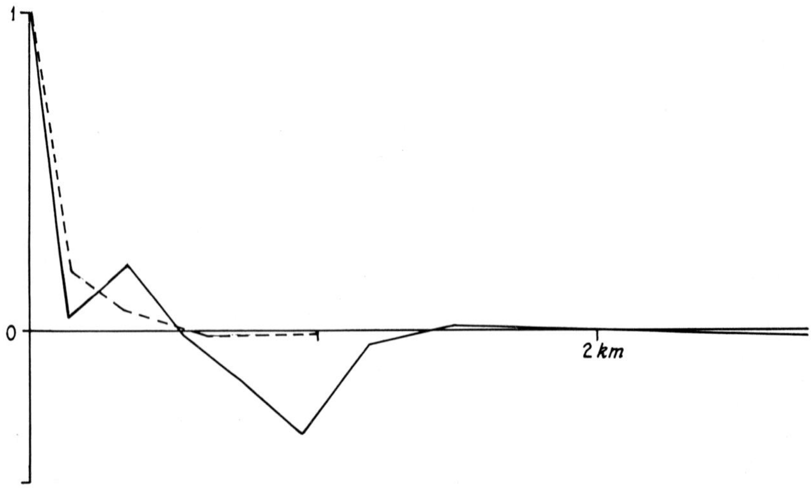
1
0
2 km

an underdominant mutation if $Nm > 1$, but may aid its spread through infection of neighboring demes.

Nagylaki (1978b) has developed a method which gives the covariance between gene frequency fluctuations at different points $[C(x_1, x_2)]$, and has applied it to analyze **fluctuations in clines**. The method extends the basic diffusion approximation [Eq. (39)] to give a differential equation for the covariance of allele frequencies, as well as the expected frequencies. It applies to arbitrarily large fluctuations, and gives results that agree well with simulations (Nagylaki and Lucier 1980). However, it has two limitations. First, it applies only to one dimension. In two dimensions, one has a differential equation in four variables (x_1, y_1, x_2, y_2), which is hard to solve even numerically. Moreover, the diffusion approximation cannot be rigorously justified, and breaks down over small distances (see Sect. 2.3). Second, many quantities—for example, the probability of fixation of a chromosome rearrangement—cannot be derived from the means and variances of allele frequencies. One could extend the method to give differential equations for the higher moments, but this would rapidly become impractical.

Some progress can be made if one considers a **polygenic character**, with constant additive genetic variance, v, rather than a discrete locus. The assumption of constant variance is reasonable if many polygenes are involved, and if there is little linkage disequilibrium between them (Bulmer 1980b; see Slatkin 1978). The effects of gene flow, selection, and drift on the mean z are then given by:

$$\frac{\partial z}{\partial t} = \frac{\sigma^2}{2}\frac{\partial^2 z}{\partial x^2} + \frac{v \partial \ln \bar{w}}{2 \partial z} + \varepsilon. \tag{42}$$

(Sampling drift produces the white noise ε; this has variance $(v/2d)$, where d is the effective population density). Equation (42) can be represented in another way by defining potential function H:

$$H = \int [\ln \bar{w} - (\sigma^2/v)(\partial z/\partial x)^2]\, dx. \tag{43}$$

Then,

$$\partial z/\partial t = (v/2)(\partial H/\partial z).$$

The character mean changes in such a way that H increases (Rouhani and Barton 1987). One can imagine an adaptive landscape, where H is plotted against different configurations z(x); the population moves up gradients in the landscape,

---

Fig. 10. a A transect through the chromosomal cline in the grasshopper *Podisma pedestris* (area B in Barton and Hewitt 1981). Distance is measured relative to the characteristic scale l. This varies along the cline, between 180 m and 440 m. **b** The correlation between fluctuations in chromosome frequency at points different distances apart (118 samples). **c** The average correlation between fluctuations in five polymorphic enzyme systems, taken from the same area (54 samples). *Dotted lines* indicate theoretical expectations

towards local maxima. The first term in H represents the effects of gene flow, which reduces the magnitude of gradients dz/dx, and the second term represents the effects of selection, which tends to increase mean fitness. In the straightforward case of stabilizing selection, there is a single equilibrium. However, if there is disruptive selection, towards peaks at -1 and $+1$, say, the population will be

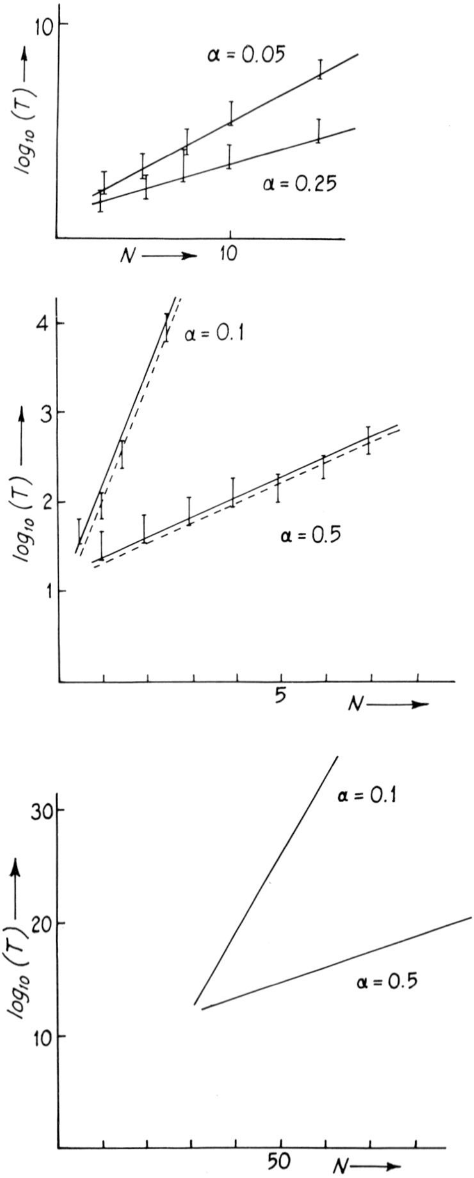

Fig. 11. The expected time *(T)* taken for random drift to knock a population to a superior adaptive peak, plotted against the dimensionless parameter *N* which determines the importance of drift relative to selection. (Rouhani and Barton 1987, Fig. 4). Disruptive selection acts on a polygenic character z, such that the log mean fitness of the population is $\log_e(\bar{W}) = -s\ [(1-z^2)^2 - (4\alpha/3)(z-2)(z+1)^2]/2$. The character is measured relative to the additive genetic variance, which is assumed constant. The expected time is scaled relative to the characteristic distance $l = \sigma/\sqrt{2s}$, and the characteristic time $\tau = 1/s$. The *bars* in **a** and **b** give the 95% confidence interval calculated from simulations. **a** For a single deme, $N = n_e s$, where n_e is the effective population size. **b** for a one-dimensional population, $N = \rho\sigma\sqrt{s/2}$. **C** In two dimensions, $N = 4\pi\rho\sigma^2$, the neighborhood size. Here the selection pressure influences the results, even after scaling: $s = 10\%$

pulled towards either state. If the population is initially at the lower peak ($z = -1$, say), there is some chance (per unit time and per unit area) that sampling drift will knock it towards the higher peak. This probability can be calculated using methods adapted from analogous problems in physics (Rouhani and Barton 1987); the calculation is based on the stationary distribution of $z(x)$, which is proportional to $\exp(2NbH)$, where Nb is the neighborhood size. Surprisingly, the chance that sampling drift will knock the population from the lower peak to the higher can be appreciable, provided that the new peak is substantially higher than the old, and provided that neighborhood size is small (< 30, say) (Fig. 11). The results do not depend strongly on the strength of selection. The analogous problem of finding the probability of fixation of an underdominant allele is much harder, because there is no potential function; however, simulations show a qualitatively similar pattern.

Group Selection, Kin Selection, and the "Shifting Balance"

Natural selection will cause evolutionary change if there is heritable variation in fitness. Selection is usually thought of as acting on individuals: if some character varies between individuals, if it is reliably passed on to offspring, and if it is correlated with the number of offspring per individual, then that character will increase. The process can be neatly summarized by this equation (Robertson 1966; Li 1967; Price 1970):

$$\bar{W}\Delta p = \mathrm{cov}(W_i, p_i) + E(W_i \partial p_i). \tag{44}$$

The ith individual has character value p_i; this might represent a **continuous trait**, or the proportion of genes at a discrete locus which carry a particular allele. ($p_i = 0$, 1/2, or 1 in a diploid; notation follows Grafen 1985). Natural selection changes the population mean if the character covaries with individual fitness (W_i). The mean will also change if there is, on average, a difference δp_i between parent and offspring [$E(W_i \partial p_i)$, where E denotes the average over the population]. With Mendelian inheritance, we expect the second term to be zero; however, there may be random changes due to sampling drift, and systematic changes due to mutation or meiotic drive.

Can this process act at other levels? One can focus on the **gene** rather than the **individual**. The same formula holds, but W_i now represents the number of copies of the ith gene successfully passed on to the next generation. δp_i can now only contribute systematically through recurrent mutation. Dawkins (1982) argues that it is safest always to follow the reproduction of genes; selection on the whole organism, and every kind of population structure, can be taken into account through interactions between the fitnesses of different genes. But one can still choose to apply Eq. (44) to individuals, as Darwin did in his ignorance of Mendel's Laws, and as we may be forced to do in ignorance of the genetic basis of continuous variation (e.g., Slatkin 1979).

Selection can also act on groups of individuals. Group structure can be introduced explicitly using Price's notation (Wade 1980, 1985; Grafen 1985):

$$\bar{W}\Delta p = \mathrm{cov}(W_g, p_g) + E[\mathrm{cov}_g(W_{gi}, p_{gi})] + E[E_g(W_{gi}\partial p_{gi})]. \qquad (45)$$

The first term represents selection between groups, which arises if the average fitness in each group (W_g) covaries with the character mean (p_g). The second term represents the effects of natural selection (averaged over groups), and the third the change caused by biased inheritance [averaged over groups (E), and over all individuals within groups (E_g)]. This formal decomposition into separate components is quite general, and applies to any arbitrary subdivision of the population.

It is clear that if some groups are more successful than others, the genes they carry will increase in frequency. However, there is much disagreement over whether it is useful to think in terms of "group selection", and if so, over what exactly the term should mean. Amidst these semantic arguments, one can lost sight of the main question, which is whether group selection causes significant evolutionary change. Interest has focused on two issues: the evolution of altruism, and Wright's (1931, 1980) model of adaptation via a "shifting balance" between individual selection, random drift, and group selection.

"Altruistic" characters which increase the fitness of others at the expense of their bearers would be eliminated by direct individual selection, but could increase if the beneficiaries tend also to carry the altruistic genes. The idea that selection can act on kin as well as individuals has a long history (Darwin 1859, Chap. 7; Haldane 1932; Fisher 1930, p. 159; Lush 1945). It is particularly important in understanding the evolution of social interactions, and can be seen as a special type of group selection. Recent interest was stimulated by Wynne-Edwards' (1962) suggestion that group selection might lead populations to limit their population size. This brought into the open the implicit assumption of group selection which lay behind many evolutionary arguments. It raised the general question of whether selection can act directly and adaptively on properties of whole groups (such as population size, or genetic variability), or whether these evolve as arbitrary by-products of selection within groups. This question is important, because intrapopulation selection may not increase mean fitness or population size (Roughgarden 1979), and because no strong and convincing short-term advantage for sex and recombination has been found (Maynard Smith 1978a; Bell 1982).

Although the evolution of altruism has received more theoretical attention, empirical studies of population structure have been motivated mainly by Wright's (1931, 1980) "**shifting balance**" model of evolution. Because different combinations of genes can give the same phenotype, a population can respond to selection in a variety of ways: it may move towards alternative stable equilibria. Wright illustrated this point with the "adaptive landscape", a plot of mean fitness against allele frequencies or character means. If one neglects frequency dependence or linkage disequilibrium, mean fitness increases under selection; a population will therefore move uphill, towards an "adaptive peak". Mass selection will take a single panmictic population to the nearest peak, but not necessarily to the global optimum. Wright argued that population subdivision would allow steady evolutionary advance, because each local population could explore the adaptive landscape independently of its neighbors. Superior peaks would spread for several reasons: isolated demes are more likely to shift from low peaks to high; migration from high peaks might shift neighboring demes to that peak; and demes at superior peaks might send out more emigrants, and might be less likely to go extinct. This last part of the "shifting balance" is a form of group or "intrademic" selection.

In Wright's model, variation between demes is generated as they shift to different equilibria. The differential proliferation of the various equilibria may involve group selection, and allows the possibility of direct selection for group characters. However, the issues are different from those raised by the evolution of altruism. Selection between individuals within demes does not necessarily oppose group selection, because each deme approaches one or other stable equilibrium. The problem is to find whether populations do shift between "adaptive peaks", whether this provides significant opportunity for selection, and whether group selection is an important part of the process. All these processes will depend strongly on population structure. The alternative, argued by Fisher (1930), is that steady progress occurs through the accumulation of individually advantageous genes. If this view is correct, then large populations will adapt most rapidly, regardless of how they are structured.

Definitions

In "traditional" models, group selection works through the extinction and recolonization of discrete demes (Wright 1945b; Lewontin 1962; Maynard Smith 1964). Here, there is a direct analogy between individual and group selection: each deme has the properties of reproduction and heredity, and so can be thought of as a 'unit of evolution". Maynard Smith (1976, p. 279) has argued that "group selection should be confined to processes that require the extinction of partially isolated groups which can reproduce and go extinct."

This restricted view of group selection is challenged by "intrademic" models (Wade 1978), which are extensions of the Levene model. In Wilson's (1975, 1977, 1982) models, "trait groups" form from a single gene pool at the start of each generation. If these differ in allele frequency, and if the output from each trait group depends on local allele frequency, alleles which raise group output will increase. This scheme differs from the "traditional" models, because selection arises from differences in group output, rather than from extinction. Trait groups do not reproduce to give genetically similar offspring groups, and so this is not "group selection" in Maynard Smith's (1964, 1976) sense. A wider definition, which includes both the "traditional" and the "intrademic" models, is that "group selection is ... that process of genetic change which is caused by the differential extinction or proliferation of groups of organisms" (Wade 1978; cf. Wright 1932, 1945b; Uyenoyama and Feldman 1980, p. 393).

Group selection can be described in the notation used above for the Levene model, and for gene flow:

$$p^* = \frac{\int d(y,p)\,[p(y)V_p(y)/\bar{V}(y)]\,g(y,x)\,dy}{\int d(y,p)\,g(y,x)\,dy}, \tag{46}$$

where $\bar{V}(y) = p(y)V_p(y) + q(y)V_Q(y)$.

Here, $d(y, p)$ is the output from a patch at y, which may depend on the allele frequency as well as the location. $[p(y)V_p(y)/\bar{V}]$ is the allele frequency after selection in each patch, and $g(y, x)$ is the proportion of genes at y which move to x; g is assumed to be independent of genotype. Migration changes density as well as gene frequency, and soi the frequency after one round of selection and migration must be normalized by the final density, $\int d(y,p)g(y,x)\,dy$. With free migration [$g(y, x)$ independent of y], Eq. (46) is just a generalization of the Levene model (though note that it gives p after migration). If selection is "soft", so that density is independent of p, it reduces to Eq. (34), whilst if selection is "hard", so that density is proportional to mean viability ($d \sim \bar{V}$), it reduces to Eq. (33). If allele frequencies vary randomly between patches (for example, because relatives clump together), and if selection is "soft", Eq. (46) reduces to Eq. (36); in this case, we saw that variation in allele frequency makes polymorphism more difficult.

Here, we are concerned with the net effect of group selection in a uniform environment, rather than with the maintenance of polymorphism in a heterogeneous environment. Suppose that there is free migration, and that $d(y, p) = d_0[1 + \alpha p(y)]$ everywhere. Then $p^* = (\bar{p} + \alpha\overline{p^2})/(1 + \alpha\bar{p}) \approx \bar{p} + \alpha \mathrm{var}\,(p)$ if $\alpha \ll 1$. The effect of group selection is proportional to the variance between groups, a result which follows directly from Eq. (45). The analogy with the "fundamental theorem of natural selection", that response is proportional to heritability, was introduced by Wade and McCauley (1980). The idea of "populational heritability" is further developed by Slatkin (1981b).

Although the general definition, based on the "differential extinction or proliferation of groups" gives a simple analysis into group and individual components [Eq. (45)], it leads to the paradoxical conclusion that group selection can occur without any population structure at all. Suppose that haploid genotypes are reproducing independently of each other, with no interactions between individuals, and hence no density regulation. If individuals are allocated to arbitrary groups, then some groups will have higher mean fitness than others, simply because they happen to include fitter individuals (Grafen 1985). This criticism applies to Wilson's models, in the case where binomial between-group variation arises solely from the random allocation of individuals to small "trait groups" (Treisman 1983; Nunney 1985). Using the notation of Eq. (46), one can see that in an unstructured population, selection is "hard", and so density is proportional to mean fitness ($\alpha = (V_P - V_Q)/\bar{\Delta} \approx s$]. Group selection causes a change $\approx s\,\mathrm{var}\,(p)$, even though there are no interactions within the arbitrarily defined groups. Nunney (1985) suggests that the term "group selection" should exclude such effects, and defines it as "a process by which a trait spreads or is maintained in a population because of the differential reproduction of genotypes that arises from the positive associations of individuals exhibiting that trait". As well as excluding the fortuitous benefit gained by groups that happen to include fit individuals, this definition has the advantage that it applies to continuous populations that do not contain discrete groups. Nunney develops models of "neighborhood selection", in which the fitness of individuals depends on the genotypes of others nearby. Equation (46) would represent such models if d depended on the allele frequency over the local neighborhood $[d(y,\bar{p})$, where $\bar{p} = \int w(\delta)p(y+\delta)d\delta]$, rather than at exactly the same point $\{d[y,p(y)]\}$. Because any model of a truly continuous population must involve density-dependent interactions between individuals in different places, such "neighborhood selection" arises naturally if these interactions depend on genotype (cf. Ellstrand and Antonovics 1984, 1985; Kelley et al. 1988).

Kin Selection

Hamilton's (1963, 1964a, b) idea that the evolution of social interactions could be understood by finding their effect on "**inclusive fitness**", together with E.O. Wilson's (1975) work on the social insects, led to a resurgence of interest in "kin selection". Hamilton (1963) defined the inclusive fitness of an individual as the "animal's production of adult offspring...stripped of all components...due to the individual's social environment, leaving the fitness he would express if not exposed to any of the harms or benefits of that environment...and augmented by certain fractions of the quantities of the harms and benefit the individual himself causes to the fitnesses of his neighbors. The fractions...are simply the coefficients of relationship". This rather complicated definition is necessary if the inclusive fitness is to be useful in predicting whether or not a trait will evolve (Grafen 1982). The definition focuses on the effects of the individual on the propagation of the genes carried by the individual and his relatives. For example, suppose that some action decreases the fitness of the actor by c, but increases a relative's fitness by b. The inclusive fitness of the actor is changed by $-c + rb$, and the trait will evolve if $c/b < r$. One can derive the same rule in another way by following individual fitnesses, or in other words, individual offspring numbers (see Cavalli-Sforza and Feldman 1978; Grafen 1982). If the gene frequency is changing slowly (an assumption which we will see is necessary for Hamilton's rule to be valid), an individual exhibiting the altruistic trait will have relatives who also exhibit the trait, and so will benefit from their altruism by (on average) rb. The ordinary fitness it equal here to the inclusive fitness. Although inclusive fitness is hard to define, and cannot be measured directly, it is sometimes easier to use it to understand kin selection. The problem is to know when the coefficient r is in fact equal to the coefficient of relationship, and so is independent of unknown genetic details of the trait.

Kin selection can be seen as a special kind of group selection, where the groups consist of families, and variation between groups arises from the relationship between family members. Thus, if the fitness of altruists is reduced by c, but the altruist increases the fitness of everyone in the family by b, the group mean fitness is $(b-c)p$; the change due to group selection is therefore $(b-c)\,\mathrm{var}(p)$, as compared with a change $-cpq$ due to selection within the family. If we define F as the standardized variance of allele frequency $[F = \mathrm{var}(p)/\bar{p}\bar{q}]$, then $\overline{pq} = \bar{p}\bar{q}(1-F)$. The net change in $\bar{p}$ is $\bar{p}\bar{q}[F(b-c)-(1-F)c]$, and so the allele will increase if $F > b/c$. If we ignore the effects of selection, and assume that all variation between families is due to their relatedness (or in other words, to sampling drift), then F is equal to the coefficient of relatedness, and we recover Hamilton's Rule. This rough argument shows the link between group and kin selection: we now summarize the rigorous work which shows exactly when Hamilton's Rule is useful.

As with any kind of group selection, the change in mean of some **additive character** (or allele frequency), p, is given by Eq. (46). Wade (1980, 1985) and Grafen (1985) show that this leads to an expression with the same form as the inclusive fitness, but with the coefficient of relationship replaced by the ratio between two covariances. [In Grafen's (1985) notation, $\mathrm{cov}_R(H_j, p_j)/\mathrm{cov}_D(H_j, p_j)$]. These covariances are taken over all the encounters that occur in the population: H_j scores 1 if an altruistic act takes place (reducing the donor's fitness by c, and raising the recipient's by b), and 0 otherwise. $\mathrm{Cov}_R(H_j, p_j)$ is the covariance between the recipient's p value and the performance of the act, and $\mathrm{cov}_D(H_j, p_j)$ is that for the donor. This expression can be seen as the regression of recipient's p value on donor's. It can also be seen as the ratio between the deviation of the average p value in recipients from the population mean, and the corresponding deviation for the donor.

The definition of r as a regression was introduced by Hamilton (1970), and applies quite generally. However, it could only be measured if one knew the genotypes of the individuals at the relevant loci: to be useful, it must equal some quantity that can be calculated from the genealogy, and applies equally to all loci. If the genes involved were neutral, the regression coefficient would equal Malécot's (1948) coefficient of relationship: the proportion of genes in the recipient which are identical by descent with a randomly chosen gene in the donor. There is an inconsistency here, because we are interested in genes that are under selection. If genes are shared through common ancestors only a few generations back (as, for example, with interactions between siblings, or parents and offspring), selection must be quite strong to distort the genealogy significantly (e.g., Charlesworth 1978; Uyenoyama 1984; Uyenoyama and Feldman 1982). However, if there is inbreeding, so that genes are shared through distant ancestors, weak selection can have a substantial cumulative effect (Uyenoyama 1984). If there is dominance, inbreeding in small groups can reveal rare homozygotes; the conditions for increase then depend on dominance (Uyenoyama 1984). With inbreeding, Hamilton's Rule may be a poor guide to the evolution of altruism.

Group Selection in Spatially Structured Populations

Most theory assumes a **"beanbag" model** of gene flow: there is either a single panmictic population, or a collection of partially isolated demes, with migration being equally likely between any two demes. How will group selection be affected if there is some spatial structure, so that migration and colonization involve neighboring populations? The answer is not obvious: although localized gene flow may allow greater differentiation overall, differences between adjacent populations will be smaller than between distant pairs, reducing the effects of group selection. We can investigate this problem by using Eq. (46) to find a diffusion approximation which will describe group and individual selection in a continuous population.

Let $V_P/V_Q = 1 + s$, where s is small. If $d(y,p) = d_O[1 + \alpha p(y)]$, where $\alpha \ll 1$, then:

$$\partial p/\partial t = (\sigma^2/2)(\partial^2 p/\partial x^2) + spq + \alpha\sigma^2(\partial p/\partial x)^2. \quad (47)$$

This expression can be also derived from Nagylaki's (1976b) extension of Eq. (39) to include variation in density (see also Nagylaki and Moody 1980), and above). Gradients in allele frequency typically span a distance of the same order as the characteristic scale, $1 = \sigma/\sqrt{2s}$; this is so whether variation is produced by selection or drift. Equation (47) therefore shows that group selection will be weak ($\approx \alpha s$) relative to individual selection ($\approx s$).

We can take, as a tractable example, a **balanced polymorphism** with stability k [i.e., $dp/dt = -k(p - \bar{p})$], and compare the expected effect of group selection $\alpha\sigma^2 E[(dp/dx)^2]$ when sampling drift perturbs allele frequencies from their common equilibrium $\hat{p}$. In a truly continuous population, the diffusion approximation which gives Eq. (47) would break down: even in one dimension, where the covariance between allele frequencies nearby at points is well behaved [Eq. (18a)], the mean square gradient diverges. In any case, the density at a point cannot depend on allele frequency at precisely the same point (see above). However, Eq. (47) does approximate a stepping stone model: with deme spacing ε, $\sigma^2 = m\varepsilon^2$ and $dp/dx \approx (p_i - p_{i-1})/\varepsilon$, and so the change due to group selection is $\alpha m E[(p_i - p_{i-1})^2] = 2\alpha m[\mathrm{var}(p) - \mathrm{cov}(p_i, p_{i-1})]$. Using the values for the variance and covariance given in Section 2 [Eq. (18a)] gives $\alpha\bar{p}\bar{q}/N\sqrt{2}$. Similarly, the average change due to group selection in a two-dimensional population is $\alpha\bar{p}\bar{q}/N(\pi \ln(\sqrt{32}))$ (for $k \ll m \ll 1$; Kimura and Weiss 1964; Crow and Aoki 1982, 1984). It is remarkable that these results depend only on the deme size (N) and the relation between gene frequency and population size (α): they are independent of the stability of the polymorphism (k), and the migration rate (m). Increased migration reduces the overall variance, and reduces the differences between adjacent demes; however, because group selection in this model acts via excess emigration from large populations, increased migration increases the rate of group selection for a given pattern of allele frequencies. Increased stability reduces the overall variance, but (if $k \ll m$) does not affect the differences between adjacent demes.

These results can be compared with the island model. With a single gene pool ($m = 1$), as in Levene's and Wilson's models, the change due to group selection is $\alpha \, \text{var}(p)$; when there is partial isolation ($m < 1$), this becomes $\alpha m \, \text{var}(p)$. Since $\text{var}(p) \approx \bar{p}\bar{q}/4Nm$, we find $\alpha\bar{p}\bar{q}/4N$. All these results are approximations, since (just as with derivations of Hamilton's Rule), we assume that the pattern of allele frequency variation is not distorted by selection. Crow and Aoki (1982, 1984) first derived these results for the stepping stone and island models, using simulations and the "neutral" approximation. Their model of differential proliferation assumed that allele frequency affected group fitness, rather than group population size; as a result, they found a dependence on Nm, rather than N. Harpending and Rogers (1987) have made a detailed study of the island model, and show that the "neutral" approximation is reasonably accurate, even with α as large as 1.

These arguments show that group selection will be weak unless allele frequency has a strong effect on deme size, and demes are small. We have dealt with the rather peculiar case of a balanced polymorphism; however, spatial distributions in a balance between drift and gene flow do not depend strongly on allele frequency or on the nature of selection, (Slatkin 1981a; Barton and Slatkin 1986), and so the same arguments should predict the fate of a novel allele.

How Strong Is Group Selection?

Theory suggests that, regardless of what causes variation between demes, and regardless of detailed population structure, group selection is inherently weaker than individual selection. With panmixia, and hard selection, group selection will outweigh individual selection only if $sF\bar{p}\bar{q} > s(1 - F)\bar{p}\bar{q}$, and hence $F > 0.5$. For group selection to be important, there must be both strong spatial differentiaion ($F \approx 1$), and alleles must affect population density much more than relative fitness ($\alpha \gg s$). This view is challenged by the experiments of Wade (1978, 1982) on population size in *Tribolium*, and Goodnight (1987) on leaf area in *Arabidopsis*. There can be substantial genetic variation between demes, allowing a rapid response to group selection, which can give qualitatively different results from individual selection. For example, Goodnight (1987) selected for leaf area either by choosing the largest (or smallest) 8 of 16 individuals within each deme, or by choosing the three demes out of nine with largest (or smallest) mean leaf area. Individual selection caused a consistent decline in leaf area, regardless of the direction of selection. In contrast, threre was considerable variation between demes (the populational heritability averaging 0.3). This allowed a response to group selection, giving a realized heritability of 0.19 for treatments with group selection alone. Wade (1977, 1982) found a similar pattern, with high populational heritabilities, and a significant effect of group selection regardless of whether individual selection acted in the same or opposite direction.

There may be several reasons for the unexpected success of these experiments. Although some involved considerable **migration**, the high populational heritabilities were in rough agreement with predictions made using Slatkin's (1981b) neutral model. For example, Wade (1982) found a significant response to group selection with a migration rate of 12.5% in an island model; the effective deme size was 14.1 adults. The realized populational heritability for this case was 0.45 over 10 generations, compared with an average realized heritability of 0.72 with no migration (Wade 1982 Table 2). The effect of migration on variation produced by drift depends on Nm, which equals 2 in this case, which is large enough to reduce between-deme variance, but not so large as to make it insignificant. In general, though there is a good deal of random variation in the data, the measured populational heritabilities do not seem too large to be explained by Slatkin's (1981b) additive model.

Other factors could cause greater differentiation between groups. Sampling drift during **founder events** can convert dominance and epistatic variance components into additive variation within and between demes (Bryant et al. 1986a, b; Goodnight 1985). Random differences in selection between demes, or the attraction of demes towards different equilibria may also contribute: this is suggested by the diverse responses to selection described by Wade (1978).

Although this laboratory work shows that moderate amounts of drift and migration can produce substantial variation on which group selection could act, it is hard to know whether such variation is common in nature. Evidence from electrophoresis (Slatkin 1981a; see above), and from the steady response to selection in small populations (Lande 1982b; Hill 1982), suggests that few natural populations are sufficiently small and isolated for drift to have generated substantial allele frequency variation between demes, and eliminated substantial amounts of variation within demes. For example, Lewontin (1962) argued that t-alleles were kept at low frequency in mouse populations by group selection, despite strong segregation distortion in males: small groups of mice that accumulated t-alleles might go extinct. However, there is no clear evidence that mice actually live in small groups of close relatives (Baker 1986); t-alleles may be kept at low frequency because females prefer not to mate with heterozygous males (Lenington 1983). In general, it is unusual to find $F_{ST} > 0.1$ within a region (Slatkin 1981a, 1985); this implies that drift alone is unlikely to generate enough populational heritability to sustain much group selection. Of course, variation between groups might be generated by selection rather than by drift (see next section). However, since most of the experiments which show strong group selection seem to involve no more variation than can be generated by drift alone, the absence of strong differentiation of neutral characters in nature reduces their relevance.

Group selection could be relatively strong if alleles had little effect on relative fitness, but had large effects on **group fitness** (i.e., $\alpha \gg s$; Fig. 4f in Wilson 1977). This assumption is built into several of the theoretical models which demonstrate strong effects of group selection (e.g., Harpending and Rogers 1987; Wilson 1982). One can imagine mechanisms which would give strong effects on group fitness—for example, slight changes in reproductive rate could cause chaotic fluctuations in population size, and hence extinction (May 1973b); alarm calls might be of great benefit to the group, but involve the caller in little risk.

However, it is hard to see these as being of wide importance, except in tight kin groups.

Wade (1978) argues that much of the theory is biased against group selection, because it assumes group selection to be opposed by individual selection. He argues that group selection can have qualitatively different effects even when it acts in the same direction as individual selection. This idea is the basis of Wright's "shifting balance": there, group selection moves the population towards a global peak of mean fitness. Genes which reduce the fitness of their neighbors may be selected, leading to a paradoxical decrease in the population mean under individual selection (Sachs 1967, Wade 1978; Goodnight 1987). It would be interesting to know the relative success of plant and animal breeders who have used different kinds of selection: Wright's ideas were suggested by his study of selection between lines in the shorthorn cattle (Wright and McPhee 1925), yet remarkable progress has been made by simple mass selection (Dudley 1977; Hill 1982). Interactions between individuals can make group selection more efficient than individual selection; however, one must still find the relative strengths of the two kinds of selection in order to judge which will be the dominant cause of gene frequency change. Models of the evolution of altruism can be seen as a convenient device for making this comparison.

Wright's "shifting balance"

A situation where variation between demes is created by drift alone, and preserved by strict isolation, is not favorable to group selection or to adaptation via the "shifting balance" (Wright 1931, 1980). Uyenoyama (1979) has shown that suitable variation can be created by random variations in selection within different groups, rather than by drift. However, this mechanism becomes very weak at extreme allele frequencies ($\approx (pq)^2$), and so cannot establish rare alleles (Charlesworth 1979; Uyenoyama 1979). The variation between demes is proportional to the variance in selection coefficients, and so group selection can only overcome individual selection if the mean selection coefficient is smaller than the variance; this requires a delicate adjustment of the parameters, similar to that which restricts the Levene model (Hoekstra et al. 1985).

Selection can combine with random processes to produce and preserve **variation between demes**. If several adaptive peaks are available, populations may be knocked from one to another by sampling drift or by fluctuations in selection. Large differences in allele frequency will be preserved, provided that gene flow is not too great. (In the island model, $m < s$, and in stepping stone or continuum models, the area which has shifted to the new peak must be $> l^2 = \sigma^2/2s$). Wright argued that selective interactions would usually lead to multiple peaks, and that evolution would involve selection between peaks as well as between individuals. Populations will certainly come to occupy different adaptive peaks: reproductive isolation entails (almost by definition) the establishment of incompatible combinations of genes. The controversial issues are, how these incompatibilities arise, and how they then spread.

Lande (1985) has modeled the spread of an **underdominant chromosome rearrangement** through a collection of partially isolated demes (see Sect. 3.2). If demes are small enough ($Ns < 10$, say), sampling drift may establish a new mutation in a single deme. It can then spread in two ways. If the rearrangement migrates to another deme, it will usually be eliminated by selection. However, there is some chance that it will drift to fixation. Second, whole demes may go extinct, and be recolonized from demes containing a different karyotype. With "soft" selection, deme sizes and extinction rates are independent of gene frequency, and so there is no group selection. With "hard" selection, deme size is proportional to mean fitness. Group selection then favors the fitter homozygote because large demes are more likely to invade their neighbors, and also because they will produce more successful colonists. Individual selection influences the other mechanism of "invasive" spread: if one homozygote has higher relative fitness than the other, it is more likely to be fixed by drift when introduced at low frequency. At any one time, almost all demes will be fixed for one or other chromosome arrangement. This maximizes the opportunity for group selection, and also greatly simplifies the analysis. Lande (1985) finds that group selection has a much weaker effect than individual selection on the probability of fixation. This conclusion clearly depends on the parameters chosen: in the limit of low gene flow, for example, spread would be entirely by extinction and recolonization, and so individual selection would have no effect (cf. Eshel 1972). Nevertheless, this model shows that even when selection towards alternative equilibria ensures that there is much more variation between demes than within them, group selection does not necessarily dominate.

The "shifting balance" can also occur in a **continuous population:** if a sufficiently large area moves towards a superior adaptive peak, differentiation can persist despite the swamping effects of gene flow (Nagylaki 1975; Fife and McLeod 1977; Rouhani and Barton 1987). Here again, the spread of the new peak is affected by both individual and group selection. Suppose, for example, that heterozygotes are less fit. A cline can be set up, because individuals which move across the cline will mate with the opposite genotype, and so their heterozygous offspring will be eliminated (Bazykin 1969). However, because the cline is maintained by intrinsic interactions between genotypes, rather than in response to external environmental heterogeneity, its position is arbitrary, and can be changed by slight perturbations (Bazykin 1969; Barton 1979). The fitter homozygote will tend to advance (selection between individuals); any net flux of genes will push the cline towards regions of low density and dispersal; and finally, alleles which increase density or dispersal will spread (a form of group selection). However, group selection will be relatively weak: it depends on a term $\alpha\sigma^2(dp/dx)^2$, which will be of order αs. As with other models of group selection, population density must depend strongly on genotype if group selection is to be significant.

Many surveys of geographic variation have been motivated by interest in Wright's "shifting balance". We now have extensive data on typical patterns of **neutral variation**. However, Wright's model involves loci under selection:

although neutral patterns can show whether drift is strong enough to cause peak shifts, they cannot tell us whether peak shifts have occurred, or how peaks spread. To answer these questions, we must look at genes that are involved in peak shifts.

Many species are subdivided into a patchwork of geographic races, separated by narrow **hybrid zones**. It is likely that most of these zones are maintained by a balance between dispersal and selection against heterozygous or recombinant genotypes (Barton and Hewitt 1985). The abundance of such hybrid zones shows that peak shifts, of the sort required by Wright's theory, have occurred. However, biogeographic and molecular evidence shows that many are ancient (Barton and Hewitt 1985); e.g., *Bombina* (Szymura and Barton 1986), *Caledia* (Shaw et al. 1985). It is hard to know whether these hybrid zones are the relics of the rare establishment of new gene combinations, or whether they are instead the visible representatives of a rapid and continual "shifting balance". These possibilities could be distinguished by crosses between distant samples (e.g., Oliver 1979; Vetukhiv 1956): it is hard to generalize from the few surveys that have been made, but the slowness of speciation suggests that peak shifts cannot be very frequent (for example, *Drosophila* sibling species are separated by several million years; Coyne and Orr 1988). The "shifting balance" inevitably leads to reproductive isolation, and so it cannot be acting rapidly.

One might hope that the positions of hybrid zones would tell us about the factors responsible for the movement of their component genes. Zones may be trapped at physical barriers, and may simply have remained near where the two races came into secondary contact. However, they often run parallel to ecological gradients (Barton and Hewitt 1985). Unfortunately, this pattern could be explained in several ways (Endler 1982). The differences might have arisen in situ, through adaptation to the different environments. As divergence continued, each race would accumulate alleles suited to its own genetic background, and so selection would eventually act against hybrids as well as in response to the environment (Endler 1977). A second possibility is that clines maintained by selection against hybrids might move such that each gene combination occupied the habitat to which it was best suited (Key 1982). If many clines responded to the same enviromental gradient, they would all come together in a single hybrid zone. Finally, each race might have expanded from allopatric refugia; if the pattern of spread depended on evironmental differences, they might come into secondary contact at an ecotone. It is hard to distinguish these possibilities, and hard to guess the relative importance of group versus individual selection in any of the three cases.

There are other obstacles to adaptation via the "shifting balance". Adaptive peaks can gain an advantage for arbitrary reasons: dominant alleles (Mallet 1985), and alleles giving increased dispersal (Barton 1979) tend to spread. More important, gene combinations that happen to coincide with regions of population expansion will gain an unfair advantage. For example, the positions of most European hybrid zones are thought to be determined by postglacial expansion from small refugia: the gene combinations that happened to be in those refugia will be grossly overrepresented. This process is analogous to sampling drift, and may overwhelm intergroup selection, simply because of the small number of independently expanding and contracting regions. A further problem is that zones involving one set of characters are usually clumped with zones involving other characters. Once independent zones are brought together, by coadaptation or by a common response to population movements, they can only be separated by drastic events. There are cases where separate clines have passed through each other (e.g., *Warramaba viatica* (White 1978; Hewitt 1979), and several involving mitochondrial DNA (Avise 1986), but these are unusual. Thus, for the same reasons that selection on small asexual populations is ineffective, Wright's "shifting balance" is unlikely to be a major cause of adaptation.

5 Genetic Diversity and the Ecology of Asexual Populations

R.C. VRIJENHOEK

Introduction

Sexual reproduction in plants and animals involves a highly conserved set of mechanisms that result in genetic recombination. As a consequence of segregation, assortment, and crossing-over of genes during meiosis, a sexual individual has the capacity to produce an immense variety of haploid gametes. Random fusion of these gametes into new zygotes produces an even greater variety of diploid genotypes, ensuring that each offspring is essentially unique. There must be a significant evolutionary benefit to this fundamental process, because nearly all higher organisms reproduce sexually, at least during some stage of their life cycles.

The classical argument for the predominance of sex postulates that genetic recombination is good for the species. **Recombination** can accelerate adaptive evolution because beneficial mutants that arise in different individuals can come together in the same individual (Fisher 1930; Muller 1932; Crow and Kimura 1965; but see Maynard Smith 1968). Also, deleterious mutations will theoretically accumulate in nonrecombinant genomes causing a decline in fitness that has been called **"Muller's ratchet"** (Felsenstein 1974). Recombination in sexual populations facilitates purging of deleterious gene loads. Over the long term, asexual species are presumed to be evolutionary dead ends (Stebbins 1950; White 1978). Clones (i.e., groups of genetically identical individuals) might flourish in some environments, but lack of recombinational variability is believed to render them less capable of responding to new physical and biotic challenges.

Stanley (1975) suggested that the absence of recombination also inhibits speciation in asexual lineages. Of course, biological species are ordinarily defined in terms of reproductive isolation, making this concept difficult to apply to asexual organisms (Maslin 1968). Nevertheless, evidence for significant morphological or ecological diversification within monophyletic asexual lineages is generally absent. White (1978, p. 317) stated the case most succinctly:

> ..."there is very little evidence that diversification and divergence of... clones by mutation is likely to proceed to the point where they would be regarded as different species by a museum taxonomist guided solely by phenetic criteria. In all cases of... clones as different as that, it is probable that they have arisen independently from their bisexual ancestors... This indicates

> the limitation of... (asexual) reproduction; its adaptive potentiality is restricted to an extent that renders really long range progressive evolutionary change out of the question".

A high rate of extinction coupled with a low rate of diversification might easily explain why the standing crop of asexual "species" is low. Notwithstanding, the view that sex predominates because of its long-term benefits to species was challenged by a simple argument based on the primacy of selection at the level of individuals (Maynard Smith 1971, 1978a; Williams 1975). Assuming that fecundity is equal, an asexual female has twice the reproductive potential of a sexual female. Asexual females produce only daughters, but on average, half of a sexual female's progeny will be sons. Given this twofold reproductive advantage, each time an all-female lineage arises, it should rapidly replace its sexual ancestors. It is difficult to see how the long-term evolutionary benefits of sex can balance the immediate demographic cost of producing males. Nevertheless, asexual lineages generally do not replace their sexual ancestors over broad geographical scales. In most taxa where all-female cloning has arisen, the sexual ancestors are still extant, often living with, or adjacent to, derived asexual forms (Grant 1981; Suomalainen et al. 1987).

The twofold advantage of all-female reproduction assumes that everything else is equal between sexual and asexual lineages. Do sexual ancestors and their derived clones have identical fertility and survivorship? Do they share identical niches? Does genetic diversity among individuals contribute to the ecological success of a sexual population? These questions are at the core of a family of ecological models that attempt to show how the twofold advantage of all-female cloning can be balanced or surpassed by the benefits of genotypic diversity to individuals. These models are based on the assumption that natural environments are variable in space and time. For example, sib competition models focus on spatially heterogeneous environments in which sexual individuals receive an immediate reward resulting from genotypic diversity among their progeny (Williams 1975; Maynard Smith 1978a; Bulmer 1980c). Sib competition models can be broken into two types (reviewed by Young 1981). First, genetically diverse progeny might be more capable of fully exploiting a variety of patches in a heterogeneous environment, leading to higher average fitness than would obtain for a homogeneous brood (Ghiselin 1974). Second, competition within environmental patches might be similar to a lottery. Sex might be favored because genetically diverse progeny have more than one chance at producing the winning genotypes (Williams 1975). Another class of models focuses on temporal variation that results from living in a biotically complex world. Genotypic diversity may be an essential component for survival in a world full of predators, competitors, and parasites that are also evolving.

Theories concerned with the **origins and maintenance of sex** have been treated in considerable detail in several comprehensive reviews by Ghiselin (1974); Williams (1975); Maynard Smith (1978a); Lloyd (1980); Bell (1982, 1985), Shields (1982); Jackson et al. (1985); Michod and Levin (1988). This chapter focuses instead on recent studies of asexual organisms, because clones provide a useful window for viewing the broader issue of sexuality. Just as geneticists have used mutations to probe the function of normal genes, we can study reproductively deviant organisms to gain insight into the function of sex and recombination. In many cases, asexual populations are remarkably successful, sometimes outnumbering their sexual relatives or even replacing them on a local scale. We need to know more about the peculiar genetic and ecological circumstances that allow asexual organisms to flourish at some localities and not at others. Recent genetic studies of asexual populations have often identified considerable genetic variation in the form of multiple clones. In this chapter, I discuss factors contributing to the origin of genotypic diversity in asexual populations, examine the translation of this genotypic diversity into phenotypic variance, and consider this variation in the light of ecological models concerning sexual and asexual reproduction. Understanding the genotypic origins of clonal diversity and its subsequent expression in ecologically relevant phenotypes is imperative for a clear understanding of the ecological dynamics of sexual and asexual populations.

Asexual Modes of Reproduction

Asexual reproduction in plants and animals can be classified in many ways. First we must distinguish between reproductive modes that involve somatic cells (i.e., vegetative reproduction, budding, and fragmentation) versus those that involve the germ line. If germinal, we must ascertain how much of meiosis is involved or circumvented because that will determine whether asexual lineages retain or lose heterozygosity. There are about as many kinds of asexual reproduction as there are ways to circumvent or block genic recombination. The variety of cytological mechanisms has led to a bewildering proliferation of terms and very little consensus about their application in plants versus animals (Gustafsson 1946; White 1973; Bell 1982; Suomalainen et al. 1987). I restrict this discussion of asexuality to "gametic cloning" (i.e., germ line processes that clone the maternal genotype). **Gametic cloning** in plants and animals leads to the production of independent offspring that are genetically identical and free to disperse.

A distinction should also be made between clonal forms of reproduction that require sperm (**pseudogamy**) versus true **parthenogenesis**. The term parthenogenesis (virgin birth) should be restricted to autonomous forms of gametic cloning. Parthenogens are free to escape from their sexual ancestors, or to replace them altogether. Pseudogamous forms, on the other hand, are forced into a parasitic relationship with a suitable sexual host that can provide sperm. In pseudogamous forms, the sperm or pollen nucleus does not fuse with the egg nucleus (i.e., syngamy is prevented), but fertilization may be necessary to initiate other developmental processes. Pseudogamous plants need pollination to ensure proper seed development because endosperm nucleii require fertilization. The function of sperm in pseudogamous animals is not understood, but embryogenesis will not proceed without it. Pseudogamy in animals is more commonly called gynogenesis.

Several unusual cases of gametic cloning involve syngamy, and thus cannot properly be called parthenogenetic or pseudogamous. Diploid all-female fishes of the genus *Poeciliopsis* have a hemiclonal form of reproduction called hybridogenesis that transmits only a haploid maternal set of chromosomes (*M*) to the eggs (Schultz 1969). The eggs are fertilized by sperm from a related sexual species (*L*), re-establishing the diploid (*ML*) genotype in each generation. Only the *M* genome is cloned; the *L* genome is substituted from the sexual gene pool in each generation. Hybridogenesis also occurs in the European water frog *Rana esculenta* (Uzzell and Berger 1975). A similar process involving two hemiclonal genomes occurs in some perennial plants of the genus *Oenothera* (Grant 1981). Permanent translocation heterozygosity results in the production of two kinds of nonrecombinant gametes (*A* or *B*), and balanced lethals permit the survival of only *AB* heterozygotes.

Genotypic Diversity in Asexual Populations

Extensive clonal diversity is a characteristic of many asexual plants and animals (Parker 1979; Bierzychudek 1985; Ellstrand and Roose 1987; Suomalainen et al. 1987). Most clonal diversity is clearly attributable to multiple (polyphyletic) origins of clones from sexual ancestors. For many plants and invertebrate animals, new clones can arise spontaneously within a species. Each new clone effectively "freezes" a genotype that once existed in a sexual population. However, clonal vertebrates all appear to be hybrids and multiple hybridization events appear to produce most of the observed clonal diversity. Whether formed spontaneously within species or by hybridization between species, clones can accumulate additional variation through mutation. Subsequent polyploidization can further expand the genomic diversity of asexual populations. Finally, rare recombination events in a predominantly asexual lineage can generate considerable genotypic diversity. In addition to the genotypic diversity among individuals, clones can maintain considerable allelic diversity within individuals in the form of permanent heterozygosity. The sources of genotypic diversity among individuals and the maintenance of heterozygosity within individuals are discussed.

Spontaneous Origins

In many **cyclical parthenogens**, clonal reproduction during most of the year is interrupted by a discrete sexual phase that typically precedes the production of overwintering eggs. Genotypic diversity in such populations is produced by recombination during the sexual generation. Clonal diversity is high in cyclical populations of cladocerans and aphids and genotypic frequencies often closely approximate Hardy–Weinberg expectations; but in some environments, they lose their sexual phase altogether and become obligately parthenogenetic (Hebert and Crease 1983; Wöhrmann and Tomiuk 1988). In obligate parthenogens, clonal diversity is typically reduced and genotypic frequencies exhibit significant disequilibrium. Obligate parthenogenesis in *Daphnia pulex* appears to result from the spread of a sex-limited meiosis suppressor (Hebert 1981). Genetic factors also appear to be responsible for elimination of the sexual phase in the aphid *Myzus persicae* (Blackman 1972).

Parthenogenetic offspring are sometimes produced spontaneously by species that are otherwise strictly sexual. Many clones apparently have arisen independently in the parthenogenetic moth *Solenobia triquetrella* (Lokki et al. 1975). Most electrophoretic genotypes occurring in the European Alps are marked by allozymes that occur in nearby sexual populations. Similarly, **frozen variation** explains most of the genetic diversity among clonal forms of the moth *Alsophila pometaria* (Harshman and Futuyma 1985). The opportunity for freezing genotypic variation during spontaneous origins of new clones can be tested directly. Annest and Templeton (1978) synthesized 233 new automictic strains of the fly *Drosophila mercatorum*. All combinations of genetics markers found in the sexual ancestors were also represented among these synthetic clones. The clones also expressed significant life history differences that affected fitness. Within ten generations, most of the clones had been eliminated, leaving only a few genotypes with high fitness under the particular conditions of laboratory culture.

Hybrid Origins

For vertebrates, and some insects, interspecific hybridization is necessary to disrupt normal gametogenesis, thereby creating conditions for the emergence of nonrecombinant reproductive processes (Schultz 1969; Vrijenhoek 1989; Moritz et al. 1989). Considerable clonal diversity has been frozen as a consequence of multiple hybrid events. Most electromorph clones of the parthenogenetic lizard *Cnemidophorus tesselatus* and the hybridogenetic fish *Poeciliopsis monacha-lucida* can be explained by multiple hybrid events (Parker and Selander 1976; Vrijenhoek 1984a). Similarly, the parthenogenetic grasshoper, *Warramaba virgo* is a hybrid of two unnamed sexual species, and most electromorph clones apparently resulted from multiple hybrid events (Honeycutt and Wilkinson 1989). Some obligately parthenogenetic cladoceran populations had polyphyletic hybrid origins involving *D. pulex* and *D. pulicaria* (Hebert et al. 1989).

Hybrid origins of clonal diversity were demonstrated in laboratory studies of *Poeciliopsis*. Schultz (1973) synthesized several new strains of the hybridogenetic fish, *P. monacha-lucida*, through crosses of *P. monacha* females X *P. lucida* males. Subsequently, 33 new strains of this fish were produced in the laboratory (Wetherington et al. 1987). Differences between the synthetic clones in life history traits such as size at birth, juvenile growth rate, and fecundity demonstrate the ability of multiple hybrid events to generate ecologically relevant clonal diversity (Wetherington et al. 1989).

Mutation and Muller's Ratchet

Mutation also contributes to genotypic diversity among clones. Every new mutation within a monophyletic asexual lineage defines a new clonal genotype. Assume, for example, that the number of gene loci (l) in a hypothetical parthenogen is 10^5 and the mutation rate per gene (u) is 10^{-5}. Then the genome-wide mutation rate (U) equals ul, or 1.0. This value is not unreasonable given a minimum estimate of $U = 0.93$ in *Drosophila melanogaster* (Mukai et al. 1972). Using $U = 1.0$, the probability of obtaining one or more new mutations in each egg is $1 - e^{-U}$, or 0.632. Mutations within a clonal lineage, having $U = 1.0$, will rapidly generate new genotypes, unless selection removes mutants as rapidly as they arise. Thus, it is possible that mutations are sufficient to drive phenotypic evolution of an asexual population. In only 2 years, a single parthenogenetic strain of *Daphnia pulex* was able to generate genetic variation that constituted 1 to 21% of the total phenotypic variance in life history traits (Lynch 1985). However, Lynch and Gabriel (1983) concluded that the clonal mutation rates need to be twice that of a sexual lineage in order for rates of phenotypic divergence to be the same. They go on to suggest that clones may serve as a significant reservoir for mutational variation. Occasional reversions to sexual reproduction can release a burst of variation that is not likely to be matched by strictly sexual populations.

Electrophoretic studies provide evidence for mutations in animal clones. Extensive **allozyme variation** among clones of the curculionid weevil, *Otiorrhynchus scaber*, is believed to have arisen by mutations in a monophyletic lineage (Saura et al. 1976a). Most clones differ slightly, and the differences can most often be explained by single mutations. The same appears to be true for two other curculionids, *Otiorrhynchus salicis* and *Polydrosus mollis*, and a chrysomelid beetle, *Adoxus obscurus* (Saura et al. 1976b; Lokki et al. 1976a,b). Although most of the clonal diversity in *Cnemidophorus tesselatus*, *Poeciliopsis*, and *Warramaba virgo* appears to be frozen from the sexual ancestors, "unique" allozymes occur in some clones (Parker and Selander 1976; Vrijenhoek 1984a; Honeycutt and Wilkinson 1989). It is likely that these allozymes are mutants, but it is also possible that these "unique" alleles exist in sexual populations that have not yet been sampled or existed in populations that are now extinct (cf. Turner et al. 1980).

Most new mutations are mildly deleterious even in the heterozygous state (Crow and Temin 1964; Simmons and Crow 1977). Muller (1964) proposed that sex is advantageous because recombination disengages a ratchet mechanism that would otherwise cause deleterious mutations to accumulate. Maynard Smith (1978a Chap. 3) explored conditions under which the ratchet mechanism could operate. The ratchet advances if the genome-wide mutation rate (U) is high, selection (s) against new mutations is weak, and the effective population size (N) is relatively small. The number of offspring expected to carry no new deleterious mutations is

$$N_o = Ne^{-U/s}.$$

If N_o is small, there is a good chance that the optimal class of offspring will fail to be represented in the next generation, and the ratchet will advance. Furthermore, positive selection for new mutations accelerates the ratchet because each favorable mutation is likely to have a few mildly deleterious mutations "hitchhiking" along with it (Felsenstein 1974).

Kondrashov (1982, 1984) argued that clones will deteriorate as a result of mutation even without the requirement for small population size. He assumes that each new mutation subtracts disproportionately from an organism's fitness. Only individuals with fewer than a critical threshold number of harmful mutations remain viable. If each gamete contains, on average, one or more new mutations, then the ability of a sexual lineage to purge its mutational load may be sufficient to outweigh the twofold cost of sex. Kondroshov's theory takes the ratchet mechanism out of the realm of long-term advantages, giving sex an immediate benefit due to recombinational purging. To judge these models, we need to know more about genomic mutation rates of plants and animals, and we need to learn whether selection against new mutations is multiplicative or not.

Organisms such as bacteria have immense population sizes, few genes, and low mutation rates. They should be able to avoid Muller's ratchet. For example, the bacterium *E. coli.* despite its capacity for parasexual recombination, exists as strictly asexual clones in nature (Selander and Levin 1980). On the other hand, hemiclones of *Poeciliopsis*, having larger genomes and much smaller population sizes, exhibit evidence for mutational accumulation (Leslie and Vrijenhoek 1978, 1980; Spinella and Vrijenhoek 1982). Similar evidence has been found in the hybridogenetic frog *Rana esculenta* (Graf and Polls 1989). Comparative studies using specific molecular probes might provide more direct evidence for an accumulation of mutations in clonal lineages.

Polyploidy

Most polyploid clones probably arose through duplication or addition of an entire haploid genome in an existing clonal lineage (Schultz 1969). Imagine, for example, an allodiploid form *AB*. Genome duplication could produce either *AAB* or *ABB* allotriploids. Such triploids would be heterozygous for heterospecific alleles carried by the *A* and *B* genomes, but they would be homozygous for all genes carried by the duplicated genomes. Alternatively, an allodiploid might add a third genome by backcrossing with one of its progenitors ($AB \times B'B'$), producing *ABB'*. These allotriploids would have an additional source of heterozygosity deriving from different homospecific alleles. In theory, these two modes of origin should be easy to distinguish by the presence or absence of heterozygosity in the homospecific genomes. Finally, an allodiploid might mate with a third species, giving rise to a trihybrid *ABC*. As we shall see below, trihybrid triploids exhibit very high levels of heterozygosity. Thus, a single diploid parthenogen can generate considerable clonal diversity through the duplication or addition of extra genomes.

Polyploidy can contribute significantly to phenotypic diversification among clones. If an allodiploid, *AB*, gave rise to two allotriploids, *AAB* and *ABB*, through genome duplication or addition, the three types might exhibit differences as a consequence of dosage effects. Genomic dosages clearly affect the dentition and behavior of allodiploid and allotriploid forms of *Poeciliopsis*. The numbers of large and small dentary teeth in this fish form a graded series clearly matching genomic dosages in the *AA*, *AAB*, *AB*, *ABB*, and *BB* forms (Schultz 1969). Cannibalistic behavior also follows this arrangement (Thibault 1974).

Lokki (1976) suggested that polyploidy would provide a buffer against the accumulation of deleterious mutations. All three alleles at a gene locus must be destroyed before a critical gene function is eliminated. This argument is correct if new mutations are completely recessive, but most new mutations appear to be slightly deleterious in the heterozygous condition (Simmons and Crow 1977). Polyploids may simply have more genes to mutate, inflating *U*. Nevertheless, addition of a new genome to an existing allodiploid lineage (giving *ABB′* or *ABC*) will provide temporary shelter against Muller's Ratchet, but duplication of an existing genome (giving *ABB*) would not because deleterious mutations would be duplicated along with the genome. We should determine whether most triploids are of the *ABB′* or *ABB* type. If Muller's Ratchet has a significant effect on clonal lineages, *ABB′* should be more common.

Recombination

Occasional recombination can produce considerable genotypic diversity within a clonal lineage. For example, just prior to meiosis in some triploid vertebrates, the entire set of chromosomes is elevated to the hexaploid level (*AABBCC*). Synapsis occurs between replicated bivalents (*A-A*, *B-B*, or *C-C*), and normal meiosis ensues, giving rise again to *ABC* progeny. However, occassional divalent mismatches during synapsis (e.g., *A-B, A-C, B-C*) or multivalent synapses (e.g., *B-B-C-C, A-A-B-B-C-C*, etc.) would allow recombination and generate clonal diversity (Asher and Nace 1971). Such recombination may be responsible for some clonal diversity in *Cnemidophorus tesselatus* and *Warramaba virgo* (Parker and Selander 1976; Honeycutt and Wilkinson 1989). Heterozygosity should be lost in such lineages at a rate that is proportional to the frequency of recombination (Asher and Nace 1971). However, this apparently is not the case because heterospecific heterozygosity is maintained in most asexual lineages. Either such recombinational events are extremely rare, or the more homozygous recombinants are strongly selected against. Interclonal selection may favor lineages that completely suppress recombination and thereby preserve heterozygosity. This is particularly evident for many parthenogenetic insects, where a variety of cytological mechanisms acts to maintain heterozygosity even in automictic forms (see Templeton 1982).

Heterozygosity

Clonal reproduction can preserve high levels of heterozygosity. Table 1 lists multilocus heterozygosity values for some insects and vertebrates that have closely related sexual and asexual forms. Because they are hybrids, clonally reproducing vertebrates have highly elevated levels of heterozygosity. The highest heterozygosity ($H = 0.714$) occurs inatrihybrid, triploid form of *Cnemidophorus tesselatus* (Parker and Selander 1976). Allotriploid clones preserve species diagnostic alleles carried by each of the three genomes. Elevated heterozygosity also occurs in most unisexual fishes. The parthenogenetic grasshopper, *Warramaba virgo*, is a hybrid and its heterozygosity is elevated relative to its sexual ancestors. Although many parthenogenetic insects are not interspecific hybrids, polyploid forms tend to have higher heterozygosity than their diploid counterparts (Table 1). Lokki (1976) suggested that this occurs because polyploids have more genes to mutate.

When parthenogens arise within a sexual species, different clones should freeze a range of heterozygosities comparable to that of the sexual population. Heterozygosity tends to conform with Hardy–Weinberg expectations for a few generations following sexual reproduction in cyclically parthenogenetic populations of *Daphnia*, but nonrandom genetic structures often develop in populations that reproduce asexually for several years. Heterozygotes exceed expectations in *Daphnia magna* populations inhabiting relatively permanent bodies of water (Young 1979), but heterozygosity is not elevated in obligately parthenogenetic *Daphnia pulex* (Hebert and Crease 1983). Similarly, rose aphids, *Macrosiphum rosae*, have heterozygosities that are roughly equivalent to Hardy-Weinberg expectations. Although near-equilibrium conditions might be expected for cyclical parthenogens, obligately parthenogenetic populations of this aphid also showed a typical range of heterozygosity values (Tomiuk and Wöhrmann 1987). With the exception of polyploid insects, there is no overwhelming evidence that heterozygosity exceeds random expectations in spontaneous parthenogens.

Hamrick (1983) reviewed the distribution of genetic variation in plants that employed various breeding systems. Species that reproduce both sexually and clonally tend to maintain higher levels of heterozygosity than species that are strictly sexual. Usberti and Jain (1978) found that heterozygosity levels were higher in sexual ($H = 0.381$) than in asexual ($H = 0.159$) races of the grass *Panicum maximum.* Compared to species having normal recombination, permanent translocation heterozygosity in the primrose *Oenothera biennis* also maintains significantly elevated levels of allozyme heterozygosity (Levy and Levin 1975).

Table 1. Heterozygosity in asexual lineages and their sexual relatives. Mean heterozygosity values were rounded to the nearest whole percentage

Asexual biotype (ploidy)	H%	Sexual relative	H%	Reference[a]
Insects				
Adoxus obscurus (3n)	34	*A. obscurus*	18	1
Polydrosus mollis (2n)	36	*P. mollis*	14	1
P. mollis (3n)	37			1
Otiorrhynchus scaber (3n)	25	*O. scaber*	31	1
O. scaber (4n)	38			1
O. salicis (3n)	24	*O. salicis*	12	1
O. singularis (3n)	37			1
Strophosomus melanogrammus (3n)	30			1
Solenobia triquetrella (2n)	23	*S. triquetrella*	23	1
S. triquetrella (3n)	23			1
Pycnoscelus surinamensis	11	*P. surinamensis*	11	2
Warramaba virgo (2n) Standard	36	*W. virgo* var. P196	4	3
W. virgo (2n) Boulder–Zanthus	39	*W. virgo* var. P196	6	3
Fish				
Poeciliopsis monacha-lucida (2n)	43	*P. monacha*	5	4
P. 2 monacha-lucida (3n)	51	*P. lucida*	2	4
P. monacha-2 lucida (3n)	54			4
P. monacha-occidentalis (2n)	43	*P. occidentalis*	2	4
P. monacha-latidens (2n)	39	*P. latidens*	1	4
P. monacha-lucida-viriosa (3n)	52			4
Poecilia formosa (2n)	32	*P. mexicana*	3	5
Menidia clarkhubsii (2n)	18	*M. beryllina*	5	6
		M. pennisulae	6	6
Lizards				
Cnemidophorus tesselatus (2n)	56	*C. tigris*	5	7
C. tesselatus (3n)	71	*C. septemvittatus*	6	7
		C. sexlineatus	7	7
C. laredoensis (2n)	27			7
C. lemniscatus (2n)	23	*C. lemniscatus*	4	8
C. velox (3n)	37			8
C. uniparens (3n)	34			8
C. neomexicanus (2n)	39			8
C. sonorae (3n)	34			8
C. exsanguis (3n)	34			8
Lepidophyma flavimaculatus (2n)	6	*L. flavimaculatus*	4	9
Heteronotia binoei (3n)	32	*H. binoei*	7	10

[a]References: 1. Lokki (1976); 2. Parker et al. 1977; 3. Honeycutt and Wilkinson (1989); 4. Vrijenhoek (1979); 5. Turner et al. (1980); 6. Echelle et al. 1983; 7. Parker and Selander (1976); 8. Dessauer and Cole (1989); 9. Bezy, pers. comm.; 10. Moritz et al. 1989.

Clonal reproduction is often cited as a means for preserving heterotic genotypes (Stebbins 1950; Dobzhansky 1955; White 1978). For example, the strong coupling between interspecific hybridization and clonal reproduction is often invoked as evidence for **heterosis** and hence the raison d'être for unisexual vertebrates (Schultz 1971; Cole 1975). Many interspecific hybrids exhibit enhanced growth and survival which should properly be called hybrid vigor, but their luxuriance is typically associated with partial or complete sterility. It is necessary therefore to distinguish hybrid vigor from true heterosis in characters affecting fitness (Dobzhansky 1950).

Hybrid vigor has been reported in some unisexual vertebrates. Natural hemiclones of *P. monacha-lucida* are more tolerant of acute heat and cold stresses than either *P. monacha* or *P. lucida*, but broad thermal tolerance was not a general characteristic of other allodiploid and allotriploid forms of *Poeciliopsis* (Bulger and Schultz 1979, 1982). The hybridogenetic frog *Rana esculenta* exhibits broader options for overwintering than either of its sexual progenitors, *R. ridibunda* and *R. lessonae* (Tunner and Nopp 1979), but its reproductive potential may be lower on average (Graf and Polls 1988). Studies of hybrid vigor and heterosis in natural clones are potentially misleading for two reasons. First, tests of performance in the laboratory might not be indicative of fitness under field conditions (Moore 1984). Second, we observe only the successful hybrid combinations; low-fitness clones have disappeared.

Wetherington et al. (1987) designed an experiment to test whether newly synthesized hybridogenetic strains of *Poeciliopsis monacha-lucida* were spontaneously heterotic. Thirty-three new hybridogenetic strains were produced. Within three generations, half of the strains were lost, presumably because of low fitness in the laboratory. On an average, the 15 remaining strains exhibited lower fertility and survival than either of the parental strains. The hybrids also suffered from an elevated frequency of birth defects. Nevertheless, good genotypic combinations have a reasonable likelihood of arising in multiple hybrid events. Two natural hybridogenetic strains had significantly higher survival than the 15 synthetic strains. Survival of the natural hybridogens also exceeded that of *P. monacha* and *P. lucida*, but it would be erroneous to attribute their elevated survival to hybridity and heterosis. All the hybridogenetic strains, synthetic and natural, had identical levels of fixed heterozygosity due to different *monacha* and *lucida* alleles. The two natural strains simply have *manacha* genomes that form particularly good combinations with the *lucida* genome used in this experiment.

Heterosis is a valid within-species phenomenon: "heterosis is simply inbreeding depression in reverse" (Falconer 1981, p 231). Thus, clones forming within a species can freeze an array of genotypes ranging from inbred to heterotic. If heterosis contributes to the fitness of clonal organisms, mean heterozygosity should be higher in the asexual populations than in the sexual relatives from which they derived. Careful comparisons of the heterozygosity distributions of sexual versus asexual populations of the same species are warranted. As discussed above, the evidence is ambiguous on this point. Mutations that have accumulated subsequent to clonal origins and polyploidy cloud the issue (Lokki 1976).

Phenotypic Variation

Asexual taxa can exhibit considerable phenotypic variation, sometimes rivaling that of their sexual ancestors; but only a few studies attempted to assess the relative contribution of genotypic and environmental factors (see Chaps. 1, 8, and 9). In the tick, *Haemaphysalis longicornis*, a sample of field-collected parthenogens was more variable than a comparable sample of sexual individuals (Oliver and Herrin 1976). Unfortunately, it was not known whether the parthenogens were composed of one clone or multiple clones. Thus it is impossible to determine whether the high variability of the parthenogens was due to genotypic variance among clones or to extreme phenotypic plasticity within a single clone. However, laboratory-reared sexuals were more variable than a laboratory-reared clone, as expected. A field study of *Warramaba virgo* also revealed that parthenogenetic populations were at least as variable as sexual populations (Atchley 1977a, 1977b, 1978). Again, the actual composition of clones was not known. A recent electrophoretic study of *W. virgo* revealed that most populations contained multiple clones (Honeycutt and Wilkinson 1989). The partitioning of phenotypic variation between genotypic and environmental components needs to be investigated in parthenogenetic populations of the ticks and grasshoppers.

In studies that took clonal diversity into account, sexual populations were found to be more variable than individual clones. Two parthenogenetic populations of *Cnemidophorus*, the first composed of one clone and the second composed of multiple clones, were compared with a sexual population. Phenotypic variances for most meristic characters (i.e., counts of scales) could be ranked monoclonal < multiclonal < sexual, as expected; but variances of growth-related mensural traits were similar in the populations. Environmental factors affecting growth probably obscured genotypic factors affecting these traits. Individual gynogenetic clones and hybridogenetic hemiclones of *Poeciliopsis* also exhibit significantly less variance in meristic traits when compared with their sexual relatives (Vrijenhoek 1984a). Even though the (*ML*) hybridogens express genetic variation due to the substitutable (*L*) genomes, they were less variable than sexual individuals taken in the same field collections. Gynogenetic clones were, as expected, the least variable.

Hemiclonal strains of *Poeciliopsis monacha-lucida* were used to partition phenotypic variation in life history traits into genetic versus environmental effects (Wetherington et al. 1989). The substitutable *L* genomes of 14 laboratory-synthesized *ML* stains were standardized in crosses to eliminate genetic variation due to the paternal genome. All genetic variance among strains was attributable to the hemiclonal *M* genomes. For eight *ML* strains whose sexual *P. monacha* ancestors derived from the Rio Fuerte, genetic variance accounted for 30% of the variance in weight at birth, and 10% of the variance in juvenile growth rate. Very few fish produced mature eggs during the 9-week-growth study. Fish in a second experiment, involving six *ML* strains whose *P. monacha* ancestors derived from the Rio Mayo, matured earlier. In this case, genetic variance accounted for 40% of the variance in weight at birth, 13% of the variance in juvenile growth rate, and 40% of the variance in egg production by young females. Substantial genotypic and hence phenotypic variation can be frozen in the differences between clonal genomes derived from genetically variable sexual ancestors.

The Ecology of Clones

The overwhelming predominance of sexual species of plants and animals tells us that asexual lineages have not replaced their sexual ancestors on a broad phylogenetic scale, but the same cannot be said for spatial distributions on contemporaneous time scales. The existence of distinct geographical races of closely related sexual and asexual forms of insects is widely known as "**geographical parthenogenesis**" (Vandel 1928). Both noncompetitive factors and competitive exclusion have been invoked to explain this pattern of geographical separation. At the same time, both noncompetitive and competitive factors have been invoked to explain local coexistence between closely related sexual and asexual lineages. Two factors that affect opportunities for coexistence versus displacement between sexual and asexual lineages must be addressed. First, I discuss the constraint that pseudogamy places on ecological interactions. Second, I explore the role that genotypic diversity can play in competitionist versus noncompetitionist models for coexistence.

Sperm Dependence

Many clonal plants and animals have not been able to divorce themselves from the mechanical or genetic requirements of fertilization. If fecundity, survival, and niche requirements are equal in sexual and sperm-dependent asexual lineages, the twofold advantage of all-female reproduction should cause clones to replace their sexual host, thereby eliminating the males on which they depend (Clanton 1934). Notwithstanding, stable complexes of sexual and pseudogamous lineages are well known among vertebrates and insects (Moore 1976; Kiester et al. 1981; Stenseth et al. 1985).

Based on studies of *Poeciliopsis*, Moore (1976) identified three factors that control the dynamics of such a host-parasite system: (1) "primary fitness" (fecundity and survival); (2) the twofold advantage of all-female reproduction; and (3) frequency-dependent mating preference on the part of sexual males. Assuming equal primary fitnesses and identical niches, bisexual and unisexual lineages will achieve a dynamic equilibrium if the twofold reproductive advantage of unisexual females is balanced by mating discrimination against the unisexuals by males (Moore and McKay 1971). Mating preferences were frequency-dependent (McKay 1971). When the sexual species is rare, males are solitary and they prefer conspecific sexual females as mates. When males are abundant, they establish dominant hierarchies. Subordinate males, being denied access to the preferred sexual females, mate with unisexuals. By examining mating success in natural populations, Moore (1976) empirically derived a mating-succes function (Fig. 1a). Assuming equal primary fitnesses, a stable equilibrium would occur with about 85% unisexual females (Fig. 1b). This prediction clearly fits the frequency of unisexual fishes in some, but not most, rivers. Moore concluded that if the twofold reproductive advantage and the HI_Q

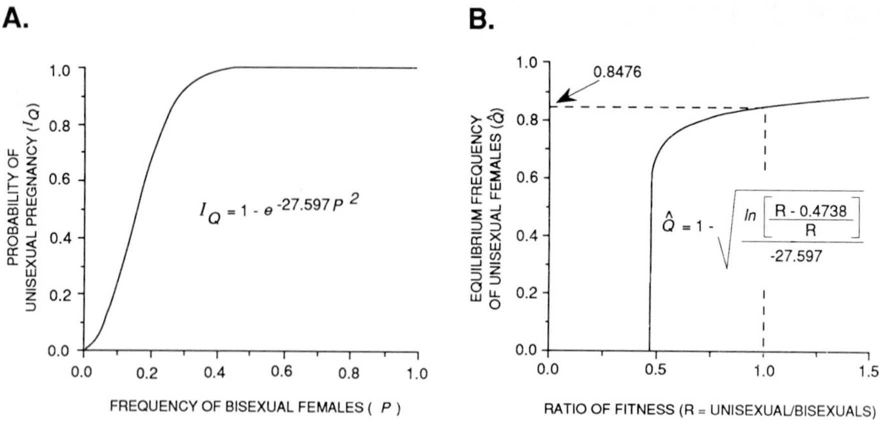

Fig. 1. Population dynamics in a mixed reproductive complex including a sperm-dependent unisexual form. **A** The probability of a pseudogamous unisexual female being inseminated (I_Q) as a function of the frequency (P) of females that are bisexual (i.e., producing male and female offspring). The equation that generated the curve portrayed here was empirically derived from the least-squares estimate of I_Q from natural complexes containing *P. monacha-occidentalis* and *P. occidentalis*. **B** The equilibrium frequency ($\hat{Q}$) of unisexual females as a function of R the ratio of primary fitness of the unisexuals: sexual females. (Both graphs after Moore 1976)

function were constants, then deviations from $\hat{Q} = 85\%$ must be due to reduced primary fitness in the unisexuals. It appears then that sperm dependence places an upper limit on the ecological success of unisexual *Poeciliopsis*. As long as the frequency-dependent mating function obtains, increasing the primary fitness of unisexuals contributes little to increasing their proportion in the population. Decreasing the primary fitness of the unisexuals to one-half that of the sexuals also has little effect. However, a decrease in the relative fitness of unisexuals below one-half of the sexual value results in elimination of the unisexuals.

Other attempts to model sperm-dependent unisexuals have recently appeared (Kiester et al. 1981; Stenseth et al. 1985). Some form of frequency-dependent fitness appears to be a general requirement for the maintenance of a sperm-dependent population. Because sperm-dependent clones are forced to coexist with a closely related sexual host, Lynch (1984) suggested that "it is clearly in a gynogen's best interest to influence the density of its host as little as possible and in a way that does not insure its own demise through competition for critical resources". Thus, he dismissed the relevance of pseudogamous organisms to controversies about coexistence of sexual and strictly parthenogenetic forms. This would be correct if all pseudogamous populations could eliminate their sexual hosts except for the limit set by sperm dependence. In most cases sperm-dependent unisexuals do not occupy the upper threshold described by Moore's $\hat{Q}$ (Fig. 1.). Other factors such as differential fitness or niche requirements often limit their success (Moore 1976; Vrijenhoek 1978). Studies of these limitations are directly relevant to the "everything else is equal" assumption.

Colonization Abilities and Geographical Parthenogenesis

Parthenogenetic reproduction provides several advantages for a colonizing species. First, only one female is needed to colonize a new habitat, whereas sexual colonists require at least a single inseminated female. Second, parthenogens have an advantage at low densities because they do not have to encounter a mate in order to reproduce. Finally, as pioneers in unsaturated habitats, parthenogens would benefit from the twofold reproductive advantage of all-female reproduction. Such factors have often been invoked to explain the elevated frequency of parthenogenetic species of insects in recently colonized areas such as northern latitudes that were recently glaciated (Suomalainen 1947; Seiler 1961; Darevsky 1966). Parthenogens also tend to occur in disturbed environments or disclimax communities (Cuellar 1977). A general characteristic of disturbed environments is the high frequency of local extinctions. In such cases, the high intrinsic rate of increase attributable to all-female reproduction may favor parthenogens in recolonizing such habitats.

Colonizing parthenogens may benefit from both heterozygosity within individuals and genotypic differences among individuals. First, sexual colonists are often subject to founder events and possible inbreeding depression. Clonal reproduction may provide "heterozygosity insurance" by protecting against inbreeding depression (Vrijenhoek 1985). I am not suggesting that cloning produces genotypes that are inherently better than outcrossed sexual individuals. That is possible, but under colonizing conditions, cloning might simply prevent loss in fitness due to increasing homozygosity for deleterious recessive alleles. The average fitness of a sexual population might decline after repeated founder events or genetic bottlenecks. Heterozygosity insurance may play a significant role in the dynamic interactions between sexual and asexual forms of *Poeciliopsis* (Vrijenhoek 1985; 1989).

Second, colonizing species often exhibit an expanded niche because of reduced interspecific competition, a phenomenon called **"ecological release"** (Pianka 1983, p. 208). Under conditions of low interspecific competition and high intraspecific competition, disruptive selection should favor extreme phenotypes that can exploit resources that would otherwise be unused. Although recombination can generate a broad diversity of genotypes in each generation, including extreme genotypes, "antagonistic recombination" (Rosenzweig 1978) will retard the rate at which a sexual population can expand or modify its niche. However, antagonistic recombination is eliminated in clones. If a diversified assemblage of clones can be drawn from a sexual gene pool, the asexual population can expand its niche rapidly through the proliferation of specialized clones (Roughgarden 1972). Baker (1965) made a similar argument for the ecological advantages of self-fertilization in colonizing plants. This idea forms the foundation for the **Frozen Niche-Variation model** (Vrijenhoek 1979, 1984b, and below).

Multiclonal niche expansion can arise in two ways: (1) Sexual founders may colonize new environments, after which new clones arise endemically. The clones would then displace their sexual ancestors from portions of the available niche, or even cause extinction of the sexuals. (2) The sexual ancestors may continuously generate new clones within their home range, and then the clones disperse and establish new colonies. Both modes have been responsible for producing unisexual *Poeciliopsis* populations. However, a significant distinction exists between the two modes of origin. Endemic populations of the unisexual fish are invariably multiclonal; whereas migrant populations are for the most part monoclonal (Vrijenhoek 1979). Founder events during colonization probably result in a considerable loss of clonal diversity.

Weedy Species

It is not necessary to invoke competitive displacement to explain geographical parthenogenesis. Just the intrinsic colonization abilities of parthenogens may be sufficient to explain their abundance in new or disturbed environments. Nevertheless, it is difficult to ignore the possibility of competition with sexual populations from which the clones arose. Since clonal genotypes are drawn from a sexual gene pool, they have significant genetic similarity and presumably ecological overlap with their ancestors (hybrids may pose a special case, see below). Competition between a clone and its sexual relatives should be highly asymmetrical. A single clonal genotype should have a small impact on the total resources available to a genetically variable sexual species. On the other hand, the resources used by sexual relatives can have a significant impact on a single clone. Asymmetrical competition should favor clones that specialize on resources at the limits of the resource use function, where competitive overlap with sexual relatives will be less severe.

The Weed hypothesis assumes parthenogens are fugitive species that are only successful in habitats where competition from the sexual relatives is weak (Moore 1984). Asexual forms are thought to be inferior competitors within the niche of the sexual ancestors. Stebbins (1958) claimed that apomictic plants are most frequently found in habitats that are unstable or extreme. However, a recent review of the plant literature found no consistent pattern in this regard (Bierzychudek 1985). Students of parthenogenetic lizards have adopted this idea; unisexual *Cnemidophorus* are thought to be the zoological equivalent of weeds, predominating in habitats described as extreme, marginal, or perpetually disturbed (Wright and Lowe 1968; Cuellar 1977). Although this hypothesis may have considerable explanatory power for the distribution of some parthenogenetic lizards, it has not been widely applied to other parthenogenetic vertebrates (cf. Uzzell and Darevsky 1975; Moore 1984). Sperm dependence limits the ability of unisexual fish and amphibia to play the role of fugitive species. Such clones cannot establish themselves in new habitats unless a suitable sexual host is available. Furthermore, ecological factors that limit the fitness of the sexual parents in marginal, ecotonal, or disclimax habitats should also reduce the density of sexual males, thereby decreasing the likelihood of unisexual inseminations.

Lynch (1984) offered an additional hypothesis to explain the peripheral or disjunct distributions of many parthenogens. In many cases, parthenogens are not protected against matings with the sexual ancestors. Such matings can disrupt the clonal mechanism leading to higher levels of ploidy and sterility. He called this "destabilizing hybridization", and suggested that such unprotected parthenogens could only persist in regions where hybridization was unlikely. This idea deserves further investigation.

Hybridization and Ecological Intermediacy

Since unisexual vertebrates are hybrids, they might be best adapted to habitats that are ecologically intermediate to those of their sexual progenitors (Moore 1984). An ecotone might provide suboptimal habitat for both parental species, leaving the unisexual hybrids a weakly contested intermediate zone. The Ecological-Intermediacy hypothesis has intuitive appeal because hybrids are often intermediate in morphology, physiology and behavior, and thus should be phenotypically well-suited for an intermediate environment. This hypothesis fails to explain the stable coexistence of multiple clones, unless the intermediate niche is itself broad and capable of being partitioned by multiple clones. Evidence for and against this hypothesis has been discussed elsewhere (Moore 1984). The local distribution of individual clones of *Poeciliopsis* across an upstream-downstream transition zone clearly does not conform to the predictions of this hypothesis (Schenck and Vrijenhoek 1986, 1988). However, the Ecological-Intermediacy hypothesis may apply on larger geographical scales that tax the physiological limits of these fish (Bulger and Schultz 1982).

General-Purpose Genotypes

The term, "General-Purpose Genotype", was coined by Baker (1965), who presented evidence for broad ecological tolerance and phenotypic plasticity in weedy plants, many of which are apomicts. Lynch (1984) suggested that:

> "As the environment changes in time, different clones will be favored in different generations, but in the long run the most successful clones will be those with the lowest temporal variance in fitness.... Relatively specialized clones will surely arise (perhaps frequently), but they will only survive as long as the narrow niche to which they are adapted remains available. Therefore, in the long term, clonal selection will promote the evolution of highly generalized (or general-purpose) genotypes".

The primary evidence in support of the General-Purpose Genotype hypothesis is the widespread distribution of some asexual organisms. Parker et al. (1977) found that a broadly dispersed parthenogenetic cockroach consists of a single electrophoretic genotype, but most studies have not determined whether the asexual population is composed of one broadly tolerant clone or an array of clones, each of which is more restricted in its distribution. Most broadly dispersed asexual plants are multiclonal (Bierzychudek 1985). For example, Lyman and Ellstrand (1984) report that, on average, North American populations of the dandelion, *Taraxacum officinale*, consist of five clonal genotypes that tend to be highly restricted in distribution. Nevertheless, one clone occurred in 19 of the 22 populations sampled, ranging from Vermont to Alaska in the north and to California in the south. This broadly distributed genotype might be an excellent candidate for having a General-Purpose Genotype. But can a broadly distributed General-Purpose Genotype coexist with a locally restricted assemblage of clones? Temporal fluctuations should favor the genotype with the least variance in fitness (i.e., the clone with the highest geometrical mean fitness). How then is clonal diversity maintained? Clonal diversity might be transient, with each population containing an old, well-established clone and some newer clones that will disappear as the environment changes (see Sebens and Thorne 1985). This is unlikely in North America where dandelions are not native; there are no sexual ancestors to generate new clones. Solbrig and Simpson (1974, 1977) reported that co-occurring North American dandelions differentially exploit recently disturbed habitats. An ***r*-selected clone** does best under low density and a **K-selected clone** is better at competing for limited space. More extensive clonal diversity occurs in Europe where the sexual ancestors still occur (Richards 1973). Further genetic and ecological studies of these dandelions would be desirable.

Frozen-Niche Variation and the Tangled Bank

Spatial heterogeneity establishes a different selective regime that is more likely to maintain genotypic polymorphism (Hedrick et al. 1976) (see also Chap. 4). Multiple niches in a spatially heterogeneous environment can maintain multiple genotypes as long as those niches are available (Strobeck 1974). The maintenance of polymorphism in a multiniche environment requires soft selection. In an ideal multiclonal population, the genotypes occupying each niche would be regulated by **density-dependent factors.** Thus, fitness becomes frequency-dependent because rare genotypes suffer the least competition in their respective niches. Sib competition models argue that sex provides an advantage under such circumstances because each recombinant genotype is rare; it is, in fact, unique. The Tangled-Bank (Bell 1982) and Frozen-Niche-Variation models (Vrijenhoek 1979, 1984b) are variations on this theme. Although similar (cf. Maynard Smith 1986), the two models have different scopes. Bell sought to explain the persistence of sex, whereas I sought to explain coexistence among multiple clones. As we shall see, both models lead to identical predictions regarding the effects of clonal diversity on coexistence with sexual relatives.

The Frozen Niche-Variation model was stimulated by field studies of unisexual *Poeciliopsis* (Vrijenhoek 1978, 1979). According to this model, clones arising from a sexual population will freeze and efficiently replicate genotypes that differ in their abilities to exploit various food and spatial resources. Multiple origins of new clones provide a base of variation upon which interclonal selection can act. The differential success of clones in acquiring resources will produce a structured assemblage of genotypes, each having high efficiency within a portion of the ancestral niche and minimal overlap with other genotypes. Recent tests of his hypothesis have found that spatial segregation on micro- and macrohabitat scales is the primary factor leading to resource partitioning among the sexual and clonal forms of *Poeciliopsis* (Schenck and Vrijenhoek 1986, 1988). Laboratory syntheses of new clones of the fish clearly demonstrate that ecologically relevant variation can be frozen from the gene pools of their sexual ancestors (Wetherington et al. 1987, 1989).

Evidence for **ecological diversification of clones** is observed in a variety of plants and animals. Different host plants influence the local distribution of clones of the geometrid moth, *Alsophila pometaria* (Mitter et al. 1979; Futuyma et al. 1981). Clone 2(B) is associated with oak stands. Compared with the sexual population, individual *Alsophila* clones appear to be more restricted in their use of host plant species and in emergence times. Different clones of the parthenogenetic fly, *Lonchoptera dubia*, appear to be favored at different latitudes and during different seasons (Ochman et al. 1980). Considerable seasonal variation in clonal frequencies occurs in rotifers and *Daphnia* (King 1972; Hebert 1974). Spatial segregation occurs at a fine environmental scale among clones of the isopod, *Trichoniscus pusilus*, and among clones of the oligochaete, *Lumbricillus lineatus* (Christensen 1979, 1980). However, environmental variables were not correlated with spatial segregation among clones of another oligochaete, *Octalasion tyrteaum* (Jaenike et al. 1980).

As long as the combined niche breadth of specialized clones does not completely eclipse that of the more generalized sexual species, coexistence of sexual and asexual forms is possible (Case and Taper 1986). Based on his Tangled Bank model, Bell (1982, p. 131) made two predictions. First, no single clone could displace the sexual lineage from much of the ancestral niche because of the clone's ecological narrowness. Second, ecological diversity among clones in a multiclonal assemblage could eclipse much of the ancestral niche, thereby displacing (or even eliminating) the sexual ancestor. I had tested these predictions in an earlier study with *Poeciliopsis* (Vrijenhoek 1979). The ecological success of the unisexual fishes was equated with their proportion of the total *Poeciliopsis* population. Unisexual success was strongly associated with clonal diversity (Table 2). In multiclonal populations, unisexual females consistently outnumbered the sexual females, sometimes comprising 85% or more of the fish (the upper threshold according to Moore's $\hat{Q}$ function; Fig. 1b). However, in monoclonal populations, unisexual fish are far less frequent, being restricted to peripheral habitats. The multiclonal assemblages predominated in all habitats, including areas of current and peripheral pools (Vrijenhoek 1984b). For *Poeciliopsis* at least, we have substantive evidence that clonal diversity contributes to the dynamic ecological interaction between sexual and unisexual portions of the fish population. Multiple clones can usurp a broad range of food and spatial resources and thereby reduce the proportion of sexual individuals in

Table 2. Ecological success of hemiclonal *Poeciliopsis*. The percentage of unisexual females is based on the total number of *Poeciliopsis* females at each locality. River systems are divided into those containing a single clone (monoclona!) versus those that contain more than one clone per locality. (multiclonal). Rivers 1-5 contain *P. monacha-occidentalis* and its sexual host *P. occidentalis*. Rivers 6-8 contain *P. monacha-lucida* and its host *P. lucida*. For a map of the rivers and a description of the original data see Vrijenhoek (1979)

River system	Number of localities sampled	Percentage of unisexuals	Range (Std. Dev.)
Monoclonal rivers			
1 Río de la Concepción	4	1.5	0–3
2 Río Sonora	10	9.6	0–30
3 Río Matape	3	6.0	0–11
4 Río Yaqui	16	7.5	0–38
Mean:	33	7.2	(9.3)
Multiclonal rivers			
5 Río Mayo	7	82.1	65–97
6 Río Fuerte	4	63.8	34–91
7 Río Sinalóa	2	45.2	40–50
8 Río Mocorito	2	36.9	32–42
Mean:	15	66.3	(21.6)

the total fish population. It is tempting to speculate that if these unisexual fish were freed from sperm dependence, an ecologically diverse multiclonal assemblage might replace sexual relatives altogether. Nevertheless, it is clear that single clones, despite their potential twofold advantage, do not have a substantial impact on the sexual population.

The Red Queen Hypothesis

Van Valen (1973) suggested that to avoid extinction in a world composed of predators, competitors, and parasites, a species must evolve as fast as it can just to stay in the same place. Thus sex, and the genotypic diversity it provides, might be essential for remaining in a coevolutionary race in which all the participants are changing (Levin 1975; Jaenike 1978; Hamilton et al. 1981). If this hypothesis is true, one would expect that sexual species would clearly predominate in regions of high biotic diversity. Asexuality, on the other hand, should be more prevalent in regions of low biotic diversity (Glesener and Tilman 1978). Asexual "species" generally are more frequent in habitats in which biotic interactions are presumed to be diminished: at higher latitudes, at higher altitude, and in disturbed communities. However, this result (i.e., geographical parthenogenesis) is also consistent with other hypotheses regarding the benefit of asexuality (e.g., colonization abilities).

Although difficult to test using the comparative approach (see Bell 1982), one must give serious consideration to the idea that **parasites** and **pathogens** can generate constant selective pressure favoring individuals simply because they are different (see also Chap. 11). For example, the periodic origin of new virus strains that evade our immune responses and the rapid evolution of antibiotic resistance by many bacteria attest to the evolutionary potential of pathogenic organisms. Long-lived hosts contend with parasites and pathogens by using highly plastic immune responses. Yet it is unlikely that a single individual could be genetically capable of resisting all the different kinds of parasites, nor is any one parasite strain likely to be capable of infecting all the potential hosts. Selection will favor parasites that can exploit the most common host phenotype. Thus, rare host phenotypes will have a temporary advantage until they too become common. By means of this frequency-dependent process, parasites (and predators) may be responsible for the maintenance of considerable genotypic diversity in their hosts (Clarke 1979). It is easy to imagine a clone that is highly resistant to a particular parasite, but it is harder to imagine a "General-Purpose" clone that is resistant to all kinds of parasites. Genotypic diversity among individuals should slow the spread of parasites between parents and offspring, through a brood, or for that matter, through a whole population. The problem of contagious spread of disease would be exacerbated in organisms that brood their offspring or are viviparous (Rice 1983). Frequency-dependent selective pressures due to diseases could be a major factor maintaining sexuality in higher organisms, as well as maintaining clonal diversity in asexual populations. If diseases are responsible for most juvenile mortality as some have suggested, then the benefits of genetic diversity might balance or even outweigh the twofold cost of sex. Bremermann (1985) provides a stimulating review of this topic and makes a plea for experiments that examine disease and the Red Queen hypothesis. Such experiments could examine the survival of populations composed of single clones, mixtures of clones, and outbred sexual populations.

Summary and Conclusions

Broad-scale extinction of biparental, sexually-reproducing species has not occurred in plant and animal taxa where all-female cloning has arisen. Therefore, we must question the assumption that "everything else is equal" between sexual and asexual lineages. The ability of clones to reproduce themselves over time and space must be severely restricted such that the potential twofold reproductive advantage is not realized. A major constraint on the evolutionary success of clones is the ecological narrowness associated with having a single genotype (Ghiselin 1974; Vrijenboek 1979; Bell 1982). Sexual individuals spread among their progeny the risk of reproduction in spatially and temporally variable environments. But can the benefits of genotypic diversity really compensate for the costs of sex? Careful examinations of fitness components in sexually versus asexually produced broods of the same species are needed at this point. For example, experiments by Antonovics and coworkers with the pasture grass

Anthoxanthum odoratum provide a model for such studies. Sexual reproduction produces unique genotypes with a frequency-dependent advantage (Antonovics and Ellstrand 1984). On average, sexual progeny appear to have reproductive rates about 1.43 times that of their asexual siblings (Kelley et al. 1988). Minority advantage in the sexual sibships appears to be due to frequency-dependent attack by parasites and predators, particularly aphids, supporting the Red Queen model (Schmitt and Antonovics 1986). No support has been found for density-dependent models involving sib competition (Ellstrand and Antonovics 1985). The experimental conditions used in these studies probably exposed sexual and clonal individuals to only a small portion of the physical and biotic variables that might affect natural populations. It is likely that the advantage of genotypically variable sexual sibships would be even greater under natural conditions. More experimental studies of this sort are needed. They are more likely to discriminate among competing hypotheses for the significance of sex than the comparative approach that has dominated past studies (cf. Bell 1982).

Studies of clonal organisms also have provided considerable insight into the **value of genotypic diversity** for natural populations. Perhaps the most significant contribution of these studies is the discovery of extensive clonal diversity in most asexual populations. In many cases, coexisting clones exhibit significant differences in their use of food and spatial resources. Most of the differences among clones appear to result from genetic variance frozen at the time of clonal origins, attesting to the variability stored in the gene pool of their sexual ancestors. Over the long term, most asexual assemblages are probably maintained by recruitment of new clonal genotypes from extant sexual ancestors (Vrijenhoek 1979). However, clonal turn-over is not just neutral flux (cf. Jaenike et al. 1982), because considerable genetic structure exists in most asexual populations. Allelic frequencies in highly diverse multiclonal assemblages of fishes and in obligately parthenogenetic clodocerans and aphids are often far removed from frequencies in the ancestral sexual population (Angus 1980; Hebert and Crease 1983; Wöhrmann and Tomiuk 1987). Furthermore, the frequencies of particular clones can show relatively long-term stability in temporally and spatially varying environments (Schenck and Vrijenhoek 1986). Significant ecological diversification exists among coexisting clones of asexual plants and animals. The ecological success of hemiclonal populations of *Poeciliopsis* appears to be a function of genotypic diversity (Vrijenhoek 1979). A diverse assemblage of hemiclones can displace their relatives from a significant portion of the available niche, but a single clone cannot. All-female reproduction alone is not sufficient to explain the success of these unisexual populations. The recombinational diversity produced by sex and frozen among these hemiclones contributes to the breadth of resources that multiclonal assemblage can exploit. At least for these fishes, sex appears to be necessary for the origin and diversification of ecologically diverse unisexual assemblages. I suspect that the ecological success and evolutionary persistence of many other asexual "species" similarly depends on the opportunity for freezing and exploiting sexually produced genotypic variation (see also Moritz et al. 1989).

A number of critical questions regarding asexual organisms need to be addressed. How often do clones go extinct? Do new clones arise so rarely that their "standing crop" never poses an ecological threat to the sexual ancestors? For cyclical parthenogens, a new crop of clones arises every year. Obligately parthenogenetic lineages arise at some localities, but we do not know how often this occurs (see Hebert et al. 1989; Wöhrmann and Tomiuk 1988). The origin of successful insect and vertebrate clones may be very rare. Severe meiotic, genetic, developmental, and ecological constraints serve as a barrier to the establishment of most new clones (Templeton 1982; Wetherington et al. 1987; Vrijenhoek 1989; Moritz et al. 1989). Nevertheless, when they arise, asexual lineages sometimes have remarkable ecological success. Special circumstances such as colonization events or environments offering a release from competitors, predators, and parasites might favor clones over sexual lineages (Glesener and Tilman 1978; Bell 1982; Moore 1984; Vrijenhoek et al. 1987).

Most sexually reproducing species exhibit direct evidence of their evolutionary antiquity. Based on allozyme variation, the genetic distance between pairs of congeneric sexual species can be used to estimate the time since **evolutionary divergence** began. Despite the uncertainty associated with estimation of evolutionary ages from Nei's D values, it is nevertheless safe to conclude that most recognized sexual species of animals are at least tens of thousands to millions of years old (Nei 1987, pp. 246–247). In contrast, allozyme surveys of unisexual vertebrates have found that most clones are about as different from one another as genomes drawn at random from contemporaneous sexual populations, although some carry allozyme variants that appear to be mutants. Analyses of mitochondrial DNA in unisexual fishes and lizards also reveal no evidence for substantive postformational divergence of clones (Brown and Wright 1979; Moritz et al. 1989; Goddard et al. 1989). Most asexual organisms have not existed long enough to accumulate genotypic differences that are even remotely comparable to the levels of divergence observed between congeneric sexual species.

Obviously, we must question the "everything else is equal" assumption of simple genetic theories about sexuality. The fact that a clone consists of only one genotype and a sexual lineage consists of a nearly limitless variety should be sufficient to refute this assumption. This assumption is comparable to believing that any one human being (i.e., one genotype) has a breadth of talents and skills equivalent to the species as a whole. Even such a prodigious talent as Leonardo da Vinci might question that assessment. The benefits of genotypic diversity among individuals must be considered in a realistic ecological context. Further attempts to model sexual versus asexual systems must consider multiple genotypes with fitnesses that vary in spatially, temporally, and biotically heterogeneous environments. That is hardly a small challenge.

Acknowledgements. I thank D. Baldwin, B. Charlesworth, N. Ellstrand, J. Quattro, and S. Weeks for helpful comments and criticisms of an earlier draft of this manuscript.

6 Variation and Selection in Plant Populations

S.K. JAIN

Introduction

It seems customary to include a chapter on plants in many population biology and evolutionary volumes with the purpose of emphasizing certain characteristics of plants and somewhat balancing the animal examples prevalent in the remaining chapters. The latter is not the case here since many chapters refer to plants, but as G.L. Stebbins (1950) wrote his masterpiece, *Variation and Evolution in Plants*, he continually tested most of the known population genetic and biosystematic ideas comparatively for plants versus animals. His chapters covered environmentally induced variation; ecotype, clinal and subspecies variation; evidence for natural selection; genetic systems involving hybridization, polyploidy, autogamy, and apomixis; and karyotype or genome evolution versus external morphology. Clearly, plant researchers have continued to emphasize these topics; more recently, demography and physiological ecology have been discussed in several volumes devoted to plant research. Stebbins (1950) established the role of genetic ideas in all of this with a wealth of succinct hypotheses and examples, combining the deductive and inductive approaches. A recent book with a very similar title (Briggs and Walters 1984. *Plant Variation and Evolution*) reaffirms many of these ideas even without covering with enthusiasm or rigor many new developments or emerging biosystematic topics in population biology. On the other hand, M. Crawley's (1986) *Plant Ecology* indeed balances the genetic treatment with such topics as life histories, dynamics of growth and form, pollination and seed dispersal ecology, and the structure of plant populations or communities. Developmental and certain molecular genetic aspects were juxtaposed with these population approaches in a recent volume (Gottlieb and Jain 1988), in which Stebbins (1988) concluded that "discoveries made during the past twenty years demand extensive revision of previous evolutionary syntheses". Here again, he suggested that plant evolutionists should clarify different species and population structure concepts since plant species are less well-defined, due to weaker isolation patterns, less complex developmental and behavioral patterns, a wider range of longevity, no locomotion, and illustrate a greater role of polyploidy, autogamy, and asexuality, but a smaller role of biotic environment. Thus, we find plants and animals treated comparatively to emphasize their common as well as distinct modes of evolution. This chapter attempts a brief survey of major findings on the patterns of variation in plant populations with emphasis on the evidence for natural selection.

Spatial Patterns of Variation

Variation patterns have been largely described in either spatial terms (ecotypes, clines, hybrid swarms, isolates, or local patchiness) or in the form of comparative species surveys (e.g., inbreeders vs outbreeders, conifers vs angiosperms). A search for pattern must clearly define the meaning and theoretical basis of such ecogeographical and biosystematic comparisons.

Ecotypes often relate to some distinct and disjunct factor of environment and plant response in terms of physiological/morphological differentiation. Clines of many types help to discover the continuous nature of variation between ecotypes or simply extremes of a gradient; a description of clinal variation might be based on one conspicuous polymorphism or a multitrait quantitative pattern. Finally, mosaic or patchwork in localized variation is based on nonrandom occurrence of two or more genotypes in a population. Note that already we are dealing with several spatial scales in variation analyses. Heslop–Harrison's (1964) classical review of genecology concluded the following:

> "1. Wide-ranging plant species show spatial variation in morphological and physiological characteristics. 2. Much of this infraspecific variation can be correlated with habitat differences. 3. To the extent that ecologically correlated variation is not simply due to plastic response to environment, it is attributable to the action of natural selection in moulding locally adapted populations from the pool of genetical variation available to the species as a whole."

More recently, Briggs and Walters (1984) thoroughly reviewed many examples of **ecotypic** and ***clinal variation*** based on largely quantitative variation. (Also see Endler 1977, for an extensive discussion of geographical variation measures and examples.)

For example, McGraw and Antonovics (1983) studied two ecotypes in *Dryas octopetala*, namely the snow-bed ecotype (larger leaf length, leaf hairs variable) and fell-field ecotype (smaller leaf with silvery hairs). Reciprocal transplant experiments and other common garden studies showed a genetic basis of ecotypic differences, seedling versus adult survivorship, and restricted (leptokurtic) gene flow, but in spite of these expected findings, the ecotypes were not perfectly matched with the environment. Results of a similar demographic study of two ecotypes in *Plantago lanceolata* are summarized in Table 1.

On the basis of morphology, reciprocal transplants and common garden experiments, two ecotypes were described in *Avena barbata* populations of the drier and warmer valley region and the cooler, humid coastal region in central California. Detailed allozyme variation studies further showed several multilocus genotypes predominant in each of four geographical regions. Clinal variation has

Table 1. Characteristic differences between two populations of *Plantago lanceolata* from a dry grassland (dg) and a wet meadow (wm) (after van Groenendael 1985)

	Wm		Dg	
	Mean	CV%	Mean	CV%
No. of leaves	3.8	25	6.6	27
Length of longest leaf (mm)	20.3	24	34	24
No. of ears	1.7	47	9.2	69
Length of scape (mm)	380	29	54	61
Length of ear (mm)	17.1	34	7.3	31
No. of seeds/ear	57.17	53	15.74	75
Seed weight (mg)	1.91	39	0.73	75
No. of side rosettes	0		3.6	
Percentage of adults flowering	20		31	
Percentage of adults with side rosettes	0		12	
No. of seeds in the soil per m^2 (1-11-1979)	500		1700	
No. of seedlings per m^2 1979	1158		779	
1980	465		486	
Percentage germinating in Spring 1979	1158		779	
1980	46		75	
Percentage of seedlings surviving 1979 → 1980	26		59	
1980 → 1981	28		45	
Half-life of adults in months	46		20	
Mean gene diversity per locus (He)	0.131		0.124	
Percentage of polymorphic loci (P)	31%		36%	

been observed in only a few of the potential areas in which populations seem to occur continuously between regions. Local polymorphic populations often show a highly patchy distribution (see Jain 1975, for description of scales of sampling). Thus, variation patterns of *A. barbata* include all three categories: ecotypes, cline, and mosaic.

Ecotype variation was described by Roy and Mooney (1982) in *Heliotropium* using desert and coastal population samples. Water-use efficiency was not as different as the response to air humidity through adaptive changes in leaf form and stomatal conductance. In *Encelia*, Ehleringer and Clark's (1988) elegant work also provided an example of adaptive variation between species. Thus, note that ecotypes are often described in relation to physiological responses to temperature extremes, moisture, nutrients, sunlight (sun vs shade), and tolerance of heavy metal toxicity. Morphological and physiological differences are readily observed and their adaptive roles are rarely in doubt. However, as noted by Quinn (1987) and others, such ecotypic variation studies do not adequately take complexity of genetic control, choice of test environments, or continuity of natural variation into account. There is increasingly greater acceptance of clinal and other complex variation patterns, as sampling methods and genetic-environmental designs for such studies have become more extensive.

Table 2. A comparison between predicted and observed natural clines.[a]

Species	Change	Width	l	b	$s = b\Delta$	Δ	$l_c = 1/\sqrt{s}$
Eucalyptus urnigera	smooth gradient	100 m	20 m	0.0036 per m	2.9	800 m	12 m
Agrostis tenuis	stepped	5–10 m	8 m	—	0.1–0.45	0.1 m	12–25 m
Agrostis stolonifera	stepped	10 m	8 m	—	0.70	0.10	9.6 m
Anthoxanthum odoratum	stepped	2–10 m	8 m	—	0.31	0.1 m	14 m
Avena barbata	smooth	5–30 m	3 m	0.25	100 m	100 m	24 m
Gaillardia pulchella	step	10–150 m	—	—	—	400 m	

[a] *Cline width*: distance between localities with gene frequencies 0.2 and 0.8; l = calculated *gene flow* distance; = calculated *selection coefficient* such that $s = b\Delta$, b = *slope of cline*, Δ = distance between the ends of cline (modified after Endler 1986).

Do we often find clinal variation? Can we establish natural selection in these examples? Theoretically, a cline results from a series of populations distributed along an environmental gradient which imposes directional selection for alternative genotypes at the two ends but selection and gene flow between adjacent populations work in balance. Accordingly, findings of clinal variation might often allow selection and gene flow to be studied quantitatively. Endler (1986) and Briggs and Walters (1984) review numerous clines (Table 2, for some examples). We must recognize at least two important complications: (1) instead of smooth clines we might see step clines, with the step-change in variation, in fact, not coinciding with the environmental transition step, and (2) the potential pseudo-clines result from founder effects followed by gene flow among the founded patches or colonies (Hedrick and Holden 1979). Barber (1965) discussed clinal variation in *Eucalyptus* in which gene flow, selective forces, and long-term evolutionary dynamics were thoroughly investigated. Selection was inferred for leaf form, bark texture, leaf waxiness, leaf pigmentation, and water relations.

Patterns of variation are also detected in comparative surveys of different species. For example, Bradshaw (1984) summarized data on copper tolerance in grasses (Table 3), such that some species as a whole might not show certain variants. He called it genostasis. Similar data are often obtained in genetic resources sampled for disease resistance genes. Mutation, drift, past selection history, and sampling methods might all be complicating such a pattern.

Table 3. The percentage of copper-tolerant individuals found in normal populations of various grass species, in relation to the presence of the species on copper-polluted waste and whether the plants collected were tolerant of copper (data of C. Ingram) (After Bradshaw 1984)

Species	Occurrence of tolerant individuals (%)	Presence of species on mines		Tolerance of collected adult plants
		On waste	Margins	
Holcus lanatus	0.16	+	+	+
Agrostis tenuis	0.13	+	+	+
Festuca ovina	0.07	−	+	−
Dactylis glomerata	0.05	+	+	+
Deschampsia flexuosa	0.03	+	+	+
Anthoxanthum odoratum	0.02	−	+	−
Festuca rubra	0.01	+	+	+
Lolium perenne	0.005	−	+	−
Poa pratensis	0.0	−	+	−
Phleum pratense	0.0	−	+	−
Cynosurus cristatus	0.0	−	+	−
Alopecurus pratensis	0.0	−	+	−
Bromus mollis	0.0	−	+	−
Arrhenatherum elatius	0.0	−	+	−

In wild coffee (*Psychotria nervosa*), Dewey and Heywood (1988) reported an elegant analysis of hierarchical spatial variation, using allozymes and statistics of genotypic dispersion (genetic similarity of neighbors). They did not find evidence for local demes or neighborhood. Whether local heterotic selection or certain dispersal patterns sufficiently account for their observations is not certain. A classical example of hierarchial pattern of variation was found in a desert annual, *Linanthus parryaii*, for which Wright (1978) had developed F-statistics for analyzing flower color polymorphism invoking a significant role of small demes. Epperson and Clegg (1986) found spatial autocorrelation at several spatial scales in the flower color polymorphism in morning glory. Turkington (1985) reported local differentiation in a perennial pasture species, *Trifolium repens*, and rather extensive and local genetic variation.

It is evident from such pattern analyses of variation that the measurement of gene flow parameters is a "hot topic" as an excellent area to bring together ecologists and geneticists. Population genetic theory is developing rapidly and perhaps also our empirical research capability. Wide use of single locus and multilocus markers in highly dynamic situations of colonization, disturbance (leading to fragmented habitat), rare species conservation programs, etc. should provide gene flow and dispersal statistics in relation to variation patterns. Likewise, ecologists are describing local dispersion, mating and dispersal parameters. Research on selection under varying degrees of controlled environments should follow in the same materials.

Genetic Variation in Breeding Systems

Some Crop Examples

Even though literature on breeding systems is rather voluminous, genetic variation in outcrossing rates has been studied in rather few species; selection experiments involving outcrossing rate and/or correlated morphological traits have been reported in even fewer cases. Breese (1966) reported success in manipulating outcrossing rate through selection for heterostathmy (a quantitative form of heterostyly) in *Nicotiana rustica*. Likewise, response to high and low outcrossing selection has been reported in field beans (*Vicia faba*) as well as some progress in breeding programs directed toward a shift to autogamy, with emphasis on alleviating inbreeding depression during the early generations. The evolution of tomato cultivars from their self-incompatible wild relatives with "pin" type stigma-anther relationships to the highly selfed "thrum" flower structure suggests a complex and gradual evolutionary genetic change. This and many similar examples of crop evolution point to various correlated changes in the morphology and function of flowers and the pollination system. Inheritance of breeding system characteristics of many legumes have been reported to be polygenic and intermediate in outcrossing expression. In field beans homozygotes which produce less pollen were more dependent on tripping and cross-pollination for seed production than the heterozygotes which produce abundant pollen. This difference appeared to be under polygenic control.

Amaranthus species vary in sex expression in both natural and domesticated stands. The maintenance of polymorphism for male sex expression in several cultivated amaranths, a situation approaching a gynodioecious system, could be explained if heterosis was an important determinant of seed yield or viability. Among major crops with a mixed mating system and outcrossing rates similar to amaranth are cotton (0.20 to 0.30) and sorghum (0.05 to 0.40). *Sorghum* exhibits wide variability in outcrossing rates; in an experiment selection for seed size was associated with an indirect response in the gynoecium morphology and a corresponding twofold increase in outcrossing rate (Foster et al. 1980). Many authors have noted apparent selection for increased inbreeding in the process of domestication and early migration of crop plants. The landrace populations of grain amaranth from Central and South America, in fact, exhibited very low levels of heterozygosity. These were below the level expected from outcrossing rates in the range of 0.10 to 0.50. In contrast, Indian landraces seemed to have significantly higher amounts of genetic variation, higher outcrossing rates, and overall a larger role of heterotic selection (unpubl. data).

Evolution of Inbreeders

Selfers theoretically gain in energetic terms for economy on flower size, nectar, and floral display features. Reproductive allocation to seed can be increased, or in many situations, the life-cycle is adapted to a shorter growing season; changes in life history and allometric changes in size variables are perhaps common. Several authors have reported on the lower energetic cost per seed in the selfing component, but greater reproductive assurance, and even a few higher-quality seed need not yield higher reproductive rates or evolve an *r*-strategy through wide seed dispersal, and so on. Thus, in theory we discover a paradox, "selfing is easier to evolve but it commonly does not; outbreeding requires stringent selection to be retained but it is." Often we have argued for such a dichotomy to evolve; however, the metaphoric balance (see model below) is not tipped easily by any single factor; complexity of adaptive/chance ecological factors and genetic/environmental components of variation together allow a dynamic wide "trough" for highly varied, mixed systems.

Plant breeding systems are discussed as a subset of genetic systems in which the rate and frequency of genetic recombination were emphasized; outcrossing, asexuality, ploidy, longevity, population size, gene flow, karyotype structure, have been discussed mostly in terms of flow of genetic variation under adaptive responses to varying environments. Details of floral biology, phenology, local population dispersion, resource allocation to male and female functions, and population density, etc. have been added from ecological studies. Richards (1986) states that "it is taken for granted that the attributes observed are under genetic

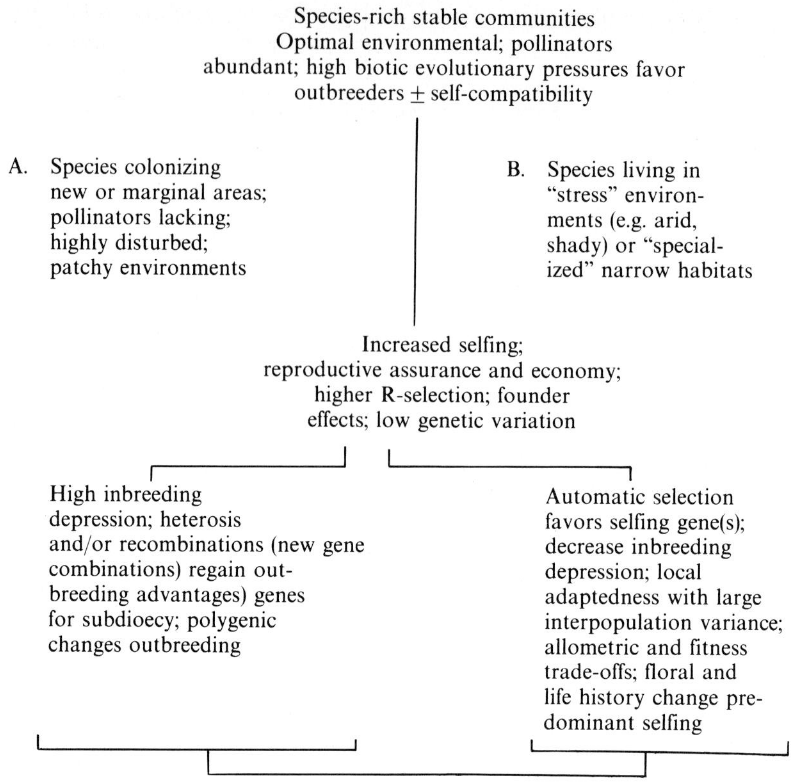

Fig. 1. A balance-tipping model (after Waller 1980) shows metaphorically some of the key factors determining changes toward outbreeding versus inbreeding

control, are heritable, and are of adaptive significance," and perhaps, taking for granted would be a folly as we leave much of the needed evolutionary research unattended. How else could we show the advantages of agamospermy in comparison with sexuals, whether inbreeding favors colonizing success, or mixed strategies are evolutionarily efficient? One must consider the concluding remarks in Richards (1986); or else we must continue to reiterate a few deduced opinions (e.g. see selfers, p. 369, Richards 1986). Selfers lack genetic variability which severely restricts the niches in time and space, and yet somehow, their "unusually high capabilities for phenotypic plasticity" and higher reproductive rates adapt to extreme, restricted, or marginal habitats. We need genetic and ecological tests which include largely deductive ideas. This would appear particularly critical if the ecology of reproduction (and resource allocation) is more important than the genetic consequences of systems such as gynodioecy, dioecy, or heterostyly.

This brief review of recent developments (also Wyatt 1988) in the evolution of breeding structure in plant populations yields two main themes: (1) mating systems are complex and dynamic features of most species, in particular those with mixed selfing and random mating, and (2) new discoveries of this complexity

Table 4. Some variables required for evolutionary hypothesis testing

Variable	*Gilia*	*Limnanthes*	*Lupinus*	*Phlox*	*Leavenworthia*	*Impatiens*
Outcrossing rate:						
Related species	+	++	++	++	+	+
Populations	+	+++	+++			+
Individuals		++	++			
Floral morphology	++	++	++	++	++	+++
Autofertility	+	+	+	+	+	+
Self-Incompatibility					++	
Sex allocation					+	+++
Pollinator visits	+	+		++		
Population size, density	+	+	+	+	++	+
Colonizing ability		+	+	++		
Genetic variation:						
allelic diversity	+	++	+	+++	+	++
hybridity	+	+	+	++		
change over time						
role of recombination						
Life history:						
reproductive effort		++	+	+	++	+++
stages of mortality		+				
longevity		+				
changes over time		+				
Phenotypic plasticity		+		+	+	

[a] +, ++, +++ Refer to the approximately increasing order of available information.

require more detailed, long-term field studies on demography, dispersal, pollination ecology, and fitness components in inbred versus outbred progenies (Table 4).

One can begin to synthesize the following findings into a coherent evolutionary model: (1) the history of population bottlenecks during colonizing episodes or simply temporally varying environments influence not only the effective population size, but also genetic assortative matings within patches and small demes. (2) Given microgeographical differentiation among demes, either due to selection or drift, interdeme gene exchange (outcrossing followed by fertilization, seed set favoring crosses) contributes significantly to the maintenance of genetic variation; the adaptive role of occasional outbursts of outcrossing in selfing species has always been considered important (Brown 1979, for a theoretical discussion). (3) Competition within sibships resulting from heterosis or increased recombination as a source of new gene combinations has revived interest in the population genetic theory involving heterozygote advantage, although optimal outcrossing rates or distances are not yet widely documented. (4) Recall that earlier studies in inbreeders like barley, lima bean and wild oats emphasized heterozygosity levels (true or pseudo-overdominance) which may have also originated more often in situations involving crosses among wide parentage; that is, heterosis could have evolved only when high levels of heterozygosity are generated (Allard et al. 1968; Brown 1984). (5) Certain populations of any selfer species could be maintaining a higher outcrossing potential, due to either genetic or environmental factors; this is unfortunately not known in most cases albeit we now have some better estimates.

Thus, various recent findings on population structure and variation in breeding systems provide clues to a model in which complete autogamy could evolve through severe bottlenecks, catastrophic selection, and chance loss of variation; translocation heterozygotes are fixed through inbreeding; only isolated stands would perpetuate complete selfing until new genetic variation accumulates in time. A mixed mating system, on the other hand, would be a dynamic balance among various population features, both adaptive and neutral; to be sure, overdominant gene combinations could play a key role here; variants affecting breeding systems would be commonly found, but the low heritability of such polygenic and even oligogenic variations must slow down any trends toward fixation of complete selfing or outbreeding systems. Primary challenges here lie in ecogenetic research designed to unravel selective forces, using population genetic studies based on well-mapped stands over several generations. Ideas based on optimal resources allocation and developmental correlations as well as plant-plant neighborhood interactions warrant new ecological approaches.

A theory of breeding systems that predicts changes only toward the extremes of selfing versus outcrossing appears too restrictive and therefore unhelpful in dealing with interdeme selection and environmental heterogeneity favoring a mixed mating system; however, models such as those of gynodioecy, heterostyly and self-incompatibility provide some interesting paradigms for further model-building. The genetics of breeding system variables is clearly not adequately known, whereas the theoretical predictions of evolutionary changes strongly depend upon the mode of inheritance and genotype-environment interactions. Heterosis and selection within a sibship framework must also consider specific kinds of variation; pooling certain sets of loci for averages has limitations for ecogenetic tests of selection. New experimental methods of population biology involving long-term field studies will hopefully replace excessive speculative writing in this area.

Breeding Systems and Variation Surveys in *Limnanthes* Spp

Limnanthes (Limnanthaceae) is a genus with 9 species comprising 13 intraspecific taxa of spring-flowering, annual, diploid (n = 5) plants. One species (*L. macounii*) is found in British Columbia; all others are native and restricted to California and southern Oregon. They usually grow in dense colonies in moist habitats such as vernal pools, or along temporary streams and wet meadows. A variety of breeding systems is found in different taxa ranging from almost complete selfing (due to cleistogamy) in *L. floccosa* to as high as 100% outcrossing due to protandry and insect pollination in *L. alba* and gynodioecy in *L. douglasii*. Outcrossing rates often vary among and within populations of most outbred taxa (McNeill and Jain 1983; Kesseli and Jain 1984). *L. alba* is the most widespread species of the section Inflexae, occurring in distinct, widely separated populations from the Coast Range east to the Sierra Foothills and from Shasta Country south to Tuolumne Country. It is also quite similar morphologically to *L. gracilis* and *L. montana* and shares a predominantly outcrossing breeding system involving showy white and protandrous flowers which are attractive to insect pollinators.

In section Inflexae, *L. floccosa* is a small-flowered species, closely related to *L. alba*, occurring in disjunct populations from Jackson County, in southern Oregon to Butte Country, California. Five subspecies have been described which range in breeding system from essentially full autogamy in *L.f* ssp. *floccosa* and *L.f* ssp. *bellingeriana* to partial outcrossing in *L.f.* ssp. *californica*, *L.f.* ssp. *grandiflora*, and *L.f.* ssp. *pumila*, with larger petals and slight protandry. On the basis of morphological similarity, geographical proximity, and certain ecological observations, *L.f.* ssp. *californica* is considered a recent derivative of *L. alba*, whereas the other subspecies of *L. floccosa* represent progressive evolution toward complete autogamy from *L. f.* ssp. *californica*. This hypothesis is of special interest in relation to the various theoretical ideas on the evolution of inbreeding in plants. Table 5 compares some floral features, breeding system parameters, and genetic variation assayed at 18 allozyme loci in four taxa. Note that *L. gracilis* is intermediate between the outbreeder *L. alba* and autogamous *L. floccosa*.

Several examples of an outcrossing species giving rise to a selfer have been suggested and documented (see Wyatt 1988, for a detailed review). *L. alba* and *L.f. floccosa* have outbreeding and high selfing systems, respectively (Brown and Jain 1979), and their morphological similarity and overlapping ranges make them a likely pair in which to look for this kind of progenitor-derivative relationship. However, the high genetic divergence between *L. alba* and *L. floccosa* argues against the postulate that *L. floccosa* was recently derived from *L. alba*. The two species share no alleles at 14–20% of loci surveyed and at 36–40% of the loci alleles of *L. floccosa* do not constitute a subset of *L. alba* alleles (McNeill and Jain 1983). None of the other taxa in this section is more closely

Table 5. Comparisons of floral, breeding system, and variability features among four *Limnanthes* taxa

Observation	*alba*	*gracilis*	*parishii*	*floccosa*
Protandry	2–3 days	1.2	1.2	0
Petal length (mm)	16 ± 1.5	11 ± 1.0	11 ± 0.7	7 ± 0.4
Anther length (mm)	7 ± 1.2	5 ± 0.3	4 ± 0.6	4 ± 0.4
Stigma lobe diam. (mm)	2 ± 0.3	1.8 ± 0.3	1.6 ± 0.4	1.0 ± 0.2
Pollen/ovule ratio	High————	————	————	————Very low
Autofertility (No. of seeds per flower)	0.08	0.33	1.65	1–5 (distinct variations)
Outcrossing (%)	70–90	40–80	45–75	0?
Polym. loci (%)	8–17	3–9	3–7	0–2
No. of alleles/locus	2.5	1.9	1.8	1.1
No. of populations	18	5	4	13

related to *L. floccosa* than *L. alba*, thus, unlike the examples in *Lycopersicon*, *Clarkia*, or *Stephanomaria* which show evidence for the evolution of autogamous species through the fixation of a subset of ancestral genotypes (Watt 1988 for review), *L. floccosa* might have either originated from some unknown taxon or diverged rapidly in its genome oranization.

Four parameters were used to quantify genetic variation at allozyme loci: percentage of loci polymorphic in a species (PLS) and per population (PLP); and average number of alleles at a locus per species (K/S) and per population (K/P). Without exception *L. floccosa* subspecies are lower than the three other species in all measures of variation. *L. alba* is the most variable taxon in all categories except PLS. *L. montana* is highly variable when populations are pooled (e.g. PLS, K/S) but lower relative to other taxa when individual populations are considered. *L.g. gracilis* is less variable than *L. alba* and more variable than *L. g. parishii* in all cases. *L.f. floccosa* is extremely homozygous but retains some variability in the form of alternatively fixed alleles in different populations at five loci. The single populations studied of *L.f. californica* had more polymorphic loci and alleles than the other four *L. floccosa* subspecies. Table 6 and 7 summarize the genetic

Table 6. Components of genetic diversity in *Limnanthes*, section Reflexae[a]

Taxon	H_T	H_S	D_{ST}	G_{ST}	D_M	R_{ST}
L.d. douglasii	0.237	0.108	0.129	0.543	0.142	1.306
L.d. nivea	0.259	0.177	0.082	0.316	0.902	0.519
L.d. sulphurea	0.093	0.086	0.007	0.075	0.014	0.163
L.d. rosea	0.174	0.133	0.041	0.237	0.048	0.363
L. striata	0.255	0.141	0.114	0.448	0.130	0.927
L. vinculans	0.119	0.110	0.990	0.077	0.012	0.111
L. bakeri	0.057	0.051	0.006	0.105	0.013	0.255
All taxa (pooled)	0.301	0.127	0.174	0.578	0.178	1.403

[a] $H_T = H_S$ (within populations) + D_{ST} (between populations), $G_{ST} = D_{ST}/H_T$, the proportion of interpopulation variation in total variation; $D_M = S(D_{ST})/S-1$ (D_{ST}), $R_{ST} = D_M/H_S$, the proportion of inter- to intra population variation.

Table 7. Components of genetic diversity in *Limnanthes*, section Inflexae

Taxon	H_T	H_S	D_{ST}	G_{ST}	D_M	R_{ST}
alba	0.214	0.160	0.054	0.252	0.059	0.370
montana	0.274	0.129	0.145	0.529	0.161	1.246
L.g. gracilis	0.250	0.177	0.029	0.140	0.038	0.217
L.g. parishii	0.120	0.101	0.019	0.159	0.022	0.221
floccosa	0.173	0.009	0.164	0.949	0.173	19.548
All taxa (pooled)	0.417	0.089	0.327	0.786	0.334	3.735
All taxa (excl. *floccosa*)	0.367	0.140	0.229	0.621	0.236	1.692

See Footenote Table 6.

variation in various *Limnanthes* taxa in terms of Nei's diversity statistics computed from the allelic frequency data. Here, intra- and interpopulation variation components are emphasized. The values of F_{ST} and R_{ST} show that inbreeder *L. floccosa* has high interpopulation variation, and as expected low H_s values; *L. bakeri* and *L. gracilis* have predominant selfing but retain rather 10–30% or even more outcrossing and accordingly show smaller R_{ST} values than the highly allogamous *L. douglasii*, *L. alba*, and *L. vinculans*, even though the latter taxon has a highly restricted distribution.

Variation in Different Species Groups

In the following sections, patterns of variation are briefly reviewed in terms of species comparisons based on ecological features such as colonizing success, geographical range, breeding system, and domestication.

Genetic Structure of Colonizing Species

The first attempt to analyze comprehensively both the genetic and ecological attributes of colonizing species was made during a symposium (Baker and Stebbins 1965). The basic approach has been to analyze successfully colonizing species as a group in an attempt to establish common genetic and ecological features. Currently, there are numerous theories and generalizations on the origin of various colonizing strategies. These are related mainly to mating systems, adaptive genetic polymorphism, ecotypic differentiation, phenotypic plasticity, and reproductive strategies (Brown and Marshall 1981).

Several authors have concluded that uniparental reproduction (selfing or apomixis) is a common feature of successful colonizers. It has been proposed that the major advantage of this mating system is that it assures reproduction even in environments with a scarcity of potential pollinators and the rapid fixation of successful genotypes. However, certain examples suggest that higher rates of outcrossing may evolve in a colonizing species (Jain and Martins 1979). In fact, Brown and Marshall (1981) concluded from literature reviews that colonizing species are highly variable in their ecological and genetic features, including the mode of reproduction and the mating system. A species may increase self-fertilization to maintain a high reproductive rate and to avoid segregational load, or alternatively may evolve higher outcrossing rates to produce genetic variability that may promote exploration of a patchy environment.

Founder effects, due to the reduced number of individuals normally involved in the establishment of new colonies, have frequently been postulated to reduce the genetic variation of these colonies and to influence the potential for successful colonization. However, certain examples suggest that the original levels of genetic diversity can be rapidly regenerated in new colonies after the initial bottleneck. Comparative studies in *Avena barbata* (Jain 1969) and *Bromus mollis* (Brown and

Marshall 1981) showed a similar average diversity between the colonial and source populations. In a comparison of the genetic structure of populations of colonizing *Lupinus succulentus* Dougl. with noncolonizing *L. nanus* Dougl., Harding and Barnes (1977) reported that the colonizing populations of *L. succulentus* have an excess of heterozygosity over the expected levels. Brown and Marshall (1981) suggested that large amounts of linkage disequilibrium between loci may even promote a postbottleneck increase in the frequency of alleles favoring outcrossing or recombination.

Ecotypic differentiation is a common feature of many plant species. Populations of colonizing plant species also show differentiation, both locally (Jain 1975) and over large geographical areas (Hancock and Bringhurst 1978). Ecotypic differentiation might evolve in colonizing species if genetic divergence in different patches is also accompanied by adequate variation within patches to allow selective changes. Snaydon and Davies (1982) reviewed several examples of rapid ecotypic differentiation on a highly localized scale. However, McWilliams et al. (1971) found that even after 90 years of cultivation over a large region of Australia, *Phalaris aquatica* L. (syn. *P. tuberosa* L.) showed no ecotypic differentiation for a wide range of morphological, developmental, and agronomic characters. Apparent genetic similarity throughout an ecologically heterogeneous region was also observed for *A. barbata* populations in central California (Jain 1975). A comprehensive study of population differentiation in colonizing plant species must include the rate and range of local gene flow through pollen and seed, estimations of neighborhood size, the spatial scales of genetic and nongenetic components of variation in both morphology and life history, and measurements of environmental heterogeneity.

Baker (1974) postulated that **phenotypic plasticity**, mainly in traits involving survival and reproduction, is a common feature of colonizing plant species. Several comparative studies on genetic versus plastic adaptive strategies have been conducted on pairs of related species. Wu and Jain (1978) found that although *Bromus rubens* L. is genetically much less variable than *B. mollis* L., *B. rubens* exhibits more individual plasticity in response to different environments. Similarly, *A. barbata* demonstrates more phenotypic plasticity than the more polymorphic *A. fatua*. Brown and Marshall (1981) suggested that opportunities may arise for successful colonization by some species in a patchy and naturally disturbed environment, leading to a concomitant increase in both genetic polymorphism and phenotypic plasticity.

Nearly 10 years of comparative studies in six annual grassland species have focused on the nature of 25 genetic structure and life history traits (Foin and Jain 1977) ranging from visual scores to responses in large, controlled experiments. These species vary widely in several ecogenetic traits, in spite of the fact that they all presumably share a similar habitat, represent a "similar" history of introduction into California, and possess the same predominantly selfing system.

Several populations of the rose clover (*Trifolium hitum*) have been studied during its recent colonization of disturbed ruderal areas adjacent to many grassland sites. Patches are established by both short- and long-range seed

Table 8. Rose clover in California: variation for flowering time (Unpub. data of F. Molina and S. Jain)

Site	No. of families	$\bar{X}$	$\hat{\sigma}^2$	V_B[c] (%)	V_W[c] (%)
Madera a	8	96.90	108.86	73.37	26.63
Auburn a	10	122.97	43.48	17.89	82.11
Nevada city a	9	129.60	37.55	18.91	81.09
Deschutes a	15	111.44	43.23	30.71	69.29
Bear Creek a	11	123.66	39.41	82.68	17.32
Madera b	14	110.97	85.04	56.32	43.68
Auburn b	17	120.15	38.17	28.39	71.61
Nevada City b	19	129.57	9.80	9.48	90.83
Deschutes b	18	114.01	10.70	38.17	61.83
Bear Creek b	11	124.75	42.41	76.45	23.55

[a] a, b refer to two planting regimes; ** $p < 0.01$.
[b] Large values of $\hat{\sigma}^2$ indicate major gene polymorphism for a flowering time locus.
[c] V_B, V_W are relative proportion of variation between and within families, respectively.

dispersal. Patch sizes were assessed by a census of seed banks and established plants over successive years, whereas polymorphisms at four morphological marker loci measured genetic variation (Jain and Martins 1979). The relationship between patchiness and patch dynamics, and polymorphism, was studied at two sites in northern California. These were an undisturbed roadside population (Auburn) of rose clover patches growing in size, and a small right-of-way area disturbed heavily by people (Placerville). Generally, neighboring patches exchanged pollen through frequent pollinator movement, and especially at the Auburn site, the larger, rapidly expanding patches showed significant increase in genetic polymorphisms. The Placerville patches generally decreased in size and showed a decrease in polymorphism. Successful patches had apparently higher outcrossing rates and higher levels of heterozygosity. New patches were continually established through animal-aided seed dispersal, but the patch size increase and seed output were not correlated. Patch extinctions are difficult to evaluate, since they must be verified by long-term observations on seed banks. Overall, the successful colonization by rose clover in disturbed patchy environments seemed to involve rapid genetic changes. For example, we have recently discovered a major gene for earliness in populations more likely to be under selective pressure from increased aridity (Table 8); these major gene polymorphisms account for a nearly two- to fivefold increase in variance among family means in populations from Madera and Bear Creek (F. Molina and S. Jain unpubl. data).

Genetic Variation in Patchily Distributed Species

There are two important questions here: Can we predict the patterns of genetic variation in patchy, i.e., spatially subdivided and temporally unstable, environments? What kind of adaptive changes result from evolution in such evironments? To answer the first question, one must look at the population genetic models in the order of increasing complexity. For instance, the loss of genetic variation within small, isolated populations is predicted by a straightforward stochastic process. The effect of random drift, particularly in relation to population size bottlenecks, has been theoretically examined. Heterozygosity is reduced, and thus average genetic variation in small populations (small, measured by harmonic N) is reduced. Motro and Thomson (1982) further emphasized these losses of variation by showing the presistent effects of bottlenecks. This, combined with the studies of restricted dispersal discussed below, provides the basis for large, persistent mosaics of genetic variation in patchy environments.

By introducing an **island model of migration**, in which all subpopulations receive migrants representing the average gene frequency of all subpopulations pooled, i.e., the metapopulation, one finds that a small amount of migration can counteract drift effects. Thus, one readily accepts the general notion that migration promotes the maintenance of genetic variation. However, as noted below, several models show that too much or too little migration may not help to retain variation. Spatial heterogeneity and temporal variation in environment, and therefore selection, patch persistence, rates of new recolonization, and rates of population growth, all need to be considered in some detail before one can trust such intuitive generalities.

Specific models of migration make an enormous difference in the distribution pattern of genetic variation. In an elegant computer simulation study, Turner et al. (1982) showed that under restricted pollen and seed dispersal, patchy genetic variation developed quickly and persisted for many generations even in large populations. Heterozygosity was distributed at the edges of such patches. This study also showed gene frequency patchiness to be greater than predicted from dispersal models. Rai and Jain (1982) empirically discovered a similar result in *A. barbata* populations. Selective forces would have dramatic local effects, as shown by Levin (1981) in his simulation of the rates of spread of favorable genes.

Several other extensions of this theory provide new insights. In a general model of polymorphism in patchy environments, Gillespie (1978) showed stochastic variation in fitnesses to be important for maintaining genetic variation under selection-migration balance. Multilocus selection was treated in the theory of the evolution of sex in heterogeneous environments by Maynard Smith (1978a). Spatial variations in the rate of outcrossing had large effects on genetic heterozygosity within populations (Brown 1979). Hedrick et al (1976) emphasized the critical role of autocorrelation between successive environments; the conditions for stable polymorphisms within demes are not determined by the temporal patterns, which, however, do yield different allelic distributions. Rates of migration might often have to be intermediate, since too much or too little migration among demes could fail to provide for polymorphism.

In small, partially isolated subpopulations, **interdeme selection** has been often proposed to lead to rapid evolutionary changes. Wade and McCauley (1980) defined "populational heritability" as the fraction of the observed between-deme phenotypic variance that is heritable in the sense that "propagules chosen at random from a "parental" deme resemble one another and that parent in their populational growth characteristics." Their research on *Trifolium castaneum* involved varying population migration rates among demes and artificial selection for productivity. According to a model by Slatkin (1981b), a response to group selection occurred even under such theoretically unlikely conditions as high rates of random extinctions and moderately high migration rates (up to 12%). Breeding and recombination systems may often evolve as group-selected traits; for example, rose clover data suggest the possible evolution of higher outcrossing rates in new colonies.

To summarize, we can offer an observation and a caveat. A theory is rapidly developing to deal with evolution under heterogeneous environments, but the results so far are rather sketchy. We still have no simple predictions on the pattern of variation, much less on the nature of adaptive outcomes.

With these theoretical ideas in mind, we might expect to find genetic variation within and among populations living in a patchy environment to provide some clues to (1) the role of genetic variation in adaptability, (2) the relative magnitudes of selective versus other factors, and (3) optimality of demographic and competitive processes in life history evolution. Prior to the development of electrophoretic assays, the literature on variation in plant populations emphasized the genecology of a few selected traits such as flower color, pubescence of leaves, leaf shape, flowering time, or seed morphs. Often such assays were aimed at ecological patterns described over large geographical scales, but were somewhat lacking in genetic analysis and microgeographical detail. Currently, we have allozyme variation data on numerous species, including colonizers, populations with different histories, and so forth. Even with no direct evidence for selection, Hamrick et al. (1979) surveyed these studies under various schemes of species groupings and found that

> "species with large ranges, high fecundities, an outcrossing mode of reproduction, wind-pollination, a long generation time, and from habitats representing later stages of succession have more genetic variation."

Brown (1979) analyzed many examples of inbreeders and outbreeders and concluded that thier patterns of interpopulation differentiation and heterozygosity within populations vary due to many complex factors affecting mating systems. Partial inbreeding due to restricted neighborhood size is important in many outbreeding species.

Variation in Inbreeding Species

Based on studies of major gene polymorphisms and quantitative traits in populations of barely, lima bean, fescues, wild oats, and on theory, Allard et al.

(1968) concluded as follows: Estimates of selective values for the simple polymorphic characters commonly found in natural and experimental populations of predominantly self-pollinated plants show that linkage blocks marked by major genes are often subject to very strong selection. In a high proportion of cases the estimated selective values show that such linkage block heterozygotes have a striking net selective advantage over homozygotes.

Natural populations of inbreeding species contain great stores of genetic variability with respect to continuously varying characters. Part of this variability takes the form of geographical differentiations, including a mosaic pattern of differentiations associated with adaptation to microniches. There is also extensive intrapopulational variability both between and within families.

When many characters are considered simultaneously, each local population is found to have a characteristic pattern of variability which distinguishes it from other local populations. The mating system is only one of many factors which affect the amount and kinds of variability in populations. The observed structure of inbreeding populations is seen to result from an integration of inbreeding into the totality of genetic and ecological factors affecting population structure. The structure of inbreeding species provides for both storage of large amounts of genetic variability and for retention of sufficient flexibility to meet the requirements imposed by heterogeneity of the local environment.

Brown (1979, 1984) reaffirmed many of these conclusions from a survey of allozyme variation in inbreeders which includes many studies on genetic resources in trees, field crops and crop relatives. Brown (1984) further argued that as we learn from molecular variation studies about novel genome organization properties, the study of multilocus organization will help in the synthesis of ecology and population genetics. More recently, Allard (1988) restated many of the same conclusions for the inbreeders. Over the 25–30-year period, our descriptions of variation patterns have become richer in detail and breadth but analyses of selective forces, both outdoors or in controlled experiments, are still meager.

Variation Under Domestication

Evolutionary studies in crop genera provide many useful examples of variation studies. Introgression, disruptive selection and shifts in mating system are common topics. Two examples are given here.

Retention of genetic variability within highly inbred populations of predominantly self-pollinating plant species suggests that response to directional selection within such populations may be significant. Ten generations of bidirectional mass selection for seed weight, plant height, and days to flower were conducted within a foundation seed lot of "Double Dwarf Yellow Milo 38" (DD38) grain sorghum [*Sorghum bicolor* (L.) Moench] to evaluate this possibility. Selection gains were evaluated by comparing nonselected and tenth-cycle selected populations grown in replicated experiments at three locations in 1978 (Foster et al. 1980). Divergence of population means of plus and minus selected strains was significant ($p \leqslant 0.01$) for all three traits, while asymmetry of response was significant ($p \leqslant 0.01$) for seed weight and days to flower only. Mean per cycle selection response expressed as a percentage of nonselected population mean over ten generations ranged from 0.25% for the early flowering population to 3.4% for increased seed weight. Correlated selection responses were observed for each character selected, but the magnitude and direction of correlated responses were inconsistent. Thus, sorghum populations may respond to directional selection despite apparently restrictive mating systems.

Allozyme studies in amaranth provided useful assays of genetic variation and the evolutionary relationships among wild, weedy, and crop species (Hauptli and Jain 1984). Thirty-four populations of cultivated New World amaranths were surveyed at nine electrophoretic enzyme loci. Eleven populations of cultivated amaranths from the Indian State of Uttar Pradesh and six from Nepal were also surveyed for a comparison. Alleles at several loci proved to be diagnostic of the crop and weed groups, and of the three individual crop species. Genetic distances based on pooled gene frequencies showed the three crop species to be generally more closely related *inter se* than they were to their putative weedy progenitor species, respectively (with the exception of the weed-crop pair *A. quitensis* and *A. caudatus*). This favors the idea of a single domestication event involving *A. hybridus* as the common ancestor rather than three separate domestication events. Close similarity between *A. caudatus* and *A. quitensis* migh have resulted from trans-domestication based on a weedy or semi-domesticated species having migrated from Meso-America to South America.

Intraspecific and interspecific crosses have been attempted between several populations of each of the cultivated grain amaranth species (P. Kulakow and S. Jain unpub. data). Success of hybridization varied depending on the populations used. A history of prior natural hybridization may explain relaxed crossing barriers between certain specific populations. A study of the inheritance of a reproductive isolation barrier between certain *A. hypochondriacus* and *A. caudatus* populations has shown that very few genes control the hybrid sterility factors.

More detailed studies of the genetic structure of landrace populations have been completed for a subset of six New World and ten Indian accessions. Variability within and between populations has been compared for both qualitative and quantitative characters. Results indicated that the New World landraces were highly differentiated between populations (Hauptli and Jain 1984). Each landrace behaved as a collection of nearly homozygous lines, although significant amounts of within population variance were also observed. Indian amaranths also showed differentiation within and between populations; these landraces showed evidence of higher levels of heterozygosity for qualitative characters.

Rare Versus Common Species

Many plant genera include both rare (and endangered) and common taxa which provide useful comparisons of population structure and levels of genetic variation. Conservation issues have made it even more timely. Rarity has been described in terms of paleoendemics, ecological specialization (close, narrow adaptedness), stress adaptation, and recent speciation events; accordingly, genetic structure of rare species can vary theoretically. A recent paper by Karron et al. (1988) showed indeed, even for congeners, a lack of correlation between commonness and genetic variation (Loveless and Hamrick 1984, found a weak correlation). *Limnanthes* taxa show a correlation in geographical range and genetic variation; range is highly correlated with the number of extant populations (islands in an archipelago, with a larger role of gene flow). Geographical range and habitat specialization appear to be reliable predictors of genetic variation in many cases; however, various other genetic and demographical variables deny us any universal rules for such associations.

Variation in Hybrid Swarms: Evidence for Introgression?

Two species of sunflower (*Helianthus bolanderi, H. exilis*) in California have been studied by several workers. Stebbins and Daly (1961) studied morphological patterns of variation in relation to the role of introgressive hybridization; one population shifted in variation patterns over 6 years. Olivieri and Jain (1977) reexamined this hypothesis in terms of a series of populations which could arise from hybridization and localized patterns of genetic drift-selection-gene flow. Rieseberg et al. (1988) provided further evidence based on mitochondrial DNA sequence variation to show that introgressive origin of *Helianthus bolanderi* is unlikely. This example reopens an entire series of questions on the role and patterns of interspecific hybridization. On the other hand, numerous examples in crop-weed complexes (e.g. rice, radish, wheat, barley) invoke introgression based on preliminary evidence of gene flow across species barriers. This entire area of plant variation research has many unexplored examples. Interspecific hybrid zones have become possibly more interesting for studying the role of selection in relation to mechanisms promoting as well as circumventing gene flow across species barriers. New opportunities exist here for studying both morphological and biochemical traits simultaneously.

Genetic Variation in a Highly Localized Context

How much does a population vary in its genetic makeup? Can we analyze the microscale distribution of genotypes (alleles, allele combinations). The spatial scale could be varied from a few centimeters to several 100 (depending on the nature of a stand). Concepts of "random mosaics" or patchwork in genetic differentiation are not new; Bradshaw (1984), Snaydon and Davis (1982), others have described them; Harper (1983), Burdon (1987), and Turkington (1985) have further reaffirmed this in *Trifolium repens* studies. *Avena barbata* provided examples of highly patchy dispersion of genotypes at two marker loci (Rai and Jain 1982). *Spartina* clones specialized for marsh, swale, and dune microhabitats, occur within small patches (Silander 1985); and Linhart (1974) reported *Veronica* populations to be differentiated along a vernal pool margin-center gradient.

The second question relates to natural selection acting on such local variation, within one or very few generations. Do different genotypes show selective advantages in patches with different ecological properties? Several studies have answered this affirmatively; in many others, the question has not been attempted adequately, which leaves the role of migration/drift/population growth variables either inadequately or speculatively yes. Natural selection requires some sort of direct evidence such as changes in genotypic proportions between life-cycle stages (e.g. Clegg et al. 1978); between generations (Snaydon and Davies 1982); performance of reciprocal transplants (Schmidt and Levin 1985); and results of colonization trials (Bradshaw 1984). Ennos (1983) has reviewed many examples of selection in reltion to the maintenance of variation, of which frequency- and/or density-dependent selection have particularly direct ecological meanings [e.g. examples of coevolving host-pathogen, or plant-herbivore systems; tests of sex role (Antonovics and Ellstrand 1984); competition and facilitation in terms of some resource use patterns]. Variation among individuals, groups of individuals (e.g. sibs) within patches or demes, and among populations within a small distance range can allow many forms of selective differentiation.

Molecular and Morphological Variation

Comparing wild and cultivated barleys using worldwide allozyme patterns, Allard (1988) concluded that wild barley has many more multilocus gene combinations, whereas rather few allelic associations in cultivated barley suggest cultivated fields to be ecologically similar worldwide. This is indeed in striking contrast to the impression from numerous morphological variation studies. There is now sufficient reason to doubt that any one set of traits or sample of gene sequences, proteins, or another kind of molecular variant would tell the same story about adaptations or phylogenies. Examples of vertebrate and invertebrate groups and also a laboratory-derived set of inbred lines in mice (Patterson 1986) clearly show varying degrees of match between the molecular and the morphological data on variation and genetic differentiation. For phenotypic evolution, it is even more likely that molecules, major gene polymorphisms, quantitative descriptors of morphology and physiological traits differ widely in the origin of genetic variation, integration into an organism, and finally into the population level expression of phenotypes.

Lande and Barrowclough (1987) simply reiterated the view that

> "most major phenotypic changes in evolution result from the accumulation of quantitative polygenic modification of existing phenotypes, rather than from single genes of major effect",

and in comparison with the lower rates of spontaneous mutations for major genes,

> "the total spontaneous rate for minor mutations at all loci influencing a particular quantitative character may often be of the order of 10^{-2} per gamete per generation or more".

An important theoretical result suggests that if we allow the expected rate of drift loss of genetic variance to be less on average than by stabilizing selection, a small population can maintain nearly as much genetic variation as a very large population. Quantitative variation at many loci (polygenes) can often be (1) large, both within and between species, (2) subject to continued, adaptive changes, (3) restored after bottlenecks fairly quickly, and (4) account for heterosis (epistatsis and buffering of multilocus heterozygotes).

Many of the pre-1970 surveys of variation were largely based on conspicuous morphological polymorphisms, some of which have large correlated effects on breeding systems, and on quantitative or phenological traits which affect life histories. During the last two decades, electrophoretic variation studies have dominated this area of research and we expect this trend to continue. However, we can also expect major developments in genome mapping using associations between molecular markers and morphological variants. Perhaps this would provide new insights into the genetic variation and covariation of many ecologically significant aspects of form and function in natural populations.

Kinds and Intensities of Natural Selection

Ennos (1983) reviewed numerous examples of plant variation in terms of the relative importance of the following kinds of selective forces: heterozygote advantage, spatially heterogeneous (multiple niche) environments, frequency-dependent selection, balancing selection favoring different genotypes at differing life-cycle stages, disassortative mating; and so on. He concluded that

> "the efficacy of various mechanisms proposed for maintaining genetic diversity depend very much on demography... as this survey makes plain, it is only rarely that we have adequate data on all parameters mentioned".

In particular, the roles of diseases and pests or herbivores as selective agents were emphasized. A recent monographic treatment of diseases in plant populations brought out newly emerging evidence for host-pathogen interactions, dynamics of vertical and horizontal resistance, changing racial composition of pathogens, and a need for detailed population analyses of so-called endemic balance (Burdon

1987). Frequency- and density-dependent selection have numerous ecological interpretations in terms of competitive and coevolutionary interactions with pollinators, pathogens, mimicry models, etc. as well as changing abiotic environments.

Endler (1986) reviewed the kinds of evidence needed to establish natural selection; manipulative genetic and ecological experiments are increasingly used in plant research (e.g. Schemske 1984; Clay and Antonovics 1985; Clegg and Epperson 1988). Levin (1984) summarized estimates of selection coefficients from reciprocal transplant experiments in which the native genotype or gene pool had as much as a 50 to 75% advantage over the alien classes as measured rather indirectly in terms of survival or biomass increases. On the other hand, Ennos (1983) and Endler (1986) reviewed examples (Table 9) of major gene polymorphisms but did not provide estimates of selective forces. Characters scored included flower and seed colors, leaf shapes and leaf pubescence, ray floret morphology, insect resistance, metal tolerance, cyanogenesis, and allozyme polymorphism. For instance, the two-locus polymorphisms foe cyanogenesis in *Trifolium repens* populations in Europe have been studied for nearly 35 years; altitudinal as well as latitudinal clines were reported but temperature, herbivory, and several other unknown factors interact and as yet, we have no estimates of the migration rates or selection coefficients. Demography and local population structure details are

Table 9. Evidence for natural selection (some examples from Endler 1986)

Species	Traits examined	Method of demonstration	Selective agent
Agrostis tenuis	Heavy metal tolerance	Genetic demography Comparisons among age classes (along a cline). Perturbation of natural populations	Metal ions, low soil soil nutrient, competition
Avena barbata	Color BB Hariness LS	Perturbation of natural populations (introduction)	Unknown
	3 Allozymes LS, BB	Genetic demography Perturbation of natural populations	Soil moisture?
	Allozymes: 3 est. loci	Genetic demography	Unknown; some fertility effects
Liatris cylindrica	Allozymes est. 3	Comparison among age classes (one sample, but ages (2-11 years)	Unknown; some fertility effects
Mimulus guttatus	Copper tolerance	Comparison among age classes Perturbation of natural populations	Copper poisoning
	Chlorophyll deficiency	Genetic demography	Differences in feritlity and rhizome growth
Senecio vulgaris	Ray floret morphology	Genetic demography	Fecundity effects; due to pollinator efficiency

not available. Several other examples in Table 9 show the lack of information required to estimate the selective forces.

Phenotypic plasticity of quantitative traits makes it even more difficult to estimate selective parameters. Moreover, several authors (Quinn 1987; Sultan 1987) have commented on the possible negative correlation between the levels and ecological roles of genetic variation and phenotypic plasticity. From a study of five *Danthonia spicata* populations representing different successional ages, Scheiner and Goodnight (1984) concluded, "there is no reason to suppose that selection for the two types of variation need necessarily be coupled." Our observations (Jain 1979) were based on three pairs of congeners, one based on an outbreeder-inbreeder comparison, and a very limited interpretation of the role of specific traits or environments. The postulate, if ever to be labeled thus, was weak but justified empirically then and has been reconfirmed in several studies. In fact, theory too does not make any strong prediction on such a postulate. We now have new definitions of plasticity (and therefore new measures) based on patterns of increased phenotypic variance, and perhaps more rigorous tests of the adaptive role.

Finally, we must ask how variation and selection enhance population well-being (ecological measures of fitness)? Wright's (1978) model refers to local population growth, dispersibility of selected gene combinations, and overall population persistence due to evolutionary changes. Here, indirectly we have genetic feedback models, or extinction of genetically depauperate species, models of variation in subdivided populations, etc. But we must ask questions directly in order to force the synthesis of population genetics and ecology e.g. which characters (phenotypic features) evolve such that the ecological view of adaptedness is enhanced by evolutionary genetic dynamics of a population under selection. How are $\bar{W}$ and equilibrium properties related to the parameters of population growth (λ, r versus K; α_{ij}) under varying densities (Ginzburg 1983) and heterogeneous environments? In order to develop ecological genetic models of adaptive strategies, we will need joint data on population numbers, life history traits, and genetic changes under various enviornmental challenges.

The experimental work on natural selection within a population has frequently been concerned with a description of the relative rates of survival of identifiable genetic units (alleles, chromosome segments, linked gene combinations, accessory chromosomes) under natural or laboratory conditions. Accordingly, the theory of population genetics has been largely developed through the use of models involving parametrization of Darwinian fitness defined in terms of the net survival and/or reproductive rates over a unit time interval. These selective or adaptive values may in fact involve a multiplicity of complex factors such as differential selection during several stages of the life cycle, developmental and genetic interactions among various traits, and so on. A detailed theoretical and experimental analysis of the role of natural selection in population dynamics would clearly require a knowledge of such complicating factors. For instance, a bulk-hybrid population of barley (CC-XXI) was investigated for changes over a period of six generations, in (1) generation means and variances for three quantitative characters, (2) phenotypic and genotypic frequencies at several marker loci, and (3) the components of selection at several life-cycle stages (germination and seedling survival, fertility, fecundity) (Jain and Marshall 1969).

The observed changes in the phenotypic frequencies at marker loci as well as in the generation means and variances suggested that there were rather weak selective forces operating in this population. Moreover, the preliminary analysis of components for selection in terms of different stages of life cycle showed clearly that (1) for different loci, the direction and relative amounts of selection pressure could be quite varied; (2) the selective advantages of one or the other allele at any locus may be reversed such as to retain allelic frequencies unchanged over short periods; and (3) the yet adaptive values computed from the individual components correspond well only in a few cases to those obtained from the frequency data.

Variation and Evolution of Life Histories

In their emphasis on the distribution and abundance of species living together in a community, ecologists have naturally focused on comparative studies on population growth, demographic patterns of regulation, and species interactions. Natural selection is frequently invoked to explain various adaptive features of life histories as well as plant-animal coevolution. Life history is defined in terms of age-dependent birth and death rates with emphasis on the first age of reproduction and the trade-offs between growth (which is related to survivorship under competition, predation, or nutrient scarcity) and reproduction. Accordingly, a population biologist measures selection intensities in terms of life-cycle components of an aggregate fitness parameter, e.g. average reproductive value or Malthusian growth rate of a genotypic class. In an optimal life history, the natural selection in a density-independent model maximizes the population growth rate (r), which also corresponds to an evolutionary stable strategy (Charlesworth 1980), or maximizes carrying capacity (K) in a density-dependent environment (Ginzberg 1983). Temporal variation in life history traits further alters the predicted fitness relationships. The concept of optimal life history is designed as a problm-solving algorithm under prescribed conditions of environmental heterogeneity and certain statistics of vital attributes. It can take many "optimal" forms depending on the genetic and nongenetic constraints as well as random events. Empirical evidence from comparative ecology of related species in general supports the optimality arguments. Since life history traits bring together information from physiological ecology and community ecology, and which we believe should include developmental biology and genetics to provide a complete fitness context, this pluralistic view of evolution seems attractive (Gottlieb and Jain 1988).

Fitness criteria have no single, universal definition; different models emphasize different notions of ecological success and/or evolutionary stability (Lewontin 1978). Accordingly, experiments on the role of life history variation in adaptive responses differ not only in the aggregate definitions of individual or population fitnesses but also in spatial-temporal scales. Causal factors of environmental heterogeneity that presumably elicit adaptive changes are not reaily guessed, let alone proved. Harper (1983) succinctly argues for long-term and ingenious field studies on genetic variation and patterns of life history together. And finally, continuous variability of most life history traits as well as aggregate fitness parameters (e.g. mean lifetime reproductive value or Wrightian $\bar{W}$) pose many uncertainties in the genetic analyses of evolutionary processes.

Mitchell–Olds and Rutledge (1986), among others, noted the advantages for life history genetic studies in plants: controlled pollinations, mapping, manipulative plantings, experimental randomization. Crop literature includes numerous quantitative genetic data and plant breeding examples based on selection, whereas field studies on natural populations are rather scarce. These authors point out many difficulties in the estimation of genetic parameters, including genetic, population, statistical, and logistic factors;

> "many features of plant life histories and reproduction can bias estimates of quantitative genetic parameters,...it is important to separate the fundamental, long-term genetic constraints which limit species evolution, from the transient genetic correlations that may be found in a particular population".

(Note that species-level evolutionary studies are necesserily long-term as noted below).

In a comparative study by Ritland and Jain (1984a), ten representative *Limnanthes alba* populations in northern California, grown in a common environment, showed significant among-population genetic variation of certain floral and life history characters (Table 10). Variation of these two sets of characters also correlated with the geographical distribution of populations, suggesting the influence of either common ancestry of neighboring populations, or gradients of natural selection. In contrast, allozyme variation at 11 polymorphic loci showed little pattern. Further study of three representative populations revealed significant differences in the sizes of dormant seed banks and several reproductive characters. These differences correlated with some field measures of soil and biotic variables, and were compared with certain predictions from reproductive strategy models in an attempt to identify any agents of selection. Alternative sets of life history characters were discernable corresponding to varieties identified on the basis of floral morphology, but these sets did not match those predicted by the r- and K- selection dichotomy.

In a parallel study the life histories of two herbaceous annuals, *Limnanthes alba* and *L. floccosa*, were studied in terms of the characters of seed dormancy, vegetative growth rate, time of reproduction, allocation to reproduction, and seed size (Ritland and Jain 1984b). Most of these characters have genetically evolved into different states in the two species, as shown by the common garden and inheritance studies. However, seed loads were highly heterogeneous in the field and seed size did not differ. In an attempt to characterize their evolution in terms of variable selection, a model, which considered both means and variances

Table 10. Means of floral and life history characters observed in ten (*Limnanthes alba* populations grown at Davis (after Ritland and Jain 1984a)

Character	Population (sampled along high to low elevation gradient)									
	1	2	3	4	5	6	7	8	9	10
Stamen length	57	68	68	69	67	59	84	94	96	98
Petal width	41	45	42	53	61	59	67	64	77	87
Corolla width	114	148	115	133	168	144	152	192	187	185
Sepal width	18	20	21	25	26	25	28	24	28	30
Petal length	127	132	124	136	142	111	131	139	140	149
Days to flower	124	119	123	129	110	114	117	116	114	109
Days to senescence	147	144	144	149	133	140	142	145	139	132
Seeds/flower (x 100)[a]	52	59	50	79	81	84	116	115	148	96
Seed number	10	12	10	8	15	15	15	20	21	19
Plant wt. (mg × 100)	292	349	270	158	267	297	282	396	423	408
Flower number	22	21	21	10	19	19	14	18	13	14
Mg × 100 seeds/plant	66	74	60	42	103	85	91	110	130	107
Flowering rate	25	27	25	30	28	30	27	25	37	42
Flowering period	13	12	12	11	14	13	12	11	11	12
Total allocation (%)	69	62	61	67	65	69	70	66	67	67
Dormancy index (%)	64	56	55	50	62	56	57	54	65	56
Seed weight (mg)	64	56	55	50	62	56	57	54	65	56
Seed allocation (%)	22	19	21	24	28	27	28	27	30	26

of character state fitness, and in which characters were considered in isolation from each other, was derived. Much of the *Limnanthes* data was consistent with such a scenario of selection acting on life history characters, although other models might equally apply. *Limnanthes floccosa*, relative to *L. alba*, displayed the postulated less risky character states of higher seed dormancy and slower germination rates, slower seedling growth, and earlier flowering. Risky versus safe strategies appeared to be valid in relation to our multiyear assessment of population persistence/extinction in numerous locations for each species.

Breeding Systems vs Life Histories

Are breeding systems and life histories correlated in terms of either sharing a set of common traits (e.g., dioecy, seed number, and quality) or being involved in simultaneous evolutionary changes? Stebbins (1958) suggested that short-lived plant species might be successful in occupying temporary habitats (as r-species) through self-fertilization whereas long-lived K-species of climax communities might have asexual or outbreeding systems. Models in life history of perennial, iteroparous species emphasize lower seedling survivorship and higher competitive ability but reveal nothing about the role of genetic variation or breeding systems. On the other hand, several recent studies on seed biology (dormancy, dispersal, timing, and rate of germination) have attempted to establish links with some variants in breeding system (e.g., cleistogamy/chasmogamy;

heterostyly/homostyly). Experimental studies on gynodioecious and dioecious species in comparison with hermaphrodites showing variable sex expression (and resource allocation) often include both genetic variation and life history variables). Evolution of selfing annuals from outcrossing progenitor species is often discussed in terms of catastrophic selection, fixation of favorable gene or chromosome variants, loss of genetic variation at least within populations, and changes toward r-strategy (higher reproductive effort; high rates of temporary population growth in heterogeneous habitats). And yet, even in this empirically rich area of plant evolutionary biology, both theoretical and empirical consensus on the mode and kinds of selection is far from satisfactory. Maybe, there is no unifying, single theme, and due to many detailed, and interacting factors in the evolution of inbreeding and a set of reproductively optimal life history features, solutions are numerous. However, as noted by Dingle (Chap. 8) and by van Noordwijk (Chap. 9), we can expect many exciting discoveries with regard to the genetic basis of life history variables and new evolutionary models involving more explicit treatments of phenotype-based selection.

Kalisz (1986) found in *Collinsia verna* the outcrossing rate to be higher in individuals which produced large seed size, thus producing a negative correlation also with the date of germination. Thus, higher selfing is associated with adult fecundity—and with "the strong directional selection for early germination." Self-incompatibility alleles may be used in *Plantago lanceolata* for monitoring a highly outbred population structure (with no or little biparental inbreeding); this can affect the measure of interdeme differentiation (F_{ST}) based on allozyme loci, and the fitness consequences of dispersal can affect the population fitness (intra- and interdeme dynamics); mating system, dispersal, and life histories evolve together.

Gynodioecy in *Limnanthes* allowed us similar opportunities. Female frequency may affect outcrossing rate (and selfing of hermaphrodites), whereas low dispersal may allow frequency-dependent sibship patterns in which the high frequency of females in certain patches is associated with more xeric (unpredictable) microsites.

Annual plants, especially weeds and successional species, have been studied commonly; their genetics, logistics of seed collection, and cultivation, etc. allow short-term population genetic studies. Seed banks in many cases have introduced an important complication: overlapping generations with low "germinability". Genetic studies year-to-year may or may not detect significant genotypic shifts even with varying seasons of drought/high rainfall or high/low mean temperatures. Evolution of seed dormancy, which might have large heritable inter- and intrapopulation components of variation, has been studied in relation to the long-term persistence strategies. Genetic analyses of variation in seed dormancy are rather few but already there is evidence for high heritability, rapid response to artificial selection, and major genes in a few cases.

Besides the percent germination (as a measure of dormancy and seed carryover over years), the timing of germination may also be under selection. Kalisz (1986) found selection to favor early germination in *Collinsia verna* but due to complex environmental heterogeneity, the overall genetic component was very low, and found also large maternal effects in the associated seed weight variance.

Mating patterns in plant populations are a subject of great interest (Hedrick, Chap. 3); several new developments have taken place in the estimation of gene frequencies in the two sexes, the paternity determination of each zygote, estimation of outcrossing rates (and pollen flow distance), computations of seed dispersal rates and distances, and the statistics of population dispersion (neighborhood structure). Already, using primarily allozyme variation (multilocus) data, several studies have begun to yield detailed insights into local population structure as well as some estimates of interpopulation gene flow. These developments will clearly allow better linking of genotypic mapping (genecology) and following their fate (in terms of mortality and fecundity, and to some degree, even their role in sexual selection).

Variation, Selection, and Adaptation

The bulk of population genetic theory deals with the maintenance of polymorphisms in a population or species; the role of natural selection is often specified by the relative fitnesses assigned to various genotypes. When these fitness sets are varied in relation to specific environmental conditions, one could argue that the genetic changes and the fate of variation are somehow linked to certain measures of adaptedness and optimal phenotypes (Schmid–Hempel, Chap. 10). The weighted mean population fitness ($\bar{W}$) has important properties relative to the local stability of equilibrium gene frequencies. However, in the ecologists' context of adaptedness, the measures of well-being of a population are interpreted in terms of reproductive success, relative abundance or persistence in time, or all of these together. Accordingly, many attempts have been made to find relations between $\bar{W}$ (mean population fitness) and values of r (intrinsic growth rate in the logistic equation) or population regulation mechanisms. In fact, this was defined as the first goal of population biology in aiming to correlate kinds and numbers within populations.

Populations living in patchy environments may allow local selection to favor different genotypes in different patches (given a rapid evolutionary rate or sufficient patch duration); this would lead to genetic differentiation (high F_{ST}, etc.). Polymorphism is assured in the metapopulation (patches pooled together), whereas within patches, polymorphism would be maintained by migration, overdominance, and other selective forces. Even then, we may not be able to specify whether this variation has an adaptive role in the ecological-demographic well-being of the species. Intuitively, we often argue that variation is useful in the continuing response to new environmental challenges. However, note that Reddingius and Den Boer's (1970) model assures the persistence of a species in a patchy environment by spreading the risk, i.e., a subdivision-dispersal-recolonization process, without any genetic shifts.

The traits likely to evolve in patchy environments include certain migration characteristics and features of the recombination-sexuality system. These apply to both nongenetic, persistence models (Reddingius and Den Boer 1970) and models of polymorphism (Slatkin 1981b). Vrijenhoek (Chap. 4) observed, on the one hand, high temporal heterogeneity, asexuality, or an *Oenothera*-type recombination system (selfing, translocation heterozygote). On the other hand, heterozygosity can be associated with sexuality or open recombination systems, according to Maynard Smith's model (1978) of sib competition under spatial heterogeneity and changing environmental conditions (including temporal heterogeneity). Likewise, a given life history feature might also be advantageous in contrasting situations: rapid population increase in newly founded patches, high r-selection, opportunism, and wide dispersal as opposed to dormancy. Interdeme selection would act on most of these features as we have discussed, i.e., through optimization to make the metapopulation more fit (see Barton and Clark, Chap 4; Schmid–Hempel, Chap. 10, for concepts and new theoretical developments in this area).

Finally, it should be noted briefly that many changes are currently taking place in evolutionary thinking. The origin of variation is subject to new molecular probes; polygenic mutation rates might be very high; regulatory genetic variation needs to be understood; a strict population genetic approach to evolutionary rate may need modification; and the origin of species (macroevolution) certainly is not just a continued gene substitution process (Singh, Chap. 7). In this changing scenario, patch dynamics, as it introduces a large role for random drift and heterogeneity of local units of evolutionary change, has an important role to play. We need to incorporate founder-flush theory; demographic genetic features of selection, migration, and drift; the role of inbreeding in subdivisions; and interdeme selection (Barton and Clark, Chap. 4). Thus, it is not business as usual—a new level of complexity is compellingly introduced. A contrasting view, however, denies the role of random drift but assumes that transitional and isolated patches show loss of phenotypic stability, leading rapidly to new evolutionary lines.

Concluding Remarks

There were at least three main points in this chapter. First, literature on plant variation studies at all levels (species populations, local neighborhoods, individuals) has grown impressively in both natural history and statistical terms. We now have an increasingly larger variety of molecular, morphological, and functional descriptors all of which should be considered together, albeit their relative information content for different processes of evolution would vary. In the past two decades, allozyme variation has provided what most would treat as valuable markers for mapping genomes, for studying population structure and gene flow, and for making "neutral-theory-based" predictions on patterns of variation. These and newly emerging DNA sequence information would also provide a valuable aid in phylogenetic and evolutionary rate determinations. It is equally important to recognize the primary interest of ecologists and most evolutionists in the morphological descriptions of individual form, function, phenology, defense, and biosystematic units.

Second, patterns of variation (spatial) are tested in terms of ecotypic, clinal, and mosaic, distributions of single and multiple character classes; statistical tests vary in the degree of sophistication and rigor but in general many good plant examples are now available for each of these categories. Depending on the scales and underlying habitat gradients (or other known environmental patterns), one could infer the role of selection in evolving these patterns through directional, disruptive, and (locally) even stabilizing selection. However, tests of adaptedness rely on manipulative experiments or simple correlations; rarely do we have a dynamic description of the changing genotypic makeup over time. Hence, most estimates of selection coefficients are based on reciprocal transplants or models of clinal variation. Likewise, we have made rather too few attempts to study the demographic genetic components of selection.

Third, evolutionary changes are complex because their consequences are not limited to the level of the biological hierarchy (molecules, characters, individuals, populations, species) in which they initially occur. The complexity is interesting and, when taken into account in the design of research, we anticipate more complete explanations of evolutionary change. Such vertical studies would entail close genetic and phenotypic analysis of individuals coupled with studies of character expression and development. Insofar as characters are studied that affect survivorship and reproductive success of individuals in natural populations (and why not study such characters?), we can also learn about character variations and their correlation to other traits, particularly in relation to concepts of optimality and other appropriately defined strategies. We expect, in fact, that these measures of fitness will find many close parallels in both morphogenetic and physiological descriptions of optimal form and function as well.

Theoretical models will also be necessary to guide population research that asks "What direction?" and "How rapidly?". The formal mathematical models that describe the dynamics of large systems with few parameters (gene frequency, outcrossing rate, dispersal rate, fecundity, etc.) have already been linked up with molecular evolution based upon gene sequence information. Similar connections, with particular relevance to plants, must now be made. The results will be critical in plants because, in the absence of dated fossils, sequence change may often provide the only estimator of evolution rate.

In a chapter appropriately titled "Variation as a genetic engineering process", Shapiro (1983) concluded that

> "the differences between species appear to have more to do with genome organization (chromosome structure, ploidy, distribution and structure of repetitive elements, sex determination) than with adaptation to different ecological niches. It may prove useful to focus on the role of variation, rather than selection, in the origin of species".

Even at the intraspecies level, we need not prove all variation to be maintained adaptively. In his elegant chapter on Darwinian plant ecology, Harper (1983) discussed variation in *Trifolium repens* for numerous traits, at even very localized spatial scales, but noted emphatically,

> "the genotypic composition of the population may in some cases dimly reflect forces that operated twenty or thirty years ago we would expect to find only a few of the many polymorphisms readily interpretable as responsive to present proximal selective forces."

Plant population biologists are increasingly concerned with detailed and highly localized processes of evolution and demography, and seem to have made a few "discoveries" such as mate selection, optimal outcrossing distance, and role of sibships. However, we also need a synthesis of "big pictures" in which qualitative results on species comparisons, using sufficient but few parameters, are desired (Levins 1968); in fact, we should note that most applications of population biology rely on a few general and short-term dynamic principles (Jain and Botsford, in press).

I close this chapter with J.B.S. Haldane's characteristically original remark that Darwin's most original contribution to biology was not the theory of evolution, but his great series of books on experimental botany. Plant biologists have indeed received a very rich legacy from Darwin, and also in a series of genetic discoveries from a great many pioneers beginning with Gregor Mendel to Barbara McClintock and with E.M. East to A.D. Bradshaw; we have great challenges ahead in plant evolutionary research.

7 Patterns of Species Divergence and Genetic Theories of Speciation

R.S. SINGH

Introduction

Species hold a central position in the hierarchical organization of biological diversity and hence the mechanism of species formation has always been and still is a central problem in evolutionary biology (Mayr 1982a). The analysis of speciation has lagged far behind that of adaptation, due to a variety of causes ranging from problems with the species concept (Ehrlich and Raven 1969; Levin 1979), to inherent difficulties in pursuing the problem of speciation experimentally (Lewontin 1974), and to doubts whether speciation is indeed a separate process of evolution independent of adaptation (Eldredge and Gould 1972). The last point has been the basis of an ongoing debate on adaptive vs non-adaptive speciation and the debate has been recently brought to an all-time high level of interest by Gould's proposition that speciation events are the major cause of biological diversity (Gould and Eldredge 1977; Goul 1980; Stanley 1979).

Darwin solved the species problem by showing that they change and in his theory the mode of change was what later came to be known as **phyletic gradualism**. It took another 80 years or so before phyletic gradualism (anagenesis) and speciation (cladogenesis) became fully recognized as two separate modes of evolution (Huxley 1942). The next 20 years (1940's–1960's), which coincide with rapid development in population genetics, saw the debate on the mechanism of speciation shift from early models framed in abstract, general terms of populations and genetically controlled phenotypic traits ("geographic" models) to formulations based on allele substitutions in gene pools ("genic" models) (Huxley 1942; Dobzhansky 1951; Mayr 1963). However, since the molecular basis of genetic variation was still unknown, very little was really said about the **nature of genetic variation** for species formation, and thus all models of speciation before the 1960's (barring of course models involving macromutations) can be characterized as basically "geographic" models. The past 20 years have seen the most wide application of molecular genetic methods to the problem of species formation and a major effort has gone into confirming or denying the allopatric model of speciation as proposed by Mayr (1963). We have learned a lot about the nature of genetic variation within species and genetic divergence between species Lewontin 1974, 1985; Bush 1975; Ayala 1975; Powell 1975; Selander 1976;

Throckmorton 1977; Nevo et al. 1984), but the problem of speciation still remains unsolved (for a most recent review of the diversity of views, see Barigozzi 1982).

In the present chapter our aim is (1) to broadly define the genetic and ecological problems of species formation, (2) to provide an in-depth review of the molecular variation data on *Drosophila* species bearing on genetic structure within species and genetic divergence between species, and (3) to review various genetic theories of speciation and see how they relate to our present knowledge of genetic variation within and between species. We will attempt to provide some new experimental approaches that can open up new avenues for research on the problem of speciation in both animals and plants.

The Problem of Species Formation

The Genetic Problem of Species Formation

There are three sources of confusion that confound the genetic problem of speciation. They are: sudden vs. gradual speciation, genic vs. genomic changes during speciation, and genetic changes vs. geographic isolation in speciation. Sudden, instantaneous, catastrophic or **quantum speciation**, in genetic terms, means speciation due to major changes in a single major gene, chromosome, or whole genome (Table 1). Possible evolution of inbreeders (selfers) from outbreeders involving major mutations in breeding systems (Jain 1976), fixation of major chromosome translocations, changes in ploidy level, and species hybridization are some well-known mechanisms of sudden speciation. In this class of

Table 1. Types of genetic variation, levels of isolation, and modes of speciation

Types of genetic variation	Levels of isolation	Modes of speciation
Primary speciation		
Major mutations (e.g., affecting breeding systems)	No isolation	Quantum speciation
Chromosomal mutations		
translocations, fusions, and fissions	Isolation and/or small populations	Quantum speciation
Polyploidy	No isolation	Quantum speciation
Allelic variation at numerous loci and minor chromosomal changes	Full isolation:	
	Isolation of a colony	Peripatric speciation
	Extrinsic barrier to gene flow	Allopatric speciation
	Semi-isolation	Parapatric speciation
	No isolation	Sympatric speciation
Secondary speciation		
Species hybridization		
Hybrid maintenance	No isolation	Hybrid speciation
Recombination	No isolation	Hybrid speciation
Amphiploidy	No isolation	Hybrid speciation

models are no unknowns about the genetic mechanisms of speciation, as reproductive isolation is a direct outcome of the genetic changes involved and they occur very rapidly if not in one step. Thus in those groups of organisms where this mechanism of speciation is most likely to occur, the genetic basis of speciation is really known; what we need to know further is, how common has this mode of speciation been? Evolution of selfers from outcrossers and speciation by polyploidy and by species hybridization are mechanisms whose occurrence can be determined with some certainty (Stebbins 1950; Grant 1981). With respect to the role of karyotypic changes other than polyploidy, White (1978) states that "over 90% and perhaps 98% of all speciation events are accompanied by karyotypic change and in the majority of these cases the structural chromosomal rearrangements have played a primary role in initiating the divergence". But all it really means is that in 90–98% of the cases closely related species differ in their karyotype; it cannot mean (because it is unknowable) that in the majority of these cases chromosomal changes have been the initiating factor in speciation (Futuyma and Mayer 1980). We need to know in how many cases the chromosomal changes and the resulting sterility barriers are secondary, superimposed on the primary sterility barriers that arose by genetic changes (Grant 1981). The knowledge of historical effective population size and genetic structure, rate of chromosomal mutations, and inbreeding depression are essential to evaluate the role of major chromosomal mutations in speciation (Walsh 1982; Lande 1985).

The second confusion is about the role of **genic vs. genomic changes** during speciation. By genomic changes here I mean the kind of macromutational changes involving the whole genome, first proposed by Goldschmidt (1940) and more recently resurrected by Gould and Wilson (Gould 1977; King and Wilson 1975; Wilson et al. 1974), effecting morphology, physiology, and function, and not changes in the karyotype that can give rise to reproductive isolation due to problems of chromosome pairing and segregation. It is, therefore, important not to confuse the usual kinds of genetic mechanism for sudden speciation mentioned above with the genetic mechanism postulated in Goldschmidt's "saltational" model of speciation. In the former case, the karyotypic changes are genetic agents of reproductive isolation; in the latter they are genetic agents of change in form and function.

The third confusion has to do with **geographic vs. genetic models of speciation**. Table 1 lists models of speciations classified according to the major types of genetic variation and levels of geographic isolation. Till the 1960's all models of speciation, barring macromutational changes, were basically geographic models and had little to say about the precise nature of the underlying genetic mechanisms. Thus such models as allopatric, sympatric, and parapatric listed in Table 1, are all models showing how one species geographically gets split into two, and the real debate has been on the relative role of geographic isolation vs. selection in this process (Mayr 1963; Maynard Smith 1966; White 1978; Thoday and Gibson 1970; Bush 1975; Felsenstein 1981). As far as the genetic basis of speciation is concerned, the main debate has been between models involving few genes (e.g., gradual sympatric models) and those involving many genes (e.g., allopatric models).

Thus we may say that the central problem in the genetic mechanisms of speciation consists of determining the genetic basis of gradual speciation involving single gene changes and minute chromosomal alterations, and the nature of the relationship between speciation and major morphological evolution. Thus in addition to identifying genes that cause reproductive isolation, it is of particular interest to find out if speciation involves a small "privileged" class of genes, in which case sympatric speciation and speciation by founder effects (in the sense of Carson and Templeton) become likely, or if, it involves a relatively large number of genes and geographic isolation, thus vindicating Mayr (1963).

The Ecological Problem of Speciation

The ecological problem of speciation is the problem of adaptation and coevolution. Ecologists treat species as adaptational units and study their evolution by identifying important ecological factors (both biotic and abiotic) and by making measurements and mapping of changes (man-made or natural) in the causal spectra with the corresponding changes in the species' response spectra. Thus the ecological study of a species system (or community) consists of the study of its ecological and coevolutionary equilibrium and its short-term (and possible long-term) evolutionary behavior in response to changes in the biotic or abiotic components of the environment. In this scenario, species are taxonomically stable but evolutionarily modifiable entities that keep adjusting their genetic makeup to keep pace with the changes in their environment; species keep evolving (or coevolving) but speciation does not enter into this picture!

Whether the problem of speciation does or does not belong in ecology would very much depend on our notion about the **origin of species**. If new speices are nothing more than "genetic flukes" and arise accidentally (e.g., see reviews by Rose and Doolittle 1982 and Krieber and Rose 1986), then the problem of speciation obviously does not belong in ecology. Ecologists can treat origin of species as a stochastic process that occasionally gives rise to new taxa and ultimately to new ecological interrelationships of species. The new species and their evolving ecological interactions can become subjects of study in ecology, but how species arise in the first place need not be. On the other hand, if species arise as a result of **genetic changes**, direct or indirect, incurred during the process of adaptation, then in ecology speciation can be seen as an escape mechanism for populations and species to rid themselves of existing species-specific ecological interactions and evolutionary pressures and for adopting new systems of ecological relationships. In this scenario, speciation becomes an ecological necessity and a part and parcel of the ecological studies of species and their communities. In this respect we might say that the concept of phenotypic adaptive landscape (Simpson 1953; Lande 1976), the concepts of niche and niche-gap (Hutchinson 1968), the principle of limiting similarity (MacArthur and

Levins 1967) and the notion of disruptive selection (Mather 1955; Thoday and Gibson 1970), together, define the ecological problem of speciation (Jain and Bradshaw 1966; Rosenzweig 1978; Van Valen 1976), and such questions as whether fecundity affects what kind of reproductive isolations develop in speciation, whether species diversity promotes or retards speciation, and whether adaptive divergence precedes or follows development of reproductive isolation are important ecological questions that need to be answered.

However, an immediate reason for the lack of interest in the problem of speciation in ecology is the same as that for the lack of interest in genetic consideration of most problems in ecology. During the last 25 years the search for the genetic basis of species formation and the molecular emphasis in population genetics has transformed the question of speciation that belonged to both genetics and ecology into one that now appears to belong only to genetics. Thus what was once considered the "genetic approach" to the problem of speciation suddenly appears to have become the *only* problem of speciation! Currently the genetic studies of speciation focus on the genetic basis of species differences and reproductive isoloation, and the question of speciation as it relates to evolution of new form and function and novel adaptations is set aside. But this cannot be for long because the ecological problems of speciation are not in place of but in addition to the genetic problem of speciation. The views that speciation involves more than creation of reproductive isolation (e.g., see Carson 1987), that regulatory gene changes affecting form and function are prominently involved in speciation (Gould 1977; Raff and Kaufman 1983), that primary reproductive traits are the key to speciation (Choudhary et al. 1988), and that major catastrophes such as species extinction and availability of new niches may lead to relaxation of selection and to rapid speciation (Wright 1982a) would ultimately bring the question of speciation back into the combined realm of genetics and ecology where it really belongs.

Theoretical Models of Speciation

To construct models of speciation we first need to know something about the nature of genetic variation within species, the levels of genetic divergence between species, and the mechanisms or processes of transformation of the former type of variation into the latter during the course of speciation. (see also Chap. 1). Since the gentic divergence observed even between closely related species includes both **speciation-related** and **phyletic genetic changes**, some efforts must be made to sort out the total genetic variation between species into different functional "classes" of genes, some of which might appear more relevant to speciation and thus worthy of further investigation. Identification of genes causing reproductive isolation is absolutely essential, but their single-minded pursuit tends to leave out some of the important problems of speciation.

Second, we need to know something about **genetic structure** and the role of various evolutionary forces that produce genetic structure within species, and about the relevance of genetic structure and isolation to speciation. Since genetic structure of species is both a consequence of evolution as well as a factor promoting evolution, a knowledge of genetic structure as specified in various geographic models of speciation without the knowledge of relevant genetic variation would not be sufficient to construct a model of speciation. It is for this reason that all models of speciation (geographic or genetic) include some features of both genetic structure and genetic variation (Mayr 1963; White 1978; Carson 1975, 1987; Carson and Templeton 1984; Templeton 1980, 1981).

Thirdly, it is also important to know how much we can rely on the present genetic structure to speculate about its role in speciation and macroevolution. Because genetic structure can hardly be assumed to be stable over the evolutionary timescale, we must have some knowledge about the **role of history**, i.e., role of population bottlenecks, colonization and range expansion, species hybridization and introgression, etc., that may have affected the speices' genetic structure in the recent past. And finally, since both adaptation and speciation together are responsible for the major morphological and physiological evolution, we would like to know if the two processes are causally connected. This is a central issue in the debate on the mechanisms of speciation and macroevolution (Eldredge and Gould 1972; Gould and Eldredge 1977; Gould 1980; Wright 1982b; Mayr 1982b; Stebbins 1982a,b). For this we need to know the genetics basis of new adaptations and its relation to the genetic basis of reproductive isolation.

Genetic Structure of Species

The Approach

Since speciation is a long-term evolutionary process and thus out of our experimental reach, we cannot experimentally proceed with this problem in the direction in which it actually occurs, i.e., from within species to between species. We are forced to resort to experimental programs which actually run in the opposite direction, and there are two such approaches.

The first approach involves genetic analysis of reproductive isolation between closely related species that can hybridize in the laboratory and produce at least partially fertile hybrids. This approach was pioneered over 50 years ago by Dobzhansky (1936) and is used to determine the number and the location of genes causing sexual isolation and hybrid sterility in **interspecific crosses** (e.g., see Zouros 1981; Coyne 1984). The advantage of this approach is that the genes discovered by this method bear directly on the genetic basis of reproductive isolation; the disadvantage is that it can be applied only with species that are at least partially fertile and the genetic analysis tends to underestimate the number of genes, which may spuriously give credence to genetic theories of speciation involving a small number of genes.

In the second approach, we start with examination of species with respect to the level of genetic divergence and try to evaluate, by **comparative analysis of species** showing varying levels of reproductive isolation, the relevance (or irrelevance) of the various types of genetic variation to speciation. Furthermore, by simple character correlation analysis with taxa during their early stages of speciation, we hope to determine whether the major morphological and physiological changes are directly associated with speciation or are a consequence of phyletic divergence after speciation. This is essentially an approach of historical reconstruction using extant closely related taxa, which, in combination with the first approach described above, can be a powerful tool for studies of speciation. This approach has been widely used with a large variety of organisms during the last 20 years and while we are still far from fully understanding the genetic mechanisms of speciation, there is a wealth of information on the genetic structure and divergence of species (Lewontin 1974; Powell 1975; Selander 1976; Hamrick 1983; Nevo et al. 1984; MacIntyre 1985; Ayala 1975; Carson and Kaneshiro 1976; Throckmorton 1977).

The data used in this review are based on electrophoretic studies of enzymes and abundant soluble proteins and come from three different sources. First, when making precise statements about levels of heterozygosity and about the relationship of genic variation within and between species, we shall use data collected in our lab with natural populations of *D. melanogaster* and its sibling species *D. simulans* (Singh and Rhomberg 1987a, b; Choudhary and Singh 1987a; Choudhary et al. 1988; Coulthart and Singh 1988a,b,c). Second, a review of genic variation between different groups of *Drosophila* will be done by analyzing the amount and the pattern of variation at homologous gene loci among species from within as well as between groups. Only those species groups where a sufficiently large number of gene loci and species have been studied are included in this analysis. These include the *melanogaster* group, the *willistoni* group, the *obscura* group, the *virilis* group, the *repleta* group, the *picture wing* group, and the *white tip* group. It is hoped that the comparison of variation at the homologous gene loci between species would bring out the effects of genetic structure, founder effect, and phylogeny on the divergence of species and species-groups in *Drosophila*. Finally, with a view to drawing parallels and making some generalizations, a set of data randomly selected from a variety of animal taxa will be presented to show that the patterns of variation between species and higher taxa do not fully conform to the gradual model of genetic transformation from **within** species to **between** species; the population bottleneck appears to have been a major factor affecting the levels of divergence between taxa.

Levels of Genetic Variation Within Populations

The summary statistics of genic variation estimated by one-dimensional electrophoresis (1DE) of enzymes and abundant soluble proteins are shown in Tables 2 and 3. The enzyme variation statistics are broken down by classes of

Table 2. Summary statistics for polymorphisms in different protein classes in *D. melanogaster* and *D. simulans*[a]

Enzymes and proteins	Species	No. of loci studied	Proportion of loci uniformly monomorphic	Proportion of loci polymorphic (mean ± SE)	Hetero-zygosity (mean ± SE)	Proportion of loci diverged (I = 0)
Enzymes	*D. melanogaster*	79	0.418	0.412 ± 0.049	0.127 ± 0.018	0.101
	D. simulans	79	0.443	0.329 ± 0.031	0.112 ± 0.016	
Larval (hemolymph)	*D. melanogaster*	8	0.375	0.550 ± 0.092	0.179 ± 0.069	0
proteins	*D. simulans*	8	0.250	0.650 ± 0.056	0.255 ± 0.081	
Larval (carcass)	*D. melanogaster*	30	0.667	0.193 ± 0.057	0.037 ± 0.015	0
and adult proteins	*D. simulans*	27	0.926	0.059 ± 0.020	0.004 ± 0.003	
Total	*D. melanogaster*	117	0.470	0.420 ± 0.070	0.102 ± 0.014	
	D. simulans	114	0.614	0.288 ± 0.022	0.094 ± 0.012	0.070

[a] Data from Singh and Rhomberg (1987b), and Choudhary and Singh (1987a).

Table 3. Proportion of loci polymorphic, average number of alleles, mean heterozygosity and mean fixation index for protein loci located on various chromosomes[a]

Chromosome	No. of loci studied (homologous)	Species	Proportion polymorphic	No. of alleles	Heterozygosity	Fixation index[b]
X	9	*D. melanogaster*	0.667	2.111 ± 1.364	0.155 ± 0.181	0.152 ± 0.165*
		D. simulans	0.444	2.111 ± 1.453	0.129 ± 0.193	0.053 ± 0.072*
2	16	*D. melanogaster*	0.562	1.937 ± 1.124	0.111 ± 0.136	0.119 ± 0.142*
		D. simulans	0.437	1.875 ± 1.147	0.077 ± 0.127	0.036 ± 0.054*
3	27	*D. melanogaster*	0.667	2.444 ± 1.577	0.155 ± 0.173	0.113 ± 0.114
		D. simulans	0.555	2.444 ± 1.601	0.174 ± 0.222	0.078 ± 0.118
2 + 3	56[c]	*D. melanogaster*	0.696	2.339 ± 1.352	0.155 ± 0.164	0.119 ± 0.131*
		D. simulans	0.536	2.286 ± 1.436	0.144 ± 0.194	0.083 ± 0.148*
Total	117	*D. melanogaster*	0.420	1.480 ± 0.110	0.102 ± 0.104	0.091 ± 0.130*
		D. simulans	0.288	1.590 ± 0.070	0.094 ± 0.012	0.026 ± 0.017*

[a] Data from Choudhary and Singh (1987a).
[b] Wilcoxon rank sum test: $* = p < 0.05$.
[c] Includes 13 additional autosomal loci that are unmapped.

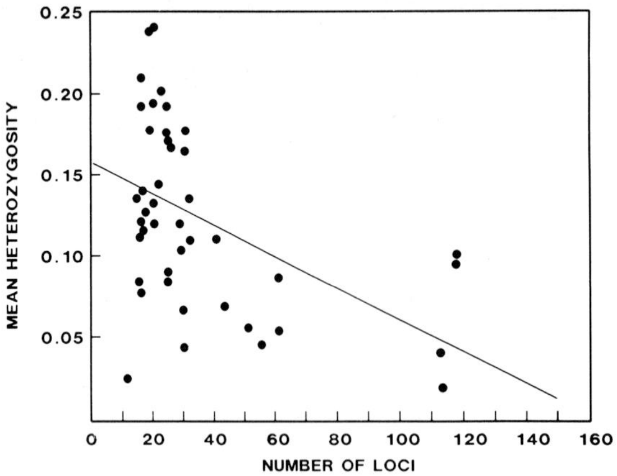

Fig. 1. Regression of mean heterozygosity on the number of loci studied in various species of *Drosophila*. $y = 0.154 - 0.000925\,x$ (regression coefficient significant: F test, $P < 0.005$). The *four points on the right* are on four sibling species of the *D. melanogaster* complex studied in our laboratory. (Singh and Rhomberg 1987b)

proteins in Table 2, and by chromosomes in Table 3. The main features of the genetic variation within populations can be summarized as follows: (1) Nearly 50% of gene-enzyme loci are uniformly monomorphic and the remaining 50% are polymorphic in one or more populations. On the 99% criterion of polymorphism, an average population is polymorphic for 40% of its loci in *D. melanogaster* and 30% in *D. simulans*. (2) Abundant water-soluble (larval hemolymph) proteins are more, and water-insoluble (larval carcass and adult) proteins much less polymorphic than the water-soluble enzymes (Table 2). (3) The autosomes and X chromosome appear to be equally heterozygous within speices, suggesting that hemizygosity of males is not a significant factor in determining the level of extant genic variation for sex-linked genes in *Drosophila*.

The overall general conclusion emerging out of our studies with enzymes and abundant proteins is that the level of genic variation in natural populations is significantly lower than what the previous studies have suggested. A large proportion of previous studies have sampled a rather small number of loci (and more often the same loci) and so the estimates of heterozygosity should be taken with some caution. If we include our own data, the levels of heterozygosity in *Drosophila* species are negatively correlated with the number of loci sampled (Fig. 1). But even our estimates of variation for enzymes may be on the high side. As well shall show later, metabolic enzymes which have been predominantly studied in the past tend to be on the average more polymorphic than nonenzymatic proteins (Coulthart and Singh 1988a).

Patterns of Genetic Variation Between Populations

Since species can vary in many respects that may affect their genetic structure, it would be ideal to compare pairs of species that are similar in all respects except the one or two factors for which they are being investigated. The sibling species of *Drosophila* provide the best material for such studies and the sibling pair *D. melanogaster*/*D. simulans* are unique in this respect. They both have a worldwide distribution yet *D. melanogaster* shows more geographic genetic differentiation than *D. simulans*. This shows up when we compare fixation indices of polymorphic loci (Fig. 2) or compare genetic distance as a function of geographic distance among populations (data not shown). Mainland populations of *D. simulans* are significantly less differentiated than those of *D. melanogaster*. The overall pattern of geographic differentiation in *D. melanogaster* is rather interesting in the sense that all temperate populations, regardless of geographic distance, are genetically more similar to each other than the tropical/tropical or the temperate/tropical comparisons (Singh and Rhomberg 1987b). However, a majority of the polymorphic loci appears to show a level of population differentiation that appears to be in agreement with the level of gene flow estimated in these speices (Singh and Rhomberg 1987a b,); thus only about 15% of the loci in *D. melanogaster* show a level and pattern of geographic differentiation that we think can not be explained without invoking natural selection (Singh and Rhomberg 1987b).

The differences in the genetic structure of *D. melanogaster* and *D. simulans* were first postulated to be due to either recent worldwide expansions of *D. simulans* or natural selection resulting from differences in their nichewidths (Choudhary and Singh 1987a). Allozyme data by themselves are incapable of distinguishing between these two alternatives but the mtDNA data gathered in our laboratory (Hale and Singh 1987) and in others (Solignac and Monnerot 1986; Baba-Aissa et al. 1988) suggest that the low geographic differentiation of *D. simulans* as shown by the enzyme data is almost surely due to its relatively more recent worldwide expansion than that of *D. melanogaster*.

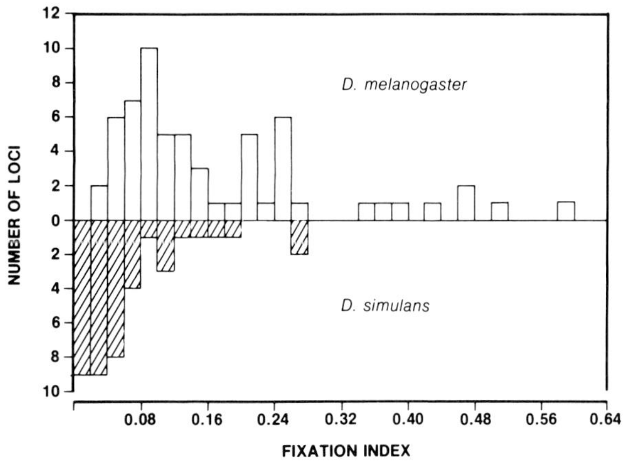

Fig. 2. Frequency distribution of polymorphic loci with respect of fixation index (F_{ST}) in natural populations of *D. melanogaster* and *D. simulans*. (Choudhary and Singh 1987a)

Some Macroscopic Features of Genic Diversity in *Drosophila*

Under the **neutral theory** of evolution (Kimura 1968b, 1983) the monomorphic state of a locus is a transient phase, as no locus remains monomorphic or polymorphic permanently. Under the modified neutral theory which incorporates slightly deleterious alleles (Ohta 1976; Kimura and Ohta 1971), some loci would have a lower probability of showing genic polymorphisms than others. The genic variation produced by mutation is then modulated by migration, natural selection, and genetic drift to give the final pattern of differentiation observed between populations and species. Thus whether the level of genic polymorphism at a locus is controlled by functional constraints, natural selection, or by population bottlenecks is not possible to be ascertained by examining population survey data from one or two species alone; for this we need comparative data on a number of related species. Such comparisons have the potential to show if the levels of variation at homologous loci are uniformly low, high, or vary from species to species and from group to group and would possibly indicate the role of purifying selection, balancing selection (or neutrality), and population bottlenecks, respectively.

A comparison of single-locus mean **heterozygosity** among different groups of *Drosophila* (Singh 1989) shows three types of loci: (1) loci that are more or less uniformly monomorphic or show little heterozygosity in all groups, e.g., α-glycerophosphate dehydrogenase, malate dehydrogenase, superoxide dismutase, fumarase, and glutamate oxaloacetate transaminase; (2) loci that consistently show high heterozygosity in all groups, e.g., xanthine dehydrogenase, acid phosphatase, aldehyde oxidase, phosphoglucomutase, and esterases; and (3) loci that show varying levels of heterozygosity from group to group. As argued above, the first group of loci can be tentatively said to be functionally constrained and under purifying selection, the second group under balancing selection and/or neutrality, and the third group affected by population bottlenecks or by natural selection that varies from group to group.

The comparison of mean heterozygosity between groups does not show if the variations in the heterozygosity levels for the third group of loci above are mainly due to differences between species within groups or between groups. This can be examined by looking at mean heterozygosity for individual species both within and between groups. The data clearly show that species belonging to the same group tend to show rather similar levels of heterozygosity for a given enzyme; in other words, the variation between species within groups is much smaller than that between groups (Singh 1989). Thus, of the two sources of variation mentioned above, the major component is the variation at the group level, and to a lesser extent the variation at the species level.

The conclusion that species belonging to the same group of *Drosophila* but which are phylogenetically separated for thousands or millions of years show similar levels of heterozygosity for individual enzymes underscores the role of phylogenetic constraints on natural selection at the level of the gene. We will come back to this point later.

Genetic Divergence Between Species

Measures of Genetic Divergence

In most genetic studies of species comparisons **Nei's genetic distance** statistic (D) has been used as a measure of genetic divergence (Nei 1975). The advantage of using D is that it combines information on both allele numbers and their frequencies, and the changes can be inferred at the level of codons. The D is also linearly related to time in closely related organisms (i.e., with $D < 1$). Another useful measure, although presently less commonly used, would be the proportion of alleles that are not shared between species, especially between closely related taxa. Unique alleles can give qualitative information about isolation of populations and their history. Still a third measure would be the proportion of loci that are alternately fixed for different alleles between two species. The three statistics together would be more useful for studying evolutionary history of populations and species than D alone, as different evolutionary processes can affect D in a variety of ways and lead to similar overall D between pairs of taxa that vary in their evolutionary history.

Patterns of Genetic Divergence Between Species and Higher Taxa

The above three measures of genetic divergence for *D. melanogaster* and *D. simulans* are shown in Table 4. The divergence statistics are shown for individual chromosomes separately. There are several interesting points to note: (1) *D. melanogaster* harbors nearly twice as many unique allleles as *D. simulans* (27% vs. 16%), and chromosome X and 2, but not 3, make a contribution to this allelic diversity. All chromosomes show an equal proportion of shared alleles. (2) Chromosome 2 has significantly more alternately fixed (monomorphic) loci than chromosome X or chromosome 3. (3) As a consequence of (1) and (2), the genetic distance estimate is nearly twice for chromosome 2 than that for chromosome 3. (4) Seven percent of all loci examined have become fully diverged and all are enzymes; none of the abundant protein loci show complete divergence (Table 2). The loci for which one species is monomorphic and the other polymorphic are also interesting. At most of these loci, one of the species apparently carries one of the alleles of the polymorphic set found in the other species (data not shown). Two simple logical possibilities exist for the origin of this situation: either one species (perhaps through one or more population bottlenecks) has lost all but one of the alleles of an ancestral polymorphism, or one species has acquired one or more alleles in addition to an ancestral monomorphism.

Another aspect of genetic divergence not shown in Table 4 is the shape of the distribution of genetic identity between species. Most sibling species show U-shape distribution of genetic identity which is also the case for *D. melanogaster* and *D. simulans* (Choudhary et al. 1988). The mean genetic distance between

Table 4. Percent shared (common) and unshared (unique) alleles, mean genetic identity, proportion of loci diverged and mean fixation index of allozyme loci for various chromosomes between *D. melanogaster* and *D. simulans*[a]

		% Alleles (number)					
Chr.	No. of loci	Unique to D.m.	Unique to D.s.	Common	Proportion loci diverged (I = 0)	Mean genetic identity (I)	Genetic distance (D)
X	9	29.0 (9)	12.9 (4)	58.1 (18)	0.11	0.709	0.344
2	17	29.4 (15)	19.6 (10)	50.9 (26)	0.29	0.631	0.460
3	27	22.7 (22)	21.6 (21)	55.7 (54)	0.04	0.774	0.256
2 + 3	57[b]	27.7 (56)	18.3 (37)	53.9 (109)	0.10	0.729	0.316
Total	112[c]	26.9 (86)	15.9 (51)	57.2 (183)	0.07	0.832	0.185

[a] Data from Choudhary and Singh (1987a) and Singh and Rhomberg (1987b).
[b] Includes 13 additional autosomal loci that are unmapped.
[c] Includes 2 additional proteins whose chromosome locations are not known.

D. melanogaster and *D. simulans* based on 112 homologous loci is 0.179 which gives a divergence time, t = 0.09 m.y. (Nei 1975). This is much less than the average divergence time of 2–3.5 m.y. based on DNA sequence data (Stephens and Nei 1985) and suggests that functional constraints and purifying selection have played an important role in the evolution of proteins.

As mentioned above, D can change by changing unique alleles, alleles frequencies at polymorphic loci, or by altering the proportion of alternately fixed (monomorphic) loci. This is reflected in the comparative genetic divergence data from a variety of *Drosophila* species graphed in Fig. 3. As expected, the D increases as a function of the proportion of completely diverged loci (I = 0) in the *obscura* and the *melanogaster* group but this is not the case in the *willistoni* group. In fact, in the *willistoni* group very little contribution to D is made by the completely diverged group of loci, and as fig. 4 shows, it is not because the completely diverged group of loci in this case have been lumped with the slightly overlapping polymorphic class due to large sample sizes; all divergence results from polymorphic loci sharing some but not all alleles. It is interesting to note (Fig. 3) that in the case of man and the chimpanzee the most contribution to D comes from nearly 50% of the loci that are completely diverged (I = 0), despite the fact that overall D is not all that different from many species of *Drosophila* (King and Wilson 1975). Clearly D alone is not a very informative statistic as far as the causes of species divergence are concerned.

To show that in all sorts of organisms, and not just in *Drosophila*, D is heavily dependent on the proportion of completely diverged loci, distribution of loci with

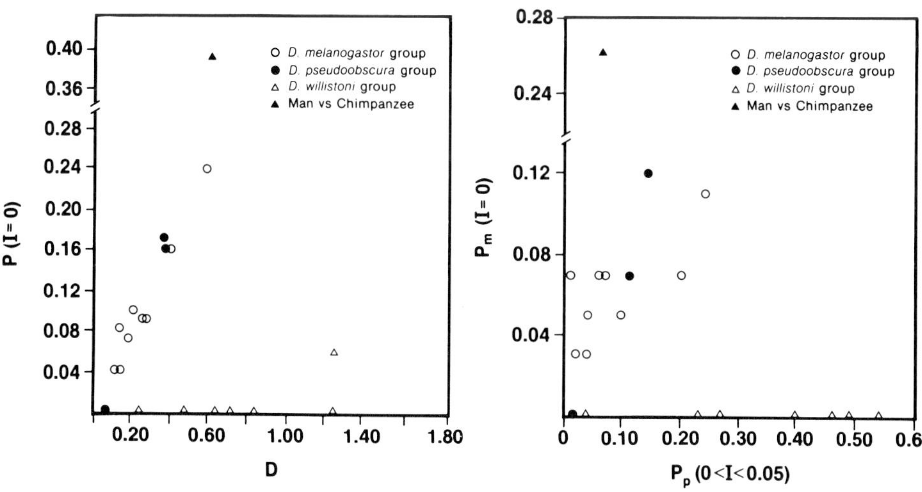

Fig. 3. *Left* relationship between genetic distance (*D*) and proportion (*p*) of loci completely diverged ($I = 0$) between pairs of closely related species in various groups of *Drosophila*. *Right* relationship between proportion of alternately fixed monomorphic loci (P_m) and polymorphic loci (P_p) showing near fixation ($o < I < 0.05$) between pairs of species. The figure for man and chimpanzee is shown for comparison. (Singh 1989)

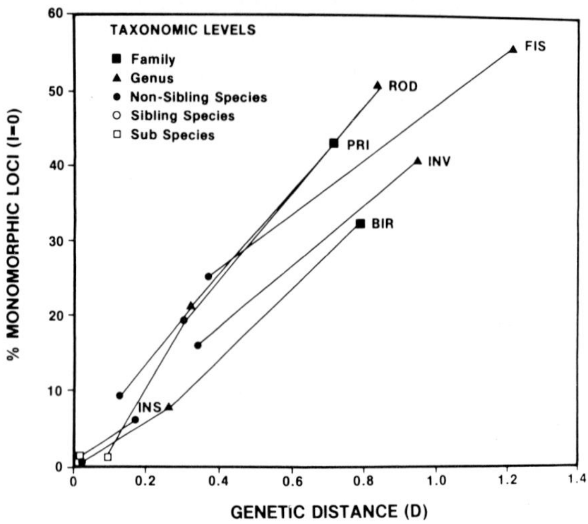

Fig. 4. A graph showing genetic distance as a function of the proportion of alternately fixed (monomorphic) loci between pairs of species at various taxonomic levels. (Singh 1989)

respect to degree of allelic divergence and the overall mean D value were examined for a variety of animals at different taxonomic levels and some relevant statistics are graphed in Fig. 4. It is interesting to note that (1) in all taxa D is almost linearly correlated with the proportion of alternately fixed (monomorphic) loci and (2) the level of divergence for a given taxonomic level varies between different groups. Of particular interest is the lower level of divergence shown by primates and birds in comparison to invertebrates, rodents, and fishes; even the family level divergence of the former taxa are lower than the genus level divergence of the latter taxa. The markedly different estimates of D values in different groups of organisms at the similar levels of evolutionary divergence mean that either the taxonomic categories between widely divergent groups of organisms are not comparable, or the population and evolutionary processes that give rise to genetic differentiation at different evolutionary stages are quite variable and hence not comparable between widely divergent groups.

Thus from the material presented in this section we reach the conclusion that population bottleneck is an important cause of genetic divergence between species. Similar conclusions have been reached through analyses of chromosome structure and evolution (Bush et al. 1977; Coyne 1983; Lande 1984, 1985).

Levels of Genetic Differentiation During the Early Stages of Speciation

During the last 20 years the technique of gel electophoresis has been applied to a wide variety of organisms at different stages of evolutionary divergence with a view to estimating the amount of genetic divergence that occurs during the **early**

Table 5. Summary of Nei genetic identity in populations of *Drosophila* at different stages of evolutionary divergence. (After Ayala 1975)

Species group or cluster	Local populations	Subspecies	Sibling species	Nonsibling species	Reference
willistoni	0.970 ± 0.006	0.795 ± 0.013	0.563 ± 0.023	0.352 ± 0.023	Ayala et al. 1974
mulleri	0.999 ± 0.001	0.878 ± 0.009	0.777 ± 0.022	—	Zouros 1973
bipectinata	—	0.907 ± 0.010	0.788	—	Yang et al. 1972
pseudoobscura	0.993	0.824	—	—	Ayala and Dobzhansky 1974
obscura	—	—	0.466 ± 0.031	0.204 ± 0.004	Lakovaara et al. 1972
Hawaiian	—	—	0.539 ± 0.054	0.164 ± 0.019	Ayala 1975

stages of speciation (e.g., see Ayala 1975; Powell 1975; Throckmorton 1977; MacIntyre and Collier 1986). The literature is too large to review here and hence only a sample of average statistics for the lower stages of evolutionary divergence in *Drosophila* is provided in Table 5. The most important finding on species divergence is the generally low level of genic differentiation associated with the early stages of speciation. The local populations of the *willistoni, mulleri,* and *pseudoobscura* group appear to be very similar and show a genetic identity of over 97%. The subspecies on the average show a genetic identity from 80–90%, sibling species from 46–78%, and nonsibling species from 16–35%. Note that while the genetic identity values for subspecies in different *Drosophila* groups are quite comparable, the values for sibling and nonsibling species vary among groups. Thus sibling species of the *mulleri* and *bipectinata* group show higher genic similarity than those of the *willistoni* and *obscura* group; and the nonsibling species of the Hawaiian *Drosophila* and the *obscura* group show genic similarities that are lower than that in the *willistoni* group. This shows that estimates of divergence above the species level are not comparable even between different groups of *Drosophila.*

The average difference of 15% in genetic identity between subspecies equates to a genetic distance $D = 0.16$, which means that on the average 1 out of 16 or about 6% of gene loci have diverged between subspecies. But even this may be an overestimate, as the percentage of completely diverged loci between *D. melanogaster* and *D. simulans*, a pair of sibling species is only 7% (Table 2). Of course, even the 6% difference must includes both relevant and irrelevant genetic changes, which leads to the unavoidable conclusion that relatively little genetic differentiation is required for speciation (Lewontin 1974; Ayala and Dobzhausky 1974; Ayala 1975; Throckmorton 1977).

As shown earlier, a second feature of the older species-divergence data on *Drosophila* appears to be a general lack of completely differentiated loci. Even species groups showing high genetic distances (e.g., *willistoni* group) lack fully differentiated loci. There is also generally a lack of species-specific (unique) alleles at the polymorphic loci. While our previous work with *D. pseudoobscura* (Singh 1983) and our present work with *D. melanogaster* and *D. simulans* with a relatively large number of loci may suggest that earlier measures of genetic divergence based on enzyme data may be underestimates, our 2D data based on male reproductive tract proteins suggest a much lower level of polymorphism in the genome and the role of a small class of genes in speciation (Choudhary et al. 1988).

Evolutionary Constraints on Species Divergence

It is a common practice in evolutionary genetics to compare genetic divergence at a given taxonomic level across organisms and to seek answers for varying rates of evolution in different organisms (e.g., King and Wilson 1975). But as pointed out above, there are various types of evolutionary constraints that operate on species divergence and that need to be taken into account in comparisons of D across widely divergent taxa (Singh 1988a). A list of these constraints and their effects on species divergence is summarized in Table 6.

The **phylogenetic constraints** can make the monomorphic or polymorphic state of a locus to be highly correlated between related species, which is the case between *D. melanogaster* and *D. simulans* (Choudhary et al. 1988). This type of positive correlation is expected for two species that arose relatively recently from the same ancestral gene pool. Another explanation, and one that need not exclude the historical one, invokes similar kinds of natural selection on homologous loci in the two species.

The **historical constraints** operate as a result of population bottlenecks (affecting historical effective species size), colonization and rapid expansion into new areas, and species hydridization and introgression. These are all essentially genetic accidents that can have profound effects on species divergence (see Choudhary and Singh 1987b and the references therein).

The effects of **functional constraints** on number of alleles and heterozygosity have been considered in the past (see Singh 1988b and the references therein), but their effects on genetic divergence between species is as yet an unexplored area. Recently we have compared genetic identity and proportion of alleles unique or shared between *D. melanogaster* and *D. simulans* for various classes of proteins and enzymes, classified according to some commonly used structural/functional constraints (Singh unpublished). Enzymes show higher divergence (or lower genetic identity) than abundant soluble proteins, and among enzymes those with externally generated substrates show higher divergence than those with internally generated substrates. The *D. melanogaster* and *D. simulans* share fewer alleles at the "null tolerant" loci than the "null nontolerant" loci, suggesting that essential enzymes are under strong purifying selection.

Table 6. Various types of evolutionary constraints and their effects on genetic divergence between species

Types of constraint	Effects
Phylogenetic constraints	Correlation between monomorphic and polymorphic state of a locus
Historical constraints	
Population bottlenecks	Increase in genetic distance due to loss or fixation of alleles
Colonization and range expansion	Increase in genetic distance due to loss or fixation of alleles
Hybridization and introgression	Decrease in genetic distance due to sharing of alleles
Functional constraints	Heterogeneity between different "classes" of gene loci in the level of genetic divergence
Ontogenetic constraints	Low divergence of early expressing genes
Selectional constraints	Lack of correlation between levels of within-species and between-species genetic divergence due to balancing selection

The **selection constraints** can affect species divergence in two opposite ways: they can retard divergence by eliminating alleles (purifying selection) or by maintaining balanced polymorphisms, or can accelerate divergence by favoring different alleles in different species (adaptive genetic divergence). Since adaptive divergence between species must use existing genetic variation within species, we should expect to find a positive association between genetic distance (D) and heterozygosity (H) at polymorphic loci. The data on *D. melanogaster* and *D. simulans* (Fig. 5) show that there is indeed a strong correlation between D and H but it is negative and not positive, as would be expected if the genetic divergence between species was based on genetic variation within species. These results lend themselves to several interpretations. First, some of the highly heterozygous loci may indeed be locked under balanced polymorphisms and consequently their divergence is slowed down, giving low D. Second, genetic divergence between species may be predominantly affected by founder effect and less so by adaptive divergence, which would explain the negative correlation between D and H (Fig. 5). And finally, high divergence of loci with low heterozygosity may suggest a role for new mutations in adaptive divergence and speciation (Wright 1982a; Choudhary et al. 1988).

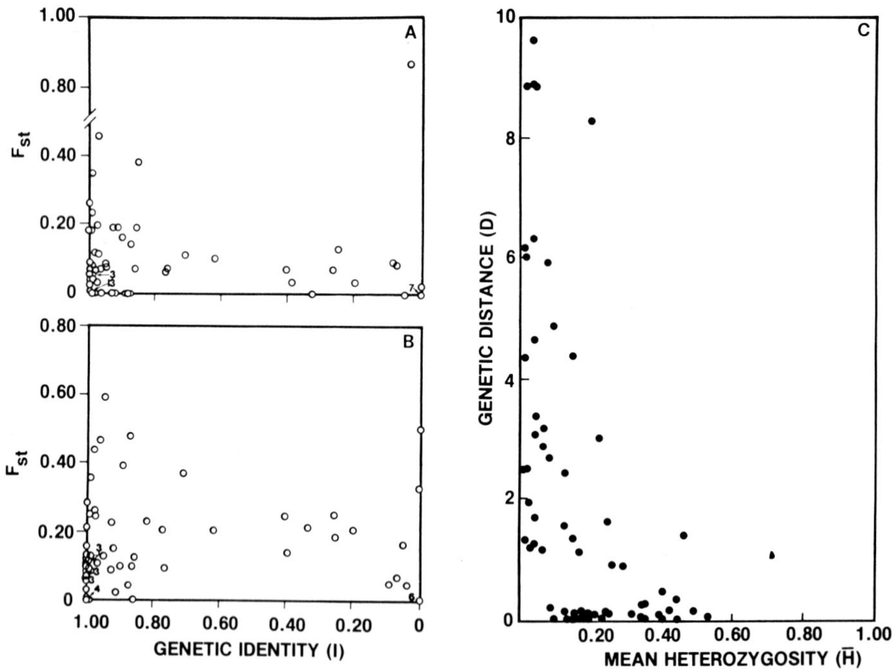

Fig. 5. Correlation between genetic identity and fixation index at polymorphic loci in *D. simulans* (**A**) and *D. melanogaster* (**B**), and between mean heterozygosity and genetic distance for all polymorphic loci in the two species (**C**). Note that six alternately fixed (monomorphic) loci between the two species are not shown in **C**. (Choudhary et al. 1988)

The **ontogenetic constraints** take into account the developmental role of genes and proteins and its effect on the level of genetic variation. There are as yet very little data which bear on this aspect of genetic variation. Our two-dimensional electrophoretic studies (see below) have shown that genes and proteins that express during the early stages of development are more likely to be conserved than those which express later. The results are compatible with the observation that in closely related species, species-distinguishing taxonomic differences first appear usually at the later stages of development.

The Nature of Genetic Variation and Theories of Speciation

Genetic Models of Speciation

Genetic models of species formation make predictions as to how many, or what proportion, of genes are involved in species formation. **Mayr's "peripatric" model** (Mayr 1954, 1963, 1982c), an elaborated version of the basic allopatric model, suggested massive reorganization of the gene pool, with extensive allele substitution or fixation, in the course of achievement of reproductive isolation. As shown in the previous section, early results from electrophoretic studies, althogh based on fewer loci, disproved this prediction in its most extreme interpretation, and our recent results based on a larger sample of loci reinforce this conclusion (Choudhary et al. 1988). However, the early studies usually failed to detect complete allelic divergence at any loci, and thus left open the question whether such divergence occurs during species formation, and if so, at what proportion of loci.

It now appears, from our results, that in a sufficiently large sample complete divergence between closely related species can be detected at a small percentage of loci. Since *D. melanogaster* and *D. simulans* have apparently been separated for at least 2 m.y. (Stephens and Nei 1985), they are probably far from their earliest stages of genetic differentiation. Thus, the figure of about 10% completely diverged loci that we obtained would seem likely to include many changes attributable to side effects of population processes (e.g., random genetic drift) accompanying species formation, or to subsequent phyletic divergence, and not causally related to the establishment of reproductive isolation per se. If as few as, say, 10% of the completely divergent loci were actually involved in reproductive isolation, this would amount to a total of only about 50–100 genes in a genome the size of that carried by *Drosophila* [5000–10000 genes (Raff and Kaufman 1983)]. Although this is still a large absolute number of genes, it is perhaps not surprising that complete interspecies allelic divergence has not been widely detected by electrophoresis, especially with species more closely related than are *D. melanogaster* and *D. simulans.*

Genetic models involving only small numbers of genes have been proposed periodically. The most important distinction between most of these models and the allopatric type is not, however, numbers of genes per se, but rather the postulation of a "privileged" class of genes that play a key role in species formation, while being constrained in their rates and/or modes of evolution within species. Two recent and rather precisely formulated examples are the **"founder-flush" model** suggested by Carson (1975, 1986, 1987) and the **"genetic transilience" model** formulated by Templeton (1980, 1981). Each of these entails special features of the genes involved; specifically, strong epistatic interactions in phenotype and fitness determination, involving a large number of polygenes (Carson's model) or a few major genes and their modifiers (Templeton's model). In addition, both authors have suggested that genes controlling specific phenotypic traits, especially those connected with mating behavior, are those most likely to fit these requirements.

These models were originally inspired by patterns of species formation in the extremely speciose drosophilid fauna endemic to the Hawaiian islands. The ecological and population-genetic contexts of the differentiation of such island species are perhaps unlikely to correspond in detail to those of the *D. melanogaster*/*D. simulans* divergence, which occurred in eastern continental Africa. Nevertheless, we think that certain important formal aspects of the models may apply, and could serve to organize further investigation. Our statistical and purely structural approach does not allow us to discriminate between direct and pleiotropic effects of diverged loci on reproductive isolation, or even between those loci that have effects and those that do not. However, within these limitations the data do contain patterns that are at least consistent with the importance of a more or less distinct class of genes in the separation between *D. melanogaster* and *D. simulans*. These genes may differ quantitatively from the rest of the genome with respect to their rates of evolution, if not qualitatively in terms of their functions or modes of evolution.

Aside from the aspect of a relatively small proportion of diverged loci, we have found evidence for the importance of selective constraint operating within species from both purifying and balancing selection (Choudhary et al. 1988). This suggests that the major source of genic divergence between species may be substitutions at loci which are maintained monomorphic within species. This supposition is in fact consistent with preponderance of within-species monomorphism among completely diverged loci. Thus, one distinguishing feature of loci with high potential for interspecies divergence may be tight constraint by purifying selection, perhaps combined with the presence or appearance of rare alleles capable of fixation by directional selection. If such genetic conditions coincided with the ecological conditions favoring interpopulation divergence, allele substitutions at essentially monomorphic loci could play an important role in species formation (Wright 1982a).

With respect to the functional aspects of loci involved in species formation, we are intrigued by the seeming tendency for **hybrid male sterility** to arise in the early stages of species differentiation in *Drosophila* (Bock 1984). This prompted us to carry out a comparison of male reproductive tract proteins by 2DE in *D. melanogaster* and *D. simulans* (Coulthart and Singh 1988 a, b, c). The results bear on the present discussion in two ways. First, male reproductive tract proteins appear to be more divergent between *D. melanogaster* and *D. simulans* than are proteins of certain "nonreproductive" tissues such as imaginal wing discs. Whether this indicates preferential genetic involvement of male-reproductive-tract functions in species formation remains to be examined. Second, an unexpected consequence of our use of 2DE, with its detection of proteins by direct staining, was that a different molecular mode of interspecies protein divergence could be easily detected; i.e., differences in protein amount and presence/absence. Since such differences were very frequent among proteins of the male reproductive tract, and most frequent among proteins with organ-specific distributions within the reproductive tract, we raise the question whether the genetic changes accompanying species differentiation are preferentially clustered along an axis representing the fundamental kinds of alterations to gene structure and function associated with primary reproductive characters. In comparison with electrophoretic data on protein structure, very few data are available on differences in expression of homologous proteins in closely related species. Exploration of the role of such factors in genetic divergence could prove to be the most fruitful approach of all.

Molecular Mechanisms of Speciation

Recent advances in molecular biology, especially the unfolding of genome structure and organization, have led to the proposition of a number of molecular mechanisms of speciation involving genomic (as opposed to genic) or extragenomic (i.e., foreign genetic elements) genetic changes (see reviews: Rose and Doolittle 1983; Krieber and Rose 1986; see also Chap. 1). These mechanisms can be grouped into three main classes: genomic disease model, mechanical incompatibility, and genome resetting.

In the **"genomic disease" model** (Rose and Doolittle 1983; Ginzburg et al. 1984), an isolated population is assumed to lose or fail to acquire mobile genetic elements (transposons) that are present in the rest of the species and the isolated population may lose property in establishing immunity to these elements. Therefore, crosses between individuals from isolated populations could lead to defects in their hybrids, and establish post-zygotic reproductive isolation. This model is based on the observations of deleterious effects of hybrid dysgenesis in *D. melanogaster* (Kidwell et al. 1977; Ginzberg et al. 1984) and parallels have been drawn between hybrid dysgenesis and species hybrid pathologies in *Drosophila* (Engels and Preston 1979; Kidwell and Novy 1979). However, there are striking differences between the dysgenic sterility and the interspecific hybrid sterility. First, dysgenic sterlity is normally observed only when flies are grown at high temperatures, while interspecific hybrid sterility occurs at normal rearing temperature. Second, both P-M and I-R dysgenesis in *D. melanogaster* have more drastic effect in females (Kidwell et al. 1977) while hybrid sterility in *Drosophila* is much more common in males (Bock 1984). Furthermore, the parallels break down when other species in the *D. melanogaster* groups are compared (Coyne 1985). The genetic divergence leading to developmental anomalies in interspecific hybrids is more advanced in the older species pair, *D. melanogaster/D. simulans*, than in the younger pair, *D. melanogaster/D. mauritiana*. Coyne (1986) found no evidence for elevated rates of male recombination in species hybrids, thus negating the role of hybriddysgenesis-like elements in reproductive isolation. In addition, genetic analyses of hybrid male sterility in *D. melanogaster* (Coyne 1984, 1985), *D. pseudoobscura* (Dobzhansky 1936, 1951; Orr 1987), and many other *Drosophila* (Naveira and Fontdevila 1986) and insects (Curtis 1982) have shown that hybrid male sterility is due to many genes, with X-linked loci making the largest contribution to sterility. These results are difficult to explain if mobile genetic elements were the leading cause of reproductive isolation.

The **"mechanical genome incompatibility" model** (Burr and Schimke 1980; Dover 1982) requires a large amount of interpopulation differences in dispersed repetitive DNA sequences. These differences could result from random amplification of preexisting sequences initially rare in the parent species, with accidental amplification of different sequences leading to incompatibility between populations. There are various molecular processes, such as transposition, unequal crossover and gene conversion, which alter the locations, numbers, and sequences of clustered and dispersed repetitive DNA's (Burr and Schimke 1980; Murray et al. 1981). Dover (1982) has integrated them into a single phenomena, called "molecular drive". These changes may disrupt chromosomal interaction (i.e., loss or reduction in meiotic recognition and not the gross chromosomal imbalance) in such a way that hybrids are sterile.

The third, the **"genome resetting" model**, also depends upon non-coding repetitive DNA element but requires at least some of these elements to be functional. They may be regulatory elements whose positions and sequences control the expression of other structural genes in developmental pathways in a complex and coordinated fashion (Britten and Davidson 1969; Davidson and Britten 1979; Raff and Kaufman 1983). The changes in positions and sequences of these regulatory elements could alter early developmental pathways, leading to rapid divergence between populations (Wilson et al. 1977; Gould 1977; Hilu 1983). In this mode of evolution the speciation is obviously a macromutational process, regulatory genes play a vital role, and relatively fewer genes are involved in speciation. This hypothesis is quite similar to the one proposed by Goldschmidt (1940) in which macromutations were supposed to be the cause of speciation except that in present-day terminology the macromutations are supposed to be changes in regulatory elements (Wilson et al. 1974, 1977). Gould (1977) thinks that it is changes in these genes that account for evolution at and beyond the species level. Both chromosomal rearrangements (White 1978) and point mutations at a critical position in the DNA sequence (Beckwith and Rossow 1974) have been proposed as mechanisms of altering regulatory genes. Many closely related species differ in their metaphase chromosome in the amount and distribution of heterochromatin (e.g., see Ahearn and Baimai 1987). Several studies have shown that rate of chromosome evolution is strongly correlated with rate of speciation (White 1978; Baker and Bickham 1986; Thompson and Sites 1986; Davenport 1986; Bush 1981; King 1981), but it has been argued that this correlation is likely to result from the role of small population sizes in both speciation and chromosome fixation and is not a causation (Charlesworth et al. 1982; Wright 1982b); there are many closely related species whose polytene chromosomes are homosequential and do not involve chromosomal alterations (Carson and Stalker 1968; Clayton et al. 1972; Carson and Kaneshiro 1976).

Genetic Mechanisms of Speciation in Plants vs. Animals

Are the genetic mechanisms of speciation different between plants and animals? The answer is clearly yes, at least with respect to the relative frequency of the various modes of speciation (see Stebbins 1950, 1982b; Grant 1981 for details). It is generally recognized that speciation by chromosomal mutations and polyploidy is more common in plants than animals. Species hybridization, followed by maintenance of hybrids or recombination leading to new species, is another mechanism which appears more likely to occur in plants than in animals. Speciation by major mutations affecting, for example, breeding systems are also thought to be more likely to occur in plants than animals. In summary, the genetic mechanisms which have a built-in capacity for producing reproductive isolation rapidly (if not in one step) and which are the basis of quantum (or instant) speciation are more likely to occur in plants than animals. It is generally thought that because of their simple development plants are more tolerant of drastic genetic changes than animals.

But what about genetic mechanisms of gradual speciation involving single gene changes or "submicroscopic structural differences", to use Stebbins' (1982a) term? Are they different between plants and animals? If one follows the history of the debate on the **sympatric** and **allopatric models** of speciation, one finds more supporters of the sympatric models among plant than animal researchers. It is argued that plants can differentiate more rapidly and counteract the homogenizing effect of gene flow without much geographic isolation (a requirement for the sympatric model) although the reasons are not always spelled out. Thus whether plants may be more prone to sympatric speciation because of limited role of gene flow (Ehrlich and Raven 1969; Levin 1979), or because their "genetic machinery" is different from that of animals in allowing rapid divergence, must depend on the existence of some fundamental difference between the biology of plants and animals bearing on these points.

It appears that there are some fundamental differences between plants and animals which would make rapid genetic divergence in plants at least in principle more likely than in animals (Walbot 1985; Stebbins 1982a). First, since animals move and plants do not, animals can reduce the immediate impacts of a large variety of environmental factors by making use of their behavioral repertoire, which plants cannot. Thus on a comparative timescale it appears likely that plants, ceteris paribus, would experience stronger natural selection than animals. Second, animals contain a germline which is generally inactive in the somatic body of the animal. The germline cells migrate to the gonads and produce gametes and thus remain unaffected by mutations in the somatic body of the animals. In plants, on the other hand, the gametes' differentiation occurs from a population of cells that have been actively dividing and differentiating in the somatic body of the plants. As Walbot (1985) puts it, "The situation would be comparable to the production of animal gametes from the skin"! Third, animal development is quite rigid in most animals and it proceeds in a "preordained" fashion after it starts. In contrast, plants produce vegetatively until an environmental signal triggers a switch for floral development. This means that while plants may be unable to modify the effect of environmental impact behaviorally, they have evolved mechanisms to be developmentally more plastic in their response to the environment. Thus plants have a much wider capacity to modify their reproductive potential than animals, which bears directly on their ability to diversify rapidly. Fourth, as mentioned above, plants appear to be more tolerant of genetic imbalances than animals, and segregation of abnormal karotypes and high occurrence of polyploidy in plants attest to this fact (Lewis 1980). Lastly, plants have a genetically active haploid phase of the life cycle which is rare among animals. The independent functioning of the haploid gametophyte in plants would eliminate many lethal genes from the populations by selection in the haploid phase of the life cycle. However, it may also mean that plants have a great redundancy of function in their genes which may buffer them against genetic damage and at the same time allow them more room to diversify genetically.

In summary, their limited movement, their developmental plasticity, their capacity for tolerance of genetic imbalance, their continuous growth and differentiation, and the potential for rapid environmental and biological (e.g., pollinators) modification of their fitness traits make plants not only more likely to favor quantum (instant) speciation (Lewis and Roberts 1956) but also, at least in principle, more likely (than animals) to favor gradual sympatric speciation. However, as we will show later the geographic axis of speciation (sympatry vs. allopatry) is not independent and has relevance only in combination with the nature of characters involved in speciation and the nature of genetic variation. When viewed together, i.e., geographic relations, types of characters, and genetic variation, the genetic theory of gradual speciation for annual flowering plants and animals may differ only quantitatively and not qualitatively (Stebbins 1982b). On the other hand, because plants do not contain a germline and generally lack sex chromosome differentiation, post-reproductive isolation in plants is usually weaker than in animals.

It is also worth noting that even in plants chromosomal sterility appears to have played a major role at the lower level of species affinity only in the annual herbs and not so much in the perennial herbs and the woody plants (Grant 1981). This would suggest that the high association of chromosomal alterations and speciation in the annual plants is only partly a property of being plants; partly it must be a property of the particular types of population dynamics, i.e., colonization and population bottlenecks and extinction to which the annual plants are subjected repeatedly.

Speciation, Adaptation, and Macroevolution

Macroevolution involves both major morphological and physiological change and reproductive isolation (see also Chap. 4). The most appealing unifying theory of adaptive evolution is the shifting balance theory (SBT) proposed by Wright (see Wright 1982 a, b for recent reviews). The shifting balance theory requires a **type of breeding structure** in which the whole of a species is subdivided into "thousands of more or less independent local populations" (Wright 1982a, p. 8). The theory also assumes the presence of genetic variation in fitness, interaction among loci, and pleiotropy. The local populations (or demes) are assumed to be adapted to their local environments and are said to be at adaptive peaks. The evolution in species with such a genetic structure consists of three phases. First, individual populations go through random genetic drift and explore the multitude of fitness terrains surrounding their adaptive peak and experience a period of maladaptation. A new geneotype with higher fitness may appear due to new associations of alleles. In the second phase, mass selection, the new genotype(s) increases in the local population and takes it to a higher fitness peak. If such a peak shift is realized in terms of increased reproductive capacity the local populations would increase in size. The third phase, interdeme selection, involves migration of individuals from the newly adapted, higher-fitness population to other demes which leads to selection among local populations (selective diffusions) and improves the fitness of the entire species.

The breeding structure required in the STB is such that local populations are independent enough to evolve and realize peak shifts individually and yet remain connected enough (through migration) for interdeme selection to occur and affect the entire species distribution. It is obvious that the breeding structure we are talking about is not a simple one made up of just numerous subdivided populations but it requires subpopulations of critical size, having critical amount of gene flow, that are *stable* over space and time. It is not known if such breeding structures exist in most organisms. The question is not whether mass selection within demes coupled with small migration between demes can lead to species improvement over time, but whether the genetic structure is stable long enough for major morphological evolutions to occur in the manner described in the shifting balance theory.

A second aspect of the shifting balance theory pertains to evolution in new environments and role of **major mutations.** Wright thought that evolution in new environments (e.g., in vacant niches after a major catastrophe) can lead to incorporation of major mutations which otherwise would not occur. Thus he states:

> "A mutation that is only very imperfectly adapted and has rather serious deleterious side effects may be the best available mutation to occupy the new niche. In many, perhaps most,

cases, the initial occupation may, however, be by a minor allelic substitution. Drastic change can then occur very rapidly in the absence of competition by a succession of such substitutions, which is essentially a continuation of essentially microevolution except in rate" (Wright 1982a, p. 14).

The above evolutionary scenario proposed by Wright for species improvement by the shifting balance process is equally likely to apply to speciation by founder effect if applied to relevant characters. This part of the shifting balance theory provides us with a general framework to explain low variation within species but high divergence between species such as that we have observed for the male reproductive tract proteins of *D. melanogaster* and its sibling species (Coulthart and Singh 1988 a, b).

The role of major mutations or accelerated evolution in speciation can be tested by comparing patterns of polymorphisms at homologous gene loci among closely related species with known phylogenetic relationships. If major mutations and/or accelerated evolution have played an important role in the evolution and speciation of *Drosophila*, the comparisons of the rates of evolution for different genes within vs. between groups should be able to provide some evidence. We have carried out such an analysis using allozyme variants by comparing the rate of allelic substitutions for the same gene in different *Drosophila* species groups. Since repeated population bottlenecks during the evolutionary history of each group would affect all loci identically, all genes within a group should show a group-characteristic fast or slow rate of evolution. On the other hand, if natural selection has played an important role, different genes would show markedly different rates of evolution within each group. Our approach was to determine the total number of alleles within each group of species for homologous enzymes. It is important to note that these comparisons employing the electrophoretic technique are based on relatively slow-evolving proteins to avoid multiple allelic substitution which, if present, would go undetected.

The results from a survey of this type using sequential gel electrophoresis on about half-a-dozen enzymes that are either invariant or have low heterozygosity show that the rates of allelic substitutions are independent of the levels of heterozygosity, and very considerably not only between enzymes but also between groups. Furthermore, for these (slow-evolving) enzymes, subunit size has no effect on either heterozygosity or rate of allelic substitution. This is what we should expect if the rate of evolution for a gene varies considerably between groups. This kind of analysis will now, of course, become more rigorous with DNA sequence data, and would allow us to test the hypothesis: that the genetic environment of various *Drosophila* species varies between groups but remains rather similar within groups so that the same gene is subjected to rather similar physiological "selection opportunities" within groups but different between groups.

The high level of heterozygosity at the ADH locus in most species of the *melanogaster* group and its rather monomorphism in most other groups cannot be explained in any other way except in the framework of the above hypothesis (to be referred to as "phylogenetic selection" hypothesis). With DNA sequence data the analysis of phylogenetic constraints and evaluation of the role of "phylogenetic selection" would become an important area of research in molecular evolutionary genetics (e.g., see Kreitman 1983).

New Directions

Measuring Genetic Variation: a Change in Emphasis from "How Much?" to "What Kind?"

All through the history of population genetics the studies of genetic variation have been mainly concerned with quantitative description and with the exception of lethal vs. quasinormal genes (Lewontin 1974; Dobzhansky 1970), very little attention has been paid to the **qualitative aspect of genetic variation**. We have shown that in *Drosophila* the amount of heterozygosity is negatively correlated with the number of loci sampled (Singh and Rhomberg 1987b), which raises the question whether enzymes give us a true picture of genic variation in the genome. Because of the variation in heterozygosity between different classes of proteins, the mean heterozygosity of a species would very much depend on the number and the kind of genes sampled (Singh 1988b). We carried out a preliminary comparative study of proteins found in male reproductive tract tissues of *D. melanogaster* and *D. simulans*, using high-resolution two-dimensional gel electrophoresis (2DE) (Coulthart and Singh 1988 a, b, c). Three important conclusions have emerged from these data (Table 7, 8).

First, the overall proportions of completely diverged loci are not very different between the soluble enzymes and other proteins surveyed by 1DE (7.0%) and the abundant solubilized proteins surveyed by 2DE (13.2%). The difference is reduced still further if the 1DE set is restricted to soluble enzymes (10.1%). The second point concerns the differences in proportions of completely diverged loci seen between, on one hand, male reproductive tract proteins that are found in both testis and gland tissue of that tract ("common" and "testis-elevated" proteins) and, on the other hand, those proteins found in testis but not gland tissue, or vice versa ("testis-specific" and "gland-specific" proteins, respectively). The former classes contain 9.8 and 3.5% completely diverged loci, respectively, and the latter 17.5 and 26.1% completely diverged loci, respectively. The figure of 9.8% completely diverged loci for "common" proteins is strikingly close to the 10.1% found with soluble enzymes. The question arises whether genes with relatively broad, or ubiquitous, patterns of expression between different tissues or stages of an organism ("housekeeping" genes) might have characteristically lower probabilities of complete interspecies divergence than do genes with relatively late or tissue/stage-specific patterns of expression.

Lastly, we draw attention to interspecies divergence of a different type, readily detectable with 2DE but not as readily with 1DE. Overall, 20–30% of *C. melanogaster* or *D. simulans* male reproductive tract **protein loci** (studied by 2DE)

Table 7. Summary statistics for polymorphism in male reproductive tract proteins detected by 2DE in *D. melanogaster* and *D. simulans*[a]

Tissue	Species	No. of loci studied	Proportion of loci uniformly monomorphic	Proportion of loci polymorphic (mean ± SE)	Heterozygosity (mean ± SE)	Proportion of loci diverged	
						Homologous (I = 0)	Nonhomologous (showing +/−)
Common	*D. melanogaster*	110	0.927	0.041 ± 0.045	0.010 ± 0.008	0.098	0.183
	D. simulans	131	0.614	0.288 ± 0.022	0.094 ± 0.012		
Testis-specific	*D. melanogaster*	74	0.878	0.101 ± 0.029	0.030 ± 0.007	0.175	0.234
	D. simulans	87	0.896	0.092 ± 0.016	0.032 ± 0.010		
Gland-specific	*D. melanogaster*	26	0.769	0.173 ± 0.082	0.065 ± 0.033	0.261	0.391
	D. simulans	25	0.680	0.280 ± 0.000	0.119 ± 0.019		
Testis-elevated	*D. melanogaster*	34	0.911	0.058 ± 0.042	0.023 ± 0.019	0.035	0.291
	D. simulans	28	0.857	0.107 ± 0.051	0.028 ± 0.030		
Total (2DE)	*D. melanogaster*	244	0.893	0.080 ± 0.044	0.024 ± 0.010	0.132	0.231
	D. simulans	271	0.900	0.085 ± 0.000	0.029 ± 0.001		

[a] Data Coulthart and Singh (1988b).

Table 8. Single-line comparisons for polypeptides of reproductive and nonreproductive tissues[a]

Species compared	Wing discs		Reproductive tracts	
	Fraction[b]	Fraction[c]	F	F_p^c
D. simulans vs. *D. melanogaster*	0.833 (0.811)	0.60	0.639	0.28
D. simulans vs. *D. sechellia*	0.856	0.68	0.791	0.40
D. melanogaster vs. *D. sechellia*	0.700	0.53	0.664	0.41
D. simulans vs. *D. mauritiana*	(0.955)	—[d]	0.848	0.43
D. melanogaster vs. *D. mauritiana*	(0.803)	—	0.636	0.30
D. mauritiana vs. *D. sechellia*	—	—	0.781	0.38

[a] Data from Coulthart and Singh (1988b), and (figures in parentheses) from Ohnishi et al. (1983).
[b] Shared fraction of total spots.
[c] Pairable fraction of unshared spots.
[d] Data not available.

show "presence/absence" differences between *D. melanogaster* and *D. simulans.* In these cases, a protein present in one of the two species lacks a detectable homologous spot in the other species; we have proposed that these seemingly species-specific proteins actually represent large interspecific differences in levels of gene expression (Coulthart and Singh 1988b).

Table 8 presents data on similarity measures for whole reproductive tract proteins and proteins of third-instar imaginal discs in randomly selected pair of isofemale lines (in *D. melanogaster* and its sibling species. The similarity measures for imaginal discs are higher in every case than they are for reproductive tracts. That is, more polypeptides in imaginal discs were identical (the F measure in Table 8) between species in each pair, and, of the nonidentical polypeptides, more could be matched (the F_P in Table 8) with a homologous one in the other species than was possible with reproductive tract polypeptides. Interestingly, Ohnishi et al. (1983), with proteins of adult whole-body homogenates in two-dimensional gels stained with Coomassie blue, obtained F values which are much closer to those for imaginal discs than those for male-reproductive-tract proteins (shown in brackets in Table 8). The proteins of tests appear to be as widely divergent between these two species as do those of whole reproductive tracts (Coulthart and Singh 1988b).

A New Genetic Model of Speciation

In this section, we first summarize some general observations bearing on genetic and phenotypic variation that we think must be taken into account in devising a general theory of speciation and then present a general model of speciation that lends itself to some experimental testing. First, the level of genetic divergence between sibling species as shown by allozyme data even in the most extensive studies appears to be not more than at 10% of the loci. Since most sibling species have undergone some level of phyletic divergence since their origin, it appears almost certain that while the absolute number of genes involved may still be high, the proportion of genes involved in speciation is much less than 10%. This strongly suggests the involvement of a rather small number of genes in speciation. Second, in our enzyme studies there is a strong negative correlation between the levels of genetic variation within species and the levels of genetic divergence between species, suggesting that gradual divergence of polymorphic loci within species is not the most important factor in species divergence. Third, a certain class of genes (e.g., the male reproductive tract proteins) appear to have low variation within species but high divergence between species. These observations again suggest the role of a rather small "privileged" class of genes in speciation whose rates if not the modes of evolution may be different than those of most genes. Fourth, in most organisms species appear to retain the ability to produce viable hybrids for much longer an evolutionary period than the ability to produce fertile hybrids. Furthermore, where species hybrids of both sexes are produced, they look phenotypically normal except in their genitalia. This would suggest that genes affecting reproductive physiology change much faster than genes affecting the organism's basic metabolic physiology (i.e., "housekeeping" genes). Lastly, any theory of geographic speciation must also deal with the issue of founder effects which may lead to reduction in fitness due to homozygosity of balanced polymorphic genes or due to fixation of deleterious alleles. Population bottleneck appears to be a dominant factor in the divergence of populations and species.

We now briefly outline a genetic model of speciation to be referred to as the **"reproductive model" of speciation** emphasizing the role of small populations (founder effects) and genes affecting primary reproductive traits (both mating and postmating types). As shown above, population bottleneck appears to be a dominant force in the divergence of species. Founder effects can lead to reduction in fitness due to both inbreeding depression resulting from homozygosity of balanced polymorphisms and fixation of numerous slightly deleterious alleles. The initial reduction in fitness is more likely to be due to inefficiency of the primary reproductive physiology affecting fertility and fecundity than to changes in the general morphology and physiology of organisms affecting viability. It is supposed that natural selection would be more sensitive to changes in the former types of characters than in the latter; this is a central presupposition of the model.

The reduction in population fitness can be counteracted in two ways, by fixation of new beneficial mutations or by selection of fitness "modifiers", presumably involving gene regulation. Thus direct evolutionary changes in genes affecting primary reproductive traits involving new mutations and indirect changes resulting from regulatory genes, are the two main sources of genetic variation relevant to reproductive isolation. The population genetic model most relevant to this kind of divergence would be one involving mass selection. Besides these, some genetic changes in the reproductive physiology would also occur due to adaptive divergence.

Depending upon the nature of genetic variation and sampling variance, the **evolution of reproductive isolation** may lead to either premating or postmating isolating barriers, but not necessarily to both, at least during the initial stages of speciation. This may lead to a negative correlation between premating and postmating reproductive character divergence during the early stages of speciation. In view of this model it is not surprising that in the Hawaiian *Drosophila*, the majority of which show strong premating isolation, relatively little genic differentiation between populations and species has been reported (Carson and Kaneshiro 1976). On the other hand, the majority of the continental species show strong postmating isloation and significant genetic divergence.

The model is also applicable to plants. Plants in general show weaker postzygotic reproductive barriers than animals probably because of lack of strong pleiotropic effects of general adaptive divergence on reproduction physiology. On the other hand, prezygotic isolating barriers develop rapidly, as in the case of annual flowering plants involving autogamy and plant-pollinator interactions.

Some Macroevolutionary Consequences of the Model

In the widely accepted "allopatric" geopraphic model, species formation is seen as a correlated side effect of independent general adaptive evolution during a period of reduced gene flow (Mayr 1963; Charlesworth et al. 1982). In recent years, however, the coupling between speciation and adaptive divergence has generally been de-emphasized, and while speciation is still seen as a gradual process, its origin is often seen as a fortuitous outcome of genetic changes not necessarily accompanied by adaptation (e.g., see Gottlieb 1982). However, there are varying views on this point, with the coupling between speciation and adaptation usually being emphasized in the "genic" models and de-emphasized in the "chromosomal" models of speciation. Gould, on the other hand, has emphasized the coupling between speciation and macroevolution to the extent that in his view major morphological and physiological adaptive changes mostly occur during speciation and the process occurs rapidly (geologically speaking) but scattered far in between long periods of evolutionary stasis (Eldredge and Gould 1972). Stanley (1982) believes that major adaptive changes occur during speciation but not the converse, i.e., not all speciation events involve major adaptive changes.

Our species divergence data, along with other genetic, chromosomal and morphological data on *Drosophila* divergence, provide evidence to suggest that speciation and major morphological evolution (major adaptation) should be viewed as two different modes of evolution—one concerned with the creation of genetic incompatibility between evolutionary lineages and the other concerned with maintenance of genetic compatibility within lineages. It is also interesting to note that in many organisms the ability to produce fertile interspecies hybrids has been lost much sooner than the ability to produce viable offspring. On the other hand, in those species where reproductive isolation is of premating type, laboratory crosses often produce fertile hybrids even when the infertile hybrids are otherwise phenotypically normal. This suggests that only the genes related to mating preference (sexual isolation) and gamete formation have diverged and the genes related to basic metabolic physiology and function are fully compatible between species. These observations suggest that during evolution the physiological compatibility of genes involved in sexual isolation (premating or postmating type) is lost much sooner than that of the "housekeeping" genes involved in the development, differentiation, and maintenance of organisms. Following this reasoning, it can be hypothesized that the major morphological and physiological changes do not occur along with speciation but follow later, initially as a result of "regulatory re-adustment" to the new gene pool (altered due to speciation), with or without new mutations, and later as a result of rapid adaptive divergence of the newly evolved lineage.

Conclusions

Speciation remains an unsolved problem in evolutionary biology partly because it is not one but a bagfull of allied problems. The central problem of speciation consists of determining the genetic basis of gradual speciation and relationship between speciation, adaptation, and major morphological evolution. The genetics basis of sudden speciation involving major gene mutations, major chromosomal mutations, and polyploidy is obviously known and what we need to know further is the frequency of their occurrence and the groups of organisms in which such mechanisms have played major roles. Population structure plays an important role in speciation but that role very much depends on the nature of genetic variation and the breeding systems of the organisms. The genetic structure of species cannot be assumed to be stable over time and hence the role of history vs. selection in the shaping of the present genetic structure of species is an important area for investigation.

Extensive one- and two-dimensional gel electrophoretic studies of species differences between *D. melanogaster* and *D. simulans* have brought three important conclusions. First, alternate fixation of different alleles at a protein coding locus appears to be a fairly general feature of divergence between the two species, occurring at a predictable frequency of about 10% for a given sample of loci. Second, there may be systematic differences between "classes" of loci, defined on the basis of their differing roles in the economy and organization of the organism, in their probabilities of undergoing complete interspecies divergence. And thirdly, at least in certain organs and tissues, a very common mode of interspecies divergence at the protein level results in presence/absence differences, as distinct from structural differences visible as electrophoretic mobility shifts. Considering the divergence time between *D. melanogaster* and *D. simulans*, our results are consistent with the importance of a small and perhaps more or less distinct class of genes in speciation. These genes may differ quantitatively from the rest of the genome with respect to their rates of evolution, if not qualitatively in terms of their modes of evolution. These observations have led us to propose a general theory of speciation emphasizing primary reproductive characters (sexual behavior, gonadal traits, and gamete formation) and the role of small populations (founder effects). It is concluded that major morphological and physiological changes do not occur along speciation but follow later, initially as a result of "regulatory re-adjustment" to the new gene pool (altered during speciation) with or without new mutations and later as a result of rapid adaptive divergence of the newly evolved lineages.

8 The Evolution of Life Histories

H. DINGLE

Introduction

The fitness of organisms is the primary focus of evolutionary biology. Since the fittest are those which provide more descendants to future generations than their competitors, and fitness is largely determined by the age-dependent schedule of behavior, birth, death, and development rates, life histories are adaptations of unique importance to the analysis of Darwinian evolution (Stearns 1976; Bell 1980). Life history characters covary and function together, and so constitute complex adaptations because they are interdependent (Frazetta 1975). Such sets of covarying traits are often referred to as "strategies" or "tactics" in the literature of ecology and evolution, meaning that they are "coadapted traits designed, by natural selection, to solve particular ecological problems" (Stearns 1976). Because of their relation to fitness, a great deal of theory, experiment, and field study has been directed at determining which associations of life history traits will evolve under particular ecological conditions.

An excursion into a field or woodlot to study natural history soon reveals the richness and variety of life histories. Some of the birds are migrants, others are residents. Hummingbirds or hawks may lay only two eggs, while hole nesters like titmice or woodpeckers may lay seven or eight or more. Many of the flowering plants are annuals which flower and produce seed only once; others are perennials, some of which, like oaks or eucalyptus, may live for over a century, flowering repeatedly. If our woodlot contains a pond, it may be filled with microcrustaceans that alternate between sexual and asexual generations or that have given up sex altogether. An exploration of the leaf litter will reveal both winged and wingless ground beetles; these may lay many eggs immediately after eclosing to adults, or wait until the following year and lay only a few. The curious naturalist will ask how these different life history patterns have evolved. Which environmental conditions, for example, have selected for which combinations of traits?

In considering this interaction of traits with environmental conditions, it is useful to begin with what Law (1979b) appropriately calls a "**Darwinian Demon**". This demon is an organism that begins reproducing at the instant of its own birth, produces infinitely large numbers of offspring at instantaneous intervals over its lifetime, experiences no mortality, and can disperse or migrate and find mates at will. Clearly such an entity is an absurdity, but a consideration of the reasons why it cannot exist should yield insight into the limits imposed on a life history. Such constraints may be endogenous to the organism itself—it can, after all, only assimilate so much or grow so fast—or they may be exogenous, since resources are limited both by their own intrinsic factors and by the density of the populations exploiting them. The partitioning of those resources to the various components of the life cycle constitutes the major element of an evolutionary strategy. The ways in which that partitioning is accomplished are a function of natural selection, genetic variance and covariance, phylogeny, and development.

Much of the theoretical effort devoted to life histories has been focused to a greater or lesser extent on determining which schedule of partitioning is "optimal". The basic question revolves around how the organism best allocates the available resources to maximize total lifetime production of progeny (Law 1979b). Most of the attention has focused on division or "trade-off" between reproductive organs (and reproductive behavior) and soma, a problem addressed early on by R.A. Fisher (1930), who was one of the first to consider the evolutionary implications of the partitioning. Other resource demands are also important to life histories, however, not the least of which are migration and dormancy, behaviors critical both for survival and for determining the time and place of breeding. It is important to consider timing, for the question is often not how to apportion between reproduction and soma, but when to switch energy and resources from one to the other. In this chapter, I shall consider briefly some of the theory concerning life history partitioning with respect to reproduction, behavior, and timing, and provide some examples of how particular life cycles may or may not fit predicted theoretical patterns.

Until quite recently these predicted patterns were developed without reference to genetics (see reviews in, e.g., Stearns 1976, 1977), even though genetic variance is necessary for evolution to occur and had been included by Fisher (1930) in his oft-cited analyses. The situation is changing rapidly; genetic variances and covariances of quantitative characters are becoming central themes in life history theory and experiments as evolutionary biologists strive to integrate genetics with ecology and phylogeny into what is becoming known as "genetical ecology" (See Chap. 9). It has become increasingly evident that as a model of evolution life history theory must be soundly based in an understanding of the relevant genetics (Lande 1982a, b; Reznick 1985). For that reason **quantitative genetics** will be the major emphasis of this chapter. It should also be apparent, however, that the roles of genetic variance and covariance are understandable only in the context of selection (ecology), ontogeny, and phylogeny, even though my discussion of these elements must perforce be more limited.

The Ecological and Demographic Background

Life Tables, Age Structure and Population Growth

The elements of the life table of a population are births, deaths, and the ages at which these occur. Given a fixed life table, a fixed sex ratio, and a fixed schedule of age-specific fecundity, the population will tend to a stable age distribution, a result demonstrated by Sharpe and Lotka in 1911 (Stearns 1976). Most subsequent discussions of demographic parameters and development of accompanying theory have assumed that age distributions are stable, and for simplicity we can make this assumption here in considering how selection might modify life tables. For fuller discussions, Stearns (1976) is still the most complete review, and briefer entertaining summaries are provided by Horn (1978) and Horn and Rubenstein (1984). I have drawn freely from these reviews in the next few sections. A good recent review is provided by Partridge and Harvey (1988).

The problem faced by the organism is how to arrange its life history most efficiently to increase its population growth rate (or of its genotype relative to the rest of the population). As we shall see, an important conclusion to come out of life history analysis is that the **schedule of reproduction**, i.e., the ages at which it occurs, is at least as important as the total number of offspring produced. It turns out that in general, population growth rate is most sensitive to variation in age at first reproduction. To see why this is so, we can begin by considering life tables.

The probability of survival to age x, 1_x, and the age-specific instantaneous birth (maternal) rate, m_x, are two primary **life table parameters.** The integral of their products $\int_\alpha^\omega 1_x m_x d_x$, is equal to R_0, the replacement rate of the population or the "net reproductive rate" (α and ω are the ages of first and last reproduction, respectively). From the 1_x and m_x terms Lotka in 1913 derived what has become known as the "characteristic equation":

$$\int_\alpha^\omega 1_x m_x e^{-rx} d_x = 1, \tag{1}$$

where r is the growth rate (or "intrinsic rate of increase") of the population when the age distribution is stable. The magnitude of r can be increased by increasing 1_x or m_x, or by earlier reproduction. This can be quite simply illustrated by considering Laughlin's (1965) equation for the approximation of r without assuming a stable age distribution which he called r_c:

$$r_c = \frac{\ln R_0}{T_c} \tag{2}$$

where T_c is the average interval between reproductive onset in two successive generations. The value of r_c can be increased by increasing R_0 (either 1_x or m_x) in the numerator or reducing T_c in the denominator, i.e., by earlier reproduction. We now return to the question of how the organism can modify the life history most efficiently to increase r.

This was the question considered by Cole (1954) in a paper which has become a classic. By means of numerical simulations, Cole demonstrated that an organism living one year could become reproductively equivalent to an immortal perennial by adding only one further young to its litter, or similarly, a perennial could become equivalently immortal by adding one young to its first litter. This is shown visually in Fig. 1. The above description assumes asexuality and no mortality for the young, but can be extended to sexual reproduction with

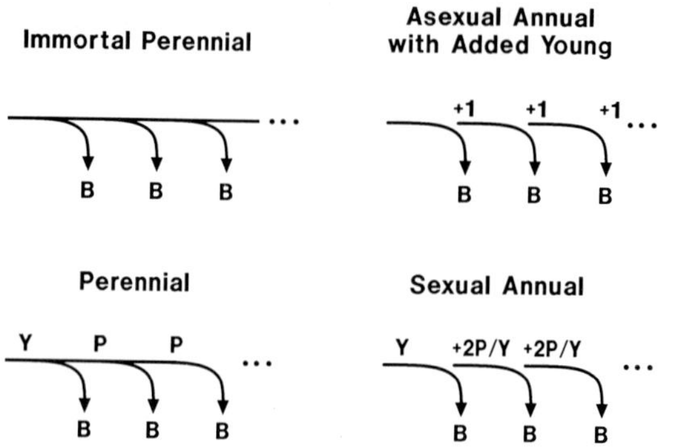

Fig. 1. Horn's models of equivalent reproductive patterns. *Upper left* is the reproduction of an immortal perennial producing litters of *B* young; this is matched by an asexual annual which simply adds one young to each litter (Cole's result). *Lower left* is a perennial with a fraction *Y*, of juveniles surviving the first season and a fraction *P*, of the parents surviving subsequent seasons. This pattern is equivalent to that of a sexual annual adding 2P/Y young per litter (or P/Y young in the case of an asexual annual). (After Horn 1978)

mortality in both parents and offspring (Horn 1978; Horn and Rubenstein 1984). The number of additional young must compensate for the average risk of mortality for both parent and offspring. If only Y offspring survive the breeding interval, then to insure one additional young surviving to parenthood, the number of young added per brood must be $1/Y$ since $(1/Y)(Y) = 1$. If parental probability of survival is P, then the current value of the parent is only the fraction P of the surviving young. In a sexual population, two surviving young are needed on average to replace the parents' genotypes. Thus a sexually reproducing organism that produces $B + 2\,P/Y$ young (where B is the original litter size) and dies is equivalent to a perennial that produces B young each year of its life. Thus the schedule of reproduction was at least as important as the amount of reproduction.

Cole's results were extended by Lewontin (1965), who was particularly concerned about life history traits in relation to **colonization**. Making the usual assumptions about stable age distribution and exponential population growth,

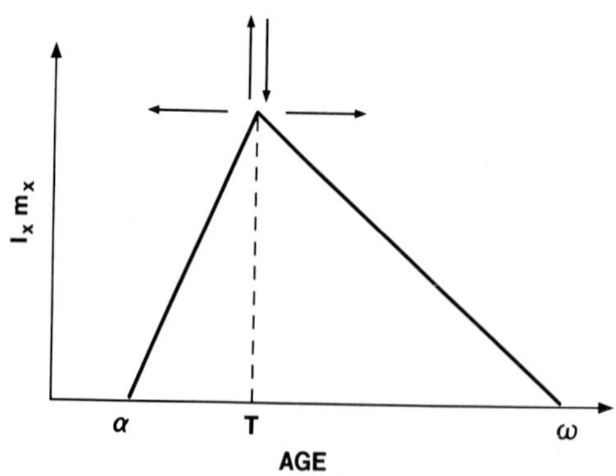

Fig. 2. Lewontin's analysis of the $l_x m_x$ curve. (Lewontin 1965). By changing values of α, ω, and T (indicated by *arrows*) and the area under the curve (offspring numbers), Lewontin demonstrated that (usually) age at first reproduction (α) had a greater influence on r than did l_x, m_x or ω(a 10% change in α was roughly equivalent to a 100% change in fertility)

he took r as a fitness measure and modeled life history variability using a triangular function of R_0 as illustrated in Fig. 2. By translating the triangle and assigning various values, he examined influences on r and demonstrated that in general r is most sensitive to changes in age at first reproduction [α in Eq. (1)]. A 10% change in α was roughly equivalent to a 100% change in fecundity. From this Lewontin predicted that α would be subject to more intense selection in colonizing species because of the importance of high r (and hence early reproduction) and should therefore show less genetic variance (removed by selection) than fecundity. I shall address the question of genetic variance in Section 3.

r-K Selection and Bet-Hedging

The interrelated consequences of varying life history traits has led evolutionary ecologists to ponder what sorts of combinations of characters should evolve under given sets of environmental circumstances. Two prominent theoretical approaches stand out, so-called "r and K selection" and "bet-hedging"; Stearns (1976) and Horn and Rubenstein (1984) give full discussions. The notion of r-K selection was fully developed by MacArthur and Wilson (1967). The model predicts that under density-independent conditions, as for example after a colonizing event by a weedy species, natural selection will favor rapid population growth, i.e., high r, while with increasing density dependence, as might occur in a stable woodland, higher carrying capacity, K, will be favored along with ability to compete and avoid predation and other sources of mortality.

The term bet-hedging was applied by Stearns (1976) to ideas which had been developed largely by Murphy (1968) and Schaffer (1974a). This argument considers especially the sources of **mortality**. If environmental variability influences juvenile mortality primarily, then reduced reproductive effort, fewer young, and longer life are favored (similar to K selection), while if adult mortality is influenced, then greater reproductive effort, more offspring, and short life spans are more likely (similar to r selection). In the first case of juvenile mortality the idea is to protect young against environmental vicissitudes with a long-lived parent, whereas in the second the idea is to get one's young out as quickly as possible before encountering further high mortality risks. Note that it is the ratio of adult to juvenile mortality that is important.

There is an extensive literature testing the predictions of r and K selection against life history data for various organisms. In some cases the predictions seem to hold, but in others they do not. The tendency for r-K to predict seems to depend in part on taxonomic level compared, with higher levels more likely to fit the predictions (Stearns 1980). This raises problems of phylogeny and ontogeny, and their influences on life histories. Other problems concern the lack of a common currency because, although r can be calculated from life table data, K cannot (Stearns 1977), although K can be estimated empirically. Finally, both bet hedging and r-K theories often predict the same result. For example, if adult mortality is variable, both theories predict late maturity, smaller reproductive effort, fewer young, and longer life in stable as opposed to fluctuating environments (Stearns 1976). Unfortunately, we still lack means of distinguishing between the differences in selection involved. One must also be aware that r and K are not simply ends of a one dimensional spectrum (Horn and Rubenstein 1984), but involve parameters and variables in multidimensional genetic and environmental "space."

Reproductive Investment and Lifespan

As part of his efforts to sort out the partitioning of growth and reproduction over the lifespan, Fisher (1930) developed the notion of reproductive value. He used the analogy of compound interest where a sum of money invested took on its own value plus that of interest compounded into the future; earlier more frequent interest payments (offspring) will compound more. Similarly the reproductive value of a female of age x, relative to that of a female at birth, is given by:

$$\frac{V_x}{V_o} = \frac{e^{rx}}{l_x} \int_x^{\infty} m_t l_t e^{rt} \, dt, \tag{3}$$

where m_t is the birth rate of individuals age t and l_t is survival from birth to age t. Following Horn (1978) this can be rewritten as

$$\int_x^{\infty} m_t (l_t/l_x) e^{-r(t-x)} \, dt, \tag{4}$$

which can be more easily interpreted as the sum into the future from age x of the product: (reproductive output at future age) × (probability of reaching the future age from the present age) × (discount on future reproduction due to growth in number of competitors between present and future ages). Note the same three elements of survivorship, reproduction, and compounding due to earlier reproduction. There are a number of limitations to reproductive value, including the fact that it is derived for populations in a stable age distribution (see discussions in Stearns 1976, and Charlesworth 1980), but for our purposes here, it is a useful way to consider the consequences of discounting and compounding.

The question of how best to partition resources between current and future young was more fully developed by Williams (1966), who also considered the question of whether the **effort** given **to offspring** should be in many small or a few large young. In general he argued that many small offspring will be favored if young can grow rapidly because they are born into a resource-rich environment, but that a few large protected progeny will be favored with scarce resources or high predation risk. Weedy plants, for example, produce many small seeds which do well on open ground with plenty of sun and nutrients, while tropical trees often produce very large seeds which can survive on nutrient poor soils (e.g., coconuts). With respect to **timing of reproduction**, Williams reasoned that since beyond a certain point reproductive value declines with age, reproductive effort (the proportion of resources mobilized for reproduction) should, in contrast, increase with age. Put in another way, older parents having lived most of their lifespans have little to lose by investing heavily in young while younger parents will invest less because they may live to reproduce again. Implicit in these (and Fisher's) arguments was the notion of cost and therefore trade-off, that is that present reproduction should reduce future reproduction, growth, lifespan, or all three. In other words there should be a negative correlation between present reproduction and the latter traits. The presumptive reasons for this are that if less allocation is made to present growth, there will be less future reproduction and/or greater risk of mortality. Various theoretical papers have considered the different aspects of this implication (reviews in Stearns 1976, 1977; Charlesworth 1980; Reznick 1985).

More recently a debate has arisen over whether it is in fact legitimate to consider **trade-off** a necessary aspect of life histories. Tuomi et al. (1983), for example, argue that organisms can compensate for costs by allocating resources according to their availability or timing reproduction to occur when resources are plentiful. In temperate-zone birds, as a case in point, photoperiod cues reproduction in the spring or early summer when resources to feed the young are readily available. Also, with respect to this line of reasoning, van Noordwijk and de Jong (1986) have developed a model demonstrating that the sign of correlations between life history traits depends on relative variation in acquisition and allocation of resources. This has been further extended by van Noordwijk and Gebhardt (1987). Bell (1983, 1984a,b) attempted in a series of papers to measures various costs of reproduction in several clonal organisms and found that with his analyses costs were generally not present. Various other papers (reviewed in Reznick 1985) also claim that rather than expected negative correlations between early and late life history variables, such correlations were often positive, quite the opposite of the predictions of a trade-off hypothesis. The whole subject has been reviewed by Reznick (1985; see also Reznick et al. 1986), who summarizes a large body of empirical evidence, pro and con. Reznick's general conclusion is that evidence contrary to the trade-off hypothesis generally comes from **phenotypic correlations** whereas any hypothesis of life history evolution requires that predicted correlations be genetic. When specifically genetic correlations are examined, trade-offs generally exist. As van Noordwijk and Gebhardt (1987) point out, however, the sign of these genetic correlations varies with varying environments (resources) thus considerably weakening Reznick's conclusion. This issue is also discussed with respect to reaction norms in Section 4.2 below.

The question of the importance of (negative) correlations between gene effects on reproduction and other life history characters also enters into considerations of senescence. Williams (1957) argued that with trade-offs due to what Rose (1983) has further developed as "antagonistic pleiotropy", i.e., negative correlations between gene influences on different life history characters, genes enhancing early life history characters might be favored in spite of later deleterious effects. One such effect would be reduced lifespan, especially since older individuals have reduced reproductive value. One of the major theories of senescence which arises from such considerations is that natural selection in acting to increase fitness will inevitably give rise to a deteriorating soma with that deterioration manifested as the aging process (see also Rose 1983 and references therein). But as with other aspects of trade-off models, the aging hypothesis has yet to be adequately tested with varying environments which in at least one case have been shown to have a profound influence on the conclusions drawn (Clare and Luckinbill 1985; Luckinbill and Clare 1985, 1986; see Sect. 4.2).

The Genetics of Life Histories

The difficulty with much of the development of life history theory as described above was that it grew without reference to the genes (see also Chapt. 1). Notions of optimality, for example, were conceived and refined with little or no consideration of whether it was even possible to evolve the traits conferring optimality. Without such considerations no evolutionary theory can be complete, so it is appropriate at this point to relate life histories to genes. Since most life history traits are metric or continuously varying traits influenced by several loci, the methods necessary for their analysis derive from quantitative genetics. Knowledge of the genetic variances and covariances provide much information about prior selection on traits and about their potential for future response, at least in the short term. I shall begin this section with a brief introduction to quantitative genetics, then indicate some of the results obtained from its methods with respect to life histories. In Section 4 I shall consider some complications arising from more detailed considerations of environmental and developmental effects.

Quantitative Genetics and Its Experimental Designs

Methods for quantitative genetic analyses are covered lucidly and in detail by Falconer (1981) and also by van Noordwijk (Chap. 9) so a brief summary to introduce the ensuing discussion will suffice here. The basic model assumes that phenotypic differences are the result of both genetic and environmental differences with the total (i.e. phenotypic) variance expressed by

$$V_P = V_A + V_N + V_E + V_{GE}, \qquad (5)$$

where V_A is the additive genetic variance, V_N is the nonadditive genetic variance (due to dominance and epistasis), V_E is the environmental variance, and V_{GE} includes the variance due to genotype x environment interaction, where genes are differentially expressed across environments, and covariance due to genotype x environment correlation. The **additive genetic variance** arises from the summation of influences across all genes contributing to a character and is the variance

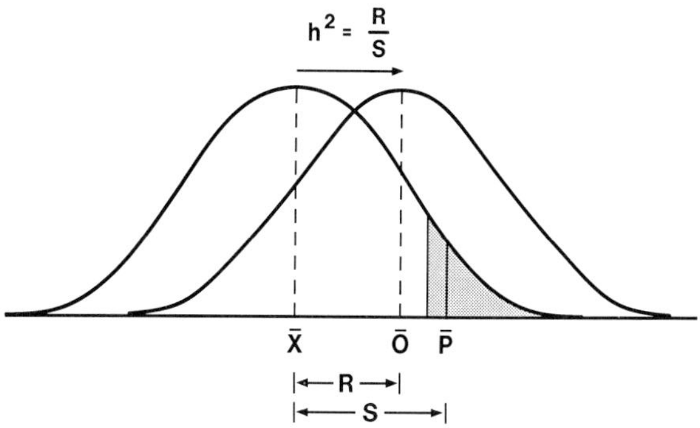

Fig. 3. Estimation of heritability by selection. Starting with a population with mean trait score $\bar{x}$ a subset is selected with mean $\bar{p}$ to serve as parents of the next generation. The difference between $\bar{x}$ and $\bar{p}$ is the selection differential, S. The difference between the next generation offspring mean, $\bar{o}$, and $\bar{x}$ is the response to selection, R, and since the response is a function of the selection differential and the heritability, h^2, $R = h^2S$ or $h^2 = R/S$

contributing specifically to resemblance among relatives. It thus determines sensitivity to selection (since variation must be heritable for selection to occur) and the maximum rate of evolutionary change in a trait.

The relation between additive genetic variance and phenotypic variance is expressed as the heritability of a trait or the ratio V_A/V_P from Eq. (5). Since V_A arise from resemblance among relatives, the heritability can be estimated in several ways involving comparisons of relatives, the most powerful being by offspring trait on parent trait least squares linear regression (because V_A contributes to resemblance) or by selection (the so-called realized heritability). Thus:

$$h^2 = V_A/V_P = b_{OP} \tag{6}$$

$$\text{or } h^2 = R/S, \tag{7}$$

where h^2 is the heritability, b_{OP} is the slope of the regression of mean offspring score (corrected for sex) on mid-parental score, R is the response to selection, and S is the selection differential. Determination of h^2 by selection is illustrated in Fig. 3. Heritability can also be estimated from components of variance among full sibs, in which case it also includes V_N (and so is often called "broad sense" heritability), or among half sibs.

Metric characters can also be correlated (e.g., height and weight) with the causes of correlation being both genetic and environmental. This can be expressed by

$$r_P = h_x h_y r_A + e_x e_y r_E, \tag{8}$$

where r_P is the **phenotypic correlation** between characters X and Y, r_A is the additive genetic correlation, r_E is the environmental correlation, h_x and h_y are the respective square roots of the heritabilities of X and Y, and e_x and e_y are the square roots of the respective 1-h^2 values (and so include variance due to nonadditive gene effects as well as environmental variance sensu stricto). The value of r_A indicates the extent to which additive genes influence that trait in common (e.g., by pleiotropy), with the result that when selection influences one trait, it will influence the other as well. The degree of influence will reflect the magnitude of the genetic correlation. The parent-offspring relationship (and selection) can also be used to estimate r_A. In this case, however, the appropriate characters are X in the parents and Y in the offspring so that

$$r_A = \frac{cov_{XY}}{\sqrt{(cov_{XX}\, cov_{YY})}}, \tag{9}$$

where cov_{XY} is the so-called cross-covariance obtained from the product of the values of X in the parents and Y in the offspring and cov_{XX} and cov_{YY} are the offspring-parent covariances for each character. Full descriptions of these and other methods of estimation of genetic parameters are given by Falconer (1981).

It is also well to keep in mind the limitations of genetic parameter estimation. In the first place these estimates are valid only for the environment in which the characters are measured (but note that this is true for any biological parameter including simple means or variances). We shall see some of the potential complications arising from genotype-environment interactions below. Secondly, genetic correlations, especially, are subject to large sampling errors and therefore often lack precision (Falconer 1981); one consequence is that estimates require large sample sizes. Furthermore these correlations are strongly influenced by gene frequences (Bohren et al. 1966) and so may differ among populations and across generations of selection which, of course, changes gene frequences. Predictions based on genetic correlation structure are thus likely to be valid only over the relatively short term. In spite of these limitations, however, considerable insight into life history evolution has been gained by employing quantitative genetic theory and methods.

Quantitative Genetic Life History Models

Two models which combine quantitative genetics and the various demographic considerations discussed in Section 2 are those of Lande (1982a,b) and Istock (1982, 1984). Lande's model uses matrix algebra and represents the life history by a column vector Z composed of age-specific life history characters (e.g. births, deaths) such that

$$Z = (z_1, \cdots z_n)\,T, \tag{10}$$

where T indicates matrix transposition. Similarly the population is represented by a mean vector $\bar{Z}$ with elements $\bar{z}_i$. Assuming that the intrinsic rate of increase, r, from Eq. (1) represents fitness, natural selection is taken as the gradient vector of r with respect to all changes in the mean values of the life history traits

$$\nabla r = \left(\frac{\delta r}{\delta \bar{z}_i}, \ldots, \frac{\delta r}{\delta \bar{z}_n}\right)^T \tag{11}$$

and the rate of change of $\bar{Z}$ will be

$$\frac{d\bar{z}}{dt} = G\nabla r, \tag{12}$$

where G is the genetic variance-covariance matrix for $\bar{Z}$ and ∇r is the gradient vector as indicated. The extent to which there is genetic covariance among traits [see Eq. (9)] thus controls the change allowable within the system since changes resulting from selection on one trait will also influence others with which it is genetically correlated. The intrinsic rate of increase, r, of the population then evolves at the rate.

$$\frac{dr}{dt} = (\nabla r)^T G \nabla r \geqslant 0. \tag{13}$$

In order to construct the model, Lande made several assumptions. These include multivariate normality of all the traits, a stable age distribution (since Lotka's characteristic equation is the demographic basis), weak selection so that the demographic processes are stronger than the imposed selection, and relative constancy of the genetic and phenotypic variance-covariance matrices, respectively. Results from wood frogs and milkweed bugs (Sect. 3.3 below) suggest that at least the latter may not be valid, and the assumption of linearity is violated when calculating quadratic components. Lande's model is useful, however, in its specific consideration of genetic covariance structure.

Istock's (1982) "**regulative value**" **model** is similar to Lande's in some ways but explicity incorporates ecological constraints acting through the phenotype. Istock assumes that at each age there is an "unrealized" maximum fecundity and survival possible in excess of what is actually seen in nature (note the similarity here to Law's "Darwinian Demon"). The population growth restriction resulting from the difference is derived as

$$c_i = l_i m_i + \sum_{x=i}^{\omega} l'_{ix} M_x, \tag{14}$$

where c_i is the "regulative value" of a specific age i, l_i is the actual average survival to age i, l'_{ix} is the "unrealized" higher survival possible for some genotypes, m_i is "unrealized" fertility for some genotypes that would otherwise express maximum fertility M_x, x is age, and ω is age at last reproduction as in Eq. (1). The loss of potential population growth and fitness is measured by c_i in terms of life history statistics at age i with the total restraint of the population across all ages

$$\sum_{x=0}^{\omega} c_x = R_{o,max} - R_{o,nature} \tag{15}$$

where R_o is the replacement rate from Eq. (2), with R_o, max being the maximal possible R_o and R_o, nature that is actually observed. Note here that the summation is across all ages, x, not just from age i as in Eq. (14). The genes are incorporated in the model by the expression

$$S_{iK} = h^2_{iK} \sum_m \sum_x c^x \rho_{i,x,m} \quad (16)$$

with S_{iK} the "selective value" operating on the kth life history trait at age i, h^2 the heritability, and ρ the phenotypic correlation of the kth trait at age i with all other traits m at all ages x. Genetic constraints occur in the heritability (the additive genetic variance available to selection) and the phenotypic correlation which includes genetic covariance. Note, however, that ρ also includes ecological constraints (since it is a phenotypic correlation it includes environmental variance) so that loss of fitness may be inhibited by purely ecological consequences. The important point to be made by both the Lande and Istock models is that the evolution of life histories depends on selection (the environment), genetic variance (permitting response to selection), and genetic covariance (interaction among traits in response to selection).

Genetical Analysis

Phenotypic differences in life history traits are frequently observed among populations within species. The simplest way to determine if gene differences contribute to these phenotypes is to rear samples from two or more populations in a common environment. This was the approach taken by Law et al. (1977) using the grass *Poa annua*. They predicted that *P. annua* from frequently disturbed environments with much bare ground would display early reproduction and shorter lifespans, as might be expected of colonizing r-selected populations, while grasses from more stable pasture situations would delay reproduction and live longer, i.e., would be K-selected. They then reared from seed individuals from the two sorts of habitats in a common garden, so that they experienced approximately identical conditions, and measured the prereproductive period, plant size at 7 and 14 months, and the number of infloresences over ages of 4 to 17 months. From these data they also calculated rates of increase. Their results clearly showed that plants from disturbed habitats were generally short-lived with rapid development to reproduction and high early seed production. Plants from pastures displayed the opposite trend, so that gene differences obviously contributed to an r and K strategy dichotomy which presumably was based on selection, although as Law et al. were careful to point out, the evidence on this point is only correlative. Other examples where gene differences contribute to intraspecific variation in plant life histories are given by Venable (1984).

A second case of life history differences persisting in a common environment was studied by Solbreck (1986). Many insects have evolved wing polymorphisms which allow alternation between migratory and sedentary life styles. Each morph is associated with a suite of life history traits which involve trade-offs in energy allocation to migration and reproduction (Dingle 1985). Typically, this is in the form of earlier and often more rapid reproduction (and hence higher population growth rates) in the sedentary micropters. An example of this, which seems to be controlled by genes at a single locus, occurs in the Mediterranean lygaeid bug *Horvathiolus gibbicollis* (Solbreck 1986). In this species the short-winged form is dominant to the long-winged with all offspring of crosses between the morphs being short-winged; the expected 3:1 ratio of short to long shows up in the F_2. Two correlated responses in the life history are noteworthy. First, the short-winged bugs reach adulthood earlier, and secondly, the temporal pattern of egg-laying is different. The short-winged bugs begin to oviposit earlier but then level off and finish later, whereas the long-winged reach a higher peak associated with wing muscle histolysis following flight. Several examples of insects in which similar trade-offs are controlled by multiple loci are discussed by Dingle (1985).

In addition to the discontinuous variation found in traits such as wing polymorphism, life history variation is often continuous. Particularly obvious are variations occurring over geographic gradients. An especially complete study of such geographic variation is that of Berven and Gill (1983) on the wood frog *Rana sylvatica*. They examined three populations for length of larval period and size at metamorphosis, traits reflecting considerable developmental plasticity; two were high and low-altitude temperate populations, and the third was from the arctic tundra. Two environmental variables, temperature and density, were included in the analysis. Results over a range of densities for the lowland temperate and the arctic tundra populations are displayed in Fig. 4. The two populations differ conspicuously: the temperate frogs metamorphosed at varying ages but relatively constant size, while the arctic ones responded with varying size but more or less constant age. Temperate frogs thus had canalized size with plastic development while the reverse was the case with the arctic population. Field observations confirmed that selection was favoring large size in this case. When heritabilities were calculated, they showed a similar pattern; the canalized traits displayed low heritabilities not significantly different from zero (implying that additive genetic

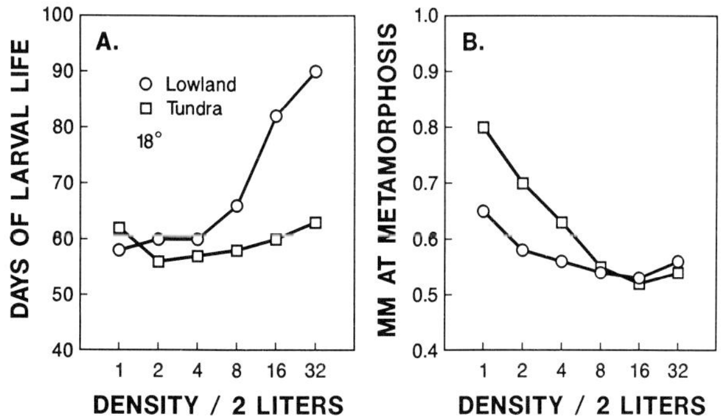

Fig. 4. The effect of tadpole density on: **A** the length of larval life and **B** the size at metamorphosis of Maryland lowland and Manitoba tundra wood frogs, *Rana sylvatica*, reared at 18°C. (After Berven and Gill 1983)

variance had been stripped away by selection), and the plastic traits displayed moderate and significant heritabilities of 0.27 in both instances. Genetic correlations between the two traits were 0.65 in the temperate frogs and 0.09 and not different from zero in the arctic. It is evident that not only do the phenotypes of the two populations differ, but both the genes (variance) and their organization (genetic covariance) also differ. The overall pattern of development was evidently the result of trade-offs between selective factors favoring large size at metamorphosis and those favoring early completion of development (see also Blakley 1981 for a case in insects).

Populations of the milkweed bug, *Oncopeltus fasciatus*, also differ with respect to both genes and their organization (Palmer and Dingle 1986; Dingle et al. 1988; Dingle and Evans 1987; Dingle 1987). In eastern North America, populations of this insect are migratory and invade in spring and summer; a short-day adult reproductive diapause permits escape in the fall. On the Caribbean island of Puerto Rico the bugs are nonmigratory and reproduce throughout the year with no dispause. Flight behavior and life history traits were analyzed using both offspring-parent regressions and correlations and laboratory selection for wing length, a morphological trait correlated with migration. Both populations displayed strong bidirectional responses to selection on wing length (confirming that heritabilities were high) and an absence of genetic correlations between age at first reproduction and other life history traits (age at first reproduction did not respond to selection on wing length). This latter result means that age at first reproduction can vary genetically without producing responses in other life history characters. Perhaps the most interesting result, however, was that in the North American population there were strong positive genetic correlations between wing length and flight and between wing length and fecundity. This implied a colonizing syndrome with migrants being both larger and more fecund ("r strategist"). In contrast, neither of these correlations could be demonstrated in the nonmigrant Puerto Rico bugs, implying a quite different genetic organization and no apparent colonizing syndrome in the latter. Differences between the populations are illustrated in Fig. 5.

Finally, we can conclude this section on genetic analysis with an example of a "coevolved" insect-plant interaction studied by Berenbaum et al. (1986) (see also Chap. 11). The parsnip webworm (*Depressaria pastinacella*) is a small moth that lays its eggs in the umbels of wild parsnip (*Pastinaca sativa*), and the larvae feed on flowers and developing seeds. As a defense against the webworm, the plant produces furanocumarins which are toxic to the insects. Plants in a central Illinois population vary in resistance to insect damage, as measured by successful seed production in controlled experiments, and resistance was largely determined by the amounts of the furanocumarins, bergapten and sphondin. It was positively related to the amounts of these chemicals in seeds and negatively related to the amounts in leaves (which influenced host finding by the moths). Each of the four factors displayed high and significant heritabilities, and therefore could respond to selection. There were trade-offs, however, because several of the resistance factors had negative genetic correlations with potential seed production, and

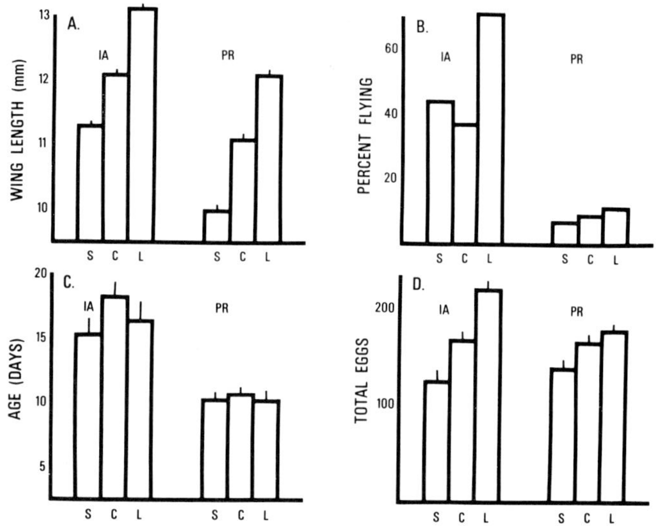

Fig. 5 The response in migrant Iowa (*IA*) and nonmigrant Puerto Rico (*PR*) populations of the milkweed bug, *Oncopeltus fasciatus*, to laboratory selection on wing length. Both populations show direct bidirectional responses (**A**), and neither displays a correlated response in age at first reproduction (**C**). Migrant Iowa bugs display correlated responses in flight and fecundity during the first 5 days of reproduction, while Puerto Rico bugs do not (**B, D**)

there were also negative genetic correlations among resistance factors. The proportion of seed bergapten, for example, was negatively correlated (−0.855) with the amount of seed sphondin. Berenbaum et al. interpret these sets of correlations as a sort of evolutionary "stalemate" in a coevolutionary "arms race".

Environmental Complications

As indicated briefly in Section 3.1 above, there are limitations to genetic parameter estimation and their interpretation. Two major complications are introduced by the influence of the parental (or even grandparental) environments and by the variability in the conditions directly experienced by individual genotypes (gene x environment interactions and correlations). The most important of the former are usually maternal effects resulting from the environment provided by the mother herself, such as uterine physiology or provision of egg yolk; other less obvious influences also abound. The external world impinging directly on the organism also quite obviously exerts an effect, and characteristically, organisms respond differentially by both phenotype and genotype. In estimating parameters such as heritabilities or genetic correlations, environments are usually held constant insofar as possible. From the perspective of evolutionary biology, however, it may be impossible, for example in field studies, or undesirable to do so, with the result that genotype x environment interactions occur. A third complication may arise when organisms are phenotypically plastic and respond to different environments with variable behavior or morphology; insect wing polymorphisms are a classic example. These three factors are discussed in the sections below.

Maternal Effects

Environmental influences on life history traits may derive not only from ambient sources such as photoperiod, temperature, or food supply acting directly on individuals, but also from factors transmitted by the parents. Usually the source of these factors is the mother, and maternal effects can result from a variety of prenatal and postnatal influences of the mother on her young, the most obvious of which occur in mammals. Paternal effects, however, are not unknown. In insects, for example, they have been reported in the dispause responses of Lepidoptera (King 1974; Danks 1987) which is interesting because, unlike most animals, in Lepidoptera, males are the homogametic sex. Maternal effects are often regarded as a nuisance in the agricultural literature because they confound efforts to estimate genetic parameters such as heritabilities or genetic correlations. These parameters are very important to plant or animal breeders trying to select for desirable traits. In contrast, from the perspective of the evolution of life histories, maternal effects are particularly interesting because they confer flexibility across generations by allowing parents to exert a measure of control over the responses of their offspring (Groeters and Dingle 1987). They provide this flexibility even when there are constraints due to across environment genetic correlations (see above) and thus allow strategic "decisions" concerning a future generation to be based on input from a current environment.

An example of this latter point is a Maryland population of the migratory milkweed bug, *Oncopeltus fasciatus* (Groeters and Dingle 1987, 1988) illustrated in Fig. 6A and B. This experiment used a "split-brood" design in which mothers are reared in two environments, and their offspring are then divided equally and reared half in the same environment as their parent and half in the opposite environment (see also van Noordwijk, Chap. 9). This design allows not only the determination of direct environmental influence and genotype (family) x environment interaction, but also the influence of parental environment on offspring performance (maternal effects). In this case both the temperature and the photoperiod experienced by the mother influence the age of first reproduction (AFR) of her offspring. Offspring reared on a LD 14:10 photoperiod (at 23°C) reproduce early (at around 33 days post adult eclosion) regardless of parental photoperiod while those reared at LD 11:13 reproduce much earlier if their mothers were also reared at LD 11:13 (Fig. 6A). (Note that this result also demonstrates an interaction between parental and offspring environments.) Since delay in AFR represents an adult reproductive diapause in this species (Dingle 1974), the maternal effect may be interpreted as an adaptation to suppress

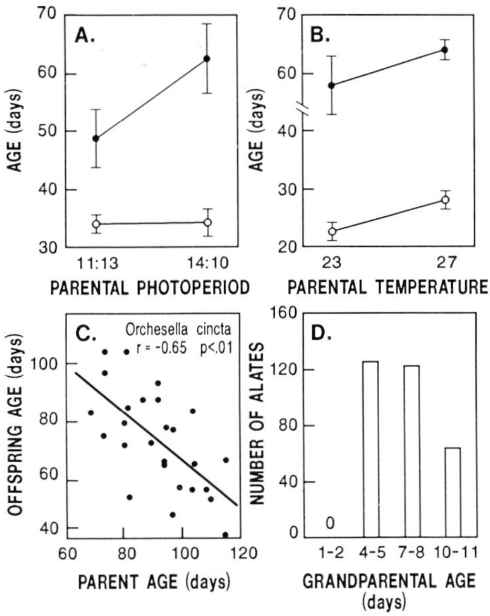

Fig. 6. A variety of maternal effects. **A** The effect of parental photoperiod on offspring age at first reproduction in milkweed bugs, *Oncopeltus fasciatus*. There is no effect on offspring reared at LD 14:10, but at LD 11:13 offspring of parents reared at 11:13 delay less than those of 14:10 parents. (Groeters and Dingle 1987). **B** Milkweed bugs delay age at first reproduction less in LD 11:13 if their parents were reared at 23 versus 27°C (Groeters and Dingle 1988). **C** In the collembolan, *Orchesella cincta*, there is a negative correlation between offspring and parent ages at first reproduction. (After Janssen 1985). **D** The effect of grandparental age on the production of winged offspring (alates) by aphid mothers. (After MacKay and Wellington 1977).

diapause in a population whose parents diapaused, as would be the case in a spring generation of bugs which might still be experiencing short days but in which diapause would be maladaptive. Similarly, when bugs are reared in LD 11:13, maternal temperature influences offspring AFR; mothers reared at 23°, and hence themselves having later AFR's, produce offspring that reproduce somewhat earlier (Fig. 6B). These results would further supplement the tendency for diapausing mothers to produce nondiapausing offspring. Parallel results have been obtained for the flesh fly, *Sarcophaga bullata*, where a maternal effect eliminates pupal diapause (Henrich and Denlinger 1982).

A particularly interesting case of maternal effects on life histories comes from a study of the collembolan, *Orchesella cincta* (Janssen et al. 1988). This is a small forest floor insect with distinct annual winter and summer generations differing in a number of life history traits. These differences result from strong negative maternal effects (Fig. 6C) so that individuals display traits opposite to their parents but resembling their grandparents. These maternal effects seem to adapt these insects to an annual phenology in which parents and offspring encounter dissimilar habitats (winter and summer) whereas those of grandparents and grandchildren are similar. Individuals thus encounter the environments not of their parents, but of their grandparents, and the negative maternal effects allow appropriate adjustments. Maternal effects thus play an important role in life history flexibility with respect to ecological fine tuning.

Janssen et al. (1988) following Falconer (1965), calculated values for maternal effects from the following model relating the phenotypic value of an individual, P, as a deviation from the population mean:

$$P = A + M + D + C + E,$$

where A is the additive genetic value of an individual (see also Sec. 3.1 above), M is the maternal influence on that individual, and D, C, and E are dominance, common environment of full-sibs other than maternal, and other environmental effects, respectively. M is defined as a linear function, m, of the mother's phenotypic value P′ (also as a deviation from the population mean) such that $M = mP'$ with m a partial regression coefficient relating daughter-mother phenotypic values. The covariance between P and P′ is $\mathrm{cov}(PP') = V_A[1/(2-m)] + mV_{p'}$. Since both mothers and grandmothers were available, Janssen was able to calculate regression coefficients for both daughter on mothers ($b_{pp'}$) and daughters on grandmothers ($b_{pp''}$) and from these obtain M since

$$b_{pp'} = h^2\left(\frac{1}{2-m}\right) + m$$

$$\text{and} \quad b_{pp''} = h^2\left[\frac{1+2m}{2(2-m)}\right] + m^2.$$

Solving for h^2 (the heritability) in the first equation yields

$$h^2 = (2-m)(b_{pp'} - m)$$

and substituting this value in the second equation gives

$$m = \frac{2b_{pp''} - b_{pp'}}{2b_{pp'} - 1}.$$

For the population of collembolans illustrated in Fig. 6C, Janssen et al. estimated $b_{pp'}$ and $b_{pp''}$ as -0.62 and 0.14, respectively, yielding a value of -0.40 for m, a sizable negative maternal effect.

There is also ample evidence for maternal effects in plants particularly with respect to seed size. Schaal (1984), for example, examined the effect of maternally influenced seed weight in the Texas bluebonnet, *Lupinus texensis*, on a number of growth and reproductive characters in the resultant plants. Seedlings which were produced from large seeds were themselves large, displayed higher survivorship early in life, and displayed higher age-specific reproduction in the middle age classes where reproduction was most intense. The fitness of the plants during the times of the life cycle most critical for success were thus in large part determined not by genotype of the individual but by the maternal plant. To the extent that the environment of the mother influences seed size, which varies across plant species, individual plants can be profoundly influenced by environments across generations.

Across generation influences may not be limited to parents and offspring. An example of grandmaternal effects is illustrated in Fig. 6D for the pea aphid *Acyrthosiphon pisum*. In this species, as in many aphids, crowding, a parthenogenetic female causes her to produce a higher proportion of winged (alate) progeny as opposed to wingless (apterous) progeny. In these clonal organisms the former serve as migrants to colonize new host plants, while the latter reproduce rapidly on suitable hosts. MacKay and Wellington (1977) demonstrated that the proportion of alates produced by a mother aphid depended on the age of her mother, i.e., the grandmother, at the time she was herself born. Mothers born to 1-2-day-old grandmothers produced no alate offspring when crowded, but those born at 4–5 days produced alates at maximum levels. The occurrence of grandmaternal effects in aphids is no doubt the result of their extremely compressed life histories because embryos actually begin to develop in the bodies of their grandmothers (presumably an adaptation to the ephemeral nature of new growth host tissue). The point to be made, however, is that "environmental" effects can span two generations and have profound effects on actual and potential rates of life history evolution. Similarly if a grandpaternal genotype influences maternal effect, as it does for size of grandoffspring in the mosquito fish *Gambusia affinis* (Reznick 1981), then expression of gene differences influencing traits may require more than one generation (see below).

Genotype-Environment Correlations and Interaction

Additional complications arising when phenotypic variance is partitioned into genotypic and environmental components concern genotype-environment correlations and interactions. Correlations arise when certain genotypes may be associated with specific environments. Good examples occur with maternal effects. In the case of Reznick's (1981) mosquito fish, cited above, large mothers produce large offspring. Both mother and offspring have genes for large size derived from the maternal grandfather so that there is a positive correlation between genotype and (maternally produced) environment. Estimates of genetic variance, $V_A + V_N$, will thus include the genotype x environment term, V_{GE} [see Eq. (5) in sect. 2.1 above) and will be inflated unless the grandparental generation is included in the analysis.

Genotype-environment interactions occur when there is differential response of genotypes across environments (see also Chap. 1). In other words, one genotype may do "best" in one environment, high temperature for example, while a second does best in another, say low temperature. The performance of a genotype across environments is known as its "norm of reaction" (Woltereck 1909; Schmalhausen 1949) and yields a great deal of information about the potential for response to selection when the selective regime encompasses more than one environment (Rendel 1967; Gupta and Lewontin 1982; Via 1984; Via and Lande 1985; Groeters and Dingle 1987, 1988; Scharloo 1987). For example, if genotypes perform well in one environment, say by displaying higher fecundity, but poorly in a second, such that there is a negative genetic correlation across the environments, then selection for improved performance in one environment may be countered by reduced performance in the other. In this sense the evolution of an optimum for both environments is constrained. This sort of problem frequently arises in selection for high yielding crops which may do poorly under uncontrolled field conditions. In contrast, if there is positive genetic correlation across environments, selection in one leads to response in the same direction in the other so that evolution toward an optimum can proceed. If both types of genotype occur, then there is genotype-environment interaction further confounding selection toward an optimum (see Via and Lande 1985), but the genetic variation results in **plasticity** leading to flexibility in life histories when environments are spatially or temporally fluctuating (Bradshaw 1965; Giesel 1986; Groeters and Dingle 1987; van Noordwijk and Gebhardt 1987).

Some examples of the **norms of reactions** of full-sib families ("genotypes") to two environments are illustrated in Fig. 7. Age at first reproduction (AFR) in two photoperiods for the milkweed bug, *Oncopeltus fasciatus* (Groeters and Dingle 1987a) is displayed in Fig. 7A, while development time on two host plants for the leaf-mining fly, *Liriomyza sativae* (Via 1984) is shown in Fig. 7B. In both cases the data were derived from a "split-brood" experimental design described in Sect. 4.1 above. Only enough families are included to illustrate the four basic types of response. First, the response in the two environments can be similar as in family 4 in Fig. 7A. Secondly, the magnitude of the response may differ, but the rank order of genotypes remains the same; a genotype high in one environment is also high in the other. This case is illustrated by *O. fasciatus* families 2 and 3 or *L. sativae* families 2 and 4. In neither case is there genotype-environment interaction, and therefore in both cases there are strong across environment genetic correlations. The third type of response involves divergence of genotype performance such that there is similar response in one environment but different responses in the other. Families 1 and 6 in the case of *O. fasciatus* and 3 and 4 in *L. sativae* are examples. Finally, the rank order of responses may differ so that lines connecting mean

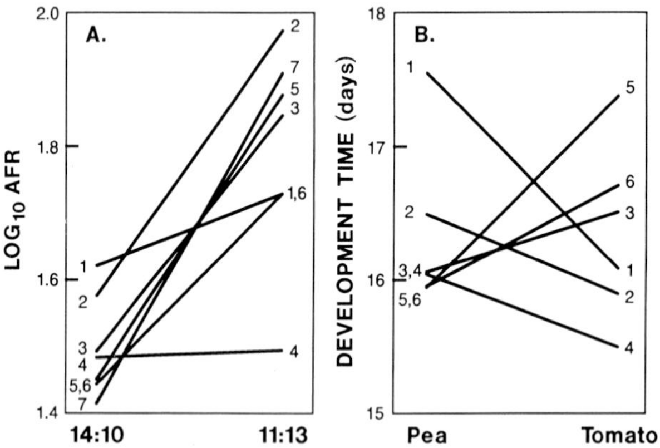

Fig. 7. Genotype-environment interactions involving age at first reproduction (*AFR*) in *Oncopeltus fasciatus* in different photoperiods (**A**) and development time in *Liriomyza sativae* (a leaf mining fly) on different host plants (**B**). Divergence of genotype (family) performance as in families *1* and *6* in **A** and *3* and *4* in **B** or changes of rank order so that lines connecting family means cross (families *3, 5,* and *7* in **A** and *1, 2, 3, 5* and *6* in **B**) constitute genotype-environment interactions. (**A** after Groeters and Dingle 1987; **B** after Via 1984)

responses cross as in the case of families 3, 5, and 7 (Fig. 7A) and families 1 and 2 and 3, 5, and 6 (Fig. 7B). In both the latter instances there is genotype-environment interaction. Note that there is considerable variation among families (genotypes) in both species in both environments. Responses to selection would thus be expected to occur in either. Evolution of a "best" or optimal genotype, however, would be constrained by the genotype-environment interactions.

Two particularly interesting cases of genotype x environment interactions have been demonstrated in species of *Drosophila* by Luckinbill and Clare (1985, 1986, Clare and Luckinbill 1985) and by Giesel et al. (1982a, b). Luckinbill and Clare worked with lifespan in *D. melanogaster* starting with conflicting observations in the literature concerning heritabilities for lifespan. A series of studies by Lints and his associates had failed to demonstrate significant heritabilities and had led to a proposal that lifespan was likely due largely to maternal effects (e.g., Lints and Hoste 1974), while Rose and Charlesworth (1981) had obtained significant heritability values. Luckinbill and Clare demonstrated conclusively that the expression of lifespan genotypes depended on larval densities. At low densities, there were no differences among genotypes, while at high larval densities, as used by Rose and Charlesworth who did not control for larval conditions, these differences were expressed, and it was possible to select for short and long lifespans by selecting for early and late reproduction, respectively. A somewhat puzzling result producing confusion about the genetics of lifespan was thus resolved.

In the second example, Giesel et al. (1982a, b) examined the influence of geographic origin of population and of temperature on heritabilities and genetic correlations, also in *D. melanogaster*. They found that both were influenced by genotype x environment interactions; in some cases, for example, genetic correlations switched from positive to negative across temperatures (see also van Noordwijk and Gebhardt 1987 for similar changes across food levels). To take a case in point, the correlation between development rate and age at death became more positively correlated with a temperature increase from 22 to 28°C in populations from Michigan and Palm Beach, Florida, but in a population from Melrose, Florida it went from positive to negative over the same range. Giesel (1986) found similar variation in genotype x environment interactions and genetic correlations when he examined the influence of photoperiod on *Drosophila*. Such effects would have profound influences on the response to natural selection acting on these traits over a range of temperatures or photoperiods, which in this case are well within the range experienced in nature.

Plasticity and Conditional Life Histories

In addition to straightforward schedules of births and deaths, life cycle syndromes may be environmentally cued for variation in timing, behavior, morphology or all three (e.g., Dingle 1986). Examples include diapause in insects, hibernation in mammals, seed dormancies in plants, migration in a variety of organisms, cyclical parthenogenesis in aphids and planktonic crustaceans, seed dispersal polymorphisms, and wing polymorphisms in insects, a list which is by no means exhaustive. We saw one example of the life history consequences of wing polymorphism in insects (*Horvathiolus gibbicolis*) in Section 3.3 above. The environmental cueing which usually occurs with these life cycle variants provides considerable flexibility by insuring breeding at the proper time and place, especially if environments are predictable. Photoperiod is the most reliable cue and is hence most frequently used. In general predictable environments result in selection for response to reliable cues such as photoperiod to produce phenotypic variation, the "**polyphenisms**" of Shapiro (1976), while unpredictable environments lead to a greater degree of genetic control and genetic polymorphisms, as seen in *H. gibbicolis* (Ziegler 1976).

Of course, the cueing itself may be genetically influenced, and Roff (1986) argues that this influence is likely to be polygenic with the heritability of the trait varying with environmental uncertainty. In the case of wing polymorphisms and polyphenisms, the trait may arise by a single gene mutation (see, e.g., Klausner et al. 1981), but selection will favor the evolution of polygenic control. The reason for this is that with a single locus system there are a very limited number of options possible, while with a polygenic system many patterns can be generated. Selection may change patterns in dispersal through shifts in genotype x environment interactions or in life cycle events such as nymphal development rates (Dingle 1986).

Variability in genetic expression may also be a consequence of **cyclical parthenogenesis**, in many ways the most interesting of life cycle polymorphisms. In aphids and cladocerans, for example, parthenogenetic spring and summer generations function to populate the summer habitat while an end of season sexual generation produces zygotes (usually eggs) to survive summer dryness or winter cold. But as demonstrated by Lynch and Gabriel (1983), during the unisexual phase there may be storage of a great deal of hidden genetic variation as a result of negative genetic correlations due to linkage. When linkages are broken by sex and recombination tremendous amounts of genetic variation may be released and made available to selection, thereby allowing considerable flexibility. There may in fact be even higher rates of per-generation evolution than would be possible with obligate sex. This observation further stresses a point made by Istock (1978) that sex in general has a direct role in fitness because genetic reassortment and recombination, along with polygenic inheritance (Roff 1986), are necessary for a "best mixed strategy" in response to environmental uncertainty.

Discussion: The Genetical Ecology of Life Histories

It should be apparent from the foregoing discussion that natural selection has molded life histories into suites of co-adapted traits which we call strategies or perhaps better, **syndromes** (Bradshaw 1986; Dingle 1986). The Darwinian Demon we discussed in the Introduction does not exist because it cannot in the face of constraints imposed by genes and ecology, and we are beginning to develop sophisticated theory concerning the form the taming of the Demon might take. Whether that theory is in the form of "regulative value" (Istock 1982) or a gradient vector and the genetic variance-covariance matrix for the traits in question [Eq. (12) and Lande 1982a, b], the importance of genes is apparent. In order for evolution to occur there must be selection acting on genetic variance and covariance so that an important question is whether such variance exists. Empirical data cited above suggests it does, and so does an extensive survey of the heritabilities of life history traits by Roff and Mosseau (1987) and Mousseau and Roff (1987). These authors demonstrate that in a variety of organisms these heritabilities, while generally lower than those of morphological or physiological characters as would be predicted if they were subject to strong selection removing genetic variance, are still more than adequate to produce rapid short-term evolution of life histories, given appropriate selection.

It is also apparent that where **environments** are predictable, organisms have evolved phenotypic responses such as wing polymorphisms so that they can produce a response suited to the environment prevailing. Such responses can even span generations in the form of parental effects that allow individuals to control the development or behavior of their offspring (Sect. 4.1). When environments are unpredictable, phenotypic switches are not possible, and adaptation resides in nonplastic genetic polymorphism without benefit of facultative adjustments to environmental cues. When organisms must deal with more than one environment, genotype x environment interactions may occur in which genotypes "best" in one condition may not be in another, thus preventing the evolution of a life history optimal under all circumstances. Similarly genetically correlated traits may evolve at unequal rates and so again constrain joint optima (e.g., Via and Lande 1985 and Sect. 3.2). Selection may thus act on an independent trait, a set of genetically correlated traits, and/or the degrees of genetic correlations among traits (Bradshaw 1986). An important additional question is how much new variation arises from mutations. We are only just beginning to unravel these sorts of complexities, but it is obvious that genetical ecology, the study of the quantitative genetics of ecologically important traits (see Chap. 9), is beginning to provide important insights into life history evolution. No longer are life histories independent of genes.

Acknowledgments. I thank A.J. van Noordwijk for his excellent critique.

9 The Methods of Genetical Ecology Applied to the Study of Evolutionary Change

A.J. VAN NOORDWIJK

Introduction

The aim of this chapter is to discuss the potential value of quantitative genetic methods in the study of micro-evolution. Micro-evolution results from the action of natural selection on the genetic variation present within populations (see also Chap. 2). It follows that an absence of micro-evolutionary change results either from the absence of net selection or from the absence of genetic variation. If we want to study micro-evolution, we therefore have to study natural selection and genetics simultaneously.

From the point of view of performing such a simultaneous study, there are three groups of traits: those where the genetics is simple but where natural selection is often difficult to study, those where the genetics is obscure but where natural selection can be studied with relative ease, and those where both the mechanisms of genetic determination and the importance of the trait for fitness are obscure. This chapter deals with the second group, that is, traits such as body size, clutch size, egg size, and bill morphology in birds, whose ecological significance is relatively well understood.

In so far as genetic variation plays a role in these traits, many different gene loci are involved and the interactions between genes and the environment tend to be complex. The genetic determination of such traits can be studied using the statistical methods from quantitative genetics.

Quantitative genetics is a theoretical construction that allows us to make statements about genetics without identifying genes. Before going into the details of this structure and before investigating the assumptions on which it is based, I want to mention a few aspects that are so obvious that they are normally forgotten.

What Do Geneticists Study?

The DNA in the chromosomes codes for proteins, many of which function as parts of enzymes, and thus regulate many aspects of life. The geneticist is interested only in certain aspects of this process. Only those genes that are not the same in all individuals are of interest to a geneticist. In other words, the geneticist needs **variation** before he can work. If one wants to understand life, the invariant gene loci may well be the more important ones, but restricting the attention to the polymorphic loci is not a serious drawback from an evolutionary point of view, where the variation between individuals is also the starting point (see also Chap. 1).

Although it is sometimes possible to study gene products directly, as in isozymes, there are often long and complex pathways between gene products and observable traits. When traits are selected for further study on grounds other than their genetic determination, most will be polygenic, that is, affected by many gene loci. There is an important distinction to be made here between all the gene loci that affect a particular trait, and the subset for which there is more than one allele present in the population. The attention of the geneticist is normally restricted to the latter. Where new variation arises through mutation, the former is the relevant set.

Phenocopies and Gene Action

The restriction of attention to variable gene loci is but one of the reasons why the relationship between the complexity of the biochemical pathways connecting the genotype to the phenotype and the variability in the trait studied is very loose. It does, however, set an upper limit in the sense that the more steps there are, the more may go "wrong" or rather may go differently. Regarding the phenotype as the result of a complex chain of reactions suggests that either genetic or **environmental influences** that disturb a particular step in the chain will have the same net effects on the phenotype. This is illustrated by our knowledge of phenocopies of many major mutants in *Drosophila*, where a particular treatment given at the right moment produces a copy of the mutant, which is then not transmitted to the next generation. In fact, in the study of gene physiology the production of phenocopies by specifically blocking certain reactions is an important method of obtaining information about gene action. Phenocopies of major developmental mutants have received special attention. The same ideas apply to more gradual and subtle effects of the environment in the expression of other traits. Scharloo (1987) gives an example where the effect of temperature on wing veination in *Drosophila* is very similar to the changes in the shape of the frequency distribution obtained as a response to selection.

Quantitative Genetics in Heterogeneous Environments

The standard methods from quantitative genetics were developed in the context of laboratory studies and of animal and plant breeding methods. It is important to uncover the assumptions underlying the standard methods because what can normally be assumed in a constant laboratory environment cannot be assumed in the more heterogeneous environments in which natural populations live.

Studying the process of evolution involves making choices. Even when using quantitative genetics applied to micro-evolutionary change, rather different approaches are possible. In recent years, the problems relating to correlations between several quantitative traits have received ample attention (see Loeschcke 1987b). I will largely ignore these methods, which are discussed by Dingle (see also Chap. 8).

Instead, I will concentrate on interactions of genes with the environment and on the consequences of heterogeneous environments. Natural environments are structured, that is the environmental variation is not random, but shows gradual changes in time and in space. In other words, two individuals living at the same time and close to each other are likely to experience largely the same environmental conditions. These patterns in the environmental conditions must be included in a study of genetic variation in natural populations. That ecology is needed for the analysis of heritable variation and the interpretation of observed selection is reflected in the term **genetical ecology** that is used for the application of quantitative genetics to data from natural populations.

In this chapter I first discuss the simplifications that are implicitly made in studying the microevolution of a particular trait. This includes such topics as: "What is a trait?", "How different is natural selection from artificial selection?", and "What are the likely effects of selection on the pool of genetic variation?". I then discuss resemblance between relatives in the context of the problems encountered in natural populations. Bill measurements in the great tit illustrate the process of inferring genetic variation from raw data. The study of phenotypic plasticity is the next major item. Two examples are given of the threshold model as a simple form of the genotype-environmental-factor-mapping function. The final section suggests directions for further research.

Some Basic Concepts

Quantitative Traits

In qualitative traits, each possible state has its own name; in quantitative traits the phenotype is expressed in a number. This definition expresses the strong element of arbitrariness—the naming part—in any distinction between qualitative and quantitative traits. Thus, as far as definitions go, the main difference is that the gradations are much finer in quantitative traits. However, these gradations require a different method of study.

The observed variation in a given situation (that is a given genetic composition and a given set of environments) is taken as the starting point. The aim is to unravel the quantitatively most important factors contributing to this variation. The major factors are genes, environment, and their interaction. Consequently, one writes the total **phenotypic variance**, that is the observed variance, as a sum of a variance component attributable to genes, a variance component attributable to the environment, and a component due to their interactions. These components may be further subdivided. One important subdivision is the distinction between additive and non-additive genetic variation. The additive genetic variance as a proportion of the total phenotypic variance is the **heritability**, which is often used to give an overall idea of the importance of the genetic component in a particular trait. The relation between variance components at population level and the phenotypic values of individuals is discussed in Section 3. For the moment it suffices that the change in population mean can be predicted from the product of heritability and the **selection differential**, which is a measure for the amount of selection that has taken place.

In quantitative genetics, there is a clear distinction between theoretical entities and observable entities. Additive genetic variance, which is the most important part of the genetic variance, is such a theoretical entity. It can never be observed directly. It can be deduced, however, from the observable resemblance between relatives. The relationship between theoretical and observable entities depends on certain assumptions. The deviations from these assumptions that are likely in natural populations are discussed in the next section.

The phenotypic variance in a given situation is the reference point in all analyses. This implies that estimates of the importance of genetic factors are valid only for the population and the conditions where measurements were made. It is for this reason that estimates made in natural populations are more interesting than heritability estimates obtained under standardized conditions, even though it is difficult to obtain the sample sizes required for precise estimates from natural populations.

Heritability estimates allow predictions of what will happen in the next few generations when a particular trait is selected up or down. The methods of quantitative genetics, however, can provide much more when seen as a general approach to problems. Starting from the total observed phenotypic variance, the relative contribution of several components can be estimated and, concentrating on the most important components, further refinements can be made if the data are detailed enough. It is thus a top-down approach, starting from the total variance. In genetically well-studied organisms it is possible to link this top-down approach to a bottom-up approach when starting from a particular mutant and studying its (pleiotropic) effects on quantitative traits. In principle, it should be possible to dissect all quantitative variation into the components provided by all the gene loci. This theoretical possibility is of the type that it is also theoretically possible to measure the distance from north to south pole with a tape measure.

This chapter cannot give a thorough introduction to quantitative genetics. A short introduction is given by Boag and van Noordwijk (1987). The standard textbook with emphasis on application to animal husbandry is Falconer's (1981) Introduction. The more theoretically inclined reader will probably find Bulmer's (1980) more formal treatment most informative.

Studying Selection Is Counting

There is a considerable body of theory on how selection leads to changes in the genetic composition of a population. This theory is discussed by Christiansen (Chap. 2) for individual gene loci and by Dingle (Chap. 8) for quantitative traits. I therefore concentrate on some ecological aspects of natural selection.

In principle, the study of natural selection is very simple. One counts individuals, according to their phenotype, at time one and then again at a later time. If the frequencies, that is the relative numbers, of the phenotypes have changed, one then tests whether the changes are likely to be chance results. If this is unlikely, there has been selection. After quantifying the change in frequencies, one knows how strong the selection has been.

Although the above is an accurate description of the majority of selection studies, there are several reasons to be unhappy with it. What fails is any indication of a causal relationship between the phenotype, the environment, and the change in frequency. Such a causal relationship is always implied in natural selection. If one does not require a cause, then genetic drift, changes due to linkage, and selection on pleiotropic effects become indistinguishable from gene frequency changes due to selection. This then makes natural selection equivalent to change in gene frequency, which is an utterly unattractive position.

Of the many definitions of natural selection, I will use a pragmatic one: We understand a particular instance of natural selection if, for a given range of phenotypes, we can predict from a knowledge of the environmental conditions, which phenotypes will increase, and which will decrease in frequency. Well-documented examples include the selection for large bills in the medium ground finch (*Geospiza media*) in the absence of small seeds (see Grant 1986 for a review) or the relationship between number of fledglings produced by starlings (*Sturnus vulgaris*) from different brood sizes and the density of their main prey, leather jackets, in the foraging area (Tinbergen 1981). In the latter example broods of six produce the highest number of fledglings in years with a high food density and produce fewer fledglings than broods of four in years with the lowest food density. Although there are a few cases where natural selection for single locus traits, including isoenzymes, are understood, most traits where natural selection has been or could easily be studied are quantitative traits.

Are the Environmental Conditions Unchanged?

There is an interesting difference between natural and artificial selection. In artificial selection it is quite feasible to apply selection without changing the environmental conditions. With natural selection, it is the change in conditions that often causes the selection in the first place. These altered conditions can also have a direct effect on the phenotype, in the same or in the opposite direction. It may then be difficult to apportion a change in population mean to the effects of

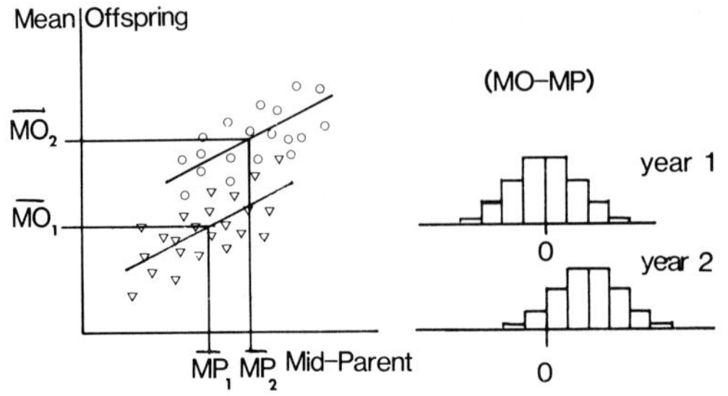

Fig. 1. The difference between a shift in population mean due to both selection and a change in environment. On the *left*, the offspring values are plotted against the parental values for 2 years, the means for parents and offspring are given. On the *right*, a frequency distribution of the differences between the offspring values and their own parents are given. The positive mean shows that in year 2 the offspring are bigger than their parents, which indicates an environmental shift

selection and to the change in environmental conditions respectively, especially if the environmental shift is not constant.

When the environmentally caused shift is either constant or proportional, it can be detected. Let us assume that we have data on the size of parents and two cohorts of offspring, and that selection has operated in the meantime. Due to selection one expects a shift in the values of the parents, but parents with the same average value should produce offspring of the same value, before and after selection. An environmental shift leads to offspring born to the same parents having different sizes in different cohorts (Fig. 1). The difference in size between parents and offspring is a useful statistic. In the case of an environmental shift, the mean of this difference between parents and offspring must have changed. A potential effect of scaling can be detected as a correlation between the absolute value of the difference and the parental value.

At What Rate Is Genetic Variation Depleted?

If a force of directional selection is applied for a long time in a consistent way, remarkable changes can take place. One of the classical examples is the selection on the oil content in corn (*Zea mays*) still responding to selection after more than three quarters of a century (Dudley 1977). Part of the selection response is probably due to **mutations** that have occurred since the selection experiment started. Even where the mutation rates are known, it is very difficult to estimate the relative contributions of new mutations and genetic variation already present at the start of the experiment. In this respect there is the important distinction between the set of all loci where mutants would affect the phenotype in the trait studied and a subset of these where more then one allele is actually present in a given population. If the total set is large, then the chance that a mutation occurs at one of these loci may become important and it is possible to obtain an equilibrium between new mutants arising and being eliminated by selection (e.g., Bürger 1988). Both when continued response to selection is mainly due to new mutations and when it is mainly due to genetic variation present at the start of the selection episode, population size is an important factor in the rate of depletion of genetic variation (e.g., Jones et al. 1968). When there are more individuals, it is more likely that a mutation that changes the phenotype in the selected direction will occur. Assuming that the total number of loci that could affect the phenotype is reasonably large, a larger population at the start of the experiment will also contain more different allelles affecting the phenotype in the selected trait. One should bear in mind that the effective population sizes in the classical experiments in which plateaus were reached, when there was no further response to selection after about 20 to 30 generations, were very small (in the order of 10–20 individuals). Repetitions with population sizes of about 50 individuals did not show similar plateaus even after 80–90 generations (Yoo 1980). There are few natural populations with effective population sizes of about ten pairs.

Therefore, depletion of genetic variation due to directional selection in the past is not an important concern, and if many loci are involved, it seems reasonable to extrapolate for up to 50–100 generations, without worrying too much about the depletion of additive genetic variance.

Selection on Pleiotropic Effects

Many gene loci affect several different traits. Changes in allele frequencies at these loci, brought about by selection for one trait, will also affect the phenotypes in the other traits, the so-called pleiotropic effects. This is self-evident when only one gene locus is involved. When more loci are involved the net effect depends on the extent to which the loci involved act in the same direction on both traits. With many loci, and especially with many traits, it is unlikely that all pleiotropic effects will go in one direction. In the extreme case this would lead to two groups of traits between which negative pleiotropy was found, and within which positive pleiotropy was found. Pleiotropic effects are capable of maintaining genetic variation that would otherwise be lost (see Chap. 8).

Appropriate Time-Scales

From a genetic point of view, there is but one logical unit of time, the **generation.** In the study of selection, however, this unit of time is often impractical. Especially with overlapping generations and with temporal fluctuations in environmental conditions, matching the time-scales used in studies of selection and those in genetic studies may become a problem.

The best time-scale for a selection study is a complete life cycle, which is equal to a generation in a semelparous species. Shorter time-spans have the problem that opposing selection forces at different stages in the life cycle can lead to the wrong answers for the overall net selection. For example, if one investigated the selection for clutch size in the great tit by counting the number of fledglings produced, one would find persistent selection for larger clutches (van Noordwijk et al. 1981). This would then lead one to look for energetic constraints in the laying female which prevent it from laying a larger clutch. If one also considers the survival after fledging, however, one finds considerable selection against large clutches. The net result is a pattern with selection for large clutches in some and for small clutches in other years, with a mean net selection close to nill. Thus over a fairly wide range, clutch size is neutral with respect to fitness when considered at a time-scale of decades, and undergoes variable selection at the time-scale of single years (van Noordwijk et al. 1981; van Noordwijk and van Balen 1988). It is an open question to what extent this pattern of selection contributes to the maintenance of genetic variation for clutch size. It does point out how changing the time-scale may alter the statements to be made about the presence of selection.

Resemblance Between Relatives

Relatives share genes, that is, they carry identical copies of the genes in their common ancestor. Therefore, relatives show a similar phenotype whenever genes play a role in the observed variation in a particular trait. This does not mean that all relatives have exactly the same phenotypic value. On the contrary, the process of Mendelian segregation will cause relatives to be dissimilar in as far as genes play a role. Nevertheless, the variance in phenotypic values within a group of relatives is smaller than the variance within a group of unrelated individuals. Moreover, the

Box 1. Individual Values and Population Variances.

Let us assume that each individual phenotype is affected by genes and by the environment and that these effects are independent and additive. We can then write P(henotype) = G(enotype) + E(nvironment) for each individual. We can express the population mean similarly and write each phenotype as a deviation from the population mean:

$$(P_i - P) = (G_i - G) + (E_i - E).$$

Squaring this expression, division by (n − 1), and summing over all individuals gives us:

$$\sum\frac{(P_i - P)^2}{n-1} = \sum\frac{(G_i - G)^2}{n-1} + \sum\frac{(E_i - E)^2}{n-1} + 2\sum\frac{(G_i - G)*(E_i - E)}{n-1}$$

and this is

$$V_P = V_G + V_E + 2COV_{GE}.$$

It is important to note that even though one assumes genes and environment act independently and additively in the individual, the covariance between genes and environment appears at the population level. The most straightforward interpretation is that this covariance stands for the nonrandom distribution of genotypes over environments. Such a nonrandom distribution may arise from active movements in animals, from selection, or from the fact that relatives live close to each other.

more closely related these relatives are and the more important the contribution of genes to the phenotypic variation is, the larger this difference in variance between related and unrelated individuals will be. There is, however, an upper limit to the resemblance between relatives that can be caused by sharing genes. If we take **full-sibs** as an example, they share on average half their genes. Due to Mendelian segregation, they will carry different parental genes for the other half of their genome. Thus, if all phenotypic variance is genetic, the variance among full sibs will be half the phenotypic variance. If the resemblance is greater, then part of it must have a non-genetic basis. The relation between the causal components in the observed phenotypes of individuals and components of variance at the population level is given in Box 1.

A difference in variance can be measured, and we would therefore like to reverse the direction of drawing conclusions. Can we infer the relative importance of genes in the phenotypic variation from such a difference in variance between related and unrelated individuals? This is equivalent to asking whether there are other potential causes, besides sharing of genes, for a resemblance between relatives. In general, such other causes exist and are important. In many cases the environmental conditions under which relatives live are more similar than the environmental conditions experienced by unrelated individuals. Thus, a resemblance between relatives may be caused by genes or shared microenvironments. Only when the latter cause can be excluded is it possible to draw inferences about the genetic nature of variation from a resemblance between relatives. Before dealing with techniques to separate an environmentally caused resemblance from a genetic one, we should discuss the difference between genotype-environment correlation and genotype-environment interaction. The first has to do with with non-random distribution of genotypes over environments, the second implies that different genotypes react differently to the same change in environmental conditions and will be dealt with in the next section.

Measuring the Genetic Variance

In principle, there are several ways to measure the genetic variance in quantitative traits. Within groups that are genetically identical, such as highly inbred lines or asexual offspring, all phenotypic variance must be environmental variance. It is often impractical to obtain genetically identical individuals from natural populations, and in inbred lines the reduced level of heterozygosity may affect the environmental variance by itself (see Chap. 5). The opposite procedure of trying to estimate all environmental variance through careful control of the environment, never works completely enough. In a rough form, however, this is the essence of the so-called **common garden experiments**. Organisms (usually plants) are grown in the same experimental garden(s) and phenotypic differences between populations that disappear are concluded to have been environmentally induced, while those that persist are concluded to have been genetic in origin. This is a method that works reasonably well when the variation within

Box 2. Reaction Norms

In thinking about interactions between genotypes and environment (see also Chap. 1), the reaction norm is an important concept. It is defined as the set of all phenotypes that a given genotype produces depending on the environmental conditions. It is usually depicted as a function. In the figure, two hypothetical reaction norms are given for the genotypes *A* and *B*. The curves give a phenotypic value for each environmental value, which is illustrated for the environments *1*, *2*, and *3* resulting in the phenotypes *A1*, *A2*, and *A3* for genotype A and in *B1*, *B2*, and *B3* for genotype B.

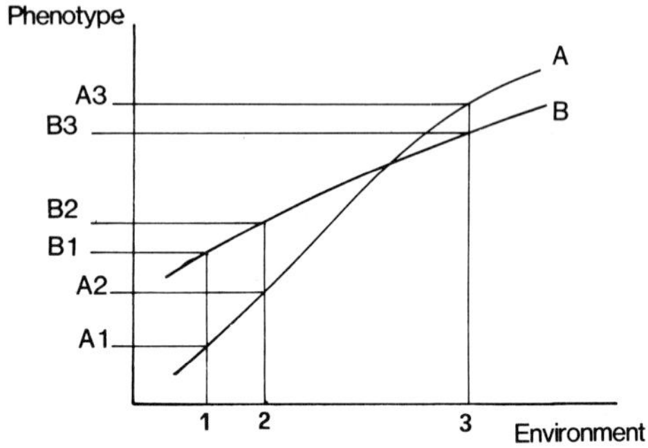

For a distribution of environmental values one obtains a distribution of phenotypes. The variance of these distributions is environmental variance. The total phenotypic distribution is the sum of the phenotypic distributions of all genotypes.

A good general introduction of quantitative genetics through reaction norms is given in Suzuki et al. (1986). Extensions to the treatment of several traits are given by van Noordwijk and Gebhardt (1987) and examples where the reaction norm is interpreted as genotype-phenotype mapping functions are given by Scharloo (1987, 1988). Another use of reaction norms is the construction of optimal reaction norms (see Stearns and Koella (1986).

populations (both genetic and environmental) is small compared to the variation between populations. It may make quite a difference in which environment the common garden is situated (see Box 2). In common garden experiments the variance within populations is not investigated, and therefore it does not give heritability estimates that can be used to estimate the effects of directional selection.

A second way of measuring the genetic variance is to measure the **response to directional selection** (see Fig. 3b in Dingle, Chap. 8). The selection can be characterized by the difference in mean between the selected individuals and the total population. The response to selection is measured as the difference in mean value between generations (but see Sect. 2.2). The heritability is equal to the response divided by the selection differential. In order to eliminate the effects of fluctuations in population mean due to environmental causes, it is useful to express all values relative to an unselected control population that was cultured in parallel. Even under standardized laboratory or agricultural conditions this use of a control population is often necessary. If selection is carried out in opposite directions simultaneously, the difference between them is similar to the difference between a selected line and an unselected control.

A third way to measure the genetic variance is through the resemblance between relatives, which is expressed either as a correlation or as a regression coefficient. Although this is probably the most widely used technique, several problems are especially important in natural populations and warrant an extensive discussion.

Problems in Natural Populations

In natural populations there are two groups of deviations from the assumptions underlying quantitative genetics. First, it cannot be assumed that genotypes are distributed randomly over environments and second, the population structure may differ from assumptions. In particular, nonrandom mating may occur, the variance in family size is often larger, and strongly fluctuating population size may cause problems.

In establishing whether genetic variation is present, it is therefore important to make two separate steps; first to quantify any resemblance between relatives and second to test explicitly whether this resemblance has a genetic cause.

Checking for Nongenetic Resemblance of Relatives

Let us assume that we have been able to calculate a resemblance between relatives from an intra-class correlation between (half-) sibs or from **regression of offspring on parental values.** How can we exclude the possibility that this resemblance is not genetically caused, but is environmental? First of all, we must distinguish between the types of relatives. The most common types will be full sibs, half sibs and parent-offspring. Of these, full sibs are not very useful, unless they are raised under controlled conditions and maternal effects, e.g., through egg size, can also be controlled. The use of paternal half-sibs enables one to exclude maternal effects, as well as creating conditions where the half-sibs to be compared are raised under environmental conditions that are as different within groups of half-sibs as they are between groups of half sibs. In some natural populations, for example herds with a single bull, or in a geographically structured plant population, the use of paternal half sibs does not rule out greater resemblance between half sibs in the environmental conditions experienced and therefore may still lead to an environmentally caused resemblance. Such problems are usually obvious once detected, but one needs a truly sceptical attitude to uncover them.

In natural populations and in field experiments, variation in timing may be an important factor that is not easily controlled. For example in a cross-fostering experiment, or in plants providing pollen, differences in timing between the cross-fostered groups or between the pollen donors may contribute to a resemblance between relatives. An example in which cross-fostering was insufficient to eliminate common environment was found by Henrich and van Noordwijk (in

Box 3. Separating Genetic and Environmental Familial Resemblance

There are several different ways to check whether a resemblance among relatives has an environmental cause. Which of these methods can be used in a particular instance depends on population structure and the nature of the environmental variation.

1. Comparing the repeatability of:
   Pairs
   Males with different females
   Females of different males
   Territories with different individuals.

This method is especially useful for reproductive traits to find out what the relative contributions of female, male, and local environment are. Different answers have been obtained for the same trait in different speices.

2. Calculating "control resemblances":
   e.g., Regression of offspring on neighbors.

This method assumes that neighbors experience almost the same environment. The general idea is to measure resemblance among unrelated individuals sharing the same environment. It is less useful if (a) neighbors tend to be related, (b) neighbors have a strong influence on each other's environmental conditions, or (c) when temporal variation is more important than spatial variation.

3. Calculating different resemblances:
   Correlation of maternal versus paternal half-sibs
   Regression on maternal versus paternal grandparents
   Compare sibs from a single brood to sibs from different broods

This method assumes that the environmental similarity is different for the two groups. Maternal effects are included in this category.

4. Checking different kinds of resemblances:
   Compare parent-offspring estimates to sib-estimate
   Compare full-sib to half-sib estimates

Similar to (3) except for differences in degree of relatedness.

5. Experimental manipulation:
   Cross-fostering
   Manipulating the environment

At first sight manipulation may seem superior. In field experiments, however, correlations between foster parents and biological parents may arise from temporal environmental variation or correlations between the trait investigated and a (genetic) correlation with timing of reproduction. Thus, field experiments need as many checks and controls as a posteriori control comparisons.

preparation), who studied body weight in the great tit with a partial cross-fostering experiment with simultaneous manipulation of brood size. They found a significant regression for weight of cross-fostered offspring on their foster-parents in enlarged broods when weights taken during the breeding season were used. The regression was absent if winter weights were used for the foster-parents instead of the breeding season weights. In these data there is an effect of brood size on the difference in weight between the breeding season (when the birds are weighed when their nestlings are 8–12 days old) and winter when the birds were weighed during roosting inspections. The obvious explanation is that during part of the breeding season, it was difficult to find sufficient food. During this period, the adults lost weight and their nestlings grew much more slowly. In this case the slope of the regression was considerably, greater than one.

A list of the various possibilities to check for an environmental component in the resemblance between relatives is given in Box 3.

Nonrandom Mating and Single-Parent h^2-Estimates

Random mating is a nice theoretical concept, but in natural populations nonrandom mating is likely to occur through various mechanisms. Random mating must be looked at for a specific trait or attribute. For traits that can be measured in both sexes, the degree to which mating is nonrandom can be measured as a correlation between partners. Traits can be defined for the express purpose of measuring the randomness of matings. To investigate the **effects of limited dispersal,** van Noordwijk (1984) has suggested using the distance between the site of birth and the center of the study area as a useful trait that will show a positive correlation between partners if individuals born close to each other are more likely to mate than individuals born far apart.

A special case that has received quite some attention is random mating with respect to relatedness. There are two levels at which conclusions can be drawn. The level of heterozygosity, given allele frequencies, tells us indirectly about the mating pattern. This method gives little information about the **mate selection** of individuals. Aspects of mate selection can be inferred from pedigrees over a number of generations.

Known pedigrees in natural populations are very rare, but those of the great tit population on the Dutch island of Vlieland are well enough known to detect inbreeding. It was shown that within the main wood on this island, the relatedness of partners is equal to the relatedness of two randomly chosen breeding birds that lived simultaneously and hence that mating is random with respect to relatedness (van Tienderen and van Noordwijk 1988).

There are cases where assortative mating will occur automatically, e.g., variation in flowering time between plants will lead to assortative mating, simply because the set of potential mates changes through a season (see also Chap. 2). Assortative mating causes serious problems when one wants to estimate a heritability based on single-parent regression. In the case of positive assortative mating, the resemblance between parent and offspring will in part be due to the genes received from the other parent and will therefore lead to an over-estimate of the heritability.

In the analysis of **nonrandom mating** it is helpful to consider the set of potential mates as an important quantity. It will not always be possible to define this set exactly, but it will help to avoid problems due to combination of data before analysis. Let us, for example, assume that the proportion of immigrants in a population of birds fluctuates between 25 and 75% and that these birds only pair at random with birds that are present in the same year. In one extreme year one would find 37.5% mixed pairs and 6.25 and 56.25% of the like pairs, while in another extreme year one would find the same proportion of mixed pairs and opposite proportions of the two types of pure pairs. In the two years together one would find 37.5% mixed pairs and a frequency of 31.25% for each of the pure types. This would be a significant deviation from random mating if one had 50 pairs in each year. This result is biologically uninteresting with respect to mate choice, at the same time the result is important in pointing out potential problems in a heritability estimate based on single-parent regression for a trait associated with migration if a similarly combined data-set is used. This is a temporal equivalent of the **Wahlund effect**, that is the deficiency of heterozygotes in a subdivided population with unequal allele frequencies (see also Chap. 4).

Variance in Family Size on How to Take Samples

One consequence of genetic variation is that measurements on relatives cannot be treated as fully independent observations. This problem is not easily quantifiable, but it is worth considering before collecting any data. The essence of all statistical testing is that the observations form a representative sample consisting of **independent measurements.** What observations can be treated as independent, depends entirely on the question asked. It is perfectly acceptable to measure a single egg 100 times to evaluate the measurement error. Ninety-nine of these same 100 measurements are useless if we want to make statements about egg-size variation in the female that laid the egg, the population from which it came or the species to which it belongs. The consequences of undetected lack of independence between observations is an over-estimate of the degrees of freedom in a statistical test which leads to reports of statistical significance in cases where the chances of finding the same results through random variation is still substantial, to "significant results" that are either not reproducible or do not make any biological sense.

Two examples, in which the presence of genetic variation for clutch size causes the repeated measurements of clutch size in the same females to the interdependent are given by van Noordwijk (1987). The first concerns the classical relationship between density and clutch size in the great tit (Lack 1966; Perrins 1965). The population densities are strongly correlated from one year to the next, and on average half the females in two subsequent years are also the same, leading to a correlation between mean clutch size in subsequent years. In later years (1962–1982), there is no correlation between population density and clutch size (McCleery and Perrins 1985). A similar effect can be found in the lack of reproducibility of microhabitats in which larger or smaller than average clutches are found (Krebs 1970). Again, the fact that the same females lay in the same area for several years has an important effect on the definition of these areas, but they are qualities of the female and not of the microhabitat.

The practical problem is of course difficult. If one catches a flock of 100 birds, does one have a sample size of 100, or is that flock composed of 10 families, the trait highly heritable and therefore the sample size only the 20 parental birds, assuming that these are unrelated? In a different example, are all the trees in a 1-ha plot descendants of a single tree that was left standing for that purpose after clearcutting, or its equivalent in terms of high variation in family size in a natural case? Who knows? However, an awareness of the problem may help in the design of sampling schemes and therefore even when the problem cannot be fully quantified, its effects can be diminished.

Fluctuating Population Sizes

Fluctuating population sizes are also potentially important in sampling and the number of independent observations that one is able to make. Especially with **cyclically parthenogenetic organism**, a sample of individuals may contain a much smaller sample of genotypes. In itself this has potential advantages for studying processes involved in genotype-environment intercations, because it is there very useful to be able to subject individuals with the same genotype to different environmental conditions.

Bill Measurements in the Great Tit as an Example

It is often difficult to collect sufficient data in a reasonable time. I therefore want to illustrate the logic that underlies a quantitative genetic analysis with a limited data-set. Nevertheless, the data led to some conclusions which surprised us (van Noordwijk and Klerks 1984), but have since been demonstrated in other studies (Gosler 1987a,b).

The bills of adult great tits on the island of Vlieland in the Netherlands were photographed during the breeding season and during a roosting inspection in December 1979. This population has been studied for a long time, and about 85% of the breeding birds were ringed as nestlings, so that their family trees are known (see van Tienderen and van Noordwijk 1988). This made it possible to compare many different types of relatives. Photos of the bill were taken from above and from the side and nine traits were measured (see Fig. 2). Two of the measurements were problematic (see Box 4) and were not analyzed. Tarsus length was also

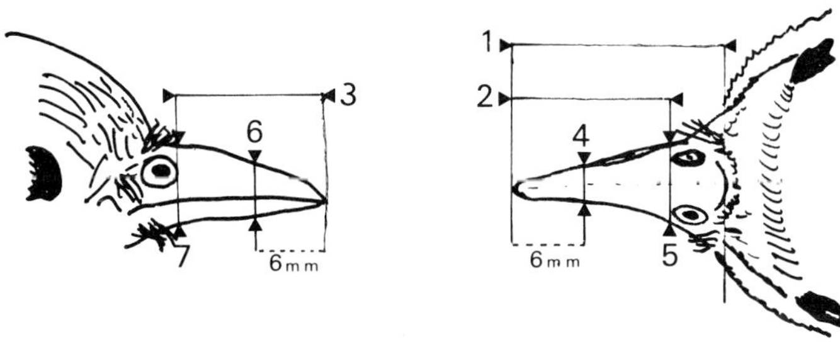

Fig. 2. The bill measurements used in Table 1: *1* Total length from above (LB); *2* length till nostril from above (LNt); *3* length till nostril from the side (LNs); *4* breadth at 6 mm from the tip (B6); *5* breadth at the nostril (BN); *6* height at 6 mm from the tip (H6); *7* height at nostril (HN). All measurements were made from photos

Box 4. Measurement Error

> Two of the measurements made on the bills gave problems. When all photos had been measured, remeasurement of the first photographs showed a significant difference with the first measurements. If it had not been identified as such, this measurement problem would have been included in the environmental variance, which would then have represented the measurement environment rather than the biologically meaningful environment. Furthermore, it is very likely that relatives often would have been measured at almost the same time, so that a gradual shift in how measurements were made would also have resulted in a resemblance between relatives. This could have been detected from a correlation between unrelated individuals that were often measured close together in time, e.g., the male and female of a pair. This correlation, which is very informative if zero, will be discussed in the context of common environment. The potential problem of measurement error, or a shift in definition, is not normally mentioned in publications. It is implicitly assumed that such problems were absent, or have been solved prior to any analysis.
>
> The problems of measurement error, and their treatment, are similar to those caused by common environment. In both cases no general rules can be given in which variance components they will be included; this depends on how environmental conditions or measurements are distributed.

measured on the same individuals, and several traits were defined as ratios (and one product) of the primary traits.

Before looking at the data, let us consider what one could observe from a series of correlations and regressions that measure the resemblance between various kinds of relatives. It is important to remember that bills are horny structures that wear and grow continuously, just as nails. Variation in bill size could have (a) an additive genetic component (as in Darwin's finches, cf. Grant 1986a), (b) it could mainly reflect the conditions during growth (as for tarsus length, cf. van Noordwijk et al. 1988), (c) it could mainly reflect the wear over the past few weeks (as in oystercatchers, cf. Hulscher 1985; Swennen et al. 1983), or any combination of these. In this analysis, it is possible to detect genetic variation (a) and lasting effects of growth conditions (b). For a proper analysis of (c) observations on diet are also needed. One aspect of wear, temporal variation, will be considered.

In Table 1, the following correlations and regressions are given: The correlation between the male and female in a pair, the repeatability, separately for females and males, the regression of offspring on mid-parent value and regression on single parent values, and finally the intraclass correlation of full sibs. In order to facilitate the comparison between the values obtained from different analyses, all resemblance between relatives were assumed to be genetic and are presented as a heritability estimate, which is definitely false for full-sibs. The major conclusions that can be drawn are indicated in the last three columns.

The correlation within pairs is not significant for any of the traits, and there are as many positive as negative values. Two conclusions can be drawn: There is no evidence for assortative mating, and there is no indication that the variation in bill size mainly reflects the wear pattern associated with a particular micro-habitat. For example, if birds foraging in pine cones mainly chafed and wore the

Table 1. Intermediate results in an analysis of bill dimensions of great tits[a] (*Parus major*)

Trait	Correl. within pairs	repeat. females	repeat. males	h^2 mid-parent	h^2 single parent	"h^2" full-sibs	Conclusions[b] ABC
Length B	0.22	−0.74	−0.06	−0.04	−0.02	1.02	− + +
Length Nt	−0.03	0.39*	0.17	0.54*	0.40*	0.82*	+ + +
Length Ns	0.19	0.39*	0.22	0.43*	0.44*	0.80*	+ + +
Breadth 6	−0.06	0.32	−0.06	0.68*	0.70*	0.92*	+(+)+
Breadth N	−0.31	0.69*	0.26	0.08	0.12	−0.04	− − −
Height 6	0.07	0.52*	0.62*	0.71*	0.92*	0.92*	+ − −
Height N	0.17	0.46*	0.02	0.05	0.12	0.50	− + −
Ba6*Ha6	−0.01	0.60*	0.18	0.71*	0.80*	0.94*	+ − −
Ba6/Ha6	−0.07	0.03	0.12	0.76*	0.75*	0.83*	+ − +
Tarsus	0.02	0.94*	0.86*	−0.07	−0.03	0.92*	− + −
N	18–35	12–22	7–19	29–63	107–165	52–85	

[a] Data from Vlieland (Klerks and van Noordwijk unpublished). see text.
[b] Conclusions: A, evidence for a genetic component in the phenotypic variance; B, evidence for effects of a common juvenile environment, C, evidence for temporal variation.

sides of their bills and if birds hammering on the branch junctures of broadleaved trees had heavy wear on the tips, affecting length but not height and width, this would result in a positive correlation within pairs, since both live and forage in the same habitat. This is not entirely theoretical; Gosler (1987a,b) found a relation between diet, social status, and bill wear in winter for male great tits in Wytham wood near Oxford. In the data presented here, the repeatabilities—which measure the constancy of individuals and normally give an upper limit to the heritability—for males are also substantially lower than those for females. It could still be that there are sex-specific effects of microhabitat. If these are important and if relatives share their microenvironments, they will be reflected in the resemblances between offspring and the parent of the same sex.

A positive significant resemblance between parents and offspring is taken as evidence for genetic variance. In all cases, the estimates based on regression on mid-parent value and on single parents are very close to each other. In the regression on single parents, there is no difference between same sex estimates (daughter-mother and son-father) and opposite sex estimates (data not presented). This implies, together with the absence of correlations within pairs, that the resemblance is not caused by sharing microenvironments.

The sample sizes allow us to detect a genetic component only if it is substantial. At the smallest sample size, a heritability of 0.40 and at the largest sample size a heritability of 0.25 would have an associated P-value of 5%. For six out of nine bill traits, significant amounts of genetic variance are found.

The resemblance between full-sibs is caused by common genes and by the effect of a common juvenile environment. The latter effect should, when it is important, lead to a stronger resemblance between full sibs than between parents and offspring. This is clearly the case for all three bill length measurements, for bill height at the nostril and for tarsus length. In some of these traits there is no evidence for genetic variation, in others there is. The third conclusion, that there is temporal variation within individuals, can be drawn when the repeatability estimate is lower than the heritability. The reason is that the two measurements on the same individual were always one from summer and one from winter. The intra-individual variance therefore always contains the temporal component. In the total variance, and in the heritability estimates based on parent-offspring regressions, only half the temporal component is represented. The small sample sizes did not show any differences among separate heritability estimates based on only summer or only winter measurements, and those based on the total dataset presented here (but see Gosler 1987a).

I have chosen these bill measurements as an example because they show many different combinations of the three most likely causes of variation in bill size in a single data-set. Some of the patterns might not be real because the sample sizes are rather small. Nevertheless, it is clear that genetic variation is present. In these analyses, bill measurements have been treated as "static" traits. A next step would include the wear and growth patterns in the analysis. One could imagine that the growth pattern has been selected to provide an adaptiveness to the wear patterns, that is, the wear caused by a particular type of food makes the bill more efficient for processing that kind of food. Such studies require both more detailed physiological knowledge about the growing process, more ecological knowledge about variability and frequency distributions of food items taken in relation to bill wear, and experimental data on the relation between bill shape and handling efficiency of different food types. One consequence of the presence of genetic variation is that experiments in which one bird is fed different diets is likely to be more informative than comparisons of different birds having different diets. In the course of such further investigations, it should also become clear how the observed heritable differences between individuals come about.

The Dynamics of Heritability Estimates

It is an axiom that a heritability estimate is valid only for the population and the environment for which it was calculated. A corollary is that the same trait in the same population could have very different heritabilities depending on the environmental conditions. Relatively little systematic work has been done on the dependence of genetic variability on the environmental conditions. Two aspects should be clearly: How the environmental variance differs between environments, and how the net effect of genes varies with conditions; i.e., both components of the phenotypic variance can vary with environmental conditions. It is therefore even possible that change in one component precisely compensates change in the other, so that the heritability remains the same. One should therefore examine either the combination of heritability and phenotypic variance or the absolute values for the genetic and the environmental variances.

In the data in Table 1, tarsus length showed no resemblance between offspring and parents. In data from other populations and indeed from other years in the same population, considerable amounts of genetic variance have been demonstrated. A similar fluctuation in heritability estimates for body weight was found in data for the great tits from the Hoge Veluwe population (van Noordwijk 1989; van Noordwijk et al. 1988). An interpretation in terms of the interaction of body size genes with the environment during the growth process is given elsewhere (van Noordwijk 1988).

Recommendations

Several of the questions posed in the previous paragraphs can only be solved if the results from many case studies are combined. It is therefore important to report results in such a way that they are comparable with other studies.

Because heritability is a proportion, one should always report phenotypic mean and variance as well. Sample size should be specified: number of families, number of individuals, number of observations. For traits that can be measured in both sexes a table with six regression estimates (female offspring on female parent, on male parent and on mid-parent and the same for the male offspring) is very informative. The correlation within pairs, gives information about nonrandom mating and/or direct effects of the microhabitat.

Heritability estimates pertain to particular environmental conditions. It is important to include information about the environmental conditions to allow comparisons to be made. In the case of a wild-caught population being kept under controlled conditions to collect the data, both the original and the laboratory conditions should be given in quite some detail.

Large numbers are needed to get precise estimates. The standard error of a regression, if the resemblance is genetic, is about $1/\sqrt{n}$; and thus about $2/\sqrt{n}$ for a heritability estimate based on regression of offspring on single parents. It is thus likely that conclusions on the presence of genetic variation will be mainly qualitative, especially those concerning the conditions under which genetic variation is more or less visible.

Genotype-Environment Interactions

In the elementary approach, the phenotype was defined as a sum of genetic affects and environmental effects that were thought to be independent. In order to maintain some realism it was necessary to state that all estimates hold only for a given population and a given environment. Now let us broaden this view and take interactions between genes and environment into account.

Genetic information can be considered as a developmental program. Genotype-environment interactions can be seen as either conditional statements in that **developmental program** (IF...THEN...) or as functions where quantities from previous steps or from the environment play a role. It is an area where the dichotomy of theoretical entities and observable entities is useful, as in quantitative genetics.

The basic theoretical idea is that there is a function relating a gene product to a phenotype. The gene product may be thought of as an amount of enzyme or even better as an amount of the endproduct of the reaction catalyzed by that enzyme. Since the environmental effects can also be expressed in terms of concentrations of chemical products, this allows us to combine gene effects and environmental effects on the phenotype. This is expressed in the name **Genotype Environmental factor Phenotype Mapping Function** (GEPMF) (see also Chap. 1). The form of a GEPMF may be based on known aspects of gene-physiology, of kinetics of gene products or may be inferred from particular kinds of observations.

The GEPMF can be seen as an extension of reaction norms (see Box 2). Reaction norms are defined for a specific genotype, the mapping function describes the effects of differences in genotype as well as differences in environment. There are a few examples in which similarity between genetic effects and environmental factors has been demonstrated (e.g., Scharloo 1987). Although the idea, then mainly based on phenocopies, is at least half a century old (Dobzhansky 1937), little work has been done in this field. It is therefore impossible to comment on the generality of a correspondance between genetic and environmental effects on the phenotype.

The GEPMF allows us to describe several phenomena in a single framework. These mapping functions are themselves thought to be under genetic control (see Scharloo 1987), e.g., some gene loci may affect temperature sensitivity without affecting the phenotype under the culturing conditions. In variable environments the distinction between direct effects on a phenotype and indirect effects through altering sensitivity to environmental conditions becomes vague and disappears.

Reaction norms can be measured directly when copies of a single genotype are available. When a single environmental factor is used, reaction norms can be plotted as the phenotypic value measured for each value of the environmental factor. In most cases, however, information on reaction norms can be obtained only indirectly.

Knowing the reaction norms for individual genotypes, one could look into their evolutionary genetics. At first sight it seems attractive to regard the reaction norm itself as a phenotype. This conflicts, however, with the definition of phenotype as that what can be observed (on an individual). A realized phenotype is a single point on the reaction norm. Where reaction norms are easily measured, with clonal reproduction, one also finds competition among clones, which affects the evolutionary dynamics.

However, even when they cannot be easily measured, we may assume that there is a reaction norm for each genotype. A first step in taking this into account is then (as always) to express every individual phenotype as a deviation from the population mean. This is easier said than done, because it is not at all clear what the relevant mean is (see van Noordwijk and Gebhardt 1987). This will depend on the ecology. If only individuals that grew up under the same, say poor, conditions have to compete with each other, then it only matters whether they are relatively

Box 5. Threshold Traits

There are many traits where variation falls in discrete classes, but that are apparently not controlled by a single locus. In most cases, there are both complex interactions with the environment and several different gene loci involved. The classical example is a genetically transmitted liability to a disease; genes play a role in the likelihood of becoming affected, but the actual infection depends on the environment.

One can also imagine that in developmental processes positive feedback loops may get started once a critical level has been reached. The result of such a positive feedback loop will then be a big step in phenotype depending on whether the critical level was reached at the critical time or not. The figure illustrates such a threshold in terms of a reaction norm.

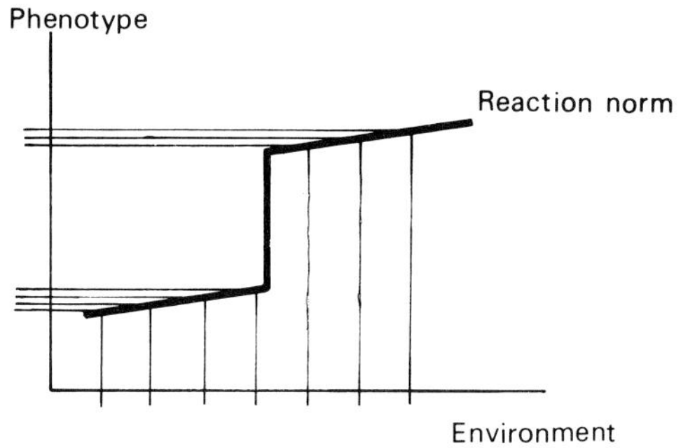

These traits can be analyzed by assuming a normally distributed hypothetical substance. At a certain threshold level of this substance, a qualitative change takes place. The position of the threshold can be deduced from the proportion affected. If one assumes that the threshold has a constant absolute value, changes in the normally distributed substance can be calculated in terms of standard deviation units. A concise description of these methods is given in Chapter 18 in Falconer (1981).

The migration data in Fig. 3 illustrate the usefulness of a threshold model. At first sight, the response to selection going from 24% nonmigrants to 52 or to 17% seems very asymmetrical. The threshold model shows, however, that both responses are precisely what would be expected with a heritability of 0.6 and a symmetrical response.

big and strong. The relevant mean is here the mean of all genotypes under that particular environmental value. If, on the other hand, individuals that grew up under different conditions compete with each other, then the environmental component plays a role as well and the relevant mean is a mean phenotype taken over different genotypes and over different environmental values.

Reaction norms and GEPMF's can have all possible shapes. A special case is the so-called threshold model (see Box 5), in which the GEPMF transforms a continuous quantity into a discrete one. The next two sections give an example of a straightforward analysis of a threshold character, migratory behavior in blackcaps and then a more speculative example, without any genetics, but where the same model gives a comprehensive summary of otherwise confusing data on clutch size in Ural owls. This is followed by a more general discussion of using methods and models from quantitative genetics in the absence of any genetic information.

Migratory Behavior in Blackcaps

Blackcaps (*Sylvia atricapilla*) are migratory birds in part of their range, nonmigratory in other parts, and in some populations only a part of the population is migratory (see Berthold 1988a,b; Berthold and Schlenker 1988). It was therefore chosen as a system to investigate the role of genes in determining whether a bird is migratory or not. Under standardized conditions a large selection experiment was performed by Berthold (1988a) and co-workers.

The results for the first generation of selection in both directions are illustrated in Fig. 3. In the middle, the distribution of a hypothetical quantitative trait in the parental generation is given. The means of the migratory and the non-migratory portions of the population are indicated. Knowing that 24% were nonmigratory and assuming a normal distribution, we can find that the threshold is 0.7 standard deviation units below the population mean and that the mean values of the nonmigratory and the migratory part of the population are 1.3 units below and 0.4 units above the population mean respectively.

To allow comparisons between generations, we assume that the value of the threshold is constant and that the standard deviation units remain unaltered. We recalculate the means setting the threshold at zero and obtain 0.7 for the

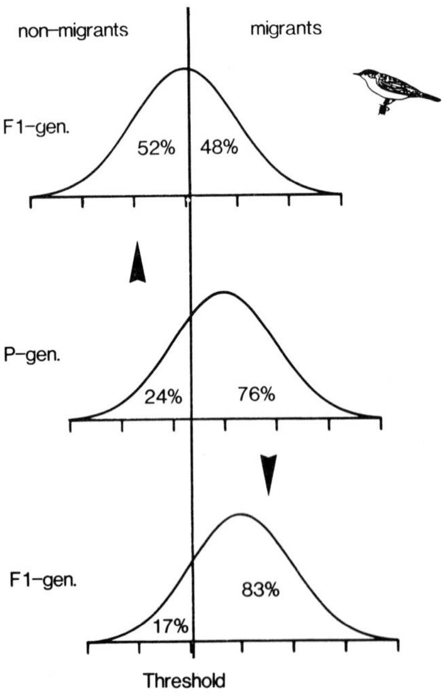

Fig. 3. The results of one generation of selection in both directions for migratory behavior in the blackcap. The P-generation is given in the middle, the F_1 selecting for nonmigrants at the top (that is offspring of two nonmigrant parents from the P-generation) and the F_1 selecting for migrants at the bottom. (Data from Berthold 1988b)

Table 2. Summary of selection for migratory behavior in the blackcap

	Proportion nonmigr.	Population migr.	mean	Mean nonmigr.	Mean migr.
P-generation	24%	76%	+0.7	−0.6	+1.1
a) F_1 selection for nonmigrants	52%	48%	−0.05	−0.8	
b) F_1 selection for migants	17%	83%	0.95		+1.25

$$\text{Heritability estimate a)}\ \frac{\text{Response}}{\text{Sel. diff.}} = \frac{(-0.05-0.70)}{(-0.60-0.70)} = \frac{-0.75}{-1.30} = 0.577$$

$$\text{Heritability estimate b)}\ \frac{\text{Response}}{\text{Sel. diff.}} = \frac{(0.95-0.70)}{(1.10-0.70)} = \frac{0.25}{0.40} = 0.625$$

population mean, and −0.6 and + 1.1 for the two fractions of the population. The differences between these means and the population mean are the selection differentials if like parents are used to produce the two F_1 generations. These are $(-0.6-0.7) = -1.3$ in selecting for non-migrants and $1.1-0.7=0.4$ in selecting for migrants. The difference in mean between the parental and the F_1 generations is the response to selection. From 46% migrants one deduces a population mean of − 0.05 and from 83% migrants a population mean of 1.25 standard deviation units from the threshold. The realized heritability estimate from the selection for nonmigratory birds is now $(-0.05–0.70)/(-1.3)=0.577$ and the estimate from the selection in the other direction is $(0.95–0.7)/0.4=0.625$. These two estimates are very close.

Both the selection differentials and the responses are three times as large when selecting for nonmigrants compared to the selection for migrants. The threshold model shows that a change from 24 to 52% and a change from 24 to 17% after one generation of selection is exactly what one would expect with a heritability of 60%. These calculations are summarized in Table 2.

The threshold model assumes an underlying normally distributed quantity. Could this provide a starting point for a connection with the ecophysiology of migratory behavior? Being migratory or not is a binary classification. Within the migratory birds, however, there is quantitative variation in both the amount and the timing of migratory restlesness. On a global level, there is a correlation between the average amount of restlessness and the proportion of migratory birds (Berthold 1988b). It is thus conceivable that for migratory restlessness, the underlying quantitative trait can be identified and measured. This would imply a threshold in the sense that a small amount of restlesness is suppressed.

Reproductive Effort in Ural Owls

In a large population study of Ural owls (*Strix uralensis*) Pietiäinen et al. (1986) found that in years in which a large proportion of the population bred, the

Table 3. The frequency of clutch sizes in the Ural owl

	No clutch	Clutch size 2	3	4	5	6 & 7
Poor years[a]	179	39	9	1		
in %	78.5	17.1	3.9	0.44		
in sd		0.79	1.71	2.62		
Intermediate years[b]	94	43	77	45	10	
in %	34.9	16.0	28.6	16.7	3.7	
in sd		−0.38	0.03	0.083	1.78	
Good years[c]	52	16	37	54	34	15
in %	25.0	7.7	17.8	26.0	16.3	7.2
in sd		−0.67	−0.45	0.01	0.72	1.46

Data from Pietiäinen (et al. 1986, in prep., and pers. comm)
[a] Data from 1977, 1981, 1984 and 1987.
[b] Data from 1978, 1980, 1982 and 1985.
[c] Data from 1979, 1983, 1986.

average clutch size was larger than in years in which a small proportion of the population was producing a clutch. Based on independent rough estimates of rodent density and the severity of the winter, they classified years as being good, intermediate, or poor in quality. The data for each of these three groups of years can be described almost perfectly with a threshold model, assuming a minimum threshold of a two-egg clutch and further thresholds at regular intervals. The energy that individual females (pairs) have available for reproduction is thought to be normally distributed. The mean of this distribution depends on the food situation in that particular year.

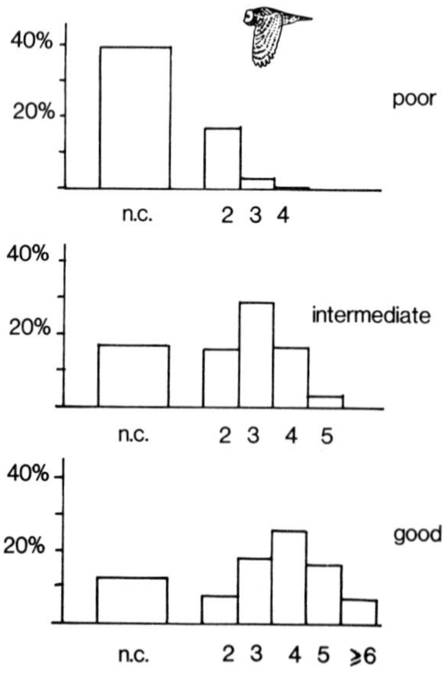

Fig. 4. Clutch size of Ural owls in Finland in poor, intermediate, and good years, *n.c.* no clutch produced. (After Pietiäinen et al. 1986)

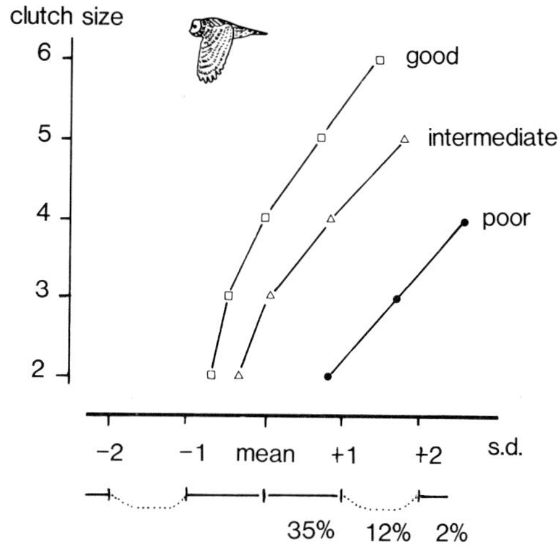

Fig. 5. Probit transformation of the data on clutch size from Fig. 4 gives the mapping functions from a hypothetical energy available for reproduction to clutch size (see text)

The data in Fig. 4 show the observed distributions for the three year groups. A casual inspection of the figure shows that the distributions resemble the cut tails from normal distributions.

Using the data on clutch size frequency (Table 3), one can reconstruct the mapping functions of the hypothetical reproductive energy into clutch size (Fig. 5). The functions for the three quality classes are constructed by plotting the probit values, the distance in standard deviation units that correspond to the observed frequency in a normal distribution. The three curves are almost parallel—the curve for good years is slightly steeper—to the right of the mean, but there are fewer two-egg clutches than expected in the good and intermediate years. There are at least three potential causes: (a) It is an effect of combining the data from 2 or 3 years in each case. (b) A certain proportion of the population does not fit into a normal distribution, for example, because a breeding attempt in the previous year has drained a disproportionate amount of energy available for reproduction, either in the fat reserves of the owls themselves or through the increased predation on the prey in their territory. (c) There is no compelling reason other than being the simplest solution that the mapping functions should be parallel straight lines.

This example illustrates the use of a threshold model in the absence of any genetic information. Apart from giving a concise description of the data, it suggests a link to the physiological mechanisms involved.

Quantitative Genetics Methods Used for Ecological Questions

The methods of quantitative genetics (QG) partition the total variance into a genetic and into an environmental component. For two reasons, they are more frequently used by people interested in the genetic component than by ecologists, interested in the environment component. Most ecologists are still unaware of the extent of **genetic variation** present in most populations for many traits, and many of the questions asked in ecology deal with mean effects rather than with variances. However, interest in the role of variation between individuals in population dynamics is growing (see Hassel 1987; Lomnicki 1988).

Genetic variance reduces the variance to be explained by ecologists. The theory of reaction norms and threshold characters does provide statements about the behavior of the mean value and can readily be integrated in ecology, as the previous examples demonstrate. In fact, the response curves measured by physiological ecologists can be seen as mean reaction norms. The main difference with the treatment of similar data by a geneticist is that the physiologist takes a mean for every environment rather than constructing a curve per individual or genotype.

There are two major reasons to advocate the use of QG methods to study the **environmental variance** as well as the **genetic variance**. First, the methods describe the interactions of individuals with their environment. In their genetic context, they make two important steps from one level of explanation to the next, namely from individual gene to phenotype and from individuals to the population level. More importantly, mapping functions allow inferences to be drawn in reverse, that is from observations at the population level to the physiological action of genes. Similar relations between processes at the level of individuals and at population level play an important role in ecology. The theoretical framework to connect the observations at the two levels is less well developed in ecology.

A second reason to study the effects of environmental conditions on individuals with QG methods is that it helps the needed integration of the environmental components with the genetic ones. Ultimately, it is the combination of genes and environment that determines microevolution.

Discussion

By about 1975 it was obvious that variation in **alloenzymes** was ubiquitous. It was (and is) not so clear what this type of genetic variation "means" to the organism. The prevailing idea was still that little genetic variation would exist in natural populations for traits closely related to fitness. Thus, the investigation of heritability for ecologically important traits in natural populations would lead either to a new appraisal of the amount of generally available genetic variation, or to a demonstration that the allozymes were not representative for all genes. It turned out that the first was the case: there are considerable amounts of genetic variation for traits closely related to fitness in many natural populations. This observation raises the classical question of how this variation is (can be) maintained. Several theoretical analyses suggest that only limited amounts of genetic variation can be maintained in a population, but observations contradict these theoretical results. The obvious conclusion is that some important premise of the theoretical framework is not fulfilled. A closer inspection suggests that part of the problem lies in the definition of basic concepts. At first it seems very strange that whereas directional selection for clutch size can be demonstrated, the same trait in the same population is neutral—not correlated with fitness—over a wide range of its values if considered at a different time-scale. What is the appropriate **time-scale for fitness**? Once the question is explicitly asked, it becomes obvious how rarely people are aware of transitions in time-scale within arguments. To connect processes at different time-scales, the speeds at which processes occur become important. These comparative rates are absent from many of the derivations in population genetics, for example where the change in gene frequency due to two different processes are equated to derive equilibrium conditions.

To list all the premises underlying population and quantitative genetics, would go far beyond the scope of this chapter. I have outlined some of the practical problems and suggested ways to get around them. One should take as little as possible for granted and start from the basic ideas, such as the two main causes for resemblance between relatives are genetic variation and common microenvironment. Common microenvironments can be caused by spatial heterogeneity, temporal heterogeneity, habitat selection, change of the habitat by the organism and as an interesting border case by selection at a previous life stage.

A further general guideline is to try and formulate explicit "mechanistic" models. These allow one to remove such absurdities as considering the size of an egg to be a property of the genes of the zygote within the egg. It is not always obvious whether traits in juveniles are mainly an expression of their own or of their parents'—especially mother's—phenotype (e.g., Reznick 1981). This will also lead to questions about the ontogeny of traits, and of the various components in the ontogeny that may be more or less likely to contain genetic variation or maternal effects.

Evolutionary Biology Is Not for Specialists

This chapter has explored the connections between genetics, ecology, and (developmental) physiology. There are few biologists who fully grasp all three fields. Specialization has been the winning strategy in science, the evolution of specialization is an interesting problem. In as far as we understand it, two elements play a major role; the predictability of future environmental conditions and the relative fitness of generalists in competition with specialists. In theoretical models, it is generally assumed that specialization automatically reduces the fitness outside the special environment. The empirical evidence for this phenomenon in nature is so far not very compelling. Could this be a reflection of the well-known costs of scientific specialization?

In the introduction to this chapter, I characterized quantitative genetics as an approach that concentrates on the most important components. As such, it can only provide explanations at an intermediary level. Quantitative genetics leaves too much hidden in a black box to be satisfactory as an ultimate explanation of phenomena. For the time being, however, it provides many starting points for the integration of knowledge about genetics, about the selective forces operating in the environment and the physiological pathways that translate the gene products and environmental influences into the phenotype. We know something about each of these processes separately, but we know next to nothing about the interactions among them. These interactions can only be revealed in simultaneous study of the elements. At present, we can only guess how important these interactions are quantitatively. I suspect that some of the interactions are important, and that explanations for the maintenance of genetic variation may be found there.

Evolution and Society

Evolutionary biology is almost exclusively seen as a cultural undertaking, furthering general knowledge. However, the demonstration that significant amounts of genetic variation are present for several ecologically important traits in several species, makes **microevolutionary change** possible within a time-scale of five to ten generations. We are familiar with the changes brought about through artificial selection in cultivated species. Changes of similar magnitudes are to be expected as a result of selection pressures caused by human environmental impact. Understanding evolutionary processes is therefore a prerequisite for understanding environmental impact.

At the same time, many of our best (studied) examples of evolutionary changes in natural populations, industrial melanism or pesticide resistance, are related to anthropogenic changes in the environment of the species concerned. These examples illustrate changes in the species or populations where sufficient genetic variation was present. For the majority of species the only evolutionary aspect studied is their extinction. We cannot (yet) predict how much genetic variation will be present for which traits under which circumstances. I suspect that the interactions between the ecology, the genetics, and the physiology are important elements of such an understanding. We will, however, need many more case studies before there is any hope of generalization.

Conclusions

What we want is a better understanding of evolutionary processes at several different levels. At the level of the individual, we want to know what the phenotype will be, given a genotype and given the environmental conditions experienced during its life-time. At the same time we want to make inferences about the genotype from observations on the phenotype and on the environment. At the population level, we want to know how much genetic variation there is, how the environmental variation is structured and how the genetic variation is maintained. Empirical studies could be grouped under the following headings:

1. the dynamics of heritability,
2. the genetics of ecophysiological response curves,
3. the frequency distributions of environmental effects on the phenotype,
4. the frequency distributions of selection pressures.

These groups are not mutually exclusive. They reflect different starting points. Studies that provide information on all these aspects simultaneously will be especially helpful in furthering our understanding of evolutionary processes.

Acknowledgments. I am indebted to Drs. Peter Berthold, Paul Klerks and Hannu Pietiäinen who freely allowed me the use of their data on migration in blackcaps, bill size in great tits, and clutch size in Ural owls respectively. I further thank Drs. Hugh Dingle and Steve Stearns for their thoughtful comments on an earlier version.

Note added in proof: A collection of articles on reaction norms by Slearus, Dodson, von Noordwijk, Schlichting and Scharloo has appeared in Bioscience 39(7) 1989.

10 In Search of Optima: Equilibrium Models of Phenotypic Evolution

P. SCHMID-HEMPEL

Introduction

A central concept in the Darwinian theory of evolution is that of adaptation through natural selection. This concept, properly used, yields a powerful tool to understand why particular phenotypic traits are found in certain organisms and not in others—a question which lies at the heart of the Darwinian revolution.

A classical case of adaptation through genetic change at a single locus is that of the increase in frequency of the dark-colored morphs of the peppered moth (*Biston betularia*) (e.g., Kettlewell 1955). Avian predation was subsequently identified as a major selective force acting against the wild-type light-colored morph which once was predominant. However, it is important to realize that the mutant, dark-colored forms were not spared from bird predators because they carry the mutant gene, but because they are more difficult to detect when resting on dark bark. In fact, with the advent of the Industrial Revolution, air pollution had turned many tree barks, where the moths prefer to rest, darker than before. Natural selection thus "sees" the phenotypic characteristics in an environmental background, and as other contributions in this volume should have made clear, it is only when these characteristics possess variation and heritability that evolution can occur. Without the study of how phenotypic traits expressed in a given environmental context influence survival and reproductive success of individuals—the proper study of adaptive value—biological phenomena would make little sense. Therefore, this kind of analysis is not simply an extension of genetics, but an independent enterprise in its own right, requiring techniques and methodology from a variety of fields in organismic biology and ecology.

The most straightforward way to investigate the **adaptive value of phenotypic traits** would simply be to follow a number of different individuals over their lifetime and observe how many descendants they leave. We may also use the comparative method to study the function of, for example, canine teeth size in primates (Harvey et al. 1978) by comparing their relative sizes across a number of species and with respect to different hypotheses. So why should optimization theory be of particular interest? One answer is that it offers a potentially quicker, and in many cases perhaps the only, way to approach adaptive value. But optimization models also help to make a connection between selection and the traits under investigation by incorporating the details of the selective mechanism. For example, one could follow light- and dark-colored forms of the peppered moth over their life-time and count how many descendants they leave. This would tell us a great deal about how selection acts, by showing that bird predation is more intense in light-colored forms when most bark is dark. Furthermore, comparison among related species may reveal that moth coloration has to do with crypsis rather than with thermoregulation. However, the explanation that "dark-colored forms are more cryptic on dark bark", and thus have higher fitness, would still encompass a wide range of possible variation in coloration. If, in contrast, we hypothesize that "coloration minimizes the probability of being eaten by a predator", a much more precise statement is gained and put to the test. Since we ask for a minimum, this is a typical question that can be answered by optimization and game theory. It should also immediately become clear that to test such a hypothesis, a number of precise constraints have to be incorporated, e.g., what is the frequency of different backgrounds in the environment, what are the sensory capabilities of the bird predators, and so forth? Optimization theory thus helps to make the notion of "adaptive value" more precise and ideally we can gain a deeper insight into the details of the selective machinery.

The basic tenet of the optimization approach is that variation in a trait reflects variation in fitness of the individuals. One has nevertheless to be careful to distinguish between the notion of **"optima"** used in everyday language where it usually means the best possible, and the same term used in connection with optimization models, where it defines the best possible within a framework of constraints. Furthermore, the term "fitness" is apparently troublesome, and it is discussed at length in other chapters of this book. (see Chap. 2). Suffice it to say here that a variety of measures for fitness are available. Life-time reproductive success usually enjoys a privileged position among these, but measures going beyond single-generation considerations are also under discussion. Nevertheless, for any given problem where optimization is used, the meaning of this term is usually well defined, such as net rate of energy gain in foraging models or the Malthusian parameter in life history problems. Such measures are local tokens for "true" fitness and they are used for their practicability and plausibility. People working in the field agree that such tokens, which are usually short-term measures of abundance, must eventually be connected to life-time reproductive success. This will allow us to connect the short-term analysis to the "background" provided by the population as a whole and thus make for the transition to evolutionary issues that involve heritabilities, amount of variation, or selection coefficients. While this is recognized, the problems remain technically difficult, but they are being worked on, and progress is being made.

To further understand the use of models as preferred by most researchers, it seems important to draw the distinction, aptly summarized by Maynard Smith (1987), between **"units of evolution"**, which must possess heredity (in the sense that offspring resemble parents), and "units of selection", which need not. This is important, since we can expect that units of evolution are entities that possess traits ensuring their survival and reproduction and which are thus worth studying by optimization methods. For all we know, individuals are such entities, whereas populations may be units of selection, but hardly units of evolution, because compelling evidence demonstrates strong between-individual within-group selection such that complex adapation at the population level is not to be expected. For this reason, most workers concentrate on the individual rather than the population as their subject of interest. However, the population background usually enters the models through the definition of strategy sets, constraints and, in particular, payoff functions, as will be discussed below. It should also be mentioned that the problem of the evolution of the units of evolution, which is involved here, is in fact a very profound question (e.g., Buss 1987).

An Introductory Example

Consider the fate of an individual of a hypothetical annual plant species. This individual will grow throughout the season, produce leaves that permit photosynthesis, and eventually produce reproductive organs. Let us assume that two growth types are possible: (A) rapid growth type that results in a long stem with small leaves high off the ground, and (B) slow growth type with short stem but larger leaves near the ground. Furthermore, let the size of the reproductive organ, and thus the resulting number of offspring, be positively correlated with leaf size. This view of the plant world is of course very simplistic, but it epitomizes the approach whereby biological realism is often sacrificed for clarity in presenting some fundamental ideas.

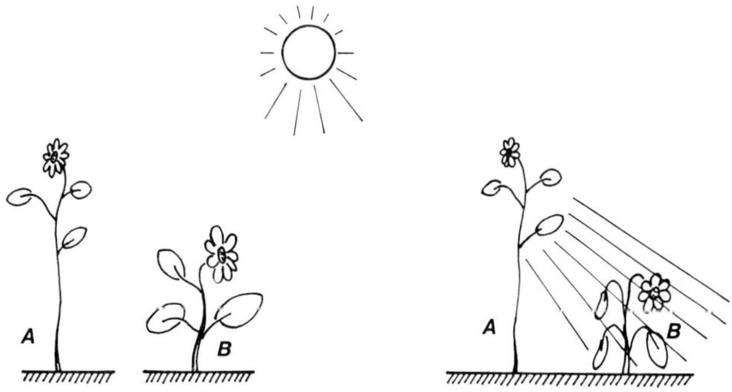

Fig. 1. Hypothetical examples with two growth types of a plant species: *A* rapid growth type; *B* slow growth type. *Left* no interaction is possible between types; *right* interactions occur, because tall plants can shade small ones. Seed production is proportional to size of reproductive organ, and a trade-off in the size of the latter and growth rate is assumed. Selection is frequency-independent in the *left* hand situation and type *B* will eventually succeed. Selection is frequency-dependent in the *right* hand situation and the outcome depends on a number of variables.

Suppose that growth type is to some degree heritable and that over generations plants reproduce under similar environmental conditions without mutual interference between neighboring individuals (Fig. 1). Which type will succeed in the course of this imaginary evolution? In the present example, the answer is simple: because the slow type is capable of producing more offspring (that carry the "slow" genes), its frequency in the population will increase generation after generation so that eventually "slow" will be the predominant growth type found in the population.

In contrast, let us now assume that interactions between neighboring plants occur. Biologically, this may be the case when sunlight is a limiting resource and plant density is very high, so that it is possible to shade a neighbor. If neighbors are of the same size, then for both plants the rate of photosynthesis, and thus the number of offspring, is reduced by a certain amount. On the other hand, if a small plant happens to be next to a tall plant, the tall plant is not affected, while the small one suffers even more than it would from competition with a small neighbor. Which type will succeed in this situation? Once again, we have to focus our attention on the average reproductive success of individuals of both types. Here, a major difficulty arises: reproductive success depends on the neighbor's growth type. It is easy to see that if there are many tall plants in the population, the small ones will do very poorly. On the other hand, if rapid growth types are rare, the slow ones are on average fairly successful. In other words, with the introduction of interactions to our fictitious world, reproductive success has become frequency-dependent. Under certain conditions it is now possible that the rapid growth type will finally be the predominant plant morph although the average speed production in the population would be maximal if only small plants were present.

The two hypothetical scenarios require different kinds of evolutionary models. However, both selective processes can be analyzed with an emphasis either on the genotypic or the phenotypic level. If the emphasis is on the phenotypic level, as discussed here, then the first scenario (i.e., without interaction) would typically require techniques from optimization theory while the second, frequency-dependent scenario (i.e., with interaction) would call for a game-theoretic treatment. Note at this point that the genetic basis assumed for the two plant types is not analyzed nor even used further, but is providing an ad hoc rationale for the approach, i.e., that one is considering phenotypic variation having microevolutionary consequences.

The attentive reader may now wonder, because interactions among individuals in the population are the rule rather than the exception, why are not game models used throughout? The answer is that optimization models can usually by made much more precise. Thus, it is often decided to sacrifice the dimension of strategic interaction as explored by game theory for improved understanding in the details of the investigated trait. Moreover, for many problems, such as foraging behavior of solitary predators, interaction with other individuals is in fact not a problem immediately impinging on the question of interest.

In contrast to this constructed example, the modeler is already confronted with the result of an evolutionary process. Hence, rather than retracing the course of evolution, the basic question is: Why do individuals of a given population possess a certain trait A and not another trait of a possible set of alternative phenotypes? Here, "traits" are usually taken to mean various phenotypic states of the same characteristic (e.g., three, four or five eggs laid) rather than different characteristics altogether (e.g., laying eggs or giving birth to independent young). The modeling approach to answer the question now focuses on an explanation for the evolutionary **maintenance of a trait** rather than on the historical evolutionary pathway that would fully describe this trait's origin. Consequently, a basic assumption is that the trait in question has been "shaped" over evolutionary time-scales long enough so that it is no longer altered by the currently prevailing forces of natural selection. This is an **equilibrium view of phenotypic evolution**, and some comments on this point will be made later. Clearly then, this approach is a modest one in the sense that it addresses only one of many aspects that would interest an evolutionary biologist. However, by modestly focusing on the problem of maintenance, the modeler gains precision in return for the loss of scope.

As it is immediately apparent, not all properties of an organism can be analyzed in this way. There will be a class of traits that might be better understood by looking at the phylogenetic history of the lineage under scrutiny than by looking at the present-day environment. In fact, the optimization approach forces the modeler to distinguish between two classes of traits, namely (1) those which can be shaped very easily by natural selection (i.e., the traits subject to stability analysis), and (2) those that remain fairly constant over evolutionary time-scales even under changing selective regimes. In mathematical language, traits of class (1) would be considered as the strategies or decision variables, whereas traits of class (2) would be constraints of the problem. However, it is usually not possible to consider all traits of class (1) at once, so that any model uses, at least implicity, a "background" in which a number of such traits are artificially kept constant and entered as constraints.

Aim and Structure of Models

The application of optimization and game theory models over the last years has added considerably to the precision and scope of seeking explanations for the existence of phenotypic traits. Developments and ideas from many fields, such as life history theory or behavioral ecology, have contributed to this success, but as a result the increasing diversity of models and mathematical techniques used has become a source of confusion even for workers in the area. I therefore present here a short survey of modeling techniques for optimization for a broad biological audience rather than for the specialized theoretician. The mathematical details are not emphasized but the primary aim is to indicate where to "shop" for the appropriate model and what the basic principles are. A number of biological examples have been selected to visualize the application of a given technique and to show what problems and assumptions are encountered. In view of the large body of literature, this survey must necessarily remain incomplete, and I must also emphasize that, in terms of sophistication, modelers have gone far beyond the basic examples illustrated here, although the principles remain the same.

While the mathematical language used in the modeling approaches discussed here originates from decision theory, it is not implied that the individual themselves "decide" about what trait they should carry (i.e., what strategy to choose). Instead, it is the process of evolution through natural selection which plays the role of the "decision-maker". Interestingly, however, this evolutionary process brings about traits which appear to an observer as if they resulted from rational decisions made by the individual animal. In this sense, the optimality or game theoretic approach leads to a quasi-rational interpretation of adaptation (see Parker and Hammerstein 1985). With regard to mathematical terminology it is, therefore, logically justified to speak of the individual as if it were itself the decision-maker.

In the case of frequency-independent selection a model consists of the following elements. Firstly, a **strategy set** contains the possible alternative phenotypic traits that are accessible to the forces of natural selection, i.e., the traits of class (1) as defined above. In the simplest case the strategy set is given by a decision variable that can assume two or more values, such as "slow" or "rapid" growth in the introductory plant example. Secondly, a **payoff function** describes how an individual's success depends on its strategy. The ultimate currency in which success should be measured is expected Darwinian fitness, e.g., the number of offspring in the plant example.

The definition of a strategy set contains assumptions about what **constrains** the set of possible traits. These assumptions are sometimes explicitly formulated, but unfortunately very often they are not. In our plant example it was explicitly stated that only two strategies, "rapid" and "slow", are possible. An implicit assumption, however, was that rapid growth cannot simultaneously lead to large reproductive organs. Biologically, this could mean that the plant is constrained to a limited amount of resources that can be allocated to either production of stem material or reproductive tissue. While justified in this hypothetical plant species, such implicit assumptions are in fact excellent pitfalls in theoretical work, since they refer to something that is eminently plausible and valid but actually may just represent large areas of unknowns.

The aim of models is to investigate what available strategy should be chosen to maximize or minimize a given payoff function, subject to constraints. The problem is thus to find an extremum of a function, and for simplicity we will mainly talk about maxima. Some applications that would fall under this definition are not considered in the following, e.g., models in community ecology which seek to maximize the probability or duration of persistence of an assemblage of species (e.g., May 1981), because such entities (i.e., communities) can hardly qualify as units of evolution. I have also omitted purely qualitative or verbal models. This should not argue against their usefulness. Quite to the contrary: such qualitative models are often a first and crucial step towards formalization of a problem; often, instead of mathematical overkill, they would be entirely sufficient.

Optimization Models in Evolutionary Biology

In the optimization models discussed here, the strategy set generally consists of a single variable, the **decision variable** that is assumed to be under the control of the individual. It can take any value within a specified interval. The optimization techniques themselves are derived from various fields of mathematical theory, and for a more thorough introduction the reader should consult one of the many textbooks (e.g., Raiffa 1968; DeGroot 1970; Intriligator 1971; Jacobs 1974; McFarland and Houston 1981; Ross 1983). For most biologists, the mathematical jargon might make it often very difficult to use these tools. Nevertheless, to appreciate what optimization models can do, it is necessary to understand some points about their structure and their technical side. I will therefore characterize the major types of models in boxes together with illustrative examples.

Life History Problems

Most life history models assume a population with stable age distribution growing in a density-independent environment (see also Chap. 8). These assumptions allow us to use the familiar Euler–Lotka equation that links important variables: survival from birth to age x, age-specific fecundity, and r, the intrinsic rate of increase or Malthusian parameter of the population. In typical life history models, the organism can decide on the amount of reproductive effort at age x (e.g., fecundity). Such effort is assumed to be negatively correlated with survival probability, while the payoff is taken to be r (e.g., Law 1979a, Schaffer 1974a; Charlesworth and Léon 1976; Schaffer and Rosenzweig 1977). Unfortunately, the mathematics quickly become too complex for analytical solutions, even with quite restrictive simplifying assumptions (such as in Law 1979a). Furthermore, many of the model parameters are usually not known for natural populations and thus quantitative predictions cannot be made. It therefore seems like a sensible research strategy to simplify the question so that models become more tractable and can thus be tested more rigorously. A number of authors have taken this approach to analyze a single but crucial life history trait: age at maturity (e.g., Bell 1980; Roff 1984; Stearns and Koclla 1986). Note that oversimplified problems can be risky and a lot of knowledge on the natural history of the problem and biological intuition is needed to avoid this caveat.

Age at Maturity

Stearns and Koella (1986), based on earlier models of Crandall and Stearns (1982) and Stearns and Crandall (1984), predict at what age and size individuals should

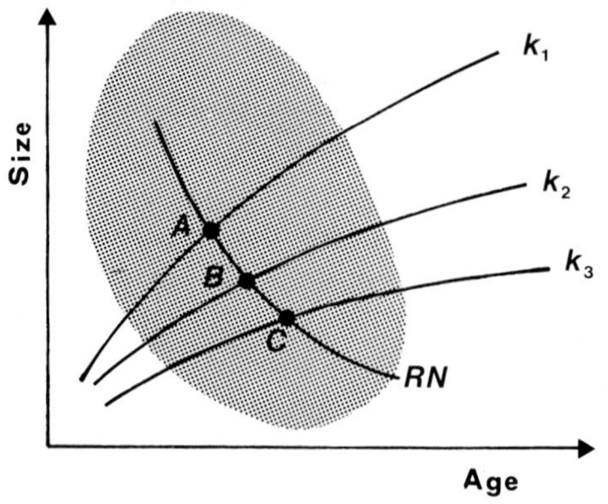

Fig. 2. A model for optimal age and size at maturity. Growth curves of the organism under consideration are k_1, k_2, k_3 and given by the environment (good-harsh). The *shaded area* is the available strategy set. *RN* denotes combinations of size and age that would maximize r, the Malthusian parameter of the population, i.e., depending on environmental conditions in which organisms find themselves, they should become mature at points *A*, *B*, or *C*. The curve *RN* is interpreted to indicate the reaction norm, given that the *shaded area* represents a single genotype. (After Stearns and Koella 1986)

reach maturity such that r for the population is maximized. The authors take growth rate of individuals as being set by the environment (e.g., slow growth in poor environments). Hence, body size is a function of age and varies among environments (Fig. 2). Furthermore, the population obeys the Euler–Lotka equation which makes it possible to express the crucial variables as a function of age at maturity alone (i.e., the decision variable). Stearns and Koella (1986) consider their model as addressing reaction norms, but note that in their model size and age at maturity are not independent decision variables as this notion might suggest. Therefore, while this model makes correct predictions on size and age at maturity for the studied cases, it does not aid in understanding how variable and conflicting needs are resolved with alternative developmental pathways.

Clutch Size

Optimal clutch size in birds is a "classical" topic in population biology (e.g., Lack 1954; Wynne-Edwards 1962; Williams 1966), and several models have been formulated (e.g., Charnov and Krebs 1974; Brockelman 1975). Here, egg number is the decision variable and eggs are often supposed to be unlimited (e.g., no constraints through nesting space). However, a crucial assumption is that large clutches entail a reproductive cost by reducing parental survival (e.g., Nur 1984). Whether such costs do really exist is a matter of dispute (e.g., DeSteven 1980; Smith 1981; Reznick 1985), but the application of models and the concomitant need to formalize the relationships have brought this problem back to the attention of researchers and stimulated the discussion.

For species without parental care, notably for parasitic wasps (for a review of the biology, see Charnov and Skinner 1984; Waage and Godfray 1985), models of clutch size have only recently been formulated (Parker and Courtney 1984; Charnov and Skinner 1984; Skinner 1985; Godfray 1986). Skinner (1985) (see Box 1) has investigated the case of *Nasonia vitripennis* (Peteromalidae,

Box 1. Simple Calculus

It is often possible to formulate a payoff function, $f = f(z)$, that depends on a single, continuous or discrete decision variable Z (Fig. B1 left). The modeler asks what value of Z maximizes f(Z). With simple calculus: $\partial f/\partial Z = 0$, and $\partial^2 f/\partial Z^2 < 0$ for a maximum, $\partial^2 f/\partial Z^2 > 0$ for a minimum (Z_1). The value of Z may be constrained to the shaded interval (a,b), the strategy set. Hence, it may be necessary to check the boundary values for maxima or minima of the function. Additional conditions might be necessary to check whether a local (Z^*, Z_2) or a global maximum (Z^*) is found. For most biological situations, however, it will be clear from the nature of the problem, that only one value, $Z = Z^*$, satisfies $\partial f/\partial Z = 0$ and that Z^* must be a maximum; the additional conditions are thus not explicitly tested. Skinner (1985) used simple calculus to predict opimal clutch size for parasitic wasps (Fig. B1 right). The wasps move around in search of hosts (e.g., eggs or larvae of lepidopterans) and lay their eggs in them. Time τ to move between hosts

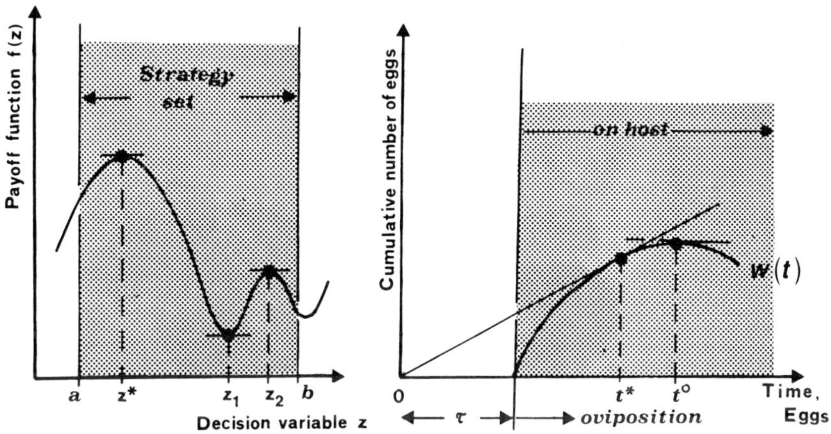

is constrained, i.e., cannot be influenced by the animal. In the model, the percentage of emerging and surviving young (a measure of fitness) is taken to decline with the number of eggs laid per host, n, according to the relationship $w' = w'(n)$. The gain from laying n eggs in a host is therefore $w = nw'$. The model's strategy set is a continuous range of oviposition times, t, that, given a constant rate of oviposition, is proportional to the number of eggs laid per host; thus $w(t) \propto w(n)$. Function w(t) is dome-shaped because of the increasing competition between eggs in the same host. If only one host were available, the wasps should choose to stay for t° to maximize w(t). With several hosts, the payoff function is $f(t) = w/(t + \tau)$, i.e., the measure of success is number of eggs laid per unit time of ovipositing and searching for hosts. Only a single maximum exists, $t = t^*$, and it is easy to show that the condition $\partial f/\partial t = 0$ leads to $\partial w/\partial t = w/(t + \tau)$ which is used for the graphical solution in Fig. B1.

Lagrange multipliers. Suppose the problem is to find a maximum of the payoff function $f = f(Z)$, subject to the constraint function $\phi = \phi(Z)$. By introducing Lagrange multipliers λ (a vector, or list of variables), this problem is simplified to finding an extremum of the unconstrained (Lagrangian) function

$$L(Z,\lambda) = f(Z) - \lambda\phi(Z).$$

By observing $\partial L/\partial Z = 0$ and $\partial L/\partial \lambda = 0$ the additional equations are gained to solve also for the unknown multipliers λ. If $\phi(Z)$ defines equality constraints, this is a problem of classical programming; with inequalities, nonlinear programming is used (see Intriligator 1971).

Chalcidoidea; Hymenoptera) attacking larvae or pupae of blowflies. Females parasitize more than one host and they have to search for them before they can lay their eggs. In this case, optimal clutch size also depends on how oviposition rate compares with rate of ovigenesis, i.e., whether the wasp is proovigenic (i.e., the life-time supply of eggs is already carried at eclosion of the female), or whether it is synovigenic (i.e., females mature their eggs while searching for hosts). Skinner's (1985) analysis thus provides an excellent example of how physiological (or phylogenetic) constraints are incorporated into a model and how they might affect the optimal solution.

Hidden Assumptions and Fitness Controversy

As already mentioned, there is hardly any model that explicitly states all its assumptions. For example, in the clutch size model of Skinner (1985), it is assumed that the wasp "knows" the properties of the environment and has an essentially unlimited life-time, so that the single, average oviposition cycle shown in Box 1 is a fair representation of the most important selective situation. If, in contrast, a female could lay only a small number of eggs throughout her life-time, then this limited capacity becomes a boundary such that oviposition decisions become contingent on the number of eggs left and dynamic models must be used (Iwasa et al. 1984) (see Boxes 2 and 3).

The use of models in life history theory has also sparked a prolonged and heated debate over the measurement of "fitness", particularly whether maximizing the Malthusian parameter is equivalent to maximizing reproductive value of individuals (Schaffer 1974b, 1981; Taylor et al. 1974; Caswell 1980; Yodzis 1981; Caswell 1981; Goodman 1982; Gleeson 1987). I shall not summarize this discussion here, but note that the very existence of this controversy is proof of the heuristic value of optimality models. In fact, the modeler has to make the difficult decision what the appropriate payoff function should be.

As mentioned in the Introduction, choice of the payoff function should reflect the essence of the problem and mimic the result of ongoing natural selection. Life-time reproductive success has often been used, but this need not necessarily be so. For example, Brown and Venable (1986) suggest that in fluctuating environments, the geometric mean of annual fitnesses should be maximized (see also Gillespie 1977; Grey 1980). Furthermore, frequency-dependent selection is an essential characteristic of many biological systems, but in this case reproductive value may be a poor choice (e.g., Charlesworth 1980).

Dynamic Problems

Life history is a pattern that unfolds over time; its analysis therefore naturally invokes models where decisions can be made more than once and where the outcome of earlier decisions affects later options. This leads to dynamic models explained in Boxes 2 and 3 (a more thorough account is given by Intriligator 1971; Jacobs 1974; Dixit 1976; Léon 1976).

In contrast to Vincent and Pulliam's (1980) continuous model of an annual

Box 2. Pontryagin's Maximum Principle-Dynamic Programming

When decisions can be made more than once, and the outcomes affect future options, methods of dynamic programming are appropriate. For problems in continuous time, the state of the system under study (a plant, a growing animal) is defined by time-dependent state variables, $\mathbf{x}(t) = [x_1(t), x_2(t), \ldots]$. State is changing under the influence of control variables, $\mathbf{u}(t) = [u_1(t), u_2(t), \ldots]$, according to the dynamic equations, $\partial\mathbf{x}(t)/\partial t = \dot{\mathbf{x}}(t), = \mathbf{f}[\mathbf{x}(t), \mathbf{u}(t)]$, that relate state and controls at time t to the state at time t + dt. Control theory views this problem as a time-dependent process unfolding from initial time t_0 to terminal time t_1, where the aim is to control this process—through adjustment of $\mathbf{u}(t)$—to maximize an objective functional (the payoff function)

$$J = \int_0^T I(\mathbf{x},\mathbf{u},t)dt + F[\mathbf{x}(T), T];$$

for convenience $t_0 = 0, t_1 = T$. I( ) is called the intermediate function, F( ) the terminal function. For many biological problems, $F(\ ) \equiv 0$, e.g., because the organism has completed its life cycle and no residual value can be assigned to final state and time. The available strategy set consists of all possible vectors $\mathbf{u}(t)$, i.e., the set of possible control functions or policies.

According to Pontryagin's maximum principle, the optimal policy is found by finding the maximum of an auxilliary function at every moment in time during (0,T). This auxilliary function is called the Hamiltonian:

$$H(\mathbf{x},\mathbf{u},t) = I(\mathbf{x},\mathbf{u},t) + \boldsymbol{\lambda}\cdot\mathbf{f}(\mathbf{x},\mathbf{u},t) \rightarrow \max \text{ for all } t \in (0, T),$$

where $\boldsymbol{\lambda}$ is a vector of costate variables (reflecting constraints similar to the Lagrange multipliers, see Box 1). The additional variables are determined with the cononical equations:

$$\partial\lambda/\partial t = \dot{\lambda} = -\partial H/\partial\mathbf{x}, \quad \lambda(T) = \partial F/\partial\mathbf{x}(T), \quad \mathbf{x}(0) = \mathbf{x}_0, \quad \partial\mathbf{x}/\partial t = \dot{\mathbf{x}} = \partial H/\partial\lambda.$$

Vincent and Pulliam (1980) used Pontryagin's principle to predict the optimal life history of an annual asexual plant. Essentially, they considered the plant to be a system that is defined as $\mathbf{x}(t) = [L(t), s(t)]$, where s(t) is biomass of reproductive tissue and L(t) biomass of the remaining, vegetative part of the plant. Over a season (0,T), the goal is to maximize $J = {}_0\!\int^T s(t)\,dt$, i.e., the amount of reproductive biomass produced. L(t) is supposed to be linearly proportional to the resources, $R(t) = R\cdot L(t)$, that can be collected, i.e., rate of photosynthesis is proportional to leaf area. The process is controlled

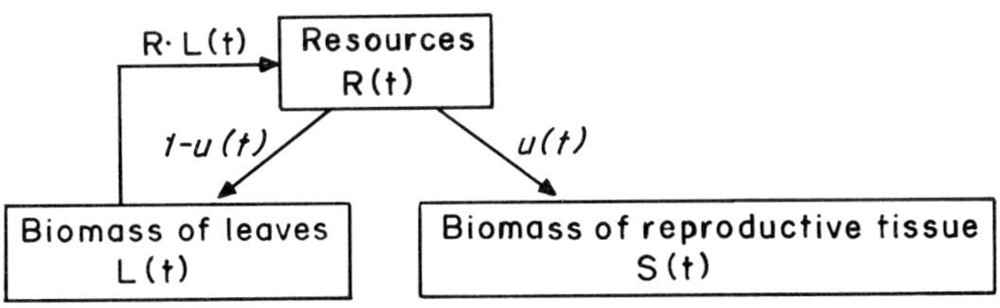

with $\mathbf{u}(t) = u(t)$, the proportion of resources allocated to produce reproductive tissue rather than vegetative parts at any one time $(0 \leq u \leq 1)$. The dynamic equations are $\mathbf{f} = (\partial L/\partial t, \ \partial s/\partial t)$, therefore $\partial L/\partial t = R[L(t)(1 - u(t)]$, and $\partial s/\partial t = \dot{s} = R\cdot L(t)\cdot u(t)$. The Hamiltonian is found from $H(t) = \dot{s}(t) + \boldsymbol{\lambda}\cdot\mathbf{f} = \dot{s}(t) + \lambda_1(t)L(t) + \lambda_2(t)s(t) = RL(t)u(t)[1 + \lambda_2(t)] + \lambda_1(t)RL(t)[1 - u(t)]$. After rearrangement, $H(t) = RL(t)[u(t)(1 + \lambda_2(t) - \lambda_1(t)) + \lambda_1]$. The canonical equations give $\lambda_1(T) = 0$, $\lambda_2(t) = 1$, $\lambda_1(t) = -\partial H/\partial L$, $\lambda_2(t) = 0$; hence $\lambda_2(t) \equiv 1$. In this simplified example, it is readily seen that to maximize H(t), u = 0 for $\lambda_1(t) > 2$ and u = 1 for $\lambda_1(t) < 2$. Because $\lambda_1(t)$ is a monotonically decreasing function, optimal control allocates all resources first into vegetative parts, then into reproductive tissue (i.e., a "bang-bang" control).

Box 3. Multistage Problems—Dynamic Programming

Time may enter as a discrete variable in a problem that is otherwise analogous to the dynamic situation of Box 2. To solve such multistage decision problems, Bellman's (1957) optimality principle is applied which states that an optimal policy (i.e. the sequence of controls, see Box 2) should have the property that whatever the initial state and decisions are, the remaining decisions must constitute an optimal policy with regard to the state resulting from the first decision. As before (Box 2), $\mathbf{x}_i = [x_1(i), x_2(i), \ldots]$ and $\mathbf{u}_i = [u_1(i), u_2(i), \ldots]$, denote state and controls at stage i respectively. The dynamics relate state and controls at stage i to state at stage $i+1: \mathbf{x}_{i+1} = \mathbf{f}(\mathbf{x}_i, \mathbf{u}_i)$. Here, the payoff function is often called the performance, V,

$$V = \sum_{i=0}^{N} B(\mathbf{x}_i, \mathbf{u}_i),$$

where B( ) is the value of a suitable benefit function evaluated at each stage i. The process lasts from initial stage $i = 0$ to terminal stage $i = N$. Now suppose, V_i^* is the value of the performance function if state at i is x_i and control is optimal (i.e., to maximize performance) from stage i until termination at $i = N$. Then, by Bellman's principle, it follows that

$$V_i^*(\mathbf{x}_i) = \max_{ui} [B(\mathbf{x}_i, \mathbf{u}_i) + v_{i+i}^*(\mathbf{x}_{i+1}, \mathbf{u}_{i+1})] = \max_{ui} \{B(\mathbf{x}_i, \mathbf{u}_i) \times v_{i+1}^*[\mathbf{f}(\mathbf{x}_i, \mathbf{u}_i), \mathbf{u}_{i+1}]\}.$$

This is the basic recurrence relation of multistage dynamic programming. Because optimal control at stage $i = N$ is often known, the problem is usually solved by backward recurrence. Thus at stage i, the recurrence relation allows to determine $\mathbf{u}_i$ that maximizes $B_i(\ )$, since V_{i+1}^* has already been fixed in the previous step. There is, however, a difficulty: while the optimal terminal decision, u_N, is often obvious, only initial state, x_0, is known (the same applies to Pontryagin's principle). The problem can usually be circumvented by iteration of the procedure: backward recurrence from a guess about terminal state, then forward recurrence with the known decisions from known initial state to determine actual terminal state, and so forth, until convergence is reached. Schaffer (1983) used dynamic programming to solve part of the generalized life history problem. The payoff to be maximized is Fisher's reproductive value at birth, R_0. The population is assumed to grow exponentially with a discrete, stable age distribution. With additional assumptions, to maximize R_0 is equivalent to maximize Λ, the exponential form of the Malthusian parameter ($\Lambda = e^{\gamma}$). Then, by definition, $J = R_0 = \Sigma_{i=0}^{N} l_i m_i / \Lambda^{i+i} \rightarrow \max$, where l_i survival probability from birth to age class i, and m_i fecundity at age class i. Applying the recurrence relation:

$$R_i = \max_{ui} [l_i m_i / \Lambda^{i+1} + R_{i+1}^*] = \max_{ui} [l_i m_i / \Lambda^{i+1} + p_i(R_{i+1}/R_0)]$$

As in box 2, u_i is the fraction of available resources allocated to reproduction; both $m_i(\ )$ and $p_i(\ )$ (survival to next age class) are functions of u_i. For $i = N$, $p_N = 0$ and the problem is solved by backward recurrence. However, Λ and R_0 depend on all controls and are thus not known in advance; thus, iteration is necessary to find the actual solution.

plant shown in Box 2, suppose the plant is a perennial, iteroparous species with discrete age classes. During each season (i.e., age class), resources must be allocated to growth or reproduction to maximize life-time production of reproductive biomass. To take such a life habit into account, Schaffer (1983) defined a value for the terminal function, F( ), of the objective functional by the residual reproductive value at the end of each age cass. F( ) also contains a value assigned to stores that help to survive to the next season. Because age classes are discrete, we deal with a discrete, multistage decision problem, and another kind of dynamic programming must be used (Box 3).

Interestingly, the same basic model can be applied to the life history of entirely different organisms. Macevicz and Oster (1976; also Oster and Wilson 1978) analyze the annual **life cycle** of a colony of **social insects** (such as found in many social wasps and bees). Colonies gain resources through the efforts of the workers that go out and gather food. Resources can be used to feed larvae that either become new workers or alate, sexual forms, i.e., new queens and males (in social insects, caste determination is usually trophogenic; Oster and Wilson 1978). Thinking in terms of optimization models shows that this decision problem is formally similar to that of the plant (Fig. 3). The predicted optimal solution is also "bang-bang": workers and sexual forms should never be produced at the same time.

It is less obvious that the similarity of plant and social insect model (Fig. 3) is also based on the assumption that both systems behave as units (see Bourke 1988; Schmid-Hempel 1990). This seems to be more likely for plants (but see Harper's 1977 modular units of construction) than for social insects where worker reproduction (which can provide infertilized eggs that develop into males) occurs (Cole 1986; Bourke 1988). Note also that the haplo-diploid sex determination system of hymenopterans (to which most of the social insects belong) can cause additional difficulties, because genetic relatedness (and thus the value of different options) to male and female offspring is different for queen and workers, so the model must make assumptions about who is in control over reproduction (e.g., Bulmer 1983).

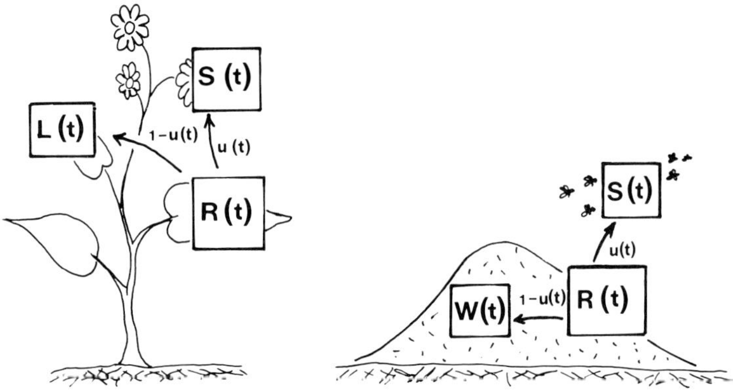

Fig. 3. Structural similarity of the life history problem for plants (*left*) and colonies of social insects (*right*). In both cases, the amount of available resources at time t, $R(t)$, is given either through production in leaves (photosynthesis) or through the activity of workers (foraging). At any time, the plant should decide on allocation of resources to produce either new leaves, $L(t)$ (an allocation that would allow for an increased rate of photosynthesis), or to reproductive tissue, e.g., to produce seeds, S(t). Fraction $u(t)$ is shunted towards reproduction. In social insects, either new workers, $W(t)$, or (alate) sexuals, $S(t)$, can be produced. The optimal life history is an allocation pattern over time that maximizes reproductive output, measured as biomass of reproductive tissue or sexuals, respectively

Life history models often predict **bang-bang control**. It is important, however, to realize that "bang-bang" is not a general principle but reflects the way problems are modeled. Bang-bang solutions are to be expected whenever the situation is deterministic and the dynamics are linear in the controls (Intriligator 1971). Indeed, if some of these assumptions are modified, graded control emerges (e.g., King and Roughgarden 1982). On the other hand, in continuous models, the control variable u(t) is allowed to jump from u = 0 to u = 1 in a discontinuous step. If instead u was forced to be a continuous function (which is more realistic), no easy solution was available. Thus, in examples of this kind, adding more realism does not appear helpful.

Observations show that bang-bang patterns in real life histories occur (e.g., Chiarello and Roughgarden 1984), but the empirical evidence is not particularly strong. For example, Green (1984) and Kolmes (1986) note that in colonies of social insects, workers and reproductives are often produced side by side rather than in a bang-bang fashion. Nonlinearity and stochasticity are most often invoked as preventing bang-bang control from being optimal. Stochasticity may also exist with respect to terminal time T (see Lucas 1985).

Pontryagin's method (Box 2) is elegant, which may explain its frequent use in solving this class of problems. However, there are two types of controls; in the commonly used open loop, control depends only on time and hence the principle behaves like a "blind robot" so that new information obtained during the process is not taken into account. This is a serious drawback and a good deal of strategic interaction is therefore lost from the model. With closed loops, controls also depend on state, but the mathematics become quickly intractable.

Optimal Timing

In some life history problems, optimal time to take action is sought (similar to age at maturity). Examples are birds that have to decide on optimal departure time for migration (Williams and Nichols 1984), appearance of mimics in a seasonal model-mimic system (Bobisud 1978), or onset of diapause before winter (Taylor 1980). This should be strictly distinguished from determining optimal switch time under bang-bang control. In the latter case, the existence of bang-bang, and thus of a switch time, is a result of solving a dynamic model, where decisions can be made repeatedly throughout the process.

In phenotypic optimality models such as discussed here, complications due to sex usually refer to the existence of two genders with different life history patterns that could necessitate a re-definition of fecundity parameters (Waller and Green 1981). In general, however, the role of sex is often better treated with a game model (e.g., Charnov and Mitchell-Olds 1981; Bulmer 1983).

Analysis of Feeding Strategies

Early attempts to use optimality models in the context of feeding strategies were made by ecologists interested in the economics of patch or diet choice (e.g.,

MacArthur and Pianka 1966; MacArthur 1972). The field has since become known as "foraging theory". Several reviews exist (Pyke et al. 1977; Krebs et al. 1983; Pyke 1984) and an up-to-date treatment of theoretical ideas is provided by Stephens and Krebs (1986).

The field, particularly when labeled **optimal foraging theory**, is surrounded by controversies that I do not wish to repeat here (see Pyke 1984; Stephens and Krebs 1986; Pierce and Ollason 1987; Stearns and Schmid-Hempel 1987). This seems, however, somewhat surprising because esentially the same logic is used as in life history theory or biomechanics (Alexander 1982), to mention just two fields that have drawn much less fire. Critiques of foraging theory often attack the payoff function that is used in most models, i.e., the net rate of energy gain. This measure is indeed further removed from Darwininan fitness than the number of eggs laid or the reproductive value so that it may seem unimportant to organisms. However, egg number or other measures such as the Malthusian parameter r, are often simply fitness tokens that may or may not reflect the forces of natural selection. The controversies should also not distract from the fact that in hardly any other domain has the balance between theoretical development and empirical test remained as good as in foraging theory. Furthermore, the experience gained through the diversity of conceptual problems that are accessible with the analysis of foraging strategies may prove useful to other domains as well.

Basic Problems

Schmid-Hempel et al. (1985) analyzed how much nectar a bee should collect from a patch of flowers. The model's decision variable is the number of flowers to be visited before returning home and unload. Formalizing the problem shows the structural similarity to the clutch size problem of Skinner (1985; Box 1). Schmid-Hempel et al. (1985) compared two models with different payoff functions. The two models could only be distinguished by their quantitative predictions, which in turn were strongly dependent on the value assigned to the metabolic costs of transporting nectar. Therefore, the outcome depends on the knowledge of (physiological) constraints, but modeling has helped to identify which of the many physiological variables should be known more precisely (consequently, subsequent research on flight energetics has been stimulated; Wolf et al. 1989). Incidentally, the example shows that in optimization models constraints and payoff functions are essentially interdependent; if costs of foraging had been heavily constrained, i.e., instantaneous costs of foraging made more expensive by discounting a limited life-time budget (which is otherwise implemented in the payoff function), the authors would have been forced to favor the alternative model.

Diet Choice in Herbivores: Linear Programing

Belovsky (1978, 1984b) used linear programming to predict optimal diet in moose and snowshoe hares (see Box 4). Linear programming is appealing because, on

Box 4. Linear programming

In linear programming, the strategy set consists of a set of several, continuous decision variables (a vector Z) and a set of explicit constraint equations that are linear in the decision variables. In generalized form: $A \cdot Z \leqq b$, where A is a matrix and b a vector of constants respectively. The goal is to maximize a linear function of the decision variables, $f = C \cdot Z$ (the scalar payoff function), where C is a vector of constants. If the strategy set consists of n decision variables, the explicit constraints define a convex polyhedron in n-dimensional space (for bounded problems). The Simplex algorithm (e.g., Intriligator 1971) is normally used to solve the problem; it makes use of the fact that possible solutions must lie on faces, edges, or vertexes of the polyhedron. For low values of n, linear programming is readily visualized. Belovsky (1978) used linear programming to predict optimal diet of snowshoe hares in winter (as sketched in the Fig.). Only two

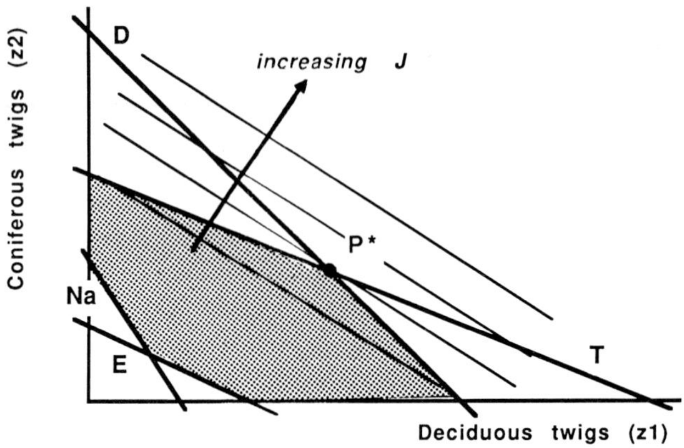

food types are available: deciduous twigs or coniferous twigs that differ in their specific energy and sodium content. The hare can decide to consume a daily amount of Z_1 and Z_2 of these two types, respectively (the two decision variables). Belovsky suggests that the hare should attempt to maximize energy consumption per day—the payoff function J—while simultaneously satisfying a minimum sodium intake. Both are linear combinations of Z_1 and Z_2. Also the constraints are linear functions of (Z_1, Z_2) and reflect maximum digestive capacity (D), maximum available feeding time (T), minimum sodium requirement (Na), and minimum energy requirement (E) (yielding the boundary lines in the Fig.). Possible solutions must therefore be part of the shaded area in the figure. A family of lines (isoclines of the payoff function J) can be drawn to find the solution P*. P* satisfies all constraints and maximizes J.

one hand, it lends itself to visualization and because, on the other, many problems can be linearized. Linear programming seems to be useful to analyze some (but not all) aspects of optimal diet composition in large, generalist herbivores (Belovsky 1984a). Whether it is also useful for other problems remains to be seen.

Models of Risk-Sensitive Foraging

Feeding animals face a problem of risk if they have to choose rewards from *known* probability distributions. Suppose a bird "knows" that it can choose to take food either from site A where ten seeds are offered half of the time and no seeds for the other half, or from site B where five seeds are always available. In this example, the expectations, μ, to receive a seed in site A (variable rewards) and site B (constant rewards) are the same, yet the two sites differ in variance.

Stephens (1981; see also Stephens and Charnov 1982) investigated such risk-sensitive foraging assuming that an organism must minimize the probability of energetic shortfall (z-score model). This need should arise when the forager has to meet a minimum requirement, R_0, in order to survive, e.g., a bird must store enough fat to survive a winter night. The model predicts that the animal should choose to forage at a site with largest variance possible whenever $\mu < R_0$, and at a site with smallest variance possible if $\mu > R_0$. Empirical evidence supporting this claim exists for small birds and shrews (Caraco et al. 1980; Barnard and Brown 1985). But foraging under risk should also be analyzed as a dynamic problem, since the actual state of a bird (e.g., its fat store) will stochastically move up and down throughout a day as a result of decisions under uncertainty (Houston and McNamara 1982; McNamara and Houston 1986).

It is also possible to approach **risk problems** with concepts from utility theory developed for human microeconomics. This approach relies on the existence of a utility function (a payoff function) that generates expectations ranking all probability distributions of rewards. Thus it must be possible to indicate preferences for the different probability distributions of outcomes (e.g. is it better to accept high variance but a low mean rather than low variance with a high mean expectation of reward?). If the utility function were known, it is possible to predict which choice would maximize expected utility (subject to constraints such as income and the price of goods). While for human economic behavior money is a useful currency, for any other organism, utility is supposed to reflect Darwinian fitness. Quite obviously, the task to derive such utility functions on a priori grounds is a formidable one. Indeed, applications of utility theory have concentrated on **descriptive models**, i.e., by observing what animals do (Caraco et al. 1980), rather than on **normative models**, i.e., what animals should do to maximize fitness. In the first case, some authors prefer to speak of **inverse optimality** (McFarland and Houston 1981; Kacelnik et al. 1981).

Information Problems

Suppose the bird "knows" little about its two-site world described before. This might occur when a male has just settled into a territory after spring migration. In this case, the animal has no exact knowledge about the probability distributions of rewards in site A or B. Previous experience may nevertheless tell the bird that it is more likely to find a constant food supply at B, because the size or structure of the place shows some characteristic features. This is a problem of incomplete information (see Box 5) and modelers use techniques of statistical decision theory to predict optimal feeding strategies (for introductions, see Raiffa 1968; DeGroot 1970; McNamara and Houston 1980; Stephens and Krebs 1986). It is important to realize that in these problems information is valuable only to the extent that it increases payoffs, here the rate of energy gain.

Box 5. Optimal statistical decisions

There is a class of problems, where the decision-maker must first gain information about important parameters to make effective decisions, because the outcome of decisions depends on the value of these parameters. The theory of Bayesian statistical decisions deals with such conditions of partial ignorance, where uncertainty can be expressed by probability distributions for the unknown parameters. The potential for applying methods of statistical decision theory in biology in large.

Consider a simple situation (McNamara 1982), where an animal can only take food in patches of two different kinds: Patches of type F contian exactly one prey item (of value 1), and patches of type N contain no food. On average, both types are equally frequent (i.e. $p = 1/2$) but otherwise indistinguishable. Travel time τ (a random variable with mean 1) is needed to get from one patch to the next. The payoff function in the model is long-term average rate of energy gain, γ, (there are no energy costs involved), i.e. $\gamma = [E(R_t)]/[E(T) + E(\tau)]$, where R_t is a random variable denoting the total reward gained by searching in the patch for time t; E(T) is expected search time in a patch. Furthermore, in patches of type F, time to find the prey is exponentially distributed with $p(t) = e^{-t}$ (mean 1); no prey is found in type N. When the animal starts foraging in a patch its a priori probability distribution that it deals with type N or F is (1/2,1/2). However, information is gained by searching that can be used to update these initial estimates to yield an a posteriori distribution. Suppose that search has lasted for time t without finding prey ("event A"). Applying Bayes theorem, the revised (a posteriori) probability that the patch is type F is

$$P(F/A) = \frac{P(F)\cdot P(A/F)}{P(F)\cdot P(A/F) + P(N)\cdot P(A/N)} = \frac{\frac{1}{2}e^{-t}}{\frac{1}{2}e^{-t} + \frac{1}{2}1} = 1/(1 + e^{t}).$$

The strategy set contains all possible search times in the patches. However, the optimal policy is of the form; leave and go on to the next patch if prey is found, otherwise search for time S. The expected gain from adopting this policy S, is E(R) = (value of prey) (probability of finding prey in an average patch within $t < S$):

$$E(R) = 1\cdot[\tfrac{1}{2}\cdot 0 + \tfrac{1}{2}\cdot P(t < S)] = \tfrac{1}{2}(1 - e^{-s}).$$

Furthermore, the expected time, E(T), needed to search an average patch with policy s is

$$E(T) = \frac{1}{2}\int_0^s te^{-t}dt + \tfrac{1}{2}(1 + e^{-s}) = \tfrac{1}{2}(1 + S - e^{-s}), \text{ thus}$$

$$\gamma = \gamma(S) = \tfrac{1}{2}(1 - e^{-s})/[E(T) + E(\tau)] = (1 - e^{-s})/(3 + S - e^{-s}).$$

To find S*, we sat $\partial\gamma/\partial S = 0$ (in this example, $S^* = 1.505$).

Statistical problems can also be formulated in discrete time. In the experiment of Krebs et al. (1978), a bird was allowed to forage either in patch A or patch B that differed in (a priori unknown) reward probabilities (i.e., probability of receiving a reward per foraging attempt). The two patches could be distinguished by their location in space, but only by trying to get food could the animal find out whether A or B was more rewarding. After each attempt, a priori estimates could therefore be updated. The problem is solved by multistage, stochastic dynamic programming, analogous to Box 3.

Estabrook and Jespersen (1974) and Bobisud and Potratz (1976) investigated the problem faced by a predator that encounters a (Batesian) model-mimic system (with the noxious model M, and a palatable mimic B). It may pay the predator to gain information about whether it is currently exploiting a M or a Blump, whenever alternative prey types are less preferable than B, yet sampling cannot be done at zero cost. This is an example for the problem of tracking a changing environment. The payoff is long-term rate of energy intake and the decision variable is the sampling interval, i.e., the time period between "looks" at the MB-system. Stephens (1987) analyzes such a general problem and gives the conditions under which sampling should occur. Although many questions remain, the analysis of information problems as pioneered in foraging theory may also prove relevant to life history analysis.

Constrained Models: Rules of Thumb

Most of the models discussed here make no particular assumption of how an animal solves its problem. For example, Skinner's (1985; Box 1) model of optimal clutch size in parasitic wasps makes no explicit account of how the animal arrives at choosing the optimal oviposition time in relation to host quality and interhost search time.

Iwasa et al. (1981; see also McNair 1982; Green 1984; Hodges 1985), in contrast, explicitly take decision mechanism into account when analyzing the related problem of a forager that exploits patchily distributed food. They assume three conceivable mechanisms, i.e., **rules of thumb**, that the animal might use to arrive at optimal decisions (leave the patch after a fixed time, after a fixed number of prey, after a fixed time since last capture). The set of possible rules, together with the range of values each variable can assume, is the strategy set of the model(s). Hence, the departure from the usual approach is simply the extent to which strategies are detailed. Iwasa et al. (1981) compared the performance of all three rules in environments that differed in prey distribution, but where each rule is used in a way that maximizes payoff (net rate of energy gain). The analysis showed that for each type of environment, another rule was best. This study makes the obvious albeit important point that the best possible solution cannot always be achieved by natural selection due to certain constraints in an organism's perception or decision-making. Nevertheless, modeling helps to answer an utterly meaningful question: which of a number of possible decision mechanisms would be best in the environment in which an organism lives?

Various Problems in Behavioral Ecology

Among the many applications of modeling in other fields of behavioral ecology we can mention only a few. The **ergonomics of caste** was intensively studied by Oster and Wilson (1978). The authors used optimization to predict, for example, what ratio of castes (workers differing in their morphology) the colony should sustain to maximize reproductive success, as measured by the number of sexual forms produced (for a critical test of the theory, see Walker and Stamps 1986).

Janetos (1980) modeled strategies of **female choice**. Females are supposed to find the best male possible, given constraints of time, memory, and mobility.

Janetos (1980) considers several conceivable ways of how a female could proceed (i.e., possible strategies or rules of thumb). This is easily explained by a familiar problem. Suppose a faculty commission has to decide who among a number of young biologists should get the job. The commission may decide to give it to candidate A without inspecting others, to the best of two candidates, or to the best of n candidates. Obviously, if n + 1 young scientists are inspected instead of just n, the certainty that choice A is best will have increased. However, the increase in certainty will level out with n while costs keep accumulating. So how many candidates should be interviewed (if the best one is sought at all)? The answer will depend on the distribution of candidate qualities, time and money constraints, and so on. Janetos (1980) finds that the female should inspect $n^* \gtrsim 5$ males. The value of n can also be taken as the memory horizon that the female should adopt. The best-of-n strategy does best in the numerical examples provided by Janetos (1980) where quality of males is normally distributed. The author claims that some support for the model predictions comes from the observation of mating behavior of a freshwater fish.

Biomechanics and Physiology

Biomechanics and Locomotion

Optimal locomotory strategies for animals were studied by Alexander (1980). For example, what gait (e.g., walk or run) minimizes energy costs of locomotion at different speeds? The change in gait for humans is observed to occur at about 2.5 m/s, exactly where walking would become more expensive than running. Examples of usage of optimality models in biomechanics are lucidly discussed in Alexander (1982).

Reichman and Aitchison (1981) ask what trail angle mammals should choose to minimize rate of metabolic expenditure when moving uphill. The best trail is shown to become shallower with an increase in body weight of the animal, but steeper with an increase in mountain slope. Reichman and Aitchison (1981) inspected snow trails of traveling animals and found both of the predicted relationships.

Physiology

The organization of metabolic processes themselves could be subject to optimality analysis (Gnaiger 1983; Watt 1986; Townsend 1987). If alternative processes (e.g., alternative enzyme sets) are distinguishable by natural selection, one should expect to find sets that maximize efficiency of resource use, or perhaps rate of power output, depending on local ecological conditions. (Watt 1986). This approach may ultimately lead to the development of an evolutionary biology of metabolic organization.

The Value, Problems, and Future of Optimization Models

The number of applications of optimality (and game theory) in various fields of population and evolutionary biology makes up an impressive body of literature (some examples are summarized in Table 1). These models share the characteristics outlined in the boxes. Not always however, is a maximum or minimum value of interest, but optimality modeling may also ask for the optimal range or variance of values (e.g., Gillespie 1977), although this is often better analyzed with game models.

In general, **modeling** appears to have been particularly successful where the question was relatively simple, well-defined (e.g., when to leave a food patch, or when to become mature), and amenable to experimental work. This should not be taken as an argument against using other methods of verification, e.g., comparative studies with data from the literature, nor should it be taken as a statement against modeling efforts that address more complex questions which otherwise cannot easily be approached. This trend reflects the art of modeling: when a simple problem is analyzed, it is more likely that the relevant constraints and objectives can be properly identified and empirical tests will be successful. Such models thus address specific cases where the biology of the system is fairly well known—a more modest approach than general models with no particular organism in mind, such as many life history models (e.g., Schaffer 1983). The value of the latter, however, consists of organizing research and identifying crucial empirical problems (e.g., what constraints are acting on the development of a leaf; how important is root mass compared to photosynthetic tissue ?).

A number of authors have criticized the approach as a whole (Lewontin 1978; Gould and Lewontin 1979; Gould and Vrba 1982; Ghiselin 1983; Gray 1986), while others have answered to the criticism (Maynard Smith 1978b; Horn 1979; Beatty 1980; Mayr 1983; Williams 1985; Stephens and Krebs 1986; Stearns and Schmid-Hempel 1987). A common source of discomfort is the question of whether natural selection always maximizes, or, in the context of game theory, whether it leads to strategies that are stable against selective forces (e.g., Maynard Smith 1981). The short answer is that we do not know. Organisms no doubt possess beautifully structured organs and behaviors to survive and reproduce—traits that impress every passionate field biologist, while it is also true that in the course of evolution old forms are replaced by new ones. However, to repeat what was mentioned in the introduction, the major task addressed with the modeling approach is not to predict evolution, but to seek explanation for the usefulness of traits in the current environment. For example, suppose a gene causes a bird to lay a smaller clutch than average, but with the beneficial effect that the number of young successfully recruited into the next generation has increased. This gene is likely to be selectively favored and hence to spread in the population. However, this finding alone is not a deep insight, because it remains to explain why the success of different strategies varies; e.g., would large clutches reduce parental survival, or are nestlings from large clutches more likely to be infested by a parasite?

Table 1. Some examples for applications of optimization models

Reference	Problem	Type of model	Payoff	Decision variable	Remarks
Law 1979	Life history with age-specific predation	Simple calculus	Max. Malthusian parameter	Reproductive effort in each age class	Predation affects age classes equally; no effect of reproductive effort on predation; examples from literature cited
Roff 1984	Opt. age at maturity	Simple calculus	Max. Malthusian parameter or max. lifetime fecundity	Age at maturity	VonBertalanffy's growth equation; cost of reproduction; quantitative test with data for teleost fisher
Schaffer et al. 1982	Life history of annual plant	Pontryagin	Max. energy invested into reproduction over season	Fraction of energy allocated stores, reproduction, or vegetative parts	Stores always retrievable at fixed cost; photosynthesis linearly dependent on leaf mass; qualitative evidence; environment varies over season
Chiariello and Roughgarden 1984	Storage allocation in annual plant	Pontryagin	Max. resource invested into seeds over season	Fraction of resources allocated to vegetative growth, seeds, or stores	Fixed season length; model tested with data on races of annual plants; uncertainty in season length invoked as cause for deviations
Brown and Venable 1986	Seed banks in temporally varying environment	Lagrange multipliers	Max. geometric mean of annual fitnesses (prob. of survival, germinating)	Seedling success granted in different years	Increase seedling success in year i decreases success in j; successive environments vary independently; general model
Bulmer 1983	Life history of social insects	Pontryagin	Max. matings through male and female offspring	Fraction of resources allocated to produce workers, males or females at any one time	Queen in control over reproduction; no eggs laid by workers; fixed season length; gives ESS-interpretation
Lucas 1985	Diet choice under time constraints	Dynamic programming	Max. net energy gain within available time	Take only good prey type or both types	Different types of time horizons (deterministic, unpredictable); handling time needed; instant recognition of prey;

Table 1. (continued)

Reference	Problem	Type of model	Payoff	Decision variable	Remarks
Belovsky 1984a,b	Diet choice in herbivores	Linear programming	Max. daily energy intake subject to linear constraints	Proportion of differrent prey types consumed, depending on season (summer/winter)	Digestive capacity, available feeding time, nutrient requirements; linearity in relationships
McNair 1982	Performance of rules-of-thumb in a patchy environment	Simple calculus	Max. net rate of energy gain	Time to give up after capture or leaving time since patch entry, depending on rule	Strategies confined to explicit rules; patch types immediately recognized; performance of rule depends on environment
McNamara and Houston 1986	Optimal course of action over winter day for small bird	Dynamic programming	Max. terminal reward function (state at end of day)	Take one of four possible actions at any one time	Daylength fixed; no direct fitness gain during day, but terminal reward related to future fitness; lethal boundary aborbs bird if state becomes zero
Green and Nunez 1986	Central-place foraging in a stochastic, patchy environment	Statistical decision problem	Max. long-term average rate of prey capture	Stopping time in patches	Average travel time between patches fixed; forager does not remember patches; only one prey retrieved per trip; no tests given
Davies and Houston 1981	Territory use	Simple calculus	Max. time available for feeding	Tolerate or expel an intruder	Time for defence, reduction in food availability constrained; quantitative test with field data for wagtails
Williams and Nichols 1984	Opt. time to take action	Dynamic programming	Max. return (unspecified)	Time to take action within fixed interval	Action can be taken only once; returns observable or unobservable, examples of applications

When trying to answer for what purpose a particular trait is designed and how well it works, the investigator first faces the difficulty of defining the problem by building a model. In fact, the trait of interest may serve several purposes simultaneously; e.g., by delivering small loads of food to nestlings frequently, a bird may simultaneously come close to maximizing net rate of energy gain (e.g., Kacelnik 1984) while also repelling intruders (e.g., Kacelnik et al. 1981), or aid thermoregulation in dependent young. Furthermore, a trait may have evolved for a certain purpose but serve another today (Gould and Vrba 1982). The modeler also has to specify a priori what set of strategies is available, and thus what constraints are assumed to act on the system. In principle, the amount of morphological or behavioral variation of the organisms under study would define the possible strategies and the constraints, while differences in the strategies should have heritable components to allow for differential reproductive success.

It appears that these difficulties are only devastating if one subscribes to the holistic approach. The picture seems too pessimistic when using a more piecemeal approach. For most problems, the available set of strategies is not impossible to know; e.g., breeding great tits may successfully incubate eight, nine or ten eggs but not 25 (Perrins 1979). Furthermore, the main purpose of laying eggs is to produce offspring, not to mark the nest site. Commercial breeders also make a living from the fact that many traits of interest can be selected for, such as increasing honey yield in bees (Bar-Cohen et al. 1978). All this should not distract from the fact that a basic knowledge of the natural history of the system under study is crucial and cannot be replaced by modeling.

However, this apparent drawback is a strength of the modeling approach: it forces the investigator to make the assumptions clear, by specifying what organisms can do and what they cannot, and by defining a payoff that is supposed to reflect the purpose of the trait. As we have said before, these payoffs are always fitness tokens. The problem is complicated when payoff and fitness are not linearly related. For example, in fluctuating environments, it may be more important to leave at least some offspring in bad years than to produce a high average number of young (see also Iwasa et al. 1984).

The modeler also does not suggest that natural selection invariably maximizes, but assumes that the traits under scrutiny have evolved to fulfil the suggested function as perfectly as possible. Formulating an explicit model helps to find out whether this hypothesis is correct. In carrying out this program, it is important to realize that optimization and game theory in fact only make sense if one seeks for local optima. For example, the best possible life history is to produce myriads of offspring within nanoseconds! Yet, to look for this global optimum is clearly nonsense, exactly because organisms are constrained in their performance. Developmental processes may require time to unfold, resources for growth must be gathered, and so forth. Moreover, the most important constraints acting on a given trait are the other traits making up an integrated organism (Stearns and Schmid-Hempel 1987).

In contrast to the discussion in several chapters of this volume, models in optimality and game theory generally do not take the **constraints** set by the **genetic system** into account; e.g., the fact that heterozygosity could prevent a trait from reaching the required optima, or that genetic variation may be absent is neglected. In fact, for almost all models appropriate heritability and genetic variation are simply assumed to exist so that strategies breed true. Yet, for virtually all problems this knowledge is not available (which also explains the frequent use of classical examples such as industrial melanism in the peppered moth). Recent studies, however, have faced this challenge and evidence is becoming available. As an example, Ritchie (1988) analyzed the ability of ground squirrels (*Spermophilus columbianus*) to achieve a diet that maximizes daily energy intake. Optimal diet was determined for each individual given constraints such as body size, feeding time, and so on. He showed that deviations from each individual's optimum were not tied to food abundance or other immediate environmental parameters, but appeared to be constant and hence suggest themselves as a behavioral trait that might be subject to natural selection. Subsequent studies now indicate that these individual deviations have heritable components (Ritchie in press).

Models of optimality and game theory therefore sacrifice realism in dealing with the genetic system for more precision in the analysis of phenotypic traits. Therefore, the focus is on the **phenotypic arena** rather than on genotypes; but this does not mean that models violate the basic principles of genetics when making simplifying assumptions where detailed knowledge is not available, or where an explicit formulation would be intractable. Rather, this choice reflects the structure of scientific enquiry with its different levels of analysis: Nobody would seriously doubt the value of doing chemistry, just because from the underlying principles of quantum mechanics it is not always possible to predict whether a certain chemical bond in a complex molecule is stable or not. Similarly, it would seem a misunderstanding of the scientific method if the study of adaptation is abandoned just because we do not know the genetics of interesting traits. In fact, if the theory of evolution through natural selection is to make sense, we must know why certain traits are better than others. Optimization is a powerful tool to help in this enterprise.

Therefore, although surrounded by controversies (partly sparked off by too many ambiguous studies), optimization will nevertheless be with us for some time to come. What future is to be expected? It would be disappointing if more and more models are published that are merely refinements of existing ones, or when critical evaluation against empirical data is not carried further. The strength of optimality models is that problems can be analyzed in much more detail and put to more stringent tests. Thus, to exploit this intrinsic strength, it seems promising to proceed "**bottom-up**", i.e., from simple problems towards more complexity, while retaining the balance between empirical evidence and theory.

There is also a need for more empirical studies on some of the basic tenets of models, such as the existence of a cost of reproduction associated with various activities. Such costs are known (e.g., Partridge and Harvey 1985). Nevertheless, the uncritical assumption of such **trade-offs** may not always hold in the light of further studies. For example, Taylor et al. (1987) expected to find a negative correlation between components of fitness at different stages of the life history of *Drosophila melanogaster*, a trade-off that is implemented in life history theories. However, larval developmental time and survival did not correlate with subsequent mating success of males. On the other hand, male mating success was positively correlated with larval competitive ability of their offspring. This provided evidence for the fitness consequences that different mate choice strategies of female *Drosophila* have (see also Partridge 1986).

Further development of the optimality approach could also contribute to a better understanding of two further categories of assumptions incorporated in any model: How do payoff functions relate to fitness, and what are the constraints acting on the system? Both problems, the meaning of fitness and the role of phylogenetic or developmental constraints, are key questions in evolutionary biology that cannot be bypassed but tackled from another perspective by the use of optimization models.

The study of adaptation would therefore greatly profit from a more general **theory of the Bauplan**, or a theory of constraints. For example, one should know what are the lineage-specific constraints on development and behavioral plasticity? This theory does not exist yet, but optimality and game theory could help in this research, precisely because it makes constraints as well as the concept of adaptation explicit. In this respect, quantification is a benefit that no other approach besides modeling can provide. Twenty years ago, it might have been enough to say that by the way parasitic wasps attack their hosts they are well adapted and maximize their fitness. Today, it would be possible to predict what number of eggs should be laid per host if rate of oviposition is to be maximized and to see whether this hypothesis about adaptive value is correct by quantitatively testing the model (Skinner 1985).

Whether or not quantification is possible, the modeling approach also has in intrinsic heuristic value in organizing research. It should have become clear by now that there is no unified theory of optima and games in evolutionary biology, but rather a set of different models borrowed from many areas of mathematics, economics, and engineering, which address specific problems. This point is important, because while the unifying theme is evolution through natural selection, models do not universally maximize fitness. Indeed, the assumption about fitness varies from model to model (Table 1) and they are subject to empirical verification just as the assumptions about what strategies are possible. Optimization theory is therefore most valuable as a tool during an intermediate stage of the inquiry into a problem: a detailed knowledge of the basic natural history is already necessary to formulate a successful and testable model. But sooner or later, a good model calls for additional knowledge about the biology of constraints and fitness tokens, and this should eventually lead to supersedure of its own assumptions.

Acknowledgments. I am greatly indebted to Peter Hammerstein, together with whom the introductory section was conceived and for the most part written.

Unfortunately, his other activities were too hectic to permit a full theoretical treatment of game models within this chapter. I am also grateful to Prof. W. Wickler, who provided generous support and a stimulating working environment at the Max-Planck facilities in Seewiesen. Helpful comments on the MS were made by S. Stearns, S. Jain, J. Koella, and J. Shykoff.

11 Coevolution at Two Trophic Levels

S.D. JAYAKAR* AND L.A. ZONTA

Introduction

In any real community of living organisms, be it on grassland weed patch by the roadside, in marshland, or in forest, there will be present a great number of species of plants, animals, and probably of bacteria and viruses. If you consider pairs of these species, they will almost all show some type of interaction. An interaction between two species can be either beneficial to both (e.g., symbiotic interactions), harmful to both (e.g., competitive interactions) or harmful to one and beneficial to the other (e.g., host-parasite or prey-predator interactions). In a diffuse community, the interaction between any two species is likely not to be direct but connected through a chain of other species.

This chapter deals with two-species systems where one of the species (P) exploits the other (H) for nutrition. The two main interactions of this kind which will be dealt with are the predator (P) and prey (H), or the parasite (P) and host (H) relationships. Even if prey-predator and host-parasite relationships have a common characteristic in that in each of them one of the species uses the other as a source of nutrition, for the details of their biology they can be very different. Most of the examples and theoretical discussion will involve one P species and one H species, but more complicated associations with more species involved are possible.

When two species have a host-parasite relationship, there can arise genetic differences in the host species which make the individuals belonging to it more (or less) susceptible to their parasite. A well-known host genetic polymorphism which is specific for defense against a parasite is the sickle-cell polymorphism in man which reduces the severity of *Falciparum malaria* when the H^S gene is in the heterozygous state. Other such polymorphisms are known in humans (Allison 1982). The complex vertebrate immune system is, of course, a defensive mechanism against parasites in general, as the production of biochemical substances in plants can be a defense against the attack of phytophagous insects

*Deceased 21.1.1988

(Feeny 1975). The genetic defense of a host against a parasite in turn can select mutants of the parasite which are counter-adapted to the genetic changes in the host. The whole process may even result in a reciprocal chain reaction, so that there is evolution in the two species which is in some way coordinated. In other words coevolution can take place.

Like evolution, coevolution between two species can only occur in the presence of **phenotypic variability** in both species, the control of this phenotypic variability by genetic factors, and fitness differences between the phenotypic variants. Moreover, also the differences in the interaction parameters between species must be genetic in nature. Under these conditions coevolution can occur but does not necessarily do so. There are **constraints on coevolution** just as there are on evolution (see also Chap. 1). Genetic variability is not unlimited and correlation between phenotypic variants could constrain the growth of one phenotye (or genotype) depending on another. Moreover, physiological and other constraints which control the rate of evolution are involved in controlling the rate of coevolution. The rates of change in coevolving species vary also according to the types of interaction between them, and the nature of their genetic control.

That coevolution can take place is seen from natural observations and from field and laboratory experiments. Much work has been carried out on prokaryotes (bacteria and bacterial viruses) in the laboratory and, since the genetics of species interactions in this group is fairly simple, coevolution is simple to observe when it occurs. For the same reason, mathematical models based on such systems are simple both to set up and to analyze, and moreover to confirm experimentally (Levin and Lensky 1983).

In many pairs of species groups, for example insects and plants, it is common to observe both host-parasite and prey-predator types of interactions. Plants are likely to be heavily infected and eaten by insects and they often produce defenses which can be of a chemical nature, like alkaloids, or which act through some morphological or other changes. The insects can counter-adapt, again through chemical, morphological, or behavioral means (Feeny 1975; Futuyma 1983).

As predators, insects can also exercise selection on seed size, since they can find more food in larger seeds, but on the other hand might find it easier to reach this food in smaller seeds with thinner coats. Observations of stored grain seeds and insect pests have confirmed that insects discriminate on the basis of the hardness of seed coats (Smith 1975). Besides insects, rodents are also predators of seeds, and much the same arguments apply to them. The attraction of a seed to its predator depends on its hardness, size, and nutritional value, and since there is discrimination by the predator for these aspects of the seed, a selection pressure is exercised on them.

Due to their evident medical and veterinary importance, many cases of coevolution involving the interaction between animals (including man) and their protozoic, bacterial, or viral parasites, have been studied in some detail. One of these, that between the rabbit and the myxomatosis virus, will be discussed later. Another case of economic importance involves plants and their fungal diseases (rusts) which can destroy entire crops. Many such cases are known and in a large proportion of these the genetics is known, and is of the gene-for-gene kind, that is every parasite gene change has a parallel gene change in the host. This is a very simple but exemplary case of coevolution and will also be discussed in some detail.

Besides the importance of host-parasite relationships for diseases of domestic animals and plants, and therefore from the point of view of agricultural and veterinary economy, prey-predator interactions can be of considerable importance in the management of game and fish resources.

Coevolutionary studies are by no means limited to living species; in fact, there is extensive fossil support for this phenomenon. Moreover, among fossil and among living species, there is also phylogenetic evidence for coevolution.

Coevolution in Action

In a few cases it has been possible to explain how coevolution occurs, following the process step by step. Two such examples are here discussed in some detail.

Myxomatosis

One of the best-known natural examples of coevolutionary changes in a host-parasite system is that of the European rabbit (*Oryctolagus cuniculus*) and the myxoma virus, which is transmitted by insects (Fenner and Ratcliffe 1965; Fenner 1971; Fenner and Myers 1978).

When the rabbit was first brought to Australia, its population size grew considerably. In an attempt to control the rabbit population, highly virulent strains of myxoma were introduced into this population, which started declining drastically. Later, less virulent strains evolved and the rabbit resistance increased. This is thought to be due to the fact that as the rabbits were reduced to very low numbers, the highly virulent virus risked eliminating itself along with its host due to the absence of possibility of infection. Then strains of intermediate virulence took over, and the strain of maximum virulence, which was the only one present in 1950–51, had already disappeared by 1959. A similar phenomenon has been observed also in France and Great Britain (Ross 1982). Table 1 shows the frequency distribution of field strains in Australia several years after the introduction of grade I virus, according to the classification of myxoma virus based on virulence proposed by Fenner and Ratcliffe (1965). The mortality rate due to myxomatosis in a standard laboratory strain of rabbits decreases from over 99% for grade I strain to less than 50% for grade V, while the mean survival time for infected rabbits increases with the decrease in virulence from less than 13 days (grade I) to over 50 days (grade V).

This is the situation which has established itself. Both the virulence of the myxoma strains and the resistance of rabbits to myxomatosis have been shown to

Table 1. Relative frequency distribution of myxoma field strains of different virulence in Australia, several years after grade I virus had been introduced into the country. (May and Anderson 1983)

	Virulence type grade					
Years	I	II	IIIA	IIIB	IV	V
1950–51	100	—	—	—	—	—
1958–59	—	25.0	29.0	27.0	14.0	5.0
1963–64	—	0.3	26.0	34.0	31.2	8.3

be genetically determined. The decreased virulence and the increased resistance were demonstrated in laboratory experiments, in which the virulence of the field virus and the susceptibility of the wild rabbit were tested against standard laboratory strains. There is another interesting aspect about myxomatosis, which involves environmental factors: the degree of virulence of the virus maintained in the field population, which is different in different countries, depends also on the insect vector of the infection. Mosquitoes (vectors in Australia and France), which are thought to be more efficient in the transmission of the virus than fleas (vectors in Great Britain) because of their bigger buccal apparatus, allow the maintenance of viruses of lower virulence in the population. In fact, while in France, like in Australia, grade IIIB and IV viruses are more frequent, in Great Britain grade II and IIIA strains are by far the most represented.

This is a case of coevolution which has been observed and studied in its various stages. It is also a clear example of coevolution in which characteristics of the parasite that limit the host damage are favored.

Mimetic Coevolution

There are several cases of morphological mimicry in two species at the same trophic level, which established as evolutionary defense against a common predator, as in the case of heliconiine butterflies. But there are also cases of mimetic coevolutionary events between two trophic levels, e.g., plants (host) which mimic the eggs of insects (parasite). The best example is offered by some heliconiine butterflies and *Passiflora* (Williams and Gilbert 1981).

The genus *Heliconius* has arisen from other heliconiines fairly recently (Brown 1981). Since cannibalistic larvae have evolved among many *Heliconius* species, females avoid oviposition on sites where eggs have already been laid. The bright yellow of *Heliconius* eggs acts as a signal for many species in the genus, since several *Heliconius* species feed on the same *Passiflora* species. When the butterflies evolved the behavior of avoiding larval crowding, the egg mimicry in *Passiflora* became an advantageous character, as a deterrent to oviposition. Because of the severe impact *Heliconius* can have on its host plants in nature, the egg mimics are a valid defense on the part of some *Passiflora* species to its insect enemy. Structures mimicking the yellow eggs of *Heliconius* have evolved independently in a number of *Passiflora* species and can derive from distinct plant structures. Egg shapes derive either from the swelling of tendril tips and of projection on stipulae, or from petiolar glands and from nectar glands which appear as yellow ocelli. Even supernumerary axillary flower buds and buds on meristem tendrils are developed as egg mimics.

Williams and Gilbert (1981) have studied the oviposition behavior of *Heliconius cydno* on *Passiflora oerstedii*, which does not have the egg mimics, and on *Passiflora cyanea*, which shows mimetic structures. In an experiment, natural yellow eggs, eggs washed in distilled water and tinted green, and yellow eggs washed with distilled water were placed on cuttings of *P. oerstedii* and then the cuttings were presented to females of *H. cydno*. Oviposition frequency (measured as ratio of number of eggs deposited to number of inspections performed by the butterfly on the plant) in the presence of a yellow egg, either natural or water-washed, was statistically lower ($P < 0.001$) than that on cuttings with green eggs or no egg at all (see Fig. 1a, b, c), while egg-free and green-egg cuttings showed equal frequency of oviposition (Fig. 1d). Moreover, when eggs were laid on a cutting with a yellow egg, they were laid several centimeters away from it. In experiments performed with *P. cyanea*, three types of plants were used, with natural mimics, with mimics in which the tip was clipped off to check for chemical stimuli (control mimics), and with all mimics removed. As shown in Fig. 2a and 2b, there was a clearly preferential oviposition on plants stripped of mimic eggs.

In a complex community containing ten or more species each of *Passiflora* and *Heliconius*, the microevolution of egg mimics by one *Passiflora* could eliminate it from the host range of several *Heliconius* species. This protected species is then an "empty niche" and would set up strong selection in some of the *Heliconius* species

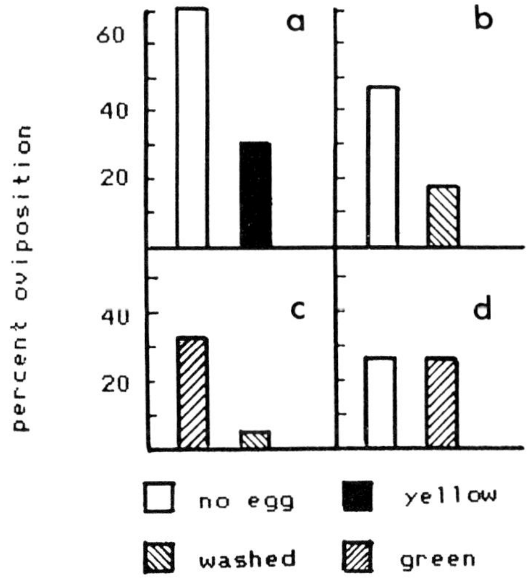

Fig. 1. Oviposition response of *H. cydno* to the presence of different types of eggs on *P. oerstedii*. (Williams and Gilbert 1981)

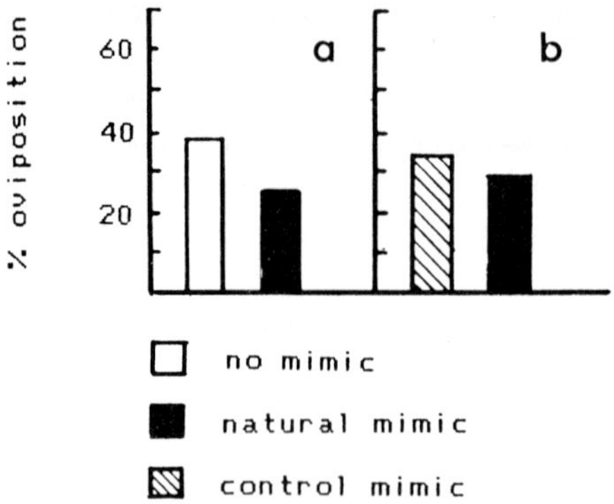

Fig. 2. Oviposition response of *H. cydno* to the presence of egg mimics on *P. cyanea*. (Williams and Gilbert 1981)

to utilize the mimic species. Further, the protected species is at an advantage with respect to expansion into new habitats, and egg mimicry is a step in the coevolution of this food web.

Indirect Evidence: Phylogenetic and Fossil

A basic question is to what extent species interacting at the present are descendants of ancestors which had the same kind of interactions in the past. Moreover, if the association through time can be demonstrated, it is important to understand the mechanisms of this reciprocal influence.

The continuous host-parasite (prey-predator) association through time can be explained by two different hypotheses. The first one is that of **association by descent**, in the sense that present H–P associations are the results of the subdivision of an ancestral parasitized host species into daughter species, which have then diverged together with their parasites. The second hypothesis is based on **colonization:** an ancestral host species subdivides and diversifies into daughter species. These new species are then invaded by the parasites, which in turn will diversify. It is likely that both events coexist, but one can be more important than the other. Additional evidence, for example geological, may turn the issue in favor of the association by descent hypothesis, as for the African ostrich and the South American rhea, which share several kinds of parasites, like lice, roundworms, and tapeworms. Association by descent is likely, since these birds are believed to have parted when South America and Africa were separated by continental drift. Moreover, it is less probable that the two birds acquired ex novo such a large proportion of parasites in common after the separation of the continents.

Systematic and phylogenetic studies have become a useful tool to understand **contemporary interactions** between species (Mitter and Brooks 1983). Evolutionary relationships among reproductively isolated groups, for example species, are generally represented as phylogenetic trees.

Several studies have been carried out on pairwise associations through time of particular host and parasite groups, by comparing their phylogenies, reconstructed within each group independently and represented as phylogenetic trees. From these studies, one of the "rules" on host-parasite interactions emerged: parasite phylogeny mirrors host phylogeny (Eichler 1948). One of the best examples is given by the study of *Enterobius* species (Nematoda pinworms) and their primate hosts, *Lemur*, Old world monkeys, great apes (Brooks and Glen 1982). A very high level of overall agreement is found with the predictions made under the model of association by descent. In some instances the systematics and phylogeny of the host have been inferred by the taxonomic relationships of their obligate parasites and vice versa. Of course, parasites must not be used as a single taxonomic character, but together with other data they can be a very helpful tool for taxonomic purposes.

Parasite-host data may even be useful in suggesting where both host and parasite originated and their dispersal routes, as for Leptodactylidae frogs and their opalinid intestinal parasites. The frog family has a geographical distribution limited to tropical and semi-tropical America, Australia, and Tasmania, and it is parasitized by similar species of opalinids (genus *Zelleriella*) in all ranges of its distribution. The hypothesis that convergent evolution has occurred in host and parasite is less likely than that of association by descent, which could have occurred in the presence of a land bridge between the geographical regions.

Sometimes even fossil records show that coevolution in exploitation interactions has occurred, as for large cursorial predators and their cursorial prey (Bakker 1983). If this kind of prey, often hoofed mammals, relied only on speed to escape predators, it should have evolved toward the longest, swiftest limb form possible; and for its predator to be successful, parallel evolution in limb form should have occurred. Fossil records show close transformation pathways of the limbs in six separate lines of ungulate prey and mammalian cursorial predator, and even suggest one common ancestral body type at the origin of the age of mammals.

Genetics of Host-Parasite Interactions

As is easy to imagine, not all hosts are equally susceptible to all parasites. Certainly some pairs show more or less susceptibility or virulence than others. There can be pairs of host-parasite genotypes which can even be incompatible. The genetics of several species pairs has been studied in detail and genotype specific incompatibility has been found in many cases. Several of these relationships are of the **gene-for-gene type**, in the sense that every parasite genotype can infect only one host genotype. A few such examples are here presented.

Crop and Pathogenic Fungi

Plant rusts, which are common parasites on plants, and particularly on crop plants, are considered an economic hazard. The rust fungi are obligate parasites and cannot exist apart from their hosts and must therefore have evolved together, each species showing a sharp limitation in host range and completing the life cycle on just one host. Rust infections have thus been controlled by the development of resistant varieties of their host plants. The resistance of a plant variety to a rust race depends upon the genotype of the variety of the plant and that of the pathogenic race, and upon the environment. The genetics of resistance to rust depends therefore on the interaction between host and parasite genotypes.

A well-known example is that of *Linum* and *Melampsora lini.* Cultivated flax *Linum usitatissimum* is economically a very important species and is not found in the wild. Differences between varieties in their resistance to a flax pathogen (*Melampsora lini*) have been found, but can vary from area to area. The relationship is of a highly specific type and complementary. Infection in a plant reveals itself as a pustule: pustule types depend both on plant variety and on rust race, and therefore define both the host resistance and the pathogenicity of the parasite. The high degree of physiological specialization of rust suggests a parallel evolution of host and parasite, with the development of complementary genic systems. There is a close correspondence between the genes for virulence in the parasite and those for resistance in the host. In fact a gene-for-gene interaction has been demostrated (Flor 1955, 1956).

Races of rust (genes for pathogenicity) are identified by the reaction to them of a series of host varieties, termed rust differentials. Genes for rust reactions are on the other hand identified by the pathogenicity of different races of rust.

While **resistance** is inherited as a dominant factor (sometimes incomplete), **virulence** in rust is inherited as a recessive factor. Table 2 and 3 show the results of some classical experiments on the genetic determination of flax resistance and rust pathogenicity, by means of two rust races (22 and 24) and two flax varieties (Ottawa 770B and Bombay), which show cross-reaction for resistance (R) and susceptibility (S). The first experiment is based on the analysis of resistance to the two rust races in the F_2 offspring of the two plants. Among the 194 F_2 plants which were tested, the number of plants resistant to both races, to just one, and

Table 2. Two-locus determination of resistant (R) or susceptible (S) response to rust races in flax varieties. (Singh 1983)

Rust race	Flax varieties parental types Ottawa	Bombay	F_2 offspring			
22	S	R	R	S	R	S
24	R	S	R	R	S	S
		Observed:	110	32	43	9
		Expected:	109	36	36	12
			(9 :	3 :	3 :	1)
Genotypes:	*LLnn*	*11NN*	*L-N-*	*L-nn*	*11N-*	*11nn*

Table 3. Two-locus determination of pathogenicity in rust races. (Singh 1986)

Flax varieties	Rust races parental 22	types 24	F_2 offspring			
Ottawa 770B (*LLnn*)	S	R	R	R	S	S
Bombay (*llNN*)	R	S	R	S	R	S
		Observed:	78	23	27	5
		Expected:	75	25	25	8
			(9 :	3 :	3 :	1)
Genotypes:	$a_La_LA_NA_N$	$A_LA_La_Na_N$	$A_L\text{-}A_N\text{-}$	$A_L\text{-}a_Na_N$	$a_La_LA_N\text{-}$	$a_La_La_Na_N$

susceptible to both was very close to that expected for a two-locus genetic determination, both loci being diallelic with a dominant allele for resistance (9:3:3:1 segregation ratio). The complementary experiment also showed a digenic determination of pathogenicity in the two rust races, the allele for avirulence being dominant (Table 3). The two loci are identified with the symbols L and N; the alleles in the plant are defined by L, l and N, n, and the complementary alleles in rust by the letters A and a for avirulent and virulent alleles respectively, with a subscript showing the locus they refer to. As usual, capital and small letters indicate dominant and recessive alleles respectively. Genes for resistance have been shown to be multiple alleles at five loci, of which three are inherited independently (K, L, M) and two belong to a linkage group (N, P).

Several other gene-for-gene interactions have been demonstrated in crop plants and their parasites (see Singh 1986, p. 72).

Plants and Insects

Parasitization of plants by insects is a very common phenomenon. An insect population faced with a range of possible host plants may be also to parasitize them all; on the other hand, due to either genetic differences or other reasons, they may be unable to use some of them as hosts. In many crop species, there is a one-to-one relationship between them and their parasites. It is therefore an ideal situation for coevolution to occur. In fact, many instances of such coevolution have been documented and, in some cases, its genetic basis is known.

Genetic analysis of this kind of series is best developed for the Hessian fly *Mayetiola destructor*) and *Triticum*, which is one definite case in insects for a gene-for-gene relationship. Different varieties of wheat can be resistant or susceptible to parasitization by the fly. Plant resistance manifests itself in that the larvae of the parasite do not survive after they start feeding on the plant

Table 4. Genotypes of the Hessian fly biotypes based on the resistant (R) or susceptible (S) response of wheat varieties. (Hatchett and Gallun 1970)

	Wheat varieties									
Biotypes	Turkey (—)		Seneca (H_7,H_8)		Monon (H_3)		Knox 62 (H_6)		Ribeiro (H_5)	
GP	S	*tt*	R	*S*–	R	*M*	R	*K*–	R	*R*–
A	S	*tt*	S	*ss*	R	*M*–	R	*K*–	R	*R*–
B	S	*tt*	S	*ss*	S	*mm*	R	*K*–	R	*R*–
C	S	*tt*	S	*ss*	R	*M*–	S	*kk*	R	*R*–
D	S	*tt*	S	*ss*	S	*mm*	S	*kk*	R	*R*–
E	S	*tt*	R	*S*–	S	*mm*	R	*K*–	R	*R*–
F	S	*tt*	R	*S*–	R	*M*–	S	*kk*	R	*R*–
G	S	*tt*	R	*S*–	S	*mm*	S	*kk*	R	*R*–

(**antibiosis**). The analysis of response to infestation among the progeny in classical genetics experiments with different wheat varieties has shown that more than a dozen nonallelic genes can confer resistance to the cultivars through dominant alleles, which are designated with the letter H and a numerical subscript. Only in one cultivar resistance is inherited as a recessive allele (Java h4). These genes can be carried on different chromosomes, or can belong to the same linkage group, as demonstrated for example for H_3, H_6, H_9, H_{10} and H_5, H_{11} which belong to two different linkage groups (Roberts and Gallun 1984).

On the other hand, crosses of different races of Hessian flies—also called biotypes—which are defined by means of the reaction to them of the wheat varieties, have revealed the existence of different genes in the parasite and that virulence is recessive to avirulence. The alleles for virulence are therefore designated with small letters. Table 4 shows how the genotype at five loci in eight Hessian fly biotypes can be described on the basis of resistance-susceptibility response of five wheat cultivars. Resistance genes identified in the cultivars are given in parentheses (Hatchett and Gallun 1970).

The one-to-one gene, however, is not the most common genetic interaction known in insect-plant systems. While resistance of rice varieties to the brown planthopper (*Nilaparvata lugens*) is based on **major genes**, which can be either dominant or recessive, the virulence of the insect seems to be **polygenic** in nature (den Hollander and Pathak 1981). Cases of resistance with polygenic determination are also known in crop species. Many pairs of crop plants and insect pests have been studied with respect to virulence and resistance. Several examples are listed and discussed in Singh (1986).

Mathematical Models

Since Mode's mathematical model based on host-parasite gene-for-gene interaction (Mode 1958), much theory has been developed (see Slatkin and Maynard Smith 1979; May and Anderson 1983).

Often two different approaches are used to study **exploited-exploiter relationships**, one derived from population genetics, in which gene frequencies are the variables (e.g., Mode 1961; Jayakar 1970; Clarke 1976; Leonard 1977; Leonard and Czochor 1980), and the other from ecology, where the variables are population densities (e.g., Hassel 1978; Pimentel et al. 1978; Schaffer and Rosenzweig 1978; Anderson 1979; Levin and Pimentel 1981; Roughgarden 1983a. More realistic models should take into account both aspects contemporarily. Here we will present a basic approach to the analysis of interactions in prokariotic species (Lenski and Levin 1985) and a general model applicable to eukaryotic species (Jayakar 1983).

Prokaryotic Systems

Bacterial viruses are dependent on bacteria for reproduction and often alter the changes of survival and reproduction of their bacterial hosts. As such they can be considered as parasites of bacteria.

The phages adsorb to sites on the bacterial membrane. Infested bacterial cells do not multiply, and after a latent period, the cell lyses and bursts, releasing a number of phage particles. The growth rate of the phage population depends on the rate of adsorption to uninfected bacteria and on the burst size.

Some bacteria are partially resistant to phage infection. If the parameters of resistance of bacteria, adsorption rates of the phage, and their burst sizes are under genetic control, we have an ideal situation for coevolution. Resistance to infection in bacteria can evolve either as alteration or as loss of a function, through several types of molecular mutations. On the other hand, the phage can mutate to counteract only bacterial alterations not loss of functions. Therefore it is difficult to find continuous coevolution in phage-bacterial systems. Lensky and Levin (1985) in their chemostat experiments with first-order resistant clones of *Escherichia coli* and several wild types of T phages have found **host-range mutants** of both T_2 and T_7, but none for either T_4 or T_5 (see examples of Fig. 3). However, they were not able to detect mutant phages active against higher-order resistant bacteria in all strains tested. It was also shown that in a culture of *E. coli* treated

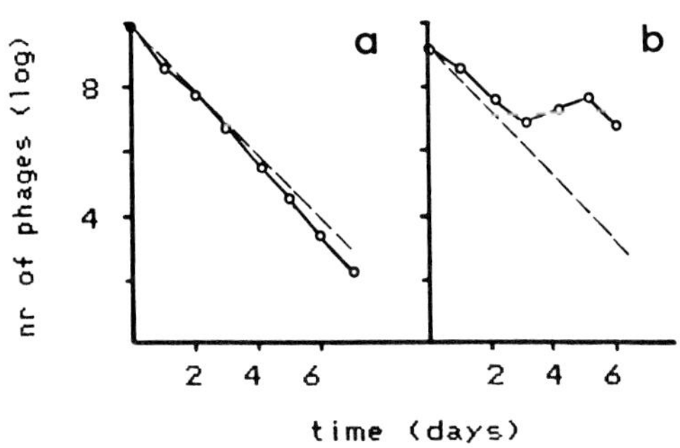

Fig. 3. In the presence of *E. coli* first-order resistant clones, phage T_4 decline is as expected (dotted line) in the absence of host-range mutants (**a**); on the contrary, phage T_2 shows the presence of host-range phages (**b**) (Lenski and Levin 1985)

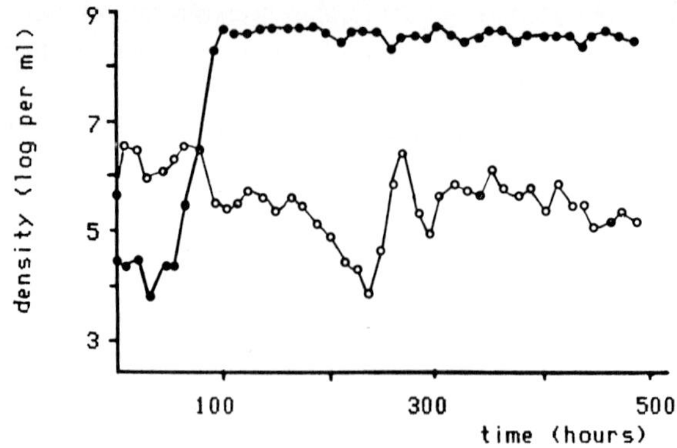

Fig. 4. Density patterns of *E. coli* (● —— ●) and phage T_4 (○ —— ○). (Lenski and Levin 1985).

with phage T_4, after the appearance of resistant bacteria and their attainment of maximum density in about 18h, T_4 phages still persisted even if no host-range mutants could be detected (Fig. 4). This situation suggests the coexistence of resistant and sensitive bacterial cells, which can be attained only if the resistant bacteria are less fit than the sensitive ones. In a competition experiment between T_4-resistant and sensitive *E. coli* strains, in the absence of T_4 phage and under resource-limiting conditions, a selective disadvantage of about 50% was estimated for the resistant bacteria with respect to the sensitive ones (Fig. 5). Therefore, even if an endless "arms race" cannot establish itself, the elimination of the phage population may be avoided whenever the resistant bacteria are less fit than the susceptibile ones, for example as regards resource utilization.

Phage-bacterial systems are of great importance because of their suitability for laboratory experiments, and a genetic study of the variation in the parameters. In a population of bacterial host and its phage, from the view point of the bacteria, favorable selection should increase the rate of cell multiplication regardless of phage density. Because adsorption of the phage leads to cell death, selection should reduce the rate of adsorption. **Selection** from the objective of the phage will depend on the burst size, and the rate of adsorption to uninfected cells; and selection should be in favor of an increase in these parameters. Selection on the phage should also reduce the latent period, namely the time between adsorption of a phage and the burst of infected cells.

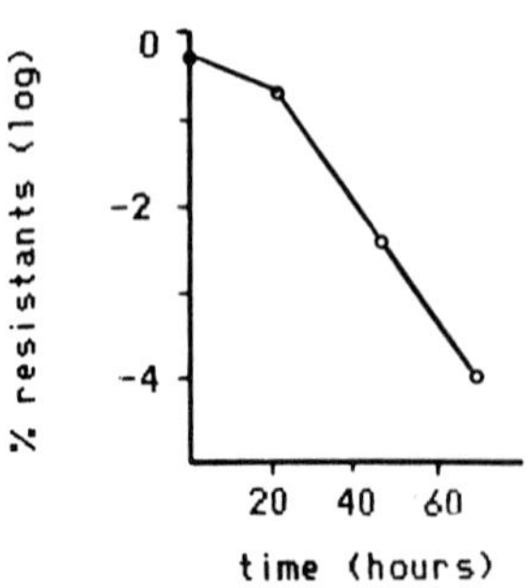

Fig. 5. Competition experiment with T_4-sensitive and T_4-resistant *E. coli* clones in the absence of phage. (Lenski and Levin 1985)

In one possible mathematical model (Levin et al. 1977) P represents the density of free phage and N the density of uninfected host cells. The parameters of the model are:
p the rate of cell multiplication;
a the rate of phage adsorption;
l the latent period between adsorption and cell burst;
b the burst size;
r the rate of flow through the habitat and the consequent dilution rate of host and phage populations.
Expressing this situation as a pair of differential equations, we have

$$\frac{dN}{dt} = pN - aNP - rN \tag{1}$$

$$\frac{dP}{dt} = be^{-rl}aN'P' - aNP - rP,$$

where N′, P′ are the densities at l time units before the present (Levin et al. 1977; Levin and Lensky 1983).

The intensity and direction of selection in this system can be inferred from the effects due to changes in the parameters defining the growth and interaction of the two populations.

At equilibrium both bacteria and phages are present in the system with the following densities

$$\hat{N} = r/[a(be^{-rl} - 1)] \text{ and } \hat{P} = (p - r)/a. \tag{2}$$

In a more recent paper (Lenski and Levin 1985), the ecological model has been expanded to incorporate mutational events which lead to bacterial resistance and host-range phage mutations, mutated phages being able to infect both types of bacteria, which are either sensitive or resistant to wild-type phages.

A Model in Ecological Population Genetics

In a system where we wish to study the changes in two species (populations), there are two important aspects to follow through time: the size of each population and its genetic composition. These two variables give us a good idea of the status of a population. The **population size** changes in time can be followed by the mathematical models of population ecology and **population dynamics**. Separately from this, the dynamics of the genetic composition can be studied by the classical methods of population genetics. However, what is necessary for our purpose is to study both the aspects simultaneously, but since this complicates considerably the mathematical modeling, and consequently its analysis, we will develop the problem here in steps. We will first discuss only the genetic changes, and the variables used will be frequencies of genes, or of gene combinations, of the species studied. The simplest of these models involves one locus and two alleles in each of the species, namely the host and the parasite, or alternatively the prey and the predator.

The interaction between host and parasite is the differential susceptibility of the host genotypes to the parasite genotypes. The parameters in the host-parasite interface will be ability to infect, to resist infection and mortality among infected hosts of different genotypes. In a prey-predator model "susceptibility" will include the ability of the prey to hide or escape from the predator. Thus the interface between prey and predator will be discussed by the different susceptibilities of prey genotypes and the different efficiencies of predator genotypes in capturing the prey.

Between a host species and a parasite, each with a series of genotypes, a pairwise compatibility of host and parasite is an ideal situation for coevolutionary changes. Let us consider the following situation: a parasite P can infect a host H. Suppose now that there is a mutation in the host genotype so that its genotype H_1 changes to H_2, and that this genotype is resistant to the parasite of genotype P_1. At this point the parasite would be exterminated unless it can mutate to a genotype P_2 which can infect also H_2. The situation would then be

$$\begin{array}{lccc} \text{Host} & \mathrm{H}_1 & & \mathrm{H}_2 \\ & \uparrow & \nwarrow & \uparrow \\ \text{Parasite} & \mathrm{P}_1 & & \mathrm{P}_2 \end{array}$$

where the arrows denote the possibility of infection. This situation could of course be extended if the host genotype undergoes a mutation H_3 which is resistant to both P_1 and P_2, and subsequently the parasite reacts through a mutation P_3 which can infect H_3. This situation can be referred to as the **"deer flees, the wolf pursues" phenomenon**, and is found in many plants which evolve defenses against insects. It is also one which is amenable to mathematical modeling.

The two host genotypes-two parasite genotypes situation has been considered by Jayakar (1970), with haploid and diploid conditions. In this model, the variables are the frequency of H_1 (H_2) in the host and that of P_1 (P_2) in the parasite, and the dynamics of these frequencies is analysed. The biological systems which inspired this kind of models are the gene-for-gene systems in host and pathogens.

In a very simple situation, with discrete generations, the only parameter utilized is the probability s of infection when a parasite encounters a susceptible host. Host and parasite fitnesses can then be written down as in Table 5.

If X and Y are the frequencies of H_1 and P_1 respectively, the recurrence equations are:

$$\mathrm{X}' = \frac{\mathrm{X}(1-\mathrm{s})}{\mathrm{X}(1-\mathrm{s}) + (1-\mathrm{X})\mathrm{Y} + (1-\mathrm{X})(1-\mathrm{Y})(1-\mathrm{s})} \qquad (3)$$

$$\mathrm{Y}' = \frac{\mathrm{XY}}{\mathrm{X} + (1-\mathrm{X})(1-\mathrm{Y})}.$$

Table 5. Fitness of the host and of the parasite for each host-parasite type pair

	Fitness				
	Host			Parasite	
	H_1	H_2		H_1	H_2
P_1	$1-s$	1	P_1	1	0
P_2	$1-s$	$1-s$	P_2	1	1

In theoretical population genetics it is referred to as a **frequency-dependent model**. But for a model which is closer to reality, in addition to the variables thus taken into account, one would have to include as variables of the system the numbers, i.e., densities, of exploiter P and of exploited H populations (Jayakar 1983; Jayakar 1984a).

In order to include a description of changes in the P and H densities, one must consider some sort of model of change for these densities and values of the parameters. The parameters may or may not be genotype-dependent, but in a general case they will depend on the genotype. Further, to describe the changes in P and H densities by the process of exploitation (either parasitization or predation), we need parameters for each pair of genotypes of host-parasite interacting populations.

The simplest form of population genetics and ecological models will be utilized in the first place, namely population genetics models with constant parameters, and the exponential growth model.

The one-generation changes will be mathematically described by first subjecting the host and parasite genotypes to growth according to the model adopted and according to the specific genotypic parameters; the next step is to subject each genotype of the host population to each genotype of the parasite population randomly and the surviving numbers will give the values of the variables after one generation. This will provide a set of difference equations (discrete generations) depending on the parameters, and the system recurrences, if it can be solved, will give us a prediction of the changes in a larger number of generations. The analysis may or may not yield a stable equilibrium, depending on the values of the parameters and their genetic control.

An example of such a system is given below.

Suppose that in one generation there are m species (types) of host and n species (types, genotypes, phenotypes) of parasites. Let us divide a generation into three phases.

a) At first we consider the growth (exponential) of each type separately.

After this phase the numbers of each type at each trophic level will change as follows:

$$H_i^* = H_i(1+r_i) \qquad i=1,\ldots,m$$

$$P_j^* = P_j(1+s_j) \qquad j=1,\ldots,n. \tag{4}$$

We have assumed exponential growth of all types of both species, with intrinsic growth rate r_i for the host types and s_j for the parasite types.
b) At the end of the first phase, we introduce the phase in which parasitization (or predation) takes place, so that the host types diminish in size and the parasites increase. After this interaction between the two trophic levels, we have:

$$H_i^+ = H_i^* - \sum_j u_{ij} H_i^* P_j^*$$
$$P_j^+ = P_j^* + \sum_i v_{ij} H_i^* P_j^*, \tag{5}$$

where u_{ij} is the average loss per individual of the type i^{th} host in an encounter with the jth parasite type, and in this same encounter the gain in parasite individuals is v_{ij}. The encounters are assumed to be of the mass action type.
c) Finally, there will be a certain natural density- and frequency-independent mortality of each host type and of each parasite type. If we assume the specific mortality rates to be b_i and d_j respectively, we obtain the following recurrence equations:

$$H_i' = H_i^+(1 - b_i) \qquad i = 1, \ldots, m \tag{6}$$
$$P_j' = P_j^+(1 - d_i) \quad j = 1, \ldots, n.$$

This is the system we have to solve.
More complicated situations can be fitted into this still simple scheme, but to solve the resulting systems of recurrence equations may become analytically too difficult as more parameters and more variables are introduced. However, sometimes the question in which one can be mainly interested is which parameters play a major role in the system and what kind of correlations can be expected between them. Sometimes specific situations, which, limiting the number of parameters, can be more easily analysed, may be even more interesting than a general model. On the other hand, when the system is too complicated, or tedious, for analytical solution, numerical analysis can help in understanding the dynamics of the system.

As an example we will discuss the model studied by Jayakar (1984b) to investigate the **evolution of resource exploitation**. This work was stimulated by the observations of Rossi and Fano (1979) on aquatic isopods which show genetic diversity within each species in the capacity of feeding on particular species of microfungi. Specialist feeders in the isopod population are found in nature at low population density, under conditions of high predation by fish. Laboratory experiments have confirmed that specialists can be maintained in the population under high predation, and that are eliminated when predation is reduced and population density increases, in other words under high or low resource availability.

The model proposed by Jayakar considers two resources Rs_1 and Rs_2 and one exploiter population in which genotype *aa* is specialist on Rs_2, while *AA* and *Aa* can utilize both Rs_1 and Rs_2. The following recurrence equations describe the dynamic of the system:

$$R_1' = R_1[1 + r_1 - \frac{r_1(R_1 + cR_2)}{k_1} - (1-p)^2 X b_{11} - 2p(1-p)X b_{12}] \quad (7)$$

$$R_2' = R_2[1 + r_2 - \frac{r_2(R_2 + cR_1)}{k_2} - (1-p)^2 X b_{21} - 2p(1-p)X b_{22} - p^2 X b_{23}]$$

$$X' = X[d + (1-p)^2 (R_1 a_{11} + R_2 a_{21}) + 2p(1-p)(R_1 a_{12} + R_2 a_{22}) + p^2 R_2 a_{23}]$$

$$p' = p[d + (1-p)(R_1 a_{12} + R_2 a_{22}) + pR_2 a_{23}]/[d + (1-p)^2 (R_1 a_{11} + R_2 a_{21}) + 2p(1-p)(R_1 a_{12} + R_2 a_{22}) + p^2 R_2 a_{23}],$$

where: R_i, r_i, k_i ($i = 1, 2$) are the density, the growth rate and the carrying capacity of Rs_i resource; c is the parameter of resource competition; X and d are the exploiter density and exponential decay parameter in the absence of resources; p and $1-p$ are the frequencies of *a* and *A* alleles; a_{ji} and b_{ji} ($j = 1,3$; $i = 1,2$) represent the increase in fitness of j exploiter type and decrease in fitness of i resource when on contact.

To find the conditions under which the specialist genotype can survive when both resources are present, the analysis of the boundary equilibrium $p = 0$ and $R_1, R_2, X \neq 0$ has been performed. The frequency p can increase when $p' > p$, that is when $\hat{R}_1 a_{12} + \hat{R}_2 a_{22} > \hat{R}_1 a_{11} + \hat{R}_2 a_{21}$, where the equilibrium values $\hat{R}_1$ and $\hat{R}_2$ are obtained by putting $R_1' = R_1$, $R_2' = R_2$ and $X' = X$, being $p = 0$. The values turn out to be

$$\hat{R}_1 = \frac{a_{21} k_1 k_2 \left(\frac{b_{21}}{r_2} - \frac{b_{11}}{r_1}\right) + (1-d)\left(k_1 \frac{b_{11}}{r_1} - ck_2 \frac{b_{21}}{r_2}\right)}{D}$$

$$\hat{R}_2 = \frac{a_{11} k_1 k_2 \left(k_1 \frac{b_{11}}{r_1} - \frac{b_{21}}{r_2}\right) + (1-d)\left(k_2 \frac{b_{21}}{r_2} - ck_1 \frac{b_{11}}{r_1}\right)}{D}, \quad (8)$$

where

$$D - a_{11}\left(k_1 \frac{b_{11}}{r_1} - ck_2 \frac{b_{21}}{r_2}\right) + a_{21}\left(k_2 \frac{b_{21}}{r_2} - ck_1 \frac{b_{11}}{r_1}\right).$$

Since it is not easy to grasp the conditions they lead to, a numerical examples has been performed, with the paramater values given in the Table 6, keeping in

Table 6. Parameter values used in the numerical example. (Jayakar 1984b)

$r_1 = 0.4$	$r_2 = 0.2$	$c = 0.1$	$d = 0.2$
$a_{11} = 0.0001$	$a_{12} = 0.0008$		$a_{13} = 0$
$a_{21} = 0.0012$	$a_{22} = 0.0016$		$a_{23} = 0.002$
$b_{11} = 0.00002$	$b_{12} = 0.00004$		$b_{13} = 0$
$b_{21} = 0.00006$	$b_{22} = 0.00008$		$b_{23} = 0.0001$

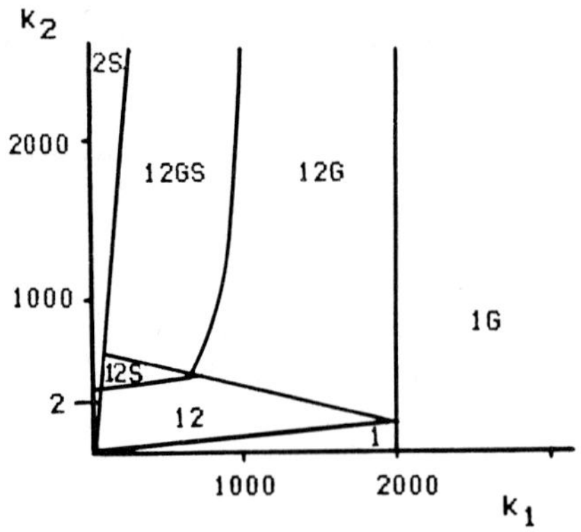

Fig. 6 k_1, k_2 parameter space subdivided into the regions where different combinations of resources (1 and 2), the generalist (G) and the specialist (S) exploiter can coexist. (Jayakar 1984b)

mind that for Rs_1 and Rs_2 to coexist the following condition must be satisfied

$$c < \mathrm{MIN}\left[\frac{k_1}{k_2}, \frac{k_2}{k_1}\right]. \tag{9}$$

Since the experimental results on isopods suggest that the existence of the specialist type is conditioned by the carrying capacities of the resource species, the above conditions have been depicted in the k_1, k_2 space as in Fig. 6, k_1 and k_2 to be interpreted as resource availability relative to the exploiter population densities. The type of community which can be maintained depends on the values of k_1 and k_2. As can be seen from the graph, there are combination of values of carrying capacities which allow the coexistence of both specialist and generalist types (1, 2, G, S), and even the existence of the specialist exploiter alone (1, 2, S).

A very important aspect of the analysis of mathematical models is that it has shown that **commensalism** or a reduction of virulence are not the obligate end points of host-parasite coevolution, with different trajectories being followed during the evolutive process (Levin et al. 1982).

Moreover, mathematical models can be of applicative importance, because they make it possible to predict the epidemiological consequencies of different strategies adopted by hosts and parasites, which could give insights toward the biological control of pests (Falcon 1982).

12 Evolution and Competition

F.B. CHRISTIANSEN AND V. LOESCHCKE

Introduction

In Darwin's theory on the mechanism of evolution, competition among living things is viewed as a major part of the "struggle for existence" and therefore as a basis for natural selection (Darwin 1872). Competition becomes more severe the more alike the competitors are, because requirements of more like individuals are in general more similar. Coexisting species from the same genus compete on average more than other coexisting species of the same family, but usually the fiercest competition is found among individuals of the same species. Competition among conspecific individuals is an important determinant of natural selection among phenotypic variants of a given species. It is distinguished as intraspecific competition from competition between species, interspecific competition.

Density Regulation and Intraspecific Competition

In a homogeneous population intraspecific competition is in its population dynamic aspects indistinguishable from effects caused by increasing population density. The idea behind simple descriptions of density-dependent growth is that all individuals in the population compete for some finite resources, which eventually will limit the growth of the population. Density-dependent growth models usually make the simplifying assumption that the individuals are equal. Phenotypic variation in a species that relates to the individual performance in intraspecific competition has a direct implication on natural selection in the species.

A population with access to unlimited resources has a propensity to grow exponentially, i.e., to double its numbers at regular intervals. If any individual survives and reproduces equally and independently of other individuals, then exponential growth results:

$$\frac{dN}{dt} = N(b-d), \text{ so } N(t) = N(0)\exp[(b-d)t], \tag{1}$$

where b is the birth rate and d is the death rate per individual in the population.

The difference, $r = b - d$, is called the **Malthusian growth rate** or just the **Malthusian parameter**. This Malthusian law of population growth (named after Malthus, who described its biological implications in 1798) led Darwin to conclude that significant intraspecific competition for limited resources must exist as the cause for the decline in individual survival and/or reproduction rates with increasing population density. This would in turn produce natural selection on variation in characters relevant to the individual performance as a response to competition.

The simplest model describing the process of decline in individual proliferation with population density is to assume that the death rate increases or the birth rate decreases as a linear function of population density (this idea is due to Verhulst 1838 and to Pearl and Reed 1920). This assumption of density regulation leads to the **logistic model** for population growth, and it was originally formulated as a simple modification of the Malthusian model. For a linearly increasing death rate it may be given as the differential equation

$$\frac{dN}{dt} = N[b - (d + \iota N)]. \tag{2}$$

Here b is the birth rate and $d + \iota N$ is the death rate per individual in the population. The Malthusian model is recovered for small populations where the per capita growth rate becomes $r = b - d$, which is equal to the initial exponential growth rate of a small population. The parameter ι is called the **density-dependent death rate coefficient**, and it is a measure of the increase in individual death rate with population size. [The birth rate could also be modified to $b - \beta N$, but the properties of the model given by Eq. (2) is the same as this more general extension, so we will consider only the simple model with density-dependent death.]

The form (2) of the logistic equation stresses the change in population size at any time. We may instead consider the variation in population size through time, which is given by the integrated equation

$$N(t) = \frac{N(0)\exp[(b-d)t]}{1 + [(b-d)/\iota]\{\exp[(b-d)t] - 1\}N(0)}. \tag{3}$$

If N(0) is small, then the population size grows initially when $r = b - d > 0$, and the population size converges to an equilibrium population size equal to $(b-d)/\iota$. This equilibrium is stable, because if $N(0) > (b-d)/\iota$, then the population size declines towards the equilibrium. For $r = b - d < 0$ the population size always declines to zero.

Gause (1935) used the logistic model of population growth to discuss data on growth of the ciliates *Paramecium aurelia, P. caudatum*, and *P. bursaria* in an experimental population started from few individuals (Fig. 1). In the analysis of the population data, Gause introduced the compound parameter $K = (b-d)/\iota$, which is convenient because a population that grows from low densities, i.e., $b - d$, will after sufficient time settle at the equilibrium population size equal to K. So the parameter K could be read from the size where the population levelled off,

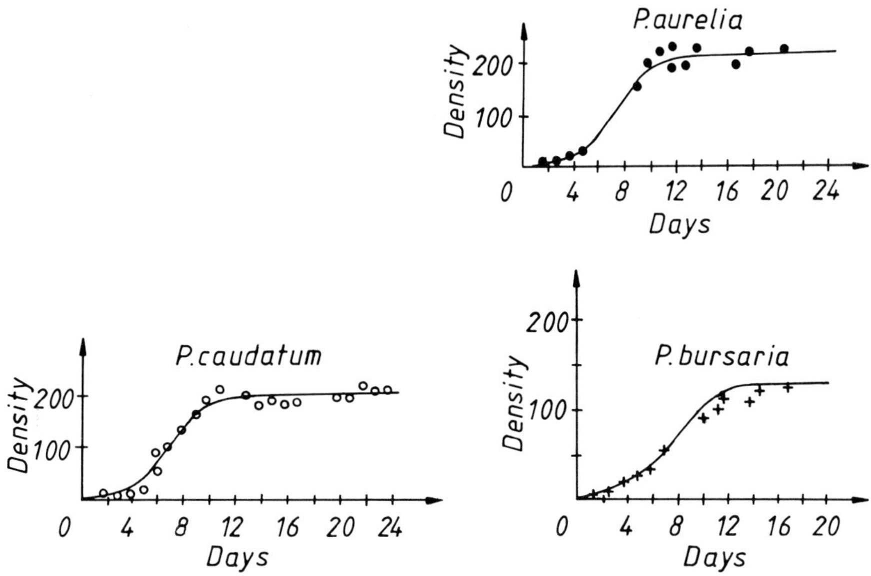

Fig. 1. Population densities in monocultures of *Paramecium*. **a** *P. aurelia*; **b** *P. caudatum*; and **c** *P. bursaria*. (Data from Gause 1934; after Begon et al. 1986)

and the initial growth rate (slope when plotted on semilogarithmic paper) gave him the parameter $r = b - d$ (Fig. 2). Using these two observations he could calculate the **intraspecific competition coefficient** which is equal to r/K. The observation of a population that grows as the curve described by Eq. (3) does not allow us to make the conclusion that the decline in the population growth rate is caused by an increase in the per capita death rate; it might as well have been a linear decline in birth rate, or both.

The observational estimates of the parameters r and K can be substituted into Eq. (2) to provide a mathematically equivalent equation for the logistic model as

$$\frac{dN}{dt} = rN(K - N)/K, \tag{4}$$

and the integrated equation is then given by

$$N(t) = \frac{N(0)Ke^{rt}}{K - N(0) + N(0)e^{rt}}. \tag{5}$$

The parameter K is called the **carrying capacity** parameter.

Intraspecific and Interspecific Competition

The logistic model [Eq. (2)] was the basis for the formulation of models describing interspecific competition by Lotka (1925, 1932) and Volterra (1927).

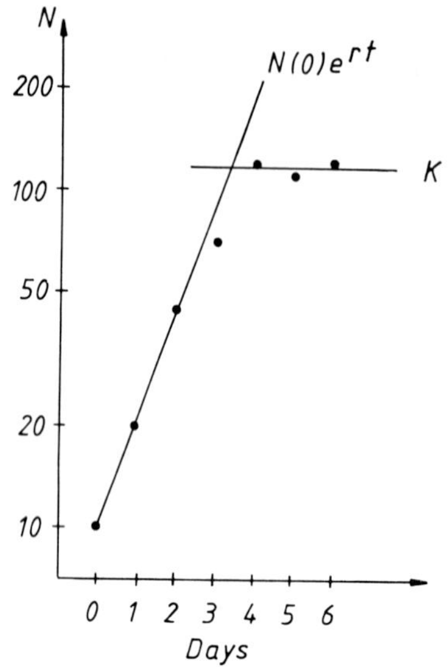

Fig. 2. Graphical estimation of the parameters r and K of the logistic model of population growth. Data on *Paramecium caudatum* provided by Arne Have

The generalization of Eq. (2) to a two-species competition model is straightforward. If we number the species as 1 and 2 we obtain the differential equations

$$\frac{dN_1}{dt} = N_1[b_1 - (d_1 + \iota_{11}N_1 + \iota_{12}N_2)] \text{ and}$$

$$\frac{dN_2}{dt} = N_2[b_2 - (d_2 + \iota_{21}N_1 + \iota_{22}N_2)], \tag{6}$$

where now ι_{ii}, $i = 1, 2$, are the density-dependent death rate coefficients due to intraspecific competition, and ι_{12} and ι_{21} are the density-dependent death rate coefficients due to interspecific competition. In this model ι_{12}, e.g., specifies the death rate of individuals of species 1 added per individual of species 2. Gause also studied the competition between different species of *Paramecium* grown in mixed experimental populations (Fig. 3) on the basis of the Lotka–Volterra Eqs. (6), and he based the analysis of data on the parameters r_i and K_i observed in the monocultures (Fig. 1). Changing to the parametrization used in Eq. (4) this provides Eqs. (6) on the form

$$\frac{dN_1}{dt} = r_1N_1(K_1 - N_1 - \alpha_{12}N_2)/K_1 \text{ and}$$

$$\frac{dN_2}{dt} = r_2N_2(K_2 - N_2 - \alpha_{21}N_1)/K_2, \tag{7}$$

where now α_{12} and α_{21} are the interspecific competition coefficients given by $\alpha_{ij} = \iota_{ij}/\iota_{ii}$, $i, j = 1, 2$.

Stable coexistence of the two competitors is only possible if both species can maintain a population in the absence of the other species, i.e., $b_i > d_i$ for $i = 1, 2$. In

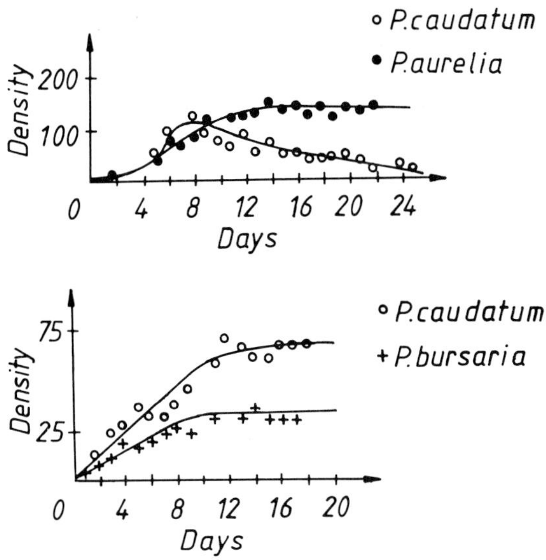

Fig. 3. Population densities in mixed cultures of *Paramecium*. **a** *P. aurelia* drives *P. caudatum* towards extinction. **b** When grown together, *P. caudatum* and *P. bursaria* coexist, though at lower densities than when alone. (Data from Gause 1934; after Begon et al. (1986)

this case, the condition for stable coexistence is that $(b_1 - d_1)/\iota_{12} > (b_2 - d_2)/\iota_{22}$ (or $K_1/\alpha_{12} > K_2$ in terms of the Gause parameterization) and that $(b_1 - d_1)/\iota_{11} < (b_2 - d_2)/\iota_{21}$ (or $K_1 < K_2\alpha_{21}$). This requires that the interspecific competition is less severe than the intraspecific competition in the sense that $\iota_{11}\iota_{22} > \iota_{12}\iota_{21}$ $(1 > \alpha_{12}\alpha_{21})$ (Fig. 4). When the interspecific competition coefficients, α_{ij}, are very close in value to the intraspecific competition coefficients, $\alpha_{ii}(=1)$, and the carrying capacities, K_i, of the competing species are of the same order of magnitude, the process of competitive exclusion or arrival at an equilibrium is very slow.

In Gause's experiments, competition between *P. aurelia* and *P. caudatum* ended always with the exclusion of one of the species. This is expected as the long-term outcome of competition if the two competing species share a common limiting resource and no heterogeneity of the experimental environment can be used to differentiate in the way resources become exploited (Gause's competitive exclusion principle). The outcome of the experiment depicted in Fig. 3a is also dependent on population growth rates as a density-independent mortality rate was introduced in the experiment by removing 10% of the populations every day for enumeration. In the experiments on competition between *P. caudatum* and *P. bursaria*, stable coexistence was found (Fig. 3b). This can be explained by a heterogeneity of the environment for these two species. *P. bursaria* is a superior competitor in one part of the environment, the sedimented yeast cells at the bottom of the test tubes, and *P. caudatum* is superior in utilizing the yeast cells suspended in the culture medium.

Interspecific competition coefficients in experimental situations can in principle be estimated from the decrease in growth as a function of density in mixed cultures compared to monocultures. An example is given in Fig. 5 on the individual growth of the mud snail, Hydrobia *ulvae*, as a function of its own growth and

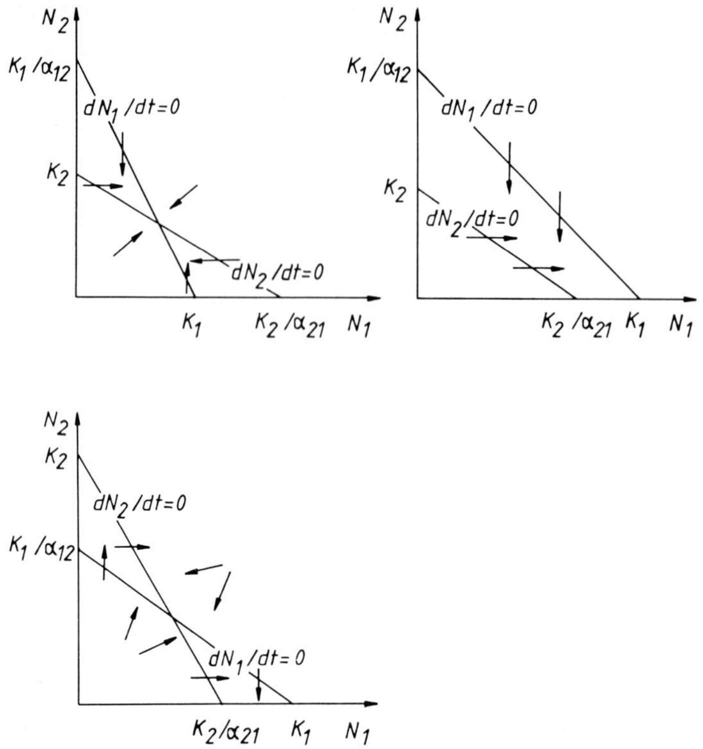

Fig. 4. A graphical analysis of a two species competition system as given by Eqs. 7. **a** Situation with stable coexistence. **b** Situation where species 1 wins. **c** Situation where the outcome of competition is dependent on the initial population sizes

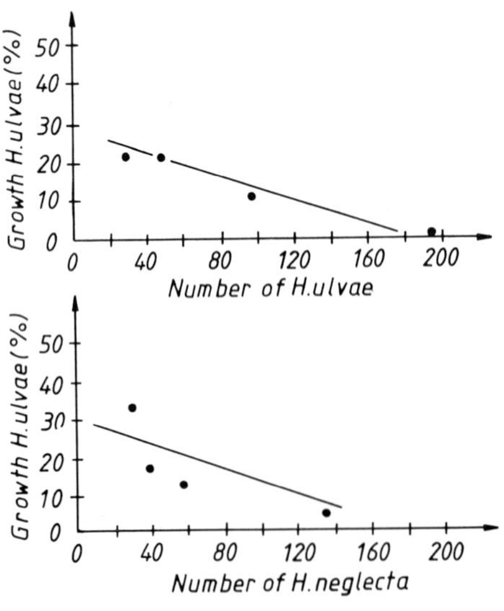

Fig. 5. The individual growth during 2 months in about 20 cm^3 sediment of the mud snail *Hydrobia ulvae* of a given size as function of its own density in monocultures and as function of the density of snails in mixed cultures with *H. neglecta* of the same size containing 20 *H. ulvae*. (After Fenchel and Kofoed 1976).

the growth of the closely related species *H. neglecta*. From the graph it can be seen that the interspecific competition coefficient is close to unity (i.e., close to the intraspecific competition coefficient).

When the interspecific competition coefficients exceed the intraspecific competition coefficients in the sense that $\alpha_{12}\alpha_{21} > 1$, then competition is called antagonistic. Antagonistic competition is for instance seen in experimental populations of the flour insect *Tribolium* (Park 1962). The predation of adults on larvae and pupae in these populations may cause the interspecific competition coefficients to exceed the intraspecific competition coefficients.

The Role of Parameters in Population Growth Models

The Malthusian model [Eq. (1)] is the basis of the theoretical studies on natural selection in continuously breeding populations by Fisher and Norton (Norton 1928; Fisher 1930; Charlesworth 1980). Genotypic variation in a population is supposed to be related to variation in the birth and death rates of individuals, and each individual, or rather each genotype, in the population is ascribed a **fitness value**, which is the Malthusian parameter as given by the difference between the birth and death rate. On this background Fisher formulated his Fundamental Theorem of Natural Selection, stating that the mean fitness increases during the evolution of the population, so the Malthusian growth rate of a population increases. However, knowing that the Malthusian growth model is only valid as a description of a population for a limited time, Fisher concluded that the increase in mean fitness at any time would result in an increased population size, because the population growth rate for any species would average zero if taken over any appreciable length of time. This conclusion was supported by the more detailed work of Kostitzin (1936, 1938), who described individual fitness based on the logistic model of population growth [Eq. (2)] using the Lotka–Volterra formulation of interspecific competition [Eq. (6)] to describe the intraspecific competition among genotypes. Thus, Kostitzin considered explicitly the effect of intraspecific competition, and he concluded that the outcome of selection is strongly dependent on the Malthusian parameter with secondary, but often qualitatively important effects of e.g., the density-dependent death rate coefficient. The results of Fisher therefore generalized to this more realistic description of population growth.

The development of theoretical population genetics proceeded using the Malthusian model as a reference for the definition of fitness, either in the form for continuously breeding populations [Eq. (7)] or more conveniently in a discrete generation version that we will consider later. The reference to the more realistic fitness description including the effects of intraspecific competition was taken up by MacArthur in a study from 1962, where he concluded that the effect of genotypic variation related to intraspecific competition overwhelmed the effect of genotypic variation in the Malthusian parameter, when evolution of a population close to the stable population size of the logistic model was considered.

MacArthur and Kostitzin addressed the same problem, namely the description of evolution in a density regulated population, and they initiated their theoretical investigations from the same growth model, namely the logistic model of population growth. They more or less performed the same calculations, and the end results of these calculations were mathematically equivalent. Yet, their biological interpretation of the results were fundamentally different. Kostitzin concluded that the effect of intraspecific competition was interesting and important, but not qualitatively new. MacArthur concluded that evolution in a density-regulated population is qualitatively determined by variation in relation to intraspecific competition with the Fisherian theory only applicable to populations where Malthusian growth prevails and population density is unimportant.

The basic biological difference between the models of Kostitzin and MacArthur is the way intraspecific competition is measured. Kostitzin, with reference to our simplified model [Eq. (2)], used the density-dependent death rate coefficient as a measure of intraspecific competition in the tradition of Lotka and Volterra. MacArthur instead used the carrying capacity of Gause's formulation as a measure of intraspecific competition. However, the carrying capacity parameter is proportional to the Malthusian parameter in the Kostitzin model, so it is only natural that Kostitzin concludes that the Malthusian parameter is important in evolution in a situation where MacArthur concluded that only the carrying capacity is of concern. The difference is therefore in the mathematical representation of the biological effect of intraspecific competition in the theoretical model, and the biological discussion of whose results are relevant in a given biological situation therefore has to center on the representation of intraspecific competition and not primarily on whether the results are reasonable or not. In this way the controversy of Kostitzin and MacArthur can contribute to our understanding of biological processes.

Exploitative Competition for Discrete Biotic Resources

Exploitation of resources leads to competition among the exploiters, but the outcome of the competition depends on the nature of the resource. A spatial resource is in principle permanent and may be occupied by the individuals for the time of their life. Other resources are consumed, and for these the dynamics of the resource has a decisive influence. A continuously renewable resource leads to a very different kind of competition than a resource which is renewed in bulk at discrete times and discrete places. However, a crow living off carrion may find food continuously replenished, whereas a carrion beetle may find the same resource renewed in bulk. Thus, like so many other evaluations in ecology, the description of the **resource dynamics** can only be done by comparing the time scales of the dynamics of the resources and that of the exploiter populations. The dynamics of a resource which is renewed in bulk will often be described as environmental patches and the arrival of consumers as colonizations, the amount of resource will be measured in terms of the persistence of and emigrant production from the patch. The continuously renewable resource is described by the rate of production of available resource and the amount of resource an individual can consume as a function of the availability of the resource. This description of individual feeding habits is simpler than the description of the episodic populations in environmental patches, so the theory of competition for a continuously renewed resource has been studied in more depth.

A simple resource dynamics occurs when the renewal of the resource can be described as growth of a prey species with a short generation time and when the consumption can be considered as due to predation by the species of interest which has a long generation time. This model is originally due to MacArthur and Levins (1964) and later discussed by, e.g., MacArthur (1972) and Christiansen and Fenchel (1977), and it has been extended to cover other kinds of resource dynamics (Abrams 1986). Let the amount, or really the concentration of resource in a given habitat at time t be R(t), and suppose that the resource grows according to the logistic equation with the carrying capacity parameter S, the resource abundance in an unexploited habitat. The rate of consumption by an individual in a habitat with the food concentration R is taken to be proportional to R independent of the number of consumers present. Thus, with x consumers in the habitat, the resource concentration changes according to the equation

$$\frac{dR}{dt} = R\xi(S - R - ux). \tag{8}$$

The consumption and growth of the resource is described in terms of the rate parameter ξ. This parameter will be used to express that the generation time of the resource is short and that the consumption of resource is on a short time scale, and this is done by assuming ξ large. The available resource will at any moment tend to stabilize at the level $R(t) = S - ux(t)$, in that a population with more than $(S - R)/u$ individuals will "graze" the resource down and a population with less than $(S - R)/u$ individuals will allow the resource to increase in density.

The competition among the consumers is in this model straightforward, in that the rate at which an individual consumes resource is equal to the new resource supplied per individual and therefore a decreasing function of the number of consumers. A more detailed model of the competition can be built by specifying the population growth rate of the consumer as a function of the per capita rate of resource consumption. This ideally requires a specification of the birth and death rate as a function of the resource consumption, but a simple model may be built by assuming a simple relationship between the population growth rate and the consumption. Assume that net reproduction occurs only when the resource is consumed at a rate larger than a threshold rate, which we may write as ξT, and that the population growth rate is proportional to the excess resource:

$$\frac{dx}{dt} = x\zeta(uR - T), \tag{9}$$

where ζ is a factor converting resource into consumers. Thus, the consumer population can increase in size only when $R > T/u$. If $R > T/u$ then the population grows until $R = T/u$ and if $R < T/u$ then the population declines until R has increased to the level T/u. When the resource level stabilize at T/u then the population ceases to grow and the population size is given by $(uS - T)/u^2$. The basic assumption of the model is that the rate parameter ξ is large so we obtain the resource concentration as approximately $R(t) - S = ux(t)$ at any time (Christiansen and Fenchel 1977), and with this assumption, Eq. (9) becomes the logistic equation with carrying capacity parameter $K = (uS - T)/u^2$.

This model may be extended to describe the **competition for the resource between two species**, in that both species grow according to the model (9), so

$$\frac{dx_1}{dt} = x_1\zeta_1(u_1R - T_1),$$
$$\frac{dx_2}{dt} = x_2\zeta_2(u_2R - T_2), \tag{10}$$

and the resource is consumed by both species as in the model (8). However, we can conclude the outcome of competition by just considering the growth of the populations of the two competitors. Species 1 grows when $R > T_1/u_1$ and species 2 grows when $R > T_2/u_2$. Thus, only the species with the lowest ratio, T_i/u_i, $i = 1, 2$, will persist in a habitat where both species are initially present, because that species can continue to grow and consume the resource at resource levels where the population size of the other species declines. This is a version of the competitive exclusion principle of Gause (1934) that says that only one species can persist on a single resource.

Coexistence of two species is possible when more resources become available. With two species and two independent resources the model may be written as

$$\frac{dx_i}{dt} = x_i\zeta_i(u_{i1}R_1 + u_{i2}R_2 - T_i),$$
$$\frac{dR_j}{dt} = R_j\xi(S_j - R_j - x_1u_{1j} - x_2u_{2j}) \tag{11}$$

for the species $i = 1, 2$ and the resources $j = 1, 2$. Again assuming that the rate parameter ξ of resource growth and consumption is large, a good approximation of the dynamics for the competitors may be reached by assuming that the resources are at any time at equilibrium corresponding to the prevailing consumer population sizes. Thus, we may assume that $R_j = S_j - x_1u_{1j} - x_2u_{2j}$, $j = 1, 2$, and obtain the equations

$$\frac{1}{x_1\zeta_1}\frac{dx_1}{dt} = (u_{11}S_1 + u_{12}S_2 - T_1) - x_1(u_{11}^2 + u_{12}^2) - x_2(u_{11}u_{21} + u_{12}u_{22}),$$
$$\frac{1}{x_2\zeta_2}\frac{dx_2}{dt} = (u_{21}S_1 + u_{22}S_2 - T_2) - x_1(u_{21}u_{11} + u_{22}u_{12}) - x_2(u_{21}^2 + u_{22}^2). \tag{12}$$

These equations have the same form as the Lotka–Volterra equation for interspecific competition [Eq. (6)]. Therefore, the Eqs. (12) can be viewed as providing the carrying capacity parameters and the intra- and interspecific competition coefficients of the Lotka–Volterra equations in terms of the parameters describing growth and consumption of the resources. The existence and coexistence of the species populations in a given habitat therefore can be discussed in relation to the availability and the exploitation of the resources (MacArthur 1972; Christiansen and Fenchel 1977). The amount of resource available to the growth of species i is described by the carrying capacity parameter $K_i = u_{i1}S_1 + u_{i2}S_2 - T_i$. The intraspecific competition in species (i = 1, 2) is measured by the term $\gamma_{ii} = u_{i1}^2 + u_{i2}^2$ and the interspecific competition between species i and species j ($\neq$ i) is measured by the term $\gamma_{ij} = u_{i1}u_{j1} + u_{i2}u_{j2}$. With this the Eqs. (12) changes into

$$\frac{1}{x_1\zeta_1}\frac{dx_1}{dt} = K_1 - x_1\gamma_{11} - x_2\gamma_{12},$$

$$\frac{1}{x_2\zeta_2}\frac{dx_2}{dt} = K_2 - x_1\gamma_{21} - x_2\gamma_{22}. \tag{13}$$

From the form of the interspecific competition coefficient it follows that $\gamma_{12} = \gamma_{21}$, and this symmetry of the matrix of competition coefficients is called **symmetric competition**. This is a general property of models for exploitative competition, and we will return to a general discussion of models for symmetric competition.

The model of MacArthur and Levins (1964) describes competition in terms of the mechanism, in that the competition coefficients and the carrying capacity parameters are given in terms of a simple model for the dynamics of food gathering. The model describes competition in terms of properties of the individual, and it has therefore the potential to describe competitive interactions in relation to observable phenotypic variation in the population. Thus, evolutionary models based on this description can advance from statements about the evolutionary change in the competitive relations as described by the competition coefficients and the carrying capacity parameters to results about the evolution of individual characteristics of the competing species.

Exploitative Competition in a Spectrum of Biotic Resources

This potential will be used in an extended version of the exploitative competition model, namely that of MacArthur and Levins (1967) and Levins (1968). This model relates the consumption of a continuum of food particle sizes to the size of trophic organs, and it is a good basis for an analysis of the evolutionary properties of exploitative competition (Schoener et al. 1986).

We will consider an example of the use of the MacArthur–Levins model to describe competition for food, namely the competition for food particles, mainly microalgae, among *Hydrobia* mud snails in Danish estuaries and fjords (Fenchel 1975; Fenchel and Kofoed 1976). The snails feed on microscopic algae at the surface of marine sediments and a snail ingests food and other particles with a distribution that depends on its size, in that the mean particle size ingested is proportional to the size of the snail (Fig. 6). On logarithmic scales the particle size distribution for the individual snail is approximately normally distributed with a variance independent of the size of the snail (Fenchel 1975). The grazing snails inflict a mortality on each of the algal species, and this predation may be described as a depression in the standing crop of each species, as their generation time is short compared to the snail predator (Fenchel and Kofoed 1976). The community of algal species, as a simplification of the model, is finally described as a continuum of sizes of food particles.

The resources are described as a continuum of qualities ρ which may be the size of the algal cell, potentially engulfed by mud snails. In the example the size ρ is measured with a logarithmic measuring stick. The abundance of the resources is described by the spectrum $S(\rho)$ that characterizes the amount of resources of quality ρ in the absence of predators (snails), and in the presence of the predators the abundance of the resource ρ is given by $R(\rho)$. The exploitation by an individual predator is described by the utilization $U(\rho)$ that specifies the rate with which the individual consumes resources of the quality ρ. The previous assumption that the resources grow and are exploited at a very fast rate provides the resource abundance as $R(\rho) = S(\rho) - xU(\rho)$ due to the exploitation of x individuals with the same utilization function. The model for population dynamics corresponding to this model of resource dynamics then specifies the individual ingestion term in Eq. (9) as $\int U(\rho)R(\rho)d\rho$ (see Christiansen and Fenchel 1977). The growth and competition of two species is then described by equations like (12) which are in form of the Lotka–Volterra Eqs. (13) with the competition coefficients

$$\gamma_{ij} = \int U_i(\rho)U_j(\rho)d\rho \tag{14}$$

and the carrying capacity parameters

$$K_i = \int U_i(\rho)S(\rho)d\rho - T_i, \tag{15}$$

where U_i is the utilization function of species i. In the following applications of the **MacArthur–Levins competition model** we will assume that the threshold T_i is so small that we can disregard this parameter.

This formulation of the classical Lotka–Volterra competition model can be considered as a formulation of competition in terms of the niche of the organism. The **niche concept** of Hutchinson (1957) can be extended to allow a differentiation between varying degrees of optimal and suboptimal conditions, defining the niche as a "fitness measure on an environmental space" (Levins 1968) in a hypervolume where each point corresponds to a specific environmental conditions. Niches are then related to the utilization functions of MacArthur and

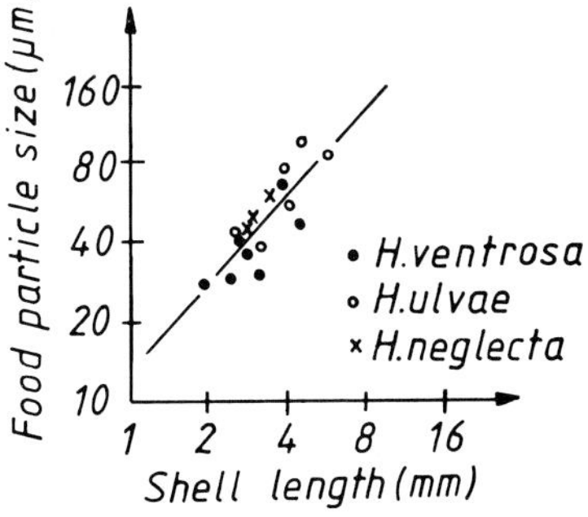

Fig. 6. The average particle sizes consumed as function of shell length in the three species of *Hydrobia*. (After Fenchel 1975)

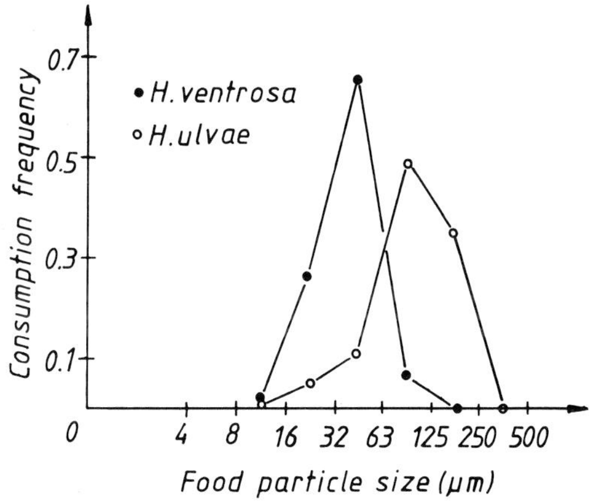

Fig. 7. The size frequency distribution of ingested mineral grains of coexisting populations of *Hydrobia ventrosa* and *H. ulvae*. (After Fenchel 1975)

Levin's model (Fig. 7). Variation in characters of the individuals that determine their resource use is through the model related to the environment in which they live, so the model specifies natural selection working on the variation in these characters. Therefore, we can study how the resource use of a species is modified by natural selection, and how adaptive changes in the realized niche occur as a response to changes in the biotic and abiotic environment.

Genotypic Variation in Resource Use

The definition of competition coefficients and carrying capacity parameters is related to the resource use of the individual in the species, so it is readily extended to definitions of competition coefficients and carrying capacity parameters related to the genotypes in a variable species. The utilization function U_{ig} for individuals of genotype g in species i is defined by taking $U_{ig}(\rho)$ as the average rate of consumption of resource ρ by individuals of genotype g. This immediately produces the carrying capacity parameter K_{ig} from Eq. (15). The competition coefficients $\gamma_{ig,jh}$ that describe the competition of genotype g in species i with genotype h in species j is given by the utilization functions U_{ig} and U_{jh} from Eq. (14).

The model simplifies when the resource abundance spectrum $S(\rho)$ and the utilization functions $U_i(\rho)$ both are assumed to be proportional to Gaussian distributions. The measurement of the resource qualities can, for simplicity, be assumed to refer to the mean of the resources spectrum, so the resource spectrum, S, has the mean zero, and the description of the environment is simplified to the specification of the variance σ^2 of the resource spectrum. The utilization function U_{ig} of an individual of genotype g in species i is specified by the mean quality D_{ig} of the exploited resources and by the variance W_{ig} of those resources. Then, apart from for evolutionary considerations unimportant proportionality constants, we get the carrying capacity parameter of genotype g in species i as

$$K_{ig} = \frac{1}{(\sigma^2 + W_{ig}^2)^{1/2}} \exp\left[-\frac{D_{ig}^2}{2(\sigma^2 + W_{ig}^2)} \right] \tag{16}$$

and the competition coefficient between genotype g in species i and genotype h in species j as

$$\gamma_{ia,jh} = \frac{1}{(W_{ig}^2 + W_{jh}^2)^{1/2}} \exp\left[-\frac{(D_{ig} - D_{jh})^2}{2(W_{ig}^2 + W_{jh}^2)} \right]. \tag{17}$$

The carrying capacity parameter [Eq. (16)] depends on the distance of the mean utilization of the genotype from the mode in the resource abundance spectrum. The competition coefficient (17) between two genotypes depends on the distance between the mean utilizations of the genotypes.The two distances are, however, evaluated with respect to different scales. The carrying capacity parameter is scaled with respect to the sum of the resource variance and the utilization variance in question, whereas the competition coefficient depends only on the utilization variances of the two genotypes.

The characters involved in this kind of exploitation of resources often show quantitative or **polygenic inheritance**, i.e., the hertitable variation of the character expresses the variation at many gene loci each with a small effect on the phenotype. Suppose the genotypic variation in species i of the model originates from the allelic variation at one such locus, so it makes sense to assume that the genotypes with respect to this particular locus differ only little. Suppose further, for simplicity, that it is an autosomal locus with two allele A and a, where the three genotypes, AA, Aa, and aa numbered 1, 2, and 3, on the average utilize resources with mean $D_i + 2d_i$, $D_i + d_i$, and D_i, respectively, i.e., we assume that the alleles have additive effects on the phenotypes, and the genotypes have the same variance in their utilization function, i.e., we assume that $W_{ig} = W_i$. The allelic effect d_i is assumed to be small, so we can address approximations of the parameters by neglecting terms of second order or higher in d_i in the power series expansion of the exponential function. The carrying capacities then become

$$K_{i1} \approx K_i\left[1 - \frac{4D_i d_i}{2(\sigma^2 + W_i^2)}\right], \tag{18}$$

$$K_{i2} \approx K_i\left[1 - \frac{2D_i d_i}{2(\sigma^2 + W_i^2)}\right], \tag{19}$$

and $K_{i3} = K_i$, where

$$K_i = \frac{1}{(\sigma^2 + W_i^2)^{1/2}} \exp\left[-\frac{D_i^2}{2(\sigma^2 + W_i^2)}\right]. \tag{20}$$

The competition coefficient between a genotype in species i and genotype h in species j (i $\neq$ j) becomes

$$\gamma_{i1,jh} \approx \gamma_{i,jh}\left[1 - \frac{4(D_i - D_{jh})d_i}{2(W_i^2 + W_j^2)}\right], \tag{21}$$

$$\gamma_{i2,jh} \approx \gamma_{i,jh}\left[1 - \frac{2(D_i - D_{jh})d_i}{2(W_i^2 + W_j^2)}\right], \tag{22}$$

and $\gamma_{i3,jh} = \gamma_{i,jh}$, where

$$\gamma_{i,jh} = \frac{1}{(W_i^2 + W_j^2)^{1/2}} \exp\left[-\frac{(D_i - D_{jh})^2}{2(W_i^2 + W_j^2)}\right]. \tag{23}$$

Evolution of Exploitative Competition

For an initial judgement of the course of evolution the genotypic fitnesses may be compared to the accuracy of the order of d_i in the above approximations, where terms of the order d_i^2 or smaller are neglected. The competition coefficient between two genotypes in species i deviates from unity by terms of the order d_i^2 or smaller [Eq. (17)], so initially we neglect variation with respect to intraspecific competition. Let the gene frequency of allele A be p and that of allele a be q = 1 − p, and let x_{jh} be the number of individuals of genotype h in species j. The rate of change in the number of A alleles is then proportional to

$$K_i\left[1 - \frac{D_i d_i(1 + p)}{\sigma^2 + W_i^2}\right] - \sum_j \sum_h x_{jh}\gamma_{i,jh}\left[1 - \frac{(D_i - D_{jh})d_i(1 + p)}{W_i^2 + W_j^2}\right] \tag{24}$$

and the comparable rate of change in the number of a alleles is

$$K_i\left[1-\frac{D_i d_i p}{\sigma^2+W_i^2}\right]$$

$$-\sum_j\sum_h x_{jh}\gamma_{i,jh}\left[1-\frac{(D_i-D_{jh})d_i p}{W_i^2+W_j^2}\right]. \tag{25}$$

The frequency of A alleles, p, will increase if and only if Eq. (24) is larger than Eq. (25). i.e., if and only if

$$0<d_i\left[\sum_j\sum_b x_{jh}\gamma_{i,jh}\frac{(D_i-D_{jh})}{W_i^2+W_j^2}-K_i\frac{D_i}{\sigma^2+W_i^2}\right]. \tag{26}$$

Thus, the increase or the decrease of the frequency of an allele will depend only on the description of the environment in terms of the resource (last term of the bracket) and in terms of the populations of the competitors (first term of the bracket). This condition depends only on the current location of the utilization functions of species i in relation to the other species and to the resource optimum, so condition (26) is in any particular circumstance either a condition saying that d_i should be positive or negative. If the genotypic variation in the other species is neglected, i.e., we say $\gamma_{i,jh}=\gamma_{i,j}$ for all h, then the result is directional selection that either favors allele A or allele a, and eventually the polymorphic population segregating the allele A and a will become monomorphic.

As we assume small differences among the genotypes, the evolution in species i is slow, so it is relevant to discuss condition (26) in the situation of ecological equilibrium, i.e., the carrying capacity parameters are given by

$$K_i=\sum_j \hat{x}_j\gamma_{i,j} \tag{27}$$

plus terms of the order d_i. Then the frequency of A alleles will increase to fixation if and only if

$$0<d_i\sum_j \hat{x}_j\gamma_{i,j}\left(\frac{D_i-D_j}{W_i^2+W_j^2}-\frac{D_i}{\sigma^2+W_i^2}\right). \tag{28}$$

This condition can by differentiation of Eq. (27) with respect to D_i be shown to be equivalent to the condition

$$0<d_i\frac{d\hat{x}_i}{dD_i} \tag{29}$$

by neglecting terms second order in d_i^2. Thus, in species i evolution will progress in the direction where its equilibrium population size increases. This may result in the extinction of one or more of the other species, but eventually it will lead to the evolution towards a location, D_i^*, of the utilization function where

$$0=\sum_j \hat{x}_j\gamma_{i,j}\left(\frac{D_i^*-D_j}{W_i^2+W_j^2}-\frac{D_i^*}{\sigma^2+W_i^2}\right). \tag{30}$$

This location D_1* is called the **optimal location** of species i with the utilization variance W_i^2 in the environment defined by the variance σ^2 of the resource spectrum and by the utilization functions of the other species, i.e., by D_j and W_j^2 for $j = 1, 2, \ldots, i-1, i+1, \ldots, n$.

For one species only, $n = 1$, the optimal location is always given by $D_1^* = 0$, i.e., the utilization mode of the species should coincide with the mode of the resource spectrum. Already for two species, $n = 2$, the characterization of the evolving species, species number 1, say, in terms of D_1^* as a function of the environment and of species number 2 becomes considerably more complex (Fenchel and Christiansen 1977; Matessi and Jayakar 1981; Loeschcke 1984a,b, 1985). Equal utilization variances, i.e., $W_1^2 = W_2^2 = W^2$, say, simplify the situation. Then for a narrow **resource spectrum,** $\sigma^2 < W^2$, the change in the utilization mode of species 1 is always in a direction towards the mode of the resource spectrum, i.e., $D_1 \rightarrow 0$ in the course of evolution, so $D_1^* = 0$. The condition for coexistence of the two species is for $D_2 > 0$, say, given by $D_1 < D_2(\sigma^2 - W^2)/(\sigma^2 + 3W^2)$ or $D_1 > D_2(\sigma^2 + 3W^2)/(\sigma^2 - W^2)$ for a wide resource, $\sigma^2 > W^2$, and by $D_2(\sigma^2 + 3W^2)/(\sigma^2 - W^2) < D_1 < D_2(\sigma^2 - W^2)/(\sigma^2 + 3W^2)$ for a narrow resource, $\sigma^2 < W^2$. Coexistence with a narrow resource therefore requires that $D_1 < 0 < D_2$, i.e., the utilization modes of the two species should be on opposite sides of the resource mode. As $D_1 \rightarrow 0$ the condition for coexistence between the two species is violated, so evolution in species 1 will cause species 2 to become extinct. Seen as coevolution in a two-species system, a narrow resource will cause an evolutionary race for staying closest to the resource optimum, and the loser in the race will eventually become extinct. In this respect, a narrow resource resembles a single discrete resource, as it leaves no opportunity for two species to coexist in the long run by differential utilization of the resources.

For a wide **resource spectrum,** $\sigma^2 > W^2$, two species may coexist even when D_1 and D_2 are of the same sign, i.e., when the utilization modes of the two species are to the same side of the resource mode. The evolutionary process also will be qualitatively different. The change in D_1 will be influenced by the location of the utilization function of species 2, and the species will stay in coexistence as $D_1 \rightarrow D_1^*$. Figure 8a show D_1^* as a function of D_2. Two optimal locations exists for each D_2. For $D_2 > 0$ and initially $D_1 < D_2(\sigma^2 - W^2)$ the location of D_1^* is given by the left arc for $D_2 > 0$, and if initially $D_1 > D_2(\sigma^2 + 3W^2)/(\sigma^2 - W^2)$ then the location of D_n^* is given by the right arc. Coevolution in a two species system in a wide resource is illustrated in Fig. 8b by a comparison between the full drawn arc (from Fig. 8a) and the broken arc. The arc drawn by a broken line show the evolutionary stable pure locations of species 2 given the location of species 1, and the intersections of the two arcs are the optimal locations of the two species. Joint evolution of the two species will cause them to eventually end up at an coevolutionary stable location, where the modes of their utilization functions are situated symmetrically around the mode of the resource spectrum, i.e., $D_1^* = -D_2^*$. The evolution of one species increases its population size (condition 29), but it may at the same time decrease the population size of the other species, so

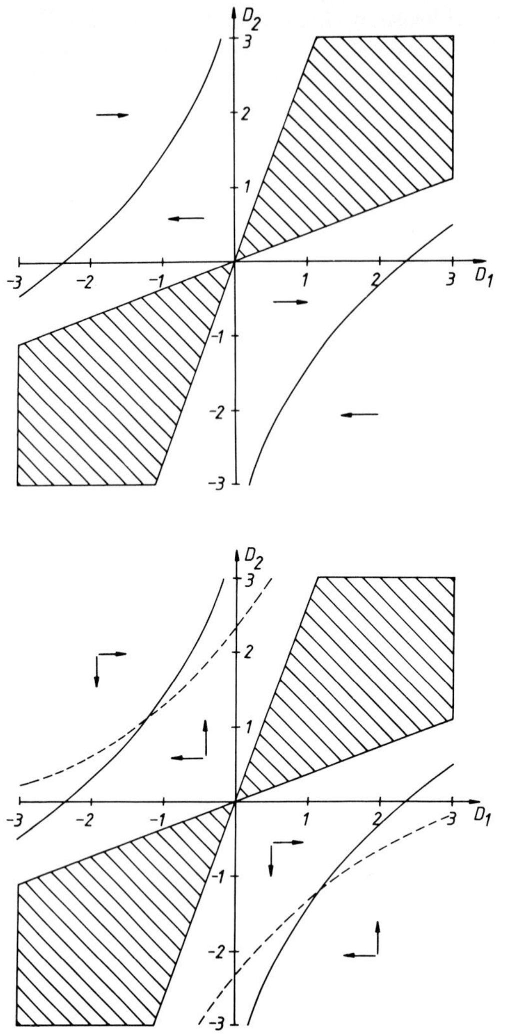

Fig. 8. The expected evolutionary shift in utilization means for two competing species with fixed utilization widths, W, in a Gaussian resource spectrum, the width of which is given by $\sigma/W = 1.75$. No coexistence is possible for pairs of utilization means, (D_1, D_2), in the *dotted* area. **a** The *solid lines* are the stable points for D_1 for a given location of D_2. Evolution in species 1 will proceed as indicated by the *arrows* under the assumption that species 2 has no variaton for the Character. **b** Stable points for D_1 (*solid lines*) and D_2 (*dotted lines*) for a given utilization mean of the competitor. The coevolutionary stable equilibrium (the intersection) will eventually be reached if enough variation in both species is accessible for evolution. (After Fenchel and Christiansen 1977)

the joint population size of the guild need to be maximal. In fact the joint population size of the two species is maximal at a point where $D_1 = -D_2$, but this point is different from the coevolutionary stable location and the species are closer together, i.e., $|D_1| < |D_1^*|$ (Roughgarden 1976, 1979).

The **coevolutionary stable locations** of the species in guilds of three, four, and five species are discussed by Loeschcke (1984b, 1985). The coevolutionary stable location of an odd number of species is characterized by the speices with a mode of the utilization function equal to the mode of the resource spectrum, and the rest of the species sorted into pairs with their utilization functions located symmetrically around the resource mode. With an even number of speices all are sorted into symmetric pairs at the coevolutionary stable locations. For a narrow resource spectrum, $\sigma^2 < W^2$, the only evolutionary stable situation is one species at the resource mode, but for a wide resource spectrum, $\sigma^2 > W^2$, all the three, four, or five species may coexist. Thus, no condition for "limiting similarity" exists, even though a comparatively low value of σ^2 may make some of the species populations very small. However, in a habitat where the residents reach coevolutionary stable locations before a new immigrant species arrives, the buildup of species diversity by this slow migration process requires a considerably wider resource spectrum. The initial equilibrium population size of an invader species before evolutionary changes has taken place is typically very small, few percent of the population size of the resident species (Fig. 9). As a very crude guide the equilibrium population size of the invader exceeds 5% of the population size of the resident species when $\sigma > 2W$ for two species, when $\sigma > 3W$ for three species, when $\sigma > 4W$ for four species, and when $\sigma > 5W$ for five species (Loeschcke 1985). Thus, low to moderate values of the resource variance, σ^2, may cause small population sizes of an invader species or of residents, and small populations are more prone to accidental extinctions than well-established resident species. These effects may well produce an effective limit to the similarity of a guild of species at a coevolutionary equilibrium.

Symmetric Competition

As mentioned earlier the exploitative competition models are all symmetric competition models: the Lotka–Volterra equations for symmetric competition are given by

$$\frac{dx_i}{dt} = v_i x_i \left(K_i - \sum_j \gamma_{ij} x_j \right), \tag{31}$$

where $\gamma_{ij} = \gamma_{ji}$ for all i and j, $i, j = 1, 2, \ldots, n$. The development of a guild of species exhibiting symmetric competition is particularly simple, and it can be characterized by the convergence of a scalar function which acts like the energy of a physical system (MacArthur 1970; Matessi and Jayakar 1976, 1981).

To construct this function define the total population size of the guild as

$$x = \sum_i x_i, \tag{32}$$

the average carrying capacity as

$$K = \sum_i K_i x_i / x \tag{33}$$

and the average competition coefficient as

$$C = \sum_i \sum_j \gamma_{ij} x_i x_j / x^2. \tag{34}$$

Now define the function ψ by

$$\psi = 2Kx - Cx^2. \tag{35}$$

This function always increases through time, i.e., $d\psi/dt \geqslant 0$, which can be seen from the following:

$$\frac{d\psi}{dt} = 2\sum_i K_i \frac{dx_i}{dt} - \sum_i \sum_j \gamma_{ij} \left[x_i \frac{dx_j}{dt} + x_j \frac{dx_i}{dt} \right]$$

$$= 2\sum_i v_i x_i \left[K_i - \sum_j \gamma_{ij} x_j \right]^2, \tag{36}$$

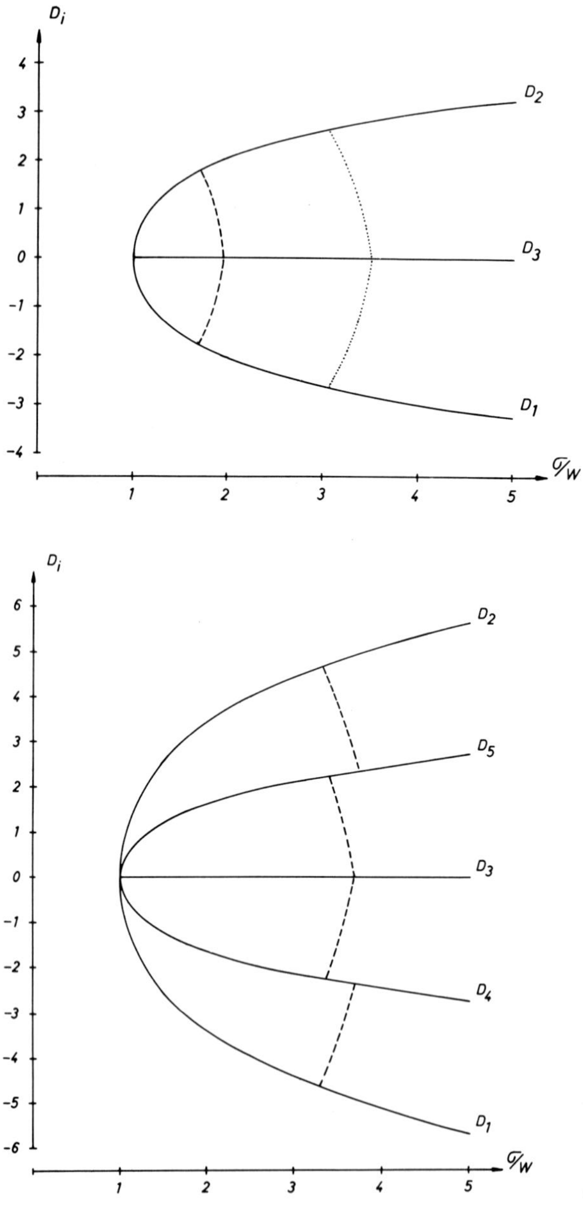

Fig. 9. Three **a** and five species; **b** coevolutionary equilibrium locations, D_j as functions of the relative resource width, σ/W. For relative resource widths to the *right* of the dashed lines (*pointed lines*) the equilibrium population size of an invader may exceed 1% (5%) of the community size. (After Loeschcke 1985).

where we substitute the differential coefficients from Eq. (31) and use the symmetry of the competition coefficients. At equilibrium all the derivates (31) are zero so the function ψ equals zero at equilibrium, and as ψ otherwise always increases the locally stable n-species equilibria of the system (31) are characterized by the local maxima of the function ψ (the Liapunov theorem on stability).

This simple dynamic behavior of the model for symmetric competition was extended by Matessi and Jayakar (1981) to a general **maximization principle for evolution** under natural selection due to symmetric intra- and interspecific competition. Consider in each species one locus that influences the parameters of the symmetric competition model (31). Suppose that in species i the locus segregate m_i alleles, A_{ig}, which occur in the frequencies p_{ig} in the population, $i = 1, 2, \ldots, n$, $g = 1, 2, \ldots, m_i$. We will describe the fitness differences among the three genotypes in each species by differences in carrying capacity parameters and in competition coefficients. Other kinds of selection will be disregarded, so we assume that the coefficient v_i is independent of the genotype and that reproduction by random mating occurs. Let the carrying capacity parameters of the genotypes in species i be given as $K_i(A_{ig}A_{ih})$, and let the competition coefficients between genotypes in species i and species j be given as $\gamma_{ij}(A_{ig}A_{ih}, A_{jk}A_{jl})$, $i, j = 1, 2, \ldots, n$, $g, h = 1, 2, \ldots, m_i$, $k, l = 1, 2, \ldots, m_j$. The average competition experienced by genotype $A_{ig}A_{ih}$ is

$$C(A_{ig}A_{ih}) = \sum_{j=1}^{n} \sum_{k=1}^{m_j} \sum_{l=1}^{m_j} \gamma_{ij}(A_{ig}A_{ih}, A_{jk}A_{jl}) p_{jk} p_{jl} x_j/x, \tag{37}$$

the average competition experienced by an individual of species i carrying allele A_{ig} is

$$C(A_{ig}) = \sum_{h=1}^{m_i} C(A_{ig}A_{ih}) p_{ih}, \tag{38}$$

and the average competition experienced by an individual of species i is

$$C_i = \sum_{g=1}^{m_i} C(A_{ig}) p_{ig}. \tag{39}$$

Averaging C_i over species provides

$$C = \sum_{i=1}^{n} C_i x_i/x, \tag{40}$$

the average competition experienced by an individual of any species in Eq. (34). Similarly we define the average carrying capacity of an individual of species i carrying allele A_{ig} as

$$K(A_{ig}) = \sum_{h=1}^{m_i} K(A_{ig}A_{ih}) p_{ih}, \tag{41}$$

and the average carrying capacity of an individual of species i is

$$K_i = \sum_{g=1}^{m_i} K(A_{ig}) p_{ig}. \tag{42}$$

Averaging K_i over species provides

$$K = \sum_{i=1}^{n} K_1 x_i / x, \tag{43}$$

the average carrying capacity of an individual of any species in Eq. (33).

We will assume that the fitness differences among the genotypes of a given species are small. The evolution of the system may then be described with good approximation by the equations

$$\frac{dp_{ig}}{dt} = v_i p_{ig} \{ [K(A_{ig}) - xC(A_{ig})] - (K_i - xC_i) \} \tag{44}$$

for $i = 1, 2, \ldots, n$ and $g = 1, 2, \ldots, m_i$. (This is the standard model for slow selection neglecting the deviations from the Hardy–Weinberg proportions brought about by the selection). The change in population sizes is given by

$$\frac{dx_i}{dt} = v_i x_i (K_i - xC_i) \tag{45}$$

for $i = 1, 2, \ldots, n$ as in Eq. (31), and with these equations the function ψ defined in Eq. (36) again monitor the development of the guild of species. The function is increasing because

$$\frac{d\psi}{dt} = 2 \sum_{i=1}^{n} v_i x_i (K_i - xC_i)^2 + 4 \sum_{i=1}^{n} \sum_{g=1}^{m_i} v_i x_i p_{ig} \{ [K(A_{ig}) - xC(A_{ig})] - (K_i - xC_i) \}^2 \tag{46}$$

is always positive. As before an equilibrium of the Eqs. (44) and (45) corresponds to a state of the population where the derivate (46) of ψ is zero, and the stable equilibria are the maxima of the function ψ.

Polymorphism and Exploitative Competition

The selection due to interspecific exploitative competition considered in condition (28) is directional selection pushing species i towards the optimal location D_i^*. However, for D_i close to D_i^* the factor of d_i in condition (28) becomes small [Eq. (30)], so its value is really unknown because we have neglected terms of the order d_i in the evaluation of this factor. Thus, close to the optimal location we have to reevaluate the selection working on characters related to the mode D_i of the utilization function. The result of this evaluation is that in many cases, in particular at coevolutionary stable locations for $\sigma^2 > W^2$, the directional selection is changed to selection that maintain the population polymorphic (Fenchel and Christiansen 1977; Loeschcke 1984a), so genetic variation will build up in the species (Fig. 10).

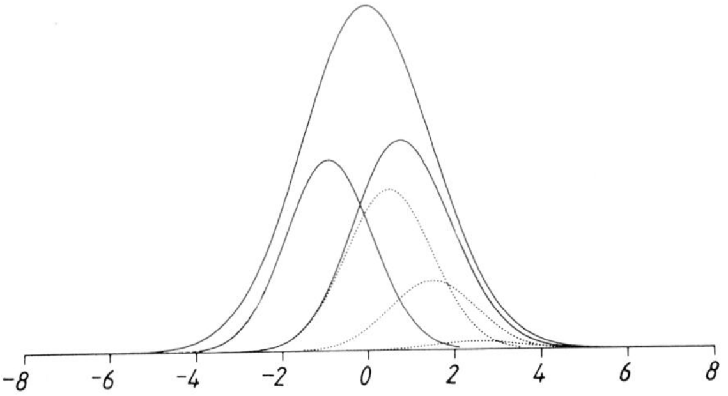

Fig. 10. Equilibrium genotype densities in one of the species in a two-species competitive situation for a relative resource width of $\sigma/W = 1.5$. The equilibrium densities of genotypes in species 2 are represented by the heights of the utilization functions (*dotted lines*). Species 1 is located at the two-species interspecific coevolutionary equilibrium location with $D_1 = -0.9$ and relative population size $x_1 = 0.55$. In species 2 there are two alleles segregating with contributions $d_1 = -0.2$ and $d_2 = +0.8$ defined as deviations from the two-species interspecific coevolutionary equilibrium location $D_2 = 0.9$. The mean location of species 2 is $D_2 = 0.96$ and the equilibrium relative population size is $x_2 = 0.67$. The utilization means of the genotypes in species 2 are $D_{21} = 0.55$, $D_{22} = 1.5$, and $D_{23} = 2.5$; and genotype frequencies are $x_{21} = 0.67$, $x_{22} = 0.30$, and $x_{23} = 0.03$. (After Loeschcke 1984b).

The essential properties of this process is captured by the consideration of a single species (Christiansen and Loeschcke 1980a). The competition coefficient between two genotypes in the species (we drop the species number) become

$$\begin{array}{c} \\ \mathrm{AA} \\ \mathrm{Aa} \\ \mathrm{aa} \end{array} \begin{array}{c} \begin{array}{ccc} \mathrm{AA} & \mathrm{Aa} & \mathrm{aa} \end{array} \\ \begin{bmatrix} \gamma & \gamma_{12} & \gamma_{13} \\ \gamma_{12} & \gamma & \gamma_{12} \\ \gamma_{13} & \gamma_{12} & \gamma \end{bmatrix} \end{array} \tag{47}$$

where

$$\gamma = \frac{1}{(2W^2)^{1/2}}. \tag{48}$$

The coefficients γ_{12} and γ_{13} are functions of the allele effect d which is assumed small so

$$\gamma_{12} \approx \gamma\left[1 - \frac{d^2}{4W^2}\right], \quad \text{and} \quad \gamma_{13} \approx \gamma\left[1 - \frac{4d^2}{4W^2}\right]. \tag{49}$$

The carrying capacities are functions of the utilization modes of the genotypes. These are all assumed to be close to the optimal location, i.e., we assume that D is small, so the disinction between D and d become unnatural. Instead define the allele effects as $d_A = d + D/2$ and $d_a = D/2$, so d is the difference between the allele effects, viz. $d = d_A - d_a$. The utilization mode of the genotypes is obtained by adding the effects of the alleles in the genotype, so the mode of AA and $2d_A = D$

$+2d$, that of Aa is $d_A + d_a = D + d$, and that of aa is $2d_a = D$. Then the carrying capacities become approximately

$$K_1 \approx K\left[1 - \frac{4d_A^2}{2(\sigma^2 + W^2)}\right], \tag{50}$$

$$K_2 \approx K\left[1 - \frac{(d_A + d_a)^2}{2(\sigma^2 + W^2)}\right], \text{ and} \tag{51}$$

$$K_3 \approx K\left[1 - \frac{4d_a^2}{2(\sigma^2 + W^2)}\right], \tag{52}$$

where

$$K = \frac{1}{(\sigma^2 + W^2)^{1/2}}. \tag{53}$$

These approximations are very accurate in that only terms which are fourth order in d and D have been neglected. Because of the symmetry of the genotypes we may assume that $d_A > |d_a|$, i.e., the genotype aa is always closest to the resource optimum and AA is always to the right of the resource optimum. With this assumption, the allele a will always increase in frequency when introduced as rare in a population monomorphic AA, because the carrying capacity of the new genotype Aa is higher than that of the resident genotype AA, and the effect of intraspecific competition on Aa is dominated by $x\gamma_{12}$ which is always less than the effect on AA which is dominated by the term $x\gamma$ describing the effect of competition between AA individuals. Thus, the equilibrium with $p = 1$ is always unstable, so the fate of the population will depend on the stability of the other monomorphic equilibrium with $p = 0$ and on the existence and stability of polymorphic equilibria.

The allele A will increase in frequency when introduced as rare at the monomorphic equilibrium aa provided

$$\frac{K_2}{\gamma_{12}} > \frac{K_3}{\gamma} \tag{54}$$

or in terms of the allele effects

$$(d_A - d_a)[d_a(\sigma^2 + 7W^2) - d_A(\sigma^2 - W^2)] > 0 \tag{55}$$

(Fig. 11). This condition is exact, i.e., it does not need the above approximations. The symmetric condition that the allele a will increase in frequency when rare in a population monomorphic AA is

$$(d_A - d_a)[d_A(\sigma^2 + 7W^2) - d_a(\sigma^2 - W^2)] < 0 \tag{56}$$

(Fig. 11). If both conditions (55) and (56) are satisfied, then both alleles will be maintained in the population, as both alleles are protected from loss due to selection. This state is called a **protected polymorphism** (Fig. 11). For small allele

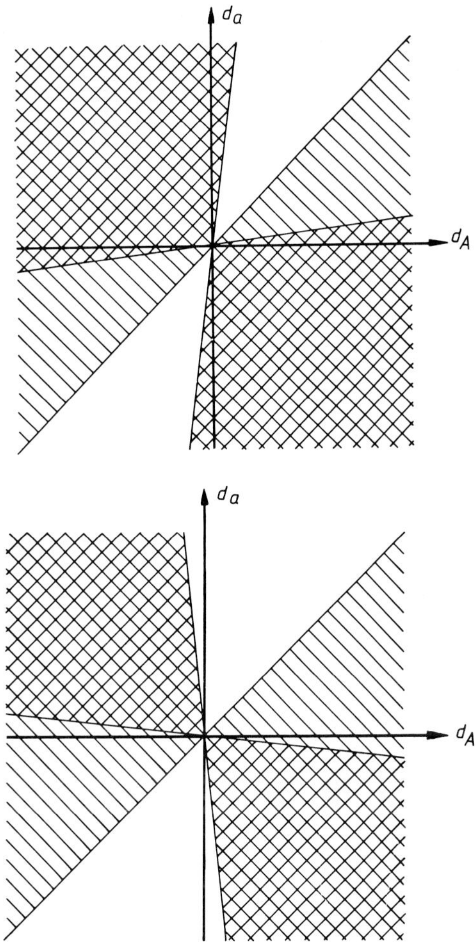

Fig. 11. The condition for increase of allele A in a population monomorphic for the allele a in terms of the contributions of the alleles. For (d_a, d_A) in the *hatched* regions the allele A will increase when rare. In the *doubly hatched* regions the alleles a and A will be maintained in the population in a protected polymorphism; **a** with the asumption that $\sigma > W$; **b** with the assumption that $\sigma < W$. (After Christiansen and Loeschcke 1980a)

effects using the approximations (48) through (54) it may be shown that the population go to a stable polymorphic equilibrium if and only if protected polymorphism prevails (Christiansen and Loeschcke 1980a). Otherwise, the population will end up monomorphic for allele a (we assumed that $d_A > |d_a|$). This completely describes the qualitative states that the population may take, because it may be shown that a population initially with three or more alleles with different effects on the utilization mode will end up in a stable state with one or two alleles.

The end state of the evolution in the population is qualitatively dependent on whether the resource spectrum is wide, $\sigma^2 > W^2$, or narrow, $\sigma^2 < W^2$. If initially the population segregates a series of alleles and if the population ends up polymorphic, then with a wide resource the two allele with the highest and the lowest effect will be maintained in the population, whereas with a narrow resource the two allele with the allele effect closest to zero (the position of the resource mode) will be kept segregating. In addition, the condition for protected polymorphism is easier to fulfill in a wide resource spectrum. Thus, given sufficient genetic variation, a population in a narrow resource spectrum will end up monomorphic with a mode of utilization function close to the resource mode, and a population in a wide resource spectrum will end up polymorphic (Christiansen and Loeschcke 1980a).

The amount of **genetic variation** that the population builds up is hard to characterize within this model, mainly because the variation builds up at all the loci that influence the character of the individual that determines the utilization mode. The model predicts increase in the genetic variation until the utilization variance W^2 and the genotypic variance roughly adds up to the resource variance σ^2. However, the model at the same time predicts that linkage disequilibrium builds up among these loci (Loeschcke and Christiansen 1984), so the standard modeling process for polygenic characters should be handled with caution. In addition, models of selection on a polygenic character need to specify the change in variance in response to selection, and from the above results on single genes it is expected that this specification will be very important for the results of selection (Bulmer 1974; Slatkin 1980b; Taper and Case 1985).

The genetic model can be generalized to account for **multidimensional resources.** Again we expect directional selection to prevail for genotypes utilizing resources closest to the resource optimum as long as resource utilization occurs far away from the resource mode (Christiansen and Loeschcke 1987). When the utilization modes of all genotypes are close to the resource optimum, polymorphic equilibria may exist. Except for special relations among the resource dimensions or utilization modes of the genotypes the system converges to a globally stable equilibrium segregating at most $m + 1$ alleles, where m is the number of dimensions of the resources.

We have used a genetic model that specifies the genotypic utilization mode as an addition of the allele effects. Deviations caused by dominance from this simple model may be important for the evolution of the population (Christiansen and Loeschcke 1980b; Christiansen 1984b). A range of types of dominance cause a breakdown of the correspondance between the condition for protected polymorphism and the requirement for the existence of polymorphic equilibria. Thus, dependent on the initial state of the population segregating two alleles (in the one dimensional resource), it may end up monomorphic by losing one of the alleles, or it may settle in a stable polymorphic state. Thus, the utilization variance of the population may depend on the course of evolution that leads the population to its present utilization mode.

Invasion and the Build-Up of Competitive Guilds

A guilds of exploiters at a coevolutionary equilibrium (Fig. 9) can be invaded by other species, so coevolved communities are in principle open to change

(Roughgarden 1983b; Case 1982; Loeschcke 1984a). Species k + 1 can invade a k-species guild at its coevolutionary equilibrium if

$$K_{k+1} > \sum_{j=1}^{k} \gamma_{k+1,j} x_j. \tag{57}$$

The fate of the invading species, however, will depend on the rate of the increase in population size of the invader. A small population is prone to random extinction, so it is unlikely that an invader can persist if its population size stays small for any extended period of time. For low population sizes the per capita rate of increase can be characterized by the relative change in population sizes in successive generations

$$\Delta_i = (x_i' - x_i)/x_i. \tag{58}$$

This rate generally will be low as long as invasion takes place within the range of the resident species (Loeschcke 1985), and the ultimate population size will be small (Fig. 9). Outside the range of residents Δ_i can be considerably higher, and the invader may reach appreciable population sizes. Thus, it seems that an invader with a utilization mode within the range of resident species has a considerably higher probability of going extinct than an invader with a mode outside this range. However, changes in the mode or variance of the resource abundance distribution are more critical to marginal invaders, and such changes may prevent long-term survival of these invader populations. So we may conclude that not too loosely packed coevolved guilds are difficult to invade by an isolated event of immigration. If, however, an invasion happens to be successful, the evolutionary response to the new competitive situation is initially slow. The fitness of rare heterozygotes in the resident species that, on the average, would reduce the competitive pressure is almost unchanged because the size of the invader population is low.

Let us now consider the **community build-up on islands** that are colonized by individuals from coevolved species on a mainland. Now assume the resource distribution on the island to be slightly different from that on the mainland that is characteristic for its equilibrium configuration. Let us consider five potential colonizing species. The outer species are named 1 and 2 and the central species 3. The species between 1 and 3 is named species 4, and the one between species 3 and 2 is species 5. Assume further that species 1 and 2 are the first to successfully colonize the island and consider the build-up of the guild when individuals from other species immigrate at random, being occasionally successful invaders. Figure 12 shows the possible paths of community build-up and corresponding species compositions in such guilds (Loeschcke 1987a). The figure shows that competition can prevent the invasion of certain species which otherwise could coexist if the competitive pressure on them is reduced by additional competitors ("diffuse" competition).

Assume now that genetic variation is introduced into the species that make up the guild of competitors. Then we expect genetic variation to build up due to

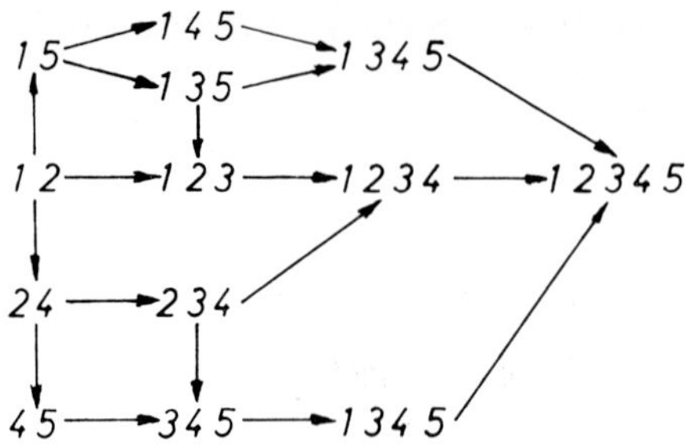

Fig. 12. Invasion-structured community build-up under exploitative competition. Species utilization means are assumed equidistant and numbered so that the outer species are called *1* and *2*, the central species is called *3*, and the intermediate ones *4* and *5*. The resource is characterized by its mean M = 0 and its variance $(\delta/W)^2 = 2.25$. It is assumed that species *1* and *2* with utilization means $D_1 = -1.5$ and $D_2 = 2.5$ are the first ones that colonize the habitat. Thereafter, the different parts of possible invasion events are depicted with respect to their effects on community composition. *Horizontal arrows* indicate an increase in species number and *vertical arrows* a substitution of a previous resident. Eventually, the evolutionary stable composition of all species is reached. (After Loeschcke 1987a).

processes more or less as described in the section on intraspecific competition. In species that are close to the optimum location with respect to the neighboring competitors polymorphism is expected as the resources available to the species are "wide" ($\sigma^2 > W^2$) when the species coexist. This genetic variation may be "disadvantageous" for long-term persistence in the guild. A species possessing genetic variation may adapt (be pushed) to a marginal position of the resource spectrum as a response to an invasion. Such a shift will occur relatively fast if an already established genotype is favored by selection. A species in a marginal location is expected to have a lower population size and therefore has a higher probability of going extinct due to demographic stochasticity, and, in addition, it becomes more vulnerable to fluctuations in the resources.

Acknowledgments. The figures were drawn by Mr. Arno Jensen.

13 Evolution and Mutualism

V. LOESCHCKE AND F.B. CHRISTIANSEN

Introduction

Mutualism is an interaction between species that conveys benefits to both. The benefits include protection, provision of nutrients, pollination, dispersal or some combination of rewards which typically differ between the partners of a mutualistic interaction. Interactions can be between a pair of species, or they can involve associations between a number of partners. The associations can be symbiotic or nonsymbiotic, and the partners can vary in their dependence on the benefits. For some species, the obligate mutualists, the presence of a mutualistic partner is necessary for their persistence. The facultative mutualist is not dependent on mutualistic partners and may participate in a mutualistic relationship when given the opportunity.

What often is referred to as **indirect mutualism** is not considered here, as we feel that the dynamics of this type of species interaction and its evolution normally are primarily determined by a predator-prey relationship (e.g., with Mullerian mimicry) or by competition (e.g., when two competitors affect each other positively through competition with a third competitor) or both types of interaction (e.g., when two consumers have a positive impact on each other through reducing competition and preventing competitive exclusion among two resource species).

Despite the widespread occurrence of mutualism, theoretical ecology has—besides a few early studies by Kostitzin (1934, 1937), and Gause and Witt (1935)—only recently focused attention on this general type of species interaction (reviews in Boucher et al. 1982; Addicot 1984; and contributions to Boucher 1985).

Population Dynamics

The simplest model of mutualistic interactions is reached from the Lotka–Volterra competition equations [Chap. 12, Eq. (7)] by changing the sign in the interaction term. For two species this Lotka–Volterra model for mutualistic interactions becomes

$$\frac{dy_1}{dt} = v_1 y_1 (K_1 - y_1 + \mu_{12} y_2),$$
$$\frac{dy_2}{dt} = v_2 y_2 (K_2 - y_2 + \mu_{21} y_1), \tag{1}$$

where μ_{ij}, for $i = 1, 2$ and $i \neq j$, is the effect of an individual of species j on the growth rate of species i due to the mutualistic interaction.

For μ_{12} and μ_{21} sufficiently large the populations of the Lotka–Volterra model [Eq. (1)] will enter into a stage where the growth rate of both species stay positive while the population size of both species increases towards infinity, so some restraints must be imposed on the model parameters to exclude this biologically unreasonable situation of unbounded growth. These restraints parallel the assumptions of the logistic model of density limited growth, and they express that the per capita benefits eventually decrease with increasing recipient density or increasing benefactor density. In the Lotka–Volterra model we are

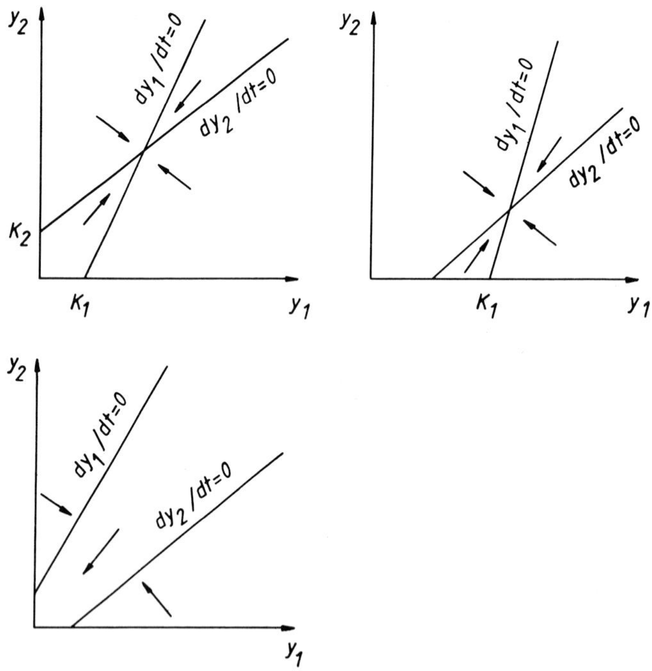

Fig. 1. The linear model [(Eq.(1)] with $\mu_{12}\mu_{21} < 1$. **a** Equilibrium values $dy_1/dt = 0$ and $dy_2/dt = 0$ for two facultative mutualists. A stable equilibrium of coexistence exists. **b** Equilibrium values $dy_1/dt = 0$ and $dy_2/dt = 0$ for one facultative (species 1) and one obligate mutualist (species 2). Stable coexistence is possible as the obligate mutualist can increase in numbers if introduced at the equilibrium population size K_1 of species 1. **c** Equilibrium values $dy_1/dt = 0$ and $dy_2/dt = 0$ for two obligate mutualists. No stable equilibrium allowing coexistence exists

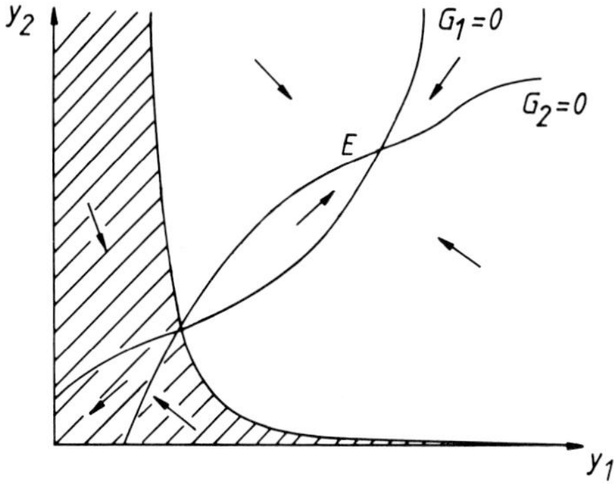

Fig. 2. A general of two facultative mutualists. Density-limited growth guarantees stable coexistence and population trajectories are attracted to the stable point *E*

forced to require that $\mu_{12}\mu_{21} < 1$ to avoid mutualism overriding density-dependent regulation in the model (Christiansen and Fenchel 1977).

For **facultative mutualists** where both species may sustain a population when alone, i.e., where $K_i > 0$ for $i = 1, 2$, the requirement that $\mu_{12}\mu_{21} < 1$ is exactly the condition for existence of a stable coexistence equilibrium in the model (Fig. 1a; Christiansen and Fenchel 1977). If neither species may sustain a population when alone or if both species are obligate mutualists ($K_i < 0$ for $i = 1, 2$; Fig. 1c), then the simple linear model does not have any equilibrium allowing coexistence. If species 2 cannot sustain a population when alone or if it is an obligate mutualist ($K_2 < 0$), then stable coexistence is possible if species 1 is a facultative mutualist that can sustain a population of its own, i.e., $K_1 > 0$, and if $\mu_{21}K_1 + K_2 > 0$, i.e., species 2 should be able to increase in numbers if introduced into a population of species 1 at the equilibrium population size K_1 (Fig. 1b).

The model [Eq. (1)] can be formulated in a more general framework as

$$\frac{dy_i}{dt} = y_i G_i(y_1, y_2) \quad \text{for } i = 1, 2. \tag{2}$$

The functions G_i need to obey some restraints for the model to qualify as a model of mutualistic interactions (Kolmogoroff 1936; Rescigno and Richardson 1967; Christiansen and Fenchel 1977). For facultative mutualists (Fig. 2) the conditions that require density-limited growth again guarantee stable coexistence. For **obligate mutualists** the same restraint on the model guarantees stable coexistence if coexistence is possible (Fig. 3). Mutualistic interactions can be very diverse, and many specific functions G_i have been suggested to fit this diversity. A recent overview on population dynamics models on mutualism is given by Wolin (1985).

Myrmecochory, Seed Predation, and Competition

The mutualistic interaction between a herbaceous plant and its ant seed-disperser may be used as an example of model (1). The benefit to the plant is clearly the dispersal, and the ant obtains food from the seeds, as an external elaisome on the

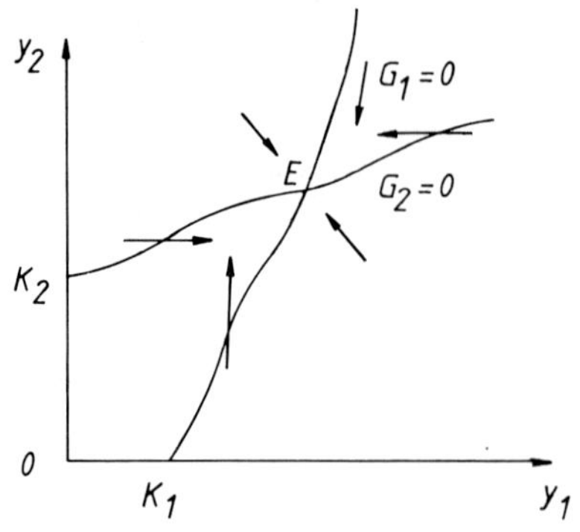

Fig. 3. A general case of two obligate mutualists. The *two curves* correspond to possible equilibrium values $G_1=0$ and $G_2=0$, respectively. Population trajectories *outside the hatched region* are attracted to the stable point *E*, while trajectories *inside the hatched area* are attracted to the origin (which implies extinction)

diaspores contains diglycerides (Marshall et al. 1979; Heithaus et al. (1980). Rodents such as *Peromyscus* predate on the seeds, and to include this into the model we add a third equation for the predator with population size y_3, so we obtain

$$\frac{dy_1}{dt} = v_1 y_1 (K_1 - y_1 + \mu_{12} y_2) \tag{3a}$$

$$\frac{dy_2}{dt} = v_2 y_2 (K_2 - y_2 + \mu_{21} y_1 - \alpha_{23} y_3) \tag{3b}$$

$$\frac{dy_3}{dt} = v_3 y_3 (K_3 - y_3 + \alpha_{32} y_2), \tag{3c}$$

where y_1 is the number of ants, which may be measured by the number of nests, y_2 the number of plants expressed as the number of seeds. The coefficients μ_{12} and μ_{21} describe the mutualistic effects of species 1 and 2 on each other, and α_{23} and α_{32} are the coefficients describing the gain and loss through predation. For a large range of parameter values which seem biologically plausible for the ant-seed-predator system a locally stable equilibrium exists (even if the mutualistic subsystem alone would not guarantee stability, i.e., when $\mu_{ij}\mu_{ji} > 1$).

Let us now introduce **competition** into the basic linear model (1) and use Eqs. (3a) and (3b) combined with a third equation describing the growth of a non-myrmecochore competitor

$$\frac{dy_3}{dt} = v_3 y_3 (K_3 - y_3 - \alpha_{32} y_2) \tag{3}$$

(Heithaus et al. 1980). Competition on myrmecochores would enhance the advantage gained through ant seed-dispersal. However, the analysis of a special case of this model ($r_i = r$ and $K_i = K$) suggests that competition, unlike predation, decreases the range of parameter values that allow stable mutualism (for stability of the whole system stability of the mutualistic subsystem is a requirement).

The Evolution of Mutualism

Competition is the necessary consequence of the exploitation of limited resources, so competition is an important cause of natural selection. Mutualism, on the other hand, has to be understood as a consequence of evolution by natural selection. Mutualism is not a property of the organism, but of its interaction with other species, and the evolution of mutualistic interactions requires prior interaction or contact between the species, e.g., in the form of some kind of symbiosis. Phenotypic variation related to the interaction with other individuals may then be subject to natural selection because of the particular interaction. There is no gene for benefiting the other. What counts is the (inclusive) fitness of a phenotype as evaluated under the interactions the individual is exposed to. Mutualistic interactions may involve a cost in addition to the benefit for the individual due to the benefit given to other individuals. For mutualism to evolve, the benefits to an individual, as measured by the (inclusive) fitness, have to exceed the costs of the individual measured in the same currency. Partners in a mutualistic interaction need not be similar either in the amount of benefit or in the amount of cost involved in the interaction, and often the interaction has to be of the sort: cheap for one and dear to the other (Janzen 1985).

The following model on the evolution of marine symbiosis and mutualism is developed by Roughgarden (1975) as a **cost-benefit model.** It explores one possible route for mutualism to evolve by focusing on one species only. The species are initially in a solitary phase with little or no interaction, and the development passes through a parasitic phase giving a symbiotic relationship that allows evolution towards a mutualistic relationship.

The formation of a symbiotic association requires the individual fitness of a symbiont to exceed that of a solitary or free-living individual. Let w_{ss} be the fitness of the original solitary phenotype within species 1, and let w_a be the fitness of an individual living in association with an individual of species 2 which we may designate the host. Suppose now that a phenotype, the symbiont phenotype, exists, which searches actively for a host. Some cost is involved in the search for a host, so individuals searching for a host have a lowered fitness, w_{sg}, in the case of failing to find a host compared to one who has not tried, w_{ss}, and the increased fitness for those being successful in finding a host is taken as w_a. An individual with the symbiont phenotype therefore is more fit than an individual with a solitary phenotype if

$$plw_a + (1 - pl)w_{sg} > w_{ss}, \tag{4}$$

where p is the probability that an individual with the symbiont phenotype finds a host, and l is the probability that the host survives while the symbiont is associated with it. The benefit may be measured as the increase in fitness $B = w_a - w_{ss}$ of a symbiont phenotype with a host and the cost may be measured by the decrease in fitness of a symbiont phenotype without a (surviving) host $C = w_{ss} - w_{sg}$. Then rearranging Eq. (5), we find that the symbiont is more fit when

$$\frac{pl}{1-pl}B > C. \tag{5}$$

So, if finding a host that survives is very likely, i.e., if pl is close to one, then condition (5) will be fulfilled even for large costs which occur, e.g., when the symbiont has a largely reduced ability to survive in a solitary state. If finding a host is very unlikely or if the association is very detrimental to the host, i.e., if p1 is small, then the benefit of the association has to be considerably larger than the cost for the symbiont to have a higher fitness than the solitary individual.

The evaluation (5) therefore focuses on some crucial factors for the evolution of symbiosis to occur. The host should be easy to find (p large) which requires that the two species are initially close in some kind of maybe fortuitous association. The host should further be likely to survive in the association with the symbiont (1 large) to secure the evolution of a beneficial symbiosis without regard to the cost. Otherwise the benefit provided to the symbiont by the host should be substantial. The evolutionary loss of an existing symbiotic relationship can also occur if one of these factors changes in size such as, e.g., if the host population becomes rare so the symbiont individual has a lower chance of finding a host individual.

Once a symbiotic (here parasitic) association has evolved, the next step is to find the conditions for the evolution of mutualism to occur. The guest (the parasite) which already receives a benefit from the host should evolve to reciprocate this benefit and become beneficial to the host. We assume again, following Roughgarden (1975), that the production of a benefit to the host involves some cost for the parasite with the only possible benefit being a higher probability that the host survives.

The evolution of mutualism from parasitism can proceed in situations where mutualistic variance of the parasite occur and are favored by natural selection. Such a phenotype spends some of the benefit it receives through the association to the advantage of the host, and this "**mutualistic parasite**" may have a higher fitness than the parasite if its cost of the conferred benefit is exceeded by the advantage gained from a higher survival probability or longer life of the host. This possibility of evolution, however, should be juxtaposed to the possibility of deepening the parasitic relationship through higher fitness of variants, who only extract an increasing benefit without giving anything in return. The cost of this behavior will be a lower survival probability or a shorter life of the host. So a symbiotic guest becomes beneficial to the host only if it is at the same time profitable in terms of his own fitness. The course of evolution from parasitism towards mutualism has to pass through a stage formally known as commensalism, i.e., when balance between the cost and benefit to the host from the parasite occurs. However, the evolution occurs in the parasite population, so it is driven by the fitness evaluation of the balance between cost and benefit for given phenotypes of the parasite. In this course of evolution in the guest species, the development of the formal relationship between the two species is quite accidental.

Consider a mutualistic variant in the guest species, and modify our previous model to account for the difference between this variant and a more parasitic variant. Let l_{hm} be the probability of survival of a host associated with a mutualistic guest and l_{hp} the probability of survival of a host associated with a nonmutualistic (parasitic) guest. Further, let w_{am} and w_{ap} be the corresponding fitness parameters of the mutualistic and parasitic guests living in association with a surviving host. From the parasitic-mutualistic contrast between the two variants we find that $l_{hm} > l_{hp}$ and $w_{ap} > w_{am}$, and the condition for the mutualistic phenotype having a higher fitness than the more parasitic guest is then

$$pl_{hm}w_{am} + (1 - pl_{hm})w_{sg} > pl_{hp}w_{ap} + (1 - pl_{hp})w_{sg}, \tag{6}$$

where again w_{sg} is the fitness value of a guest who has failed to find a host or whose dies during the association, and p is the probability of finding a host. The cost of finding a host is as before $C = w_{ss} - w_{sg}$, where w_{ss} is the fitness of the solitary phenotype which was our previous reference for measurements. The sacrifice of the mutualistic guest is $S = w_{ap} - w_{am}$ compared to the parasitic guest, and the total benefit of the symbiotic association to a mutualistic guest compared to a solitary phenotype as before can be expressed by $B_p = w_{ap} - w_{ss}$. In terms of these costs and benefit we obtain condition (6) as

$$(B_p + C)\frac{l_{hm} - l_{hp}}{l_{hm}} > S. \tag{7}$$

The term $B_p + C = w_{ap} - w_{sg}$ is really the benefit of the parasitic phenotype as compared to the mutualistic phenotype. So for mutualism to evolve, this parasitic benefit from the association times the relative improvement in host survival of the mutualistic guest has to exceed the sacrifice of the mutualistic guest. Thus, the evolution of mutualism proceeds if the cost-benefit accounting of the mutualistic guest is more favorable than that of the parasitic guest, where the comparison of the cost and the benefit is done by reducing the cost by the relative improvement in host survival. The benefit term of conditon (7) is large when the guest shows a high dependency on the association with the host, i.e., large costs if failing in finding a host that survives, so the benefit term is maximal for an obligate symbiont.

The benefit term depends only on the parasitic phenotype, so we may consider that as fixed and view condition (7) as a limit to the cost to the symbiont phenotype with which a given improvement in host survival can be bought. So, if the sacrifice of the mutualist phenotype is low compared to the gain in survival, then the mutualistic phenotype will be favored, but if the sacrifice is too high compared to the gain in survival, then the parasitic phenotype will be favored. For a given relation between sacrifice and survival gain the route towards mutualism is eased by increasing dependence of the guest on the host.

The **constraints on the evolutionary path** to mutualism through a parasitic phase become apparent in Roughgarden's model if we compare the condition for the formation of symbiosis [Eq. (15)] and that for the evolution of mutualism

[Eq. (7)]. Condition (5) is fulfilled if absolute host survival is high, while condition (7) depends only on the relative improvement in host survival. This relative improvement is necessarily low if host survival is initially high, so for high host survival rates we expect no symbiotic association to be formed and the parasitic association to remain and become more parasitic. Only for intermediate host survival rates may mutualism evolve, and for low survival rates we expect no association.

Roughgarden (1975) illustrates the use of the above outlined cost-benefit model in explaining qualitative features of the association between damselfish and sea anemones based on data from Vervey (1930) from the Batavia Bay in Indonesia.

Other cost-benefit models on the evolution of mutualism to cover different types of mutualistic interactions have been proposed by Keeler (1981, 1985). These include among others anti-extrafloral nectary mutualism, myrmecochory, mycorrhizal fungi, and mixed feeding flocks. The conclusions reached from these models are in principle close to those obtained by Roughgarden. Costs of investment and chance of failure are inversely related, and for mutualism to evolve and remain as an interaction between species costs and failure rates should be low, but the greater the benefit, the more costs can be tolerated.

In all these models genetics do not enter explicitly, in that they are models for evaluating the fitnesses of discrete phenotypes. This immediately predicts the outcome of one-locus genetic models, but if the interaction involves more complex genetics, an explicit formulation of the conditions for mutualism to evolve becomes necessary.

Acknowledgments: We are grateful to Mr. Arno Jensen for drawing the figures.

Epilog

S. JAIN AND K. WÖHRMANN

Most questions in evolutionary ecology begin with the patterns of species diversity in different kinds of communities with emphasis on adaptive responses of different species to their abiotic and biotic environments; then, evolutionary processes are introduced by allowing intraspecific variation, in physiology, behavior and interspecies interactions; and finally, population genetic and quantitative genetic models or research paradigms are increasingly acknowledged to be useful. Recent interest in the evolution of life histories and mating systems is clearly bringing together various models of evolution at the level of individuals, kinships, local interbreeding units, metapopulations, and related species. Thus, population substructures and the demographic/genetic properties of subunits along with dispersal and gene flow patterns represent one of the most active research areas. Likewise, reproductive systems (sexual, asexual; inbreeding, outbreeding; polygamy, mate choice, parental care) seem to evolve in relation to their genetic consequences (amount of recombination, fate of progeny variances, hybridity or inbreeding depression) as well as (and together with) ecological measures of reproductive success and optimal resource-use tactics.

Ecologists might be less interested at this time in the details of molecular variation, genome organization, novel variation-generating mechanisms, and perhaps even in large-scale assays of phenotypic variation, which begin with neutral theory and serve especially the phylogenetic goals. However, variation and appropiate genetic analysis of some well-chosen traits and a clear understanding of the interplay of evolutionary processes (random drift, mutation, and recombination included) is essential, and origin of species in the framework of coevolutionary outcomes and of niche evolution (both pre- and postspeciation stages) is of ecological interest.

Various contributions in this Volume have addressed many of these topics in an attempt to **promote a population biology synthesis** (or at least, collaborative research). It must be clearly recognized that adaptations to varying environments arise at all levels. Response to stress, for example, may be biochemical, cellular, organismal physiology, or behavioral plasticity before gene frequency changes might result at the population level. On the other hand, genetic systems that regulate mutational and recombinational events might also evolve in relation to the maintenance of variation in populations. It is too simplistic to label population genetic theory as beanbag genetics, devoid of interest in the ecological questions of evironmental variables or phenotypes accounting for adaptive response. The foundations of population genetics were laid by J.B. Haldane, R.A. Fisher and S. Wright, all of whom had exceptional biological insights. Formalism of theory indeed emphasized gene or genotypic changes over time in a highly abstract, albeit elegant, series of steps. Recent discussions whether population genetics is necessary or sufficient for evolutionary ecological studies have, in our view, reaffirmed its original scope. Similarly, a detailed review of various constraints on evolutionary changes has rediscovered the tenets of multivariate quantitative genetics and development, more or less as outlined by S. Wright, K. Mather, and C.H. Waddington among others.

In his book entitled *Molecules and Morphology in Evolution–Conflict or Compromise*, restricted in its scope to the discussions of systematic and phylogenetic questions, C. Patterson (1986) pointed out that molecular and morphological (phenotypic) variation might tell the same story. This is of particular concern in evolutionary ecology. It is not yet clear how robust predictions from phenotypic evolution are which are based on additive polygene models and rely upon changes in means, variances, and covariances. Conspicuous polymorphisms for pigmentation, shape, presence or absence of certain appendages or hairs, etc. along with mimicry, disease, or pest resistance, and simply inherited breeding system variants such as heterostyly or gynodioecy, all these have allowed gene frequency analyses and often succeeded in detecting natural selection, but their studies do not replace quantitative genetic studies of life history evolution. In the case of quantitative traits, phenotypic norms of reaction (representing plasticity or genotype-environment interactions) mask a direct genetic analysis of variation and selection, and the use of molecular markers, often uncertain in selection studies, offers useful methods for analyzing population substructure, mating systems, and gene flow. Clearly, different genetic variants will serve differing purposes, but it is not clear how or whether their findings will be concordant. The concept of fitness eludes any simple definition, which consequently generates many views of selective forces, and accordingly, many different views on the acceptable evidence for selection (Endler 1986); geneticists, ecologists, and biometricians indeed face a formidable challenge here.

One important goal of the population biologist is to understand the **dynamics of plant and animal populations** in terms of their growth and resource use. Many impressive theoretical developments in ecology lie in this area, which deals with the intrinsic rate of growth (r), carrying capacity (K), and environmental heterogeneity. Southwood (1976) attempted a classification of species in terms of a three-dimensional synoptic model in which evolutionary shifts between r-selection and K-selection was made dependent on a parameter of habitat duration in proportion to generation time. Emlen (1984) reviewed and also contributed a number of original ideas on the genetic and evolutionary aspects of population regulation. Within single species, this primarily depends upon the patterns of recruitment dynamics and uncertainty of environments, but in general, population dynamics depends on multispecies interactions where natural enemies (disease pattern, pests, parasites) control population numbers. Life history traits and long-term numerical analyses of population sizes together describe the processes in which monitoring of genetic variation will also be needed to detect coevolutionary changes. It is interesting to note that studies on rare and endangered taxa on one hand and the traditional interests in pest control or disease containment on the other continue to emphasize the causes of population extinction, demographic and environmental stochasticity along with some catastrophes (Lande 1988; May 1985). Genetic and evolutionary factors are yet to be integrated into these studies, except in a few examples as noted by Jayakar and Zonta (see Chap. 11; May 1985; Emlen 1984).

Both theory and experimental evidence have generated a great deal of interest in **the problem of scales** (spatial, in particular) in dealing with measuring of variation, dispersal, interaction dynamics, stability of population numbers, and community dynamics. For example, Roughgarden et al. (1988) developed a model for complex life cycles in the rocky intertidal species in which larval migration and adult stage competition together help predict population fluctuations. They suggest that a "new category of discoveries concerning scale and coupling among scales will be emerging in ecology. The same can be safely predicted from the evolutionary models discussed by Barton and Clark (see Chap. 4). Patch dynamics, nonequilibrium descriptions of community structure, local versus global stability analysis of coevolving systems, all these refer to the importance of spatial scales which raise interesting queries about the methodological choices in population studies. The time scales of micro- and macroevolution have often been treated separately, but there is considerable diversity of opinions about the ecological and population genetic time-frames for describing population dynamics. Once again, we perceive the utility of selecting an ecosystem or species group for detailed and long-term research and the need for examining several spatial and temporal scales of environmental heterogeneity and adaptive responses.

Another important development in **research methodology** is well illustrated by Hedrick's treatment of mating systems (see Chap. 3). Clearly stating the assumptions and parameters in a simple model allows the first (not necessarily trivial) few crude estimates of a parameter, say per cent outcrossing. Both in theory and in experimentation, then, one can remove assumptions one by one in order to obtain more general (biologically more realistic) estimates of several parameters jointly. Statistical and computational requirements may soon exceed an experimenter's resources, but such estimation processes are highly valuable. A clear example of nonquantitative and somewhat misleading treatments of breeding systems is found in Richards' (1986) review, which is otherwise useful narrative natural history. Many important advances in the theory of evolution of dioecy, self-incompatibility, gamete competition, assortative mating, paternity analysis, pollinators' behavior, evolution of floral characters, and so on, have been made using the combined power of estimation procedures, simple genetic and ecological experiments, and theoretical constructs of well-defined models (Clegg and Epperson 1988, for example). Behavioral ecology of mating systems in plants and animals show rapid progress with these advances in population biology.

As noted in several chapters, even the simplest building blocks of population genetic theory help us discover different kinds of selection and the potential importance of life history analyses. Models of two or more interacting species, especially those involving host-parasite or predator-prey systems, and simple genetic models are readily understood, and extending them in polygenic terms helps us to analyze competitive outcomes. As we accumulate more empirical information on life history variables and their modes of inheritance, character displacement, or convergence among coexisting species, and the so-called genetic feedback mechanism of frequency-dependent, coevolutionary systems (Pimentel 1968; Krebs and Myers 1974) will be better understood. Several epidemiological models of host-parasite systems have already shown the role of theoretical advances. Similar examples of notable theoretical contributions are found in the models of Gillespie (1978, polymorphisms under heterogeneous environments), Slatkin (1985, long-distance gene flow measures and some insight into population bottlenecks), Levin et al. (1985, strategic analysis of dispersal in a metapopulation), Karlin and Lessard (1986, importance of genetic determination models in the evolution of sex ratios), and Holsinger (1986, evolution of mixed selfing and random-mating systems under population subdivision). These are just a few of the many achievements of theory, many of them not so intuitively obvious, as noted by Ewens (1979) in his elegant review of theoretical population genetics.

One increasingly popular approach in population biology synthesis is to focus on a given geographic area or community, and here island biology has often proved attractive. **Island models of biogeography**, species level colonization-extinction processes, and island models of speciation/geographic differentiation provide a natural confluence of many interests. A special issue of Trends in Ecology and Evolution (1987) dealt with the Hawaiian islands, and compared certain ecological and evolutionary features of vertebrates, invertebrates, and plants. Here, ancestral relations and time-scales of evolutionary divergence are known, species distributions are well recorded, and speciation models are testable in various groups. However, population level information on numbers, variation, dispersal, and competition is generally inadequate. The role of sexual selection models in speciation became a subject of major polemic here again, and population studies and some manipulative experiments are needed to complement phylogenetic evidence. A recent treatment of Galapagos island finches by Grant (1986a) attempts to document the ongoing selection in terms of seasonal variation in food resources and evidence for interspecies competition, whereas genetic variation and population structure need further work.

We may argue that there is order in the living world which we might see in the way genomes are organized, mutation rates are regulated, metapopulations adapt to spatial heterogeneity in their environment, evolutionary species systems interact, and overall species diversity patterns evolve. We might not yet know how to integrate these hierachially ordered phenomena together into **central dogmas**, one or a few grand principles, and we may only guess the operation of optimality ideas in biochemical or biomechanic design terms. We also have no certain feeling about the chaotic versus nonchaotic nature of population and community changes. In fact, we seem to fumble with continued skepticism about the resolving power of our research methods, about the merging of disciplines, about the role of mathematical thinking, and so on. But we can admire the growth of our field in so many exciting ways and reaffirm the potential applications of this knowledge in solving evolutionary puzzles as well as economic problems in applied biology.

In an elegant synthesis, Anderson (1983) reviewed several topics in population biology: use of models, ecological genetics and selection, quantitative variation, and newly developing techniques, and concluded that "the original promise of a single unified biology of populations has yet to be realized", and that "large parts of population genetics and population ecology remain untouched by the other." This volume hopefully defies the latter, and denies the former. We should not expect a simple solution, a single puzzle that just needs putting together, or a large elephantine model waiting to be discovered by some miraculous recovery of vision by blind men. Levins (1968), among others, noted the complexity of an interdisciplinary flow chart of information, the need for dynamically sufficient parametrization, compromises in modeling, the search for comparative patterns, and the need for a great deal more manipulative experiments. We expect numerous patterns (puzzles) at different scales and for different groups of organisms, each one large and changing in the number and shapes of it parts over time (evolutionary). The parts become smaller (and thus more numerous) as we learn details, and imagine parts with randomly varying boundaries (due to stochasticity and phenotypic plasticity), and then, researchers focusing on how parts came into being as much as how (descriptively) they fit together. Clearly, this is a formidable challenge. Most evolutionary problems, except for cases in which molecular phylogenies can be established (Gottlieb and Jain 1988), will have a probabilistic array of "solutions", based on a rich combination of theory, field data, laboratory experiments, and large overviews (surveys) using comparisons (see Stebbins 1988 for four kinds of comparisons). Temporal trends alone would allow reconstruction of puzzle pictures along the time axis.

We make three comments in concluding this epilog: First, we find numerous developments in our field exciting and valuable in many ways. As several important early developments were also rooted in applied biology (e.g., pest control, fisheries, and forestry management), the number of applications is growing in the fields of medicine, agriculture, conservation biology, and resource management. All of the chapters here are relevant to such an applied ecological approach.

Second, it appears that our field occasionally suffers unduly from "trendiness" imposed by short-sighted problem definitions, funding matters, and by "pseudo-controversies" (e.g., importance of group selection, density-dependent, or density-independent population regulation) which arise from hypothesis-testing euphoria (and strong personalized debates) rather than careful statement of ideas and research findings. We have avoided such scenarios here. Clear definitions, good theoretical background, more emphasis on statistical rigor in both testing and estimation procedures, and more succinct and critical literature reviews help in these matters.

Finally, while several top-ranking evolutionists have proclaimed that, as we learn more about the genetics of our own nature (sociobiology, cultural evolution) and evolutionary trees, we should be able to choose a system of values on an objective basis, recent events of human history suggest the need for the rather humble realization that the ultimate wisdom will not come from science and relentness Darwinism, but from faith and wise choices based on the understanding of diversity, peaceful coexistence, and environmental ethics, all of which can be learned from population biology.

References

Ablowitz M, Zeppetella A (1979) Explicit solution of Fisher's equation for a special wave speed. Bull Math Biol 41: 835–840

Abrahamson WG (ed) (1989) Plant animal interactions. McGraw-Hill, New York

Abrams PA (1986) Character displacement and niche shift analyzed using consumer-resource models of competition. Theor Popul Biol 29: 107–160

Addicott JF (1984) Mutualistic interactions in populations and community processes. In: Price PW, Slobokchikoff CN, Gaud WS (eds) A new ecology: Novel approaches to interactive systems. Wiley, New York, pp 437–455

Ahearn JN, Baimai V (1987) Cytogenetic study of three closely related species of Hawaiian *Drosophila*. Genome 29: 47–57

Aide TM (1986) The influence of wind and animal pollination on variation in outcrossing rates. Evolution 40: 434–435

Akin E (1979) The geometry of population genetics. (Lecture Notes in Biomathematics 31) Springer, Berlin Heidelberg New York

Alexander R MN (1980) Optimum walking techniques for quadrupeds and bipeds. J Zool (Lond) 192: 97–117

Alexander R MN (1982) Optima for animals. Arnold, London

Allard R W (1988) Genetic changes associated with the evolution of adapatedness in cultivated plants and their wild progenitors. J Hered 79: 225–238

Allard RW, Jain SK, Workman PL (1968) The genetics of inbreeding populations. Adv Genet 14: 55–131

Allard RW, Babbel GR, Clegg MT, Kahler AL (1972a) Evidence for coadapatation in *Avena barbata*. Proc Natl Acad Sci 69: 3043–3048

Allard RW, Kahler AL, Weir BS (1972b) The effect of selection on esterase allozymes in a barley population. Genetics 72: 489–503

Allison AC (1982) Coevolution between hosts and infectious disease agents and its effects on virulence. In: Anderson RM, May RM (eds) Population biology of infectious diseases. Springer, Berlin Heidelberg New York, pp 245–267

Anderson RM (1979) The influence of parasitic infection on the dynamics of host population growth. In: Anderson RM, Turner BD (eds) Population dynamics. Blackwell, Oxford, pp 245–282

Anderson WW (1971) Genetic equilibrium and population growth under density-regulated selection. Am Nat 105: 489–498

Anderson WW (1983) Achieving synthesis in population biology In: King CE, Dawson PS (eds) Population biology: retrospect and prospect. Columbia University Press, New York, pp 189–205

Andrewartha HG, Birch LC (1954) The distribution and abundance of animals. In: Baker HG, Stebbins GL (eds) Genetics of colonizing species. Academic Press, London

Angus RA (1980) Geographical dispersal and clonal diversity in unisexual fish populations. Am Nat 115: 531–550

Annest JL, Templeton AR (1978) Genetic recombination and clonal selection in *Drosophila mercatorum*. Genetics 89: 193–210

Antonovics J, Ellstrand NC (1984) Experimental studies on the evolutionary significance of sexual reproduction. I. A test of the frequency-dependent selection hypothesis. Evolution 38: 103–115

Antonovics J, Bradshaw AD, Turner RG (1971) Heavy metal tolerance in plants. Adv Ecol Res 7: 1–85

Anxolabehere D, Nouaud D, Periquet G, Ronsseray S (1986) Evolution des potentialités dysnénésiques du système P-M dans des populations experimentales mixtes P, Q, M et M'de *Drosophila melanogaster*. Genetica 69: 81–95

Aquadro CF, Dene SF, Bland MM, Langley CH, Laurie-Ahlberg CC (1986) Molecular population genetics of the alcohol dehydrogenase region of *Drosophila melanogaster*. Genetics 114: 1165–1190

Asher JH Jr, Nace GW (1971) The genetic structure and evolutionary fate of parthenogenetic amphibian populations as determined by Markovian analysis. Am Zool 11: 381–398

Asmussen MA, Arnold J, Avise JC (1987) Definition and properties of disequilibrium statistics for associations between nuclear and cytoplasmic genotypes. Genetics 115: 755–768

Atchley WR (1977a) Evolutionary consequences of pathenogenesis: Evidence from the *Warramaba virgo* complex. Proc Natt Acad Sci USA 74: 1130–1134

Atchley WR (1977b) Biological variability in the parthenogenetic grasshopper *Warramaba virgo* (Key) and its sexual relatives. I. The eastern Australian populations. Evolution 31: 782–799

Atchely WR (1978) Biological variability in the parthenogenetic grasshopper *Warramaba virgo* (Key) and its sexual relatives. II. The western Australian taxa. Evolution 32: 375–388

Avery OT, MacLeod CM, McCarty M (1944) studies on the chemical nature of the substance inducing transformation of pneumococcal types. Induction of transformation by a deoxyribonucleic acid fraction isolated from pneumococcus Type III. J Exp Med 79: 137-158

Avise JC (1986) Mitochondrial DNA and the evolutionary genetics of higher animals. Philos Trans R Soc Lond Biol Sci 312: 325–342

Avise JC, Arnold J, Ball RM, Bermingham E, Lamb T, Neigl JE, Saunders NC (1987) Intraspecific phylogeography: the mitochondrial DNA bridge between population genetics and systematics. Annu Rev Ecol Syst 18: 489–522

Ayala FJ (1975) Genetic differentiation during the speciation process. In: Dobzhansky T, Hecht MK, Steere WC (eds) Evolutionary biology, Vol 8 Plenum, New York, pp 1–78

Ayala FJ, Dobzhansky T (1974) A new subspecies of *Drosophila pseudoobscura*. Pan Pac Entomol 50: 211–19

Ayala FJ, Tracey ML, Hedgecock D, Richmond RC (1974) Genetic differentiation during the speciation process in *Drosophila*. Evolution 28: 576–592

Baba-Aissa F, Solingnac M, Dennebouy N, David JR (1988) Mitochondrial DNA variability in *Drosophila simulans:* quasi absence of polymorphism within each of the three cytoplasmic races. Heredity 61: 419–426

Bailey NTS (1951) Testing the solubility of maximum likelihood equations in the routine application of scoring methods. Biometrics 7: 268–274

Baker AEM (1986) Does group selection occur in commensal house mice (*Mus domesticus*) In: Stenseth NC, Swingland I (eds) Living in a patchy environment. Oxford University Press, Oxford

Baker HG (1965) Characteristics and modes of origin of weeds. In: Baker HG, Stebbins GL (eds) The genetic of colonizing species. Academic Press, London, pp 147–172

Baker HG (1974) The evolution of weeds. Annu Rev Ecol Syst 5: 1–24

Baker HG, Stebbins GL (eds) (1965) The genetics of colonizing species. Academic press, London

Baker RJ (1981) Chromosome flow between chromosomally characterized taxa of a volant mammal, *Uroderma bilobatum* (Chiroptera: Phyllostomatidae). Evolution 35: 296–305

Baker RJ, Bickham JW (1986) Speciation by monobrachial centric fusions. Proc Natl Acad Sci USA 83: 8245–8248

Bakker RT (1983) The deer flees, the wolf pursues: incongruencies in predator-prey coevolution. In: Futuyma DJ, Slatkin M (eds) Coevolution. Sinauer, Sunderland, Massachusetts, pp 350–382

Barber HN (1965) Selection in natural populations. Heredity 20: 552–572
Bar-Cohen A, Alpern G, Bar-Anan R (1978) Progeny testing and selecting Italian queens for brood area and honey production. Apidologie 9: 95–100
Barigozzi D (1982) Mechanisms of speciation. Liss, New York
Barker JSF, East PD, Weir BS (1986) Temporal and microgeographic variation in allozyme frequencies in a natural population of *Drosophila buzzatii*. Genetic 112: 577–611
Barnard CJ, Brown CAJ (1985) Risk-sensitive foraging in common shrews (*Sorex araneus* L.). Behav Ecol Sociobiol 16:161–164
Barrett JA (1980) Pathogen evolution in multilines and variety mixtures. Z Pflanzenkr Pflanzenschutz 87: 383–396
Barton NH (1979) The dynamics of hybrid zones. Heredity 43: 341–359
Barton NH (1982) The structure of the hybrid zone is *Uroderma bilobatum*. Evolution 36: 863–866
Barton NH (1983) Multilocus clines. Evolution 37:454–471
Barton NH (1986) The effects of linkage and density-dependent regulation on gene flow. Heredity 57: 415–426
Barton NH (1987) The probability of establishment of an advantageous mutation in a subdivided population. Genet Res 50: 35–40
Barton NH (1988a) Drift in a continuous population. II local population structure. (submitted to Genetics)
Barton NH (1988b) Spatial patterns and population structure. I. Theory (submitted to Genetics)
Barton NH, Charlesworth B (1984) Genetic revolutions, founder effects, and speciation. Annu Rev Ecol Syst 15: 133–164
Barton NH, Hewitt GM (1981) A chromosomal cline in the grasshopper *Podisma pedestris*. Evolution 35: 1008–1018
Barton NH, Hewitt GM (1985) Analysis of hybrid zones. Annu Rev Ecol Syst 16: 113–148
Barton NH, Post RJ (1986) Sib competition and the advantage of mixed families. J Theor Biol 120: 381–387
Barton NH, Slatkin M (1986) A quasi-equilibrium theory of the distribution of rare alleles in a subdivided population. Heredity 56: 409–416
Bazykin AD (1969) Hypothetical mechanism of speciation. Evolution 23: 685–687.
Beardmore JA (1961) Diurnal temperature fluctuations and genetic variance in *Drosophila melanogaster*. Nature 189: 162–163
Beatty J (1980) Optimal-design models and the strategy of model building in evolutionary biology. Philos Sci 47: 532–561
Beckwith J, Rossow P (1974) Analysis of genetic regulatory mechanisms. Annu Rev Genet 8: 1–13
Begon M, Harper JL, Townsend CR (1986) Ecology. Blackwell, Oxford
Bell G (1980) The costs of reproduction and their consequences. Am Nat 166: 45–76
Bell G (1982) The masterpiece of nature: the evolution and genetics of sexuality. University Calif Press, Berkeley.
Bell G (1983) Measuring the cost of reproduction III. The correlation structure of the early life history of *Daphnia pulex*. Oecologia (Berl) 60: 378–383
Bell G (1984a) Measuring the cost of reproduction I. The correlation structure of the life table of a plankton rotifer. Evolution 38: 300–313
Bell G (1984b) Measuring the cost of reproduction. II. The correlation structure of the life tables of five freshwater invertebrates. Evolution 38: 314–326.
Bell G (1985) Two theories of sex and variation. Experientia (Basel) 41: 1235–1245
Bellman RE (1957) Dynamic programming. Princeton University Press, Princeton
Belovsky GE (1978) Diet optimization in a generalist herbivore: the moose. Theor Popul Biol 14: 105–134
Belovsky GE (1984a) Herbivore optimal foraging: A comparative test of three models. Am Nat 124: 97–115
Belovsky GE (1984b) Snowshoe hare optimal foraging and its implications for population dynamics. Theor Popul Biol 25: 235–264

Bennett JH, Binet FE (1956) Association between Mendelian factors with mixed selfing and random mating. Heredity 44: 5–13

Berenbaum MR, Zangerl AR, Nitao JK (1986) Constraints on chemical coevolution: Wild parsnips and the parsnip webworm. Evolution 40: 1215–1228

Berthold P (1988a) The control of migration in european warblers. Pro IXXth Int Orn Congr, Ottawa 1986 pp 416–423

Berthold P (1988b) Evolutionary aspects of migratory behavior in european warblers. J Evol Biol 1: 195–210

Berthold P, Schlenker R (1988) Mönchsgrasmücke. In: Glutz von Blotzheim (ed) Handbuch der Vögel Mitteleuropas, Bd 11. Verlag, Wiesbaden (in press)

Berven KA, Gill DE (1983) Interpreting geographic variation in life history traits. Am Zool 23: 85–97

Bierzychudek P (1985) Patterns in plant parthenogenesis. Experientia (Basel) 41: 1255–1264

Bijlsma PW, Allard RW, Kahler AL (1986) Non-random mating in an open-pollinated maize population. Genetics 112: 669–680

Birky CW, Maruyama T, Fuerst P (1983) An approach to population and evolutionary genetic theory for genes in mitochondria and chloroplasts, and some results. Genetics 103; 513–527

Blackman RL (1972) The inheritance of life-cycle differences in *Myzus persicae*. Bull Entomol Res 62: 281–294

Blakley, N (1981) Life history significance of size-triggered metamorphosis in milkweed bugs (*Oncopeltus*) Ecology 62: 57–64

Boag PT, Noordwijk AJ van (1987) Quantitative genetics. In Cooke F, Buckley PA (eds) Avian genetics. Academic Press, London, PP 45–78

Bobisud LE (1978) Optimal time of appearance of mimics. Am Nat 112: 962–965

Bobisud LI, Portratz CJ (1976) One-trial versus multitrial learning for a predator encountering a model-mimic system. Am Nat 110: 121–128

Bock IR (1984) Interspecific hybridization in the genus *Drosophila*. In: Hecht MK, Wallace B, Prance GT (eds) Evolutionary biology, vol 18. Plenum, New York, pp 41–70

Bodmer WF (1968) Demographic approaches to the measurement of differential selection in human populations. Proc Natl Acad Sci USA 59: 690–699

Bodmer WF, Cavalli-Sforza LL (1968) A migration matrix model for the study of random genetic drift. Genetics 59: 565–592

Bodmer WF, Felsenstein J (1967) Linkage and selection: Theoretical analysis of the deterministic two-locus random mating model. Genetics 57: 237–265

Boer PJ Den (1986), Exclusion, competition or coexistence? A question of testing the right hypothesis. Z Zool Syst Evolutionsforsch 23: 259–274

Bohren BB, Hill WG, Robertson A (1966) Some observations on asymmetrical correlated responses. Genet Res 7: 44–57

Borman SA, Levitt PR (1980) The genetics of altruism. Academic Press, London

Boucher DH (ed) (1985) The biology of mutualism. Croom Helm, London

Boucher DH, James S, Keeler KH (1982) The ecology of mutualism. Annu Rev Ecol 13: 315–348

Bourke AFG (1988) Worker reproduction in the higher eusocial hymenoptera. Q Rev Biol 63: 291–311

Bradshaw AD (1965) Evolutionary significance of phenotypic plasticity in plants. Adv Genet 13: 115–155

Bradshaw AD (1984) The importance of evolutionary ideas in ecology and vice versa. In: Shorrocks B (ed) Evolutionary Ecology. Blackwell, Oxford, pp. 1–25

Bradshaw WE (1986) Pervasive theories in insect life cycle strategies. In: Taylor F, Karban R (eds) The evolution of insect life cycles, Springer, Berlin Heidelberg New York, pp 261–275

Breese EL (1966) Selection for differing degrees of outbreeding in *Nicotiana rustica*. Ann Bot NS 23: 331–334

Bremermann HJ (1985) The adaptive significance of sexuality. Experimentia (Basel) 41: 1245–1254

Briggs D, Walters SM (1984) Plant variation and evolution. 2nd Edition. Cambridge University Press, Cambridge

Britton RJ, Davidson EH (1969) Gene regulation for higher cells: A theory. Science 165: 349–357
Britton RJ, Davidson EH (1971) Repetitive and non-repetitive DNA sequences and a speculation on the origins of evolutionary novelty. Q Rev Biol 42: 111–138
Brockelman WY (1975) Competition, the fitness of offspring and optimal clutch size. Am Nat 107: 677–99
Brooks DR, Glen DR (1982) Pinworms and primates: a case study in coevolution. Proc Helm Soc Wash 49: 76–85
Brown AHD (1979) Enzyme polymorphism in plant populations. Theor Popul Biol 15: 1–42
Brown AHD (1984) Multilocus organization of plant populations. In: Wöhrmann K, Loeschcke V (eds) Population biology and evolution, Springer, Berlin Heidelberg New York, pp 160–169
Brown AHD, Albrecht L (1980) Variable outcrossing and the genetic structure of predominantly self-pollinated species. J Theor Biol 32: 591–606
Brown AHD, Allard RW (1970) Estimation of the mating system in open-pollinated maize populations using isozyme polymorphisms. Genetics 66: 133–145
Brown AHD, Feldman MW (1981) Population structure of multilocus associations. Proc Natl Acad Sci USA 78: 5913–5916
Brown AHD, Marshall DR (1981) Evolutionary changes accompanying colonization in plants. Proc Intern Congr Syst Evol Biol pp 151–163
Brown AHD, Barrett SCH, Moran GF (1985) Mating system estimation in forest trees: models, methods, and meanings. In: Gregorius H-K (ed) Population genetics in forestry. Lecture Notes in Biomathematics 60. Springer Berlin Heidelberg, New York, p 32
Brown AHD, Feldman MW, Nevo E (1980) Multilocus structure of natural populations of *Hordeum spontaneum*. Genetics 96: 523–536
Brown CR, Jain SK (1979) Reproductive system and pattern of genetic variation in two *Limnanthes* species. Theor Appl Genet 54: 181–190
Brown JS, Venable DL (1986) Evolutionary ecology of seed-bank annuals in temporally varying environments. Am nat 127: 31–47
Brown KS (1981) The biology of *Heliconius* and related genera. Ann Rev Entomol 26: 427–456
Brown WM, Wright JW (1979) Mitochondrial DNA analysis and the origin and relative age of parthenogenetic lizards (genus *Cnemidophorus*). Science 203: 1247–1249
Bryant EH, McCommas SA, Combs LM (1986a) Morphometric differentiation among experimental lines of the housefly in relation to a boottleneck. Genetics 114: 1213–1223
Bryant EH, McCommas SA, Combs LM (1986b) The effect of an experimental bottleneck upon additive variation in the housefly. Genetics 114: 1191–1211
Bürger R (1988) The maintenance of genetic variation: A functional analytic approach to quantitative genetic models. In: Jong G de (ed) Population genetics and evolution. Springer, Berlin Heidelberg New York, pp. 63–72
Bulger AJ, Schultz RJ (1979) Heterosis and interclonal variation in thermal tolerance in unisexual fish. Evolution 33: 848–859
Bulger AJ, Schultz RJ (1982) Origins of thermal adaptation in northern vs southern populations of a unisexual hybrid fish. Evolution 36: 1041–1050
Blumer MA (1980c) The sib competition model for the maintenance of sex and recombination. J Theor Biol 87: 671–702
Bulmer MG (1972) The genetic variability of polygenic characters under optimizing selection, mutation and drift. Genet Res 19: 17–25
Bulmer MG (1974) Density-dependent selection and character displacement. Am Nat 108: 45–58.
Bulmer MG (1980a) The sib competition model for the evolution of sex and recombination. J Theor Biol 82: 335–345
Bulmer MG (1980b) The mathematical theory of quantitative genetics. Oxford University Press, Oxford
Bulmer MG (1983) Sex ratio evolution in social hymenoptera under worker control with behavioural dominance. Am Nat 121: 899–902

Bundgaard J, Christiansen FB (1972) Dynamics of polymorphisms I. Selection components in an experimental population of *Drosophila melanogaster*. Genetics 71: 439–460
Bundgaard J, Christiansen FB (1976) Dynamics of polymorphisms. II. Patterns of selection in *Drosophila*: an experimental survey. Unpublished manuscript, see Hedrick (1983)
Burdon JJ (1987) Diseases and plant population biology. Cambridge University Press, Cambridge
Burr HE, Schimke RT (1980) Intragenomic DNA sequence homologies in the chicken and other members of the class Aves: DNA re-association under reduced stringency conditions. J Mol Evol 15: 391–307
Bush GL (1975) Modes of animal speciation. Annu Rev Ecol Syst 6: 339–64
Bush GL (1981) Stasipatric speciation and rapid evolution in animals. In: Atchley WR, Woodruff DS (eds) Evolution and speciation, essays in honour of MJD White. Cambridge University Press, New York
Bush GL, Case SM, Wilson AC, Patton JL (1977) Rapid speciation and chromosomal evolution in mammals. Proc Natl Acad Sci USA 74: 3942–3946
Buss LW (1987) The evolution of individuality. Princeton University Press, Princeton
Caraco T, Martindale S, Whitham TS (1980) An empirical demonstration of risk-sensitive foraging preferences. Anim Behav 28: 820–830
Carson HL (1975) The genetics of speciation at the diploid level. Am Nat 109: 83–92
Carson HL (1986) Sexual selection and speciation. In: Karlin S, Nevo E (ed) Evolutionary processes and theory. Academic Press, London, pp 391–409
Carson HL (1987) The genetic system, the deme and the origin of species. Annu Rev Genet 21: 405–423
Carson HL, Kaneshiro KY (1976) *Drosophila* of Hawaii: Systematics and ecological genetics. Annu Rev Ecol Syst 7: 311–345
Carson HL, Stalker HD (1968) Polytene chromosome relationships in Hawaiian species of *Drosophila*. I. The grimshawi subgroup. Uni Tex Publ 6818: 335–354
Carson HL, Templeton AR (1984) Genetic revolution in relation to speciation phenomenon: the founding of new populations. Annu Rev Ecol Syst 15: 97–131
Case ML, Taper TJ (1986) On the coexistence and coevolution of asexual and sexual competitors. Evolution 40: 366–387
Case T (1982) Coevolution in resource-limited competition communities. Theor Popul Biol 21: 69–91
Caswell H (1980) On the equivalence of maximizing reproductive value and maximizing fitness. Ecology 61: 19–24
Caswell H (1981) Reply to comments by Yodzis and Schaffer. Ecology 62: 1685
Cavalli-Sforza LL, Bodmer WF (1971) The genetics of human populations. Freeman, San Francisco, California
Cavalli-Sforza LL, Feldman MW (1978) Darwinian selection and "altruism". Theor Popul Biol 14: 268–280
Charlesworth B (1978) Some models of the evolution of altruistic behavior between siblings. J Theor Biol 72: 297–319
Charlesworth B (1979) A note on the evolution of altruism in structured demes. Am Nat 113: 601–605
Charlesworth B (1980) Evolution in age-structured populations. Cambridge University Press, Cambridge
Charlesworth B (1985) The population genetics of transposable elements. In: Ohta T, Aoki K (eds) Population genetics and molecular evolution. Jap Sci Soc Press Tokyo, Springer Berlin Heidelberg New York 503 pp
Charlesworth B (1987) The heritability of fitness. In: Bradbury JW, Anderson MB (eds) Sexual selection: testing the alternatives. Wiley, New York, pp 21–40
Charlesworth B, Charlesworth D (1978) A model for the evolution of dioecy and gynodioecy. Am Nat 112: 975–997
Charlesworth B, Léon JA (1976) The relation of reproductive effort to age. Am Nat 110: 449–459
Charlesworth B, Lande R, Slatkin M (1982) A neo-Darwinian commentary on macroevolution. Evolution 36: 474–98

Charlesworth D, Charlesworth B (1987) Inbreeding depression and its evolutionary consequences. Annu Rev Ecol Syst 18: 237–268
Charlesworth D, Charlesworth B, Strobeck C (1979) Selection for recombination in partially self-fertilizing populations. Genetics 92: 237–244
Charnov EL (1982) The theory of sex allocation. Princeton University Press, Princeton
Charnov EL, Krebs JR (1974) On clutch size and fitness. Ibis 116: 217–219
Charnov EL, Mitchell-Olds T (1981) A note on sex and life histories. Am Nat 117: 814–818
Charnov EL, Skinner SW (1984) Evolution of host selection and clutch size in parasitoid wasps. Fla Entomol 67: 5–21
Cheliak WM, Morgan K, Strobeck C, Yeh FCH, Dancek BP (1983) Estimation of mating system parameters in plant populations using the EM algorithm. Theor Appl Genet 65: 157–161
Cheliak WM, Dancik BP, Morgan K, Yeh FCH, Strobeck C (1985) Temporal variation of the mating system in a natural population of jack pine. Genetics 109: 569–584
Chiarello N, Roughgarden J (1984) Storage allocation in seasonal races of an annual plant: optimal vs. actual allocation. Ecology 65: 1290–1301
Choudhary M, Singh RS (1987a) A comprehensive study of genic variation in natural populations of *Drosophila melanogaster*. III. Variation in genetic structure and their causes between *Drosophila melanogaster* and its sibling species *Drosophila simulans*. Genetics 117: 697–710.
Choudhary M, Singh RS (1987b) Historical effective size and the level of genetic diversity in *Drosophila melanogaster* and *D. pseudoobscura*. Biochem Genet 25: 41–51
Choudhary M, Coulthart MB, Singh Rs (1988) Genetic divergence and models of species formation in *Drosophila*. Science (in press)
Christensen B (1977) Habitat preferences among amylase genotypes in *Asellus aquaticus* (Isopoda: Crustacea) Hereditas 87: 21–26
Christensen B (1979) Differential distribution of genetic variants in triploid parthenogenetic *Trichoniscus pusillus* (Isopoda: Crustacea) in a heterogeneous environment. Hereditas 91: 179–182
Christensen B (1980) Constant differential distribution of genetic variants in parthenogenetic forms of *Lumbricillus lineatus* (Enchytraeidae, Oligochaeta) in a heterogeneous environment. Hereditas 92: 193–198
Christiansen FB (1975) Hard and soft selection in a subdivided population. Am Nat 109: 11–16
Christiansen FB (1980) Studies on selection components in natural populations using population samples of mother-offspring combinations. Hereditas 92: 199–203
Christiansen FB (1984a) Evolution in a temporally varying environment: density and composition dependent genotypic fitness. In: Wöhrmann K, Loeschcke V (eds) Population biology and evolution. Springer, Berlin Heidelberg New York, pp. 115–124
Christiansen FB (1984b) The definition and measurement of fitness. In: Shorrocks B (ed) Evolutionary ecology. Blackwell, Oxford, pp. 65–79
Christiansen FB (1985) Selection and population regulation with habitat variation. Am Nat 126: 418–429
Christiansen FB (1988) Epistasis in the multiple locus symmetric viability model. J Math Biol (in press)
Christiansen FB, Feldman MW (1975a) Selection in complex genetic systems. IV. Multiple alleles and interactions between loci. J Math Biol 2: 179–204
Christiansen FB, Feldman MW (1975b) Subdivided populations—review of one and two locus deterministic theory. Theor Popul Biol 7: 13–38
Christiansen FB, Feldman MW (1983) Selection in complex genetic systems. V. Some properties of mixed selfing and random mating with two loci. Theor Popul Biol 23: 257–272
Christiansen FB, Fenchel T (1977) Theories of populations in biological communities. (Ecol Studies 20) Springer, Berlin Heidelberg New York
Christiansen FB, Frydenberg O (1973) Selection component analysis of natural polymorphisms using population samples including mother-offspring combinations. Theor Popul Biol 4: 425–445
Christiansen FB, Frydenberg O (1976) Selection component analysis of natural polymorphisms

using mother-offspring samples of successive cohorts. In: Karlin S, Nevo E (eds) Population genetics and ecology. Academic Press, London, pp. 277–301
Christiansen FB, Loeschcke V (1980a) Evolution and intraspecific exploitative competition. I. One locus theory for small additive effects. Theor Popul Biol 18: 297–313
Christiansen FB, Loeschcke, V (1980b) Intraspecific competition and evolution. In: Barigozzi S (ed) Vito Volterra Symposium on mathematical models in biology. Lect Not Biomath 39. Springer, Berlin Heidelberg New York, pp. 151–170
Christiansen FB, Loeschcke V (1987) Evolution and intraspecific exploitative competition. III. One-locus theory for additive gene effects and multidimensional resource qualities. Theor Popul Biol 31: 33–46
Christiansen FB, Bundgaard J, Barker JSF (1977a) On the structure of fitness estimates under post-observational selection. Evolution 31: 843–853
Christiansen FB, Frydenberg O, Simonsen V (1977b) Genetics of *Zoarces* populations. X. Selection component analysis of the *Est III* polymorphism using samples of successive cohorts. Hereditas 87: 129–150
Clanton W (1934) An unusual situation in the salamander *Ambystoma jeffersonianum* (Green). Occas Pap Mus Zool Univ Mich 290: 1–15
Clare MJ, Luckinbill LS (1985) The effects of gene-environment interaction on the expression of longevity. Heredity 55: 19–29
Clarke B (1972) Density-dependent selection. Am Natur 106: 1–13
Clarke B (1975) The contributions of ecological genetics of evolutionary theory. Genetics 79: 101–113
Clarke B (1976) The ecological genetics of host-parasite relationships. In: Taylor AER, Muller R (eds) Genetic aspects of host-parasite relationships. Symp Br Soc Parasitol 14: 87–103
Clarke BC (1979) The evolution of genetic diversity. Proc R Soc Lond B Biol Sci 205: 453–474
Clausen J, Keck DD, Hiesey WM (1948) Experimental studies on the nature of species. III. Environmental response of climatic races of *Archillea*. Carnegie Inst Wash Publ 581: 1–129
Clay K, Antonovics J (1985) Demographic genetics of the grass *Danthonia spicata*: Success of progeny from chasmogamous and cleistogamous flowers. Evolution 39: 205–210
Clayton FE, Carson HL, Sato JE (1972) Polytene chromosome relationships in Hawaiian species of *Drosophila*. IV. Supplementary data on metaphases and gene sequences. Univ Tex publ 7213: 163–177
Clegg M (1980) Measuring plant mating systems. Bioscience 30: 814–818
Clegg MT, Epperson BK (1988) Natural selection of flower color polymorphism in morning glory populations. In: Gottlieb LD, Jain SK (eds) Plant evolutionary biology. Chapman & Hall, London pp 255–274
Clegg MT, Kahler AL, Allard RW (1978) Estimation of the life cycle, components of selection in an experimental plant population. Genetics 89: 765–792
Clegg MT, Kidwell JF, Horch CR (1980) Dynamics of correlated systems. V. Rates of decay of linkage disequilibrium in experimental populations of *D. melanogaster*. Genetics 94: 217–234
Cockerham CC (1969) Variance of gene frequencies. Evolution 223: 72–84
Cockerham CC (1973) Analyses of gene frequencies. Genetics 74: 679–700
Cockerham CC, Burrows PM, Young SSY, Prout T (1972) Frequency-dependent selection in randomly mating populations. Am Nat 106: 493–515
Cole BJ (1986) The social behaviour of *Leptothorax allardycei* (Hymenoptera: Formicidae): time budgets and the evolution of worker reproduction. Behav Ecol Sociobiol 18: 165–173
Cole CJ (1975) Evolution of parthenogenetic species of reptiles. In: Reinboth R (ed) Intersexuality in the animal kingdom. Springer, Berlin Heidelberg New York, pp 340–355
Cole LC (1954) The population consequences of life history phenomena. Rev Biol 29: 103–137
Coulthart MB, Singh RS (1988a) Low genic variation in male-reproductive tract proteins of *Drosophila melanogaster* and *D. simulans*. Mol Biol Evol 5(2): 167–181
Coulthart MB, Singh RS (1988b) High level of divergence of male-reproductive tract proteins,

between *Drosophila melanogaster* and its sibling species, *D. simulans*. Mol Biol Evol 5(2): 182–191
Coulthart MB, Singh RS (1988c) Differing amounts of genetic polymorphism in testes and male accessory glands of *Drosophila melanogaster* and *Drosophila simulans*. Biochem Genet 26: 153–164
Coyne JA (1983) Correlation between heterozygosity and rate of chromosome evolution in animals. Am Nat 123: 725–729
Coyne JA (1984) Genetic basis of male sterility in hybrids between two closely related species of *Drosophila*. Proc Natl Acad Sci USA 81: 4444–47
Coyne JA (1985) Genetic studies of three sibling species of *Drosophila* with relationship to theories of speciation. Genet Res Camb 46: 169–192
Coyne JA (1986) Meiotic segregation and male recombination in interspecific hybrids of *Drosophila*. Genetics 114: 485–494
Coyne J, Orr H (1988) Speciation in *Drosophila*. In: Otte D (ed) Speciation and adaptation. Philadelphia Academy of Sciences, Philadelphia
Crandall RE, Stearns SC (1982) Variational models of life-histories. Theor Popul Bio 21: 11–23
Crawley MJ (ed) (1983) Plant ecology. Blackwell, Oxford
Crow JF (1986) Basic concepts in population, quantitative, and evolutionary genetics. Freeman, New York
Crow JF, Aoki K (1982) Group selection for a polygenic behavioral trait: a differential proliferation model. Proc Natl Acad Sci 79: 2628–2631
Crow JF, Aoki K (1984) Group selection for a polygenic behavioral trait: estimating the degree of population subdivision. Proc Natl Acad Sci USA 81: 6073–6077
Crow JF, and Kimura M (1956) Some genetic problems in natural populations. Proc Third Berkeley Symp Math Stat Prob 4: 1–22
Crow JF, Kimura M (1965) Evolution in sexual and asexual populations. Am Nat 99: 439–450
Crow JF, Kimura M (1970) An introduction to population genetics theory. Harper and Row, New York
Crow JF, Temin RG (1964) Evidence for the partial dominance of recessive lethal genes in natural populations of *Drosophila*. Am Nat 98: 21–33
Cuellar O (1977) Animal parthenogenesis. Science 197: 837–843
Curtis CF (1982) The mechanism of hybrid male sterility from crosses in the *Anopheles gambiae* and *Glossina morsitans* complexes. In: Stein WMF (ed) The genetics of insect disease vectors. Stripes, New York
Daday H (1954) Gene frequencies in wild populations of *Trifolium repens* L. I. Distribution by latitude. Heredity 8: 61–78
Danks HV (1987) Insect dormancy: an ecological perspective. Biological Survey of Canada (Terrestrial Arthropods), Ottawa
Darevsky IS (1966) Natural parthenogenesis in a polymorphic group of Caucasian rock lizards related to *Lacerta saxicola* Eversmann. J Ohio Herpetol Soc 5: 115–152
Darwin C (1859) On the origin of species. Murray. London
Darwin C (1872) The origin of species, 6th edn. Murray, London
Davenport R (1986) Chromosomal polymorphism and speciation in the genus *Eunemobius* (Orthopetra: Grylloidea). Genetica 68: 105–107
Davidson EH, Britton RJ (1973) Organization, transcription, and regulation in the animal genome. Rev Biol 48: 565–613
Davidson EH, Britton RJ (1979) Regulation of gene expression: possible role of repetitive sequences. Science 204. 1052 —1059
Davies NB, Houston AI (1981) Owners and satellites: the economis of territory defence in the pied wagtail *Motacilla alba*. J Anim Ecol 50: 157–180
Dawkins R (1982) The extended phenotype. Oxford University Press, Oxford
Dessauer HC, Cole CJ (1989) Diversity between and within nominal forms of unisexual teiid lizards. In: Dawley R, Bogart J (eds) Evolution and ecology of unisexual vertebrates. State University New York Press, Albany pp 49–71
Dewey JS, Heywood VH (1988) Spatial gegenetic structure in a population of *Psychotrea nervosa*. Evolution 42: 834–836

Dingle H (1974) Diapause in a migrant insect, the milkweed bug *Oncopeltus fasciatus* (Dallas) (Hemiptera: Lygaeidae). Oecologia (Berl) 17: 1–10
Dingle H (1985) Migration. In: Kerkut GA, Gilbert LI (eds) Comprehensive insect physiology, biochemistry and pharmacology, Vol 9, Behavior. Pergamon, London pp 375–415
Dingle H (1986) The evolution of insect life cycle syndromes. In: Taylor F, Karban R (eds) The evolution of insect life cycles. Springer, Berlin Heidelberg New York, pp 187–203
Dingle H (1987) Quantitative genetics of life history evolution in a migrant insect. In: Jong G de (ed) Population genetics and evolution. Springer, Berlin Heidelberg New York, pp 83–93
Dingle H, Evans KE (1987) Responses in flight to selection on wing length in non-migratory milkweed bugs, *Oncopeltus fasciatus*. Entomol Exp Appl 45: 289–296
Dingle H, Evans KE, Palmer JO (1988) Responses to selection among life history traits in a non-migratory population of milkweed bugs (*Onceopeltus fasciatus*), Evolution 42: 79–92
Dixit AK (1976) Optimization in economics. Oxford University Press, Oxford
Dobzhansky T (1936) Studies on hybrid sterility. II. Localization of sterility factors in *Drosophila pseudoobscura* hybrids. Genetics 21: 113–35
Dobzhansky T (1937) Genetics and the origin of species. Columbia University Press, New York
Dobzhansky T (1950) Genetics of natural populations. XIX. Origin of heterosis through natural selection in populations of *Drosophila pseudoobscura*. Genetics 35: 288–302
Dobzhansky T (1951) Genetics and the origin of species. Columbia, New York
Dobzhansky T (1955) A review of some fundamental concepts and problems of population genetics. Cold Spring Harbor Symp Quant Biol 20: 1–15
Dobzhansky T (1970) Genetics of the evolutionary process. Columbia University Press, New York
Dobzhansky T (1973) Nothing in biology makes sense except in the light of evolution. Am Biol Teach 35: 125–129
Doolittle WF, Sapienza C (1980) The genealogy of some recently evolved vertebrate proteins. Trends Biochem Sci 10: 233–237
Dover G (1982) Molecular drive: a cohesive mode of species evolution. Nature 299: 111–117
Dudley JW (1977) 76 generations of selection for oil and protein percentage in maize. In: Pollak E, Kempthorne O, Bailey Jr TB (eds) Proc Int Conf Quant Gen, Iowa State Univ Press, Ames, pp 459–473
Dunn LC (1956) Analysis of a complex gene in the house mouse. Cold Spring Harbor Symp Quant Biol 21: 187–195
Echelle AA, Echelle AF, Crozier CD (1983) Evolution of an all-female fish, *Menidia clarkhubbsi* (Atherinidae). Evolution 37: 722–784
Ehleringer J, Clark C (1988) Evolution and adaptation in *Encelia* (Asteraceae). In: Gottlieb LD, Jain SK (eds) Plant Evolutionary Biology. Chapman & Hall, London pp 227–248
Ehrlich PR, Raven PH (1964) Butterflies and plants: A study in coevolution. Evolution 18: 586–608
Ehrlich PR, Raven PH (1969) Differentiation of populations. Science 165: 1228–1232
Eichler W (1948) Some rules in ectoparasitism. Ann Mag Nat Hist 12: 588–598
Eldredge N, Gould SJ (1972) Punctuated equilibria: an alternative to phyletic gradualism. In: Schopf TJM (ed) Models in paleobiology. Freeman, Cooper, San Francisco, pp 82–115
Ellstrand NC, Antonovics J (1984) Experimental studies of the evolutionary significance of sexual reproduction. I. A test of the frequency-dependent selection hypothesis. Evolution 38: 103–115
Ellstrand NC, Antonovics J (1985) Experimental studies of the evolutionary significance of sexual reproduction. II. A test of the density-dependent selection hypothesis. Evolution 39: 657–666
Ellstrand NC, Foster KW (1983) Impact of population structure on the apparent outcrossing rate of grain sorghum (*Sorghum bicolor*) Theor Appl Genet 66: 323–327
Ellstrand NC, Roose ML (1987) Patterns of genotypic diversity in clonal plant species. Am J Bot 74: 123–131
Emlen JM (1984) Population biology: The coevolution of population dynamics and behavior. MacMillan, New York

Endler JA (1973) Gene flow and population differentiation. Science 179: 243–250
Endler JA (1977) Geographic variation, speciation, and clines. Princeton University Press, Princeton
Endler JA (1982) Problems in distinguishing historical from ecological factors in biogeography. Am Zool 22: 441–452
Endler JA (1986) Natural selection in the wild. Princeton University Press, Princeton
Engles WR, Preston CR (1979) Hybrid dysgenesis in *D. melanogaster*: the biology of female and male sterility. Genetics 92: 161–174
Ennos RA (1982) Association of the cyanogenic loci in white clover. Genet Res 40: 65–72
Ennos R (1983) Maintenance of genetic variation in plant populations. Evol Biol 16: 129–155
Ennos RA, Clegg MT (1982) Effect of population substructuring on estimates of outcrossing in plant populations. Heredity 43: 283–292
Epling C, Dobzhansky T (1942) Genetics of natural populations. VI. Microgeographical races in *Linanthus parryae*. Genetics 27: 317–332
Epperson BK, Clegg MT (1986) Spatial autocorrelation analysis of flower color polymorphism within substructured populations of morning glory (*Ipomoea purpurea*), Am Nat 128: 840–858
Epperson BK, Clegg MT (1987) First pollination primacy and pollen selection in the morning glory, *Ipomoea purpurea*. Heredity 58: 5–14
Eshel I (1972) On the neighbor effect and the evolution of altruistic traits. Theor Popul Biol 3: 258–277
Eshel I (1978) On the mutual maintenance of a polymorphism in two nonepistatic loci with partial selfing. Theor Popul Biol 13: 99–111
Estabrook GF, Jespersen DC (1974) Strategy for a predator encountering a model-mimic system. Am Nat 108: 443–457
Ewens WJ (1972) The sampling theory of selectively neutral alleles. Theor Popul Biol 3: 87–112
Ewens WJ (1979) Mathematical population genetics (Biomathematics, 9 Springer, Berlin Heidelberg New York
Ewens WJ, Feldman MW (1976) The theoretical assessment of selective neutrality. In: Karlin S, Nevo E (eds) Population genetics and ecology. Academic Press, London, pp 303–337
Ewens WJ, Thomson G (1977) Properties of equilibria in multilocus systems. Genetics 87: 807–819
Ewing E (1980) Genetic variation in a heterogeneous environment VIII. Temporal and spatial heterogeneity in haploid populations. Am Nat 116: 885–890
Falcon LA (1982) Use of pathogenic viruses as agents for the biological control of insect pests. In: Anderson RM, May RM (eds) Population biology of infectious diseases. Springer, Berlin Heidelberg New York, pp 191–120
Falconer DS (1965) Maternal effects and selection response. In: Geerts SJ (ed) Genetics today. Proc XI Internat Congr Genetics, Vol 3. Pergamon, Oxford
Falconer DS (1981) Introduction to quantitative genetics, 2nd edn. Longman, London
Feeny P (1975) Biochemical coevolution between plants and their insect herbivores. In: Gilbert LE, Raven PH (eds) Coevolution of animals and plants. University of Texas Press, Austin, pp 3–19
Feldman MW, Christiansen FB (1975) The effect of population subdivision on two loci without selection. Genet Res 24: 151–162
Feldman MW, Christiansen FB (1984) Population genetic theory of the cost of inbreeding. Am Nat 123: 642–653
Feldman MW, Liberman U (1979) On the number of stable equilibria and the simultaneous stability of fixation and polymorphism in two locus models. Genetics 92: 1355–1360
Feldman MW, Christiansen FB, Liberman U (1983) On some models of fertility selection. Genetics 105: 1003–1010
Felsenstein J (1974) The evolutionary advantage of recombination. Genetics 78: 737–756
Felsenstein J (1975a) A pain in the torus: some difficulties with the model of isolation by distance. Am Nat 109: 359–368
Felsenstein J (1975b) Genetic drift in clines which are maintained by migration and natural selection. Genetics 81: 209–222

Felsenstein J (1976) The theoretical population genetics of variable selection and migration. Annu Rev Genet 10: 253–280
Felsenstein J (1981) Skepticism towards Santa Rosalia, or why are there so few kinds of animals? Evolution 35: 124–138
Fenchel TM (1975) Character displacement and coexistence in mud snails (Hydrobiidae). Oecologia (Berl) 20: 19–32
Fenchel TM, Christiansen FB (1977) Selection and interspecific competition. In: Christiansen FB, Fenchel TM (eds) Measuring selection in natural populations: Lect Not Biomath 19. Springer, Berlin Heidelberg New York, pp 477–498
Fenchel TM, Kofoed LH (1976) Evidence for exploitative interspecific competition in mud snails (Hydrobiidae). Oecologia (Berl) 14: 317–326
Fenner F (1971) Evolution in action: myxomatosis in the Australian wild rabbit. In: Kramer A (ed) Topics in the study of life, The bio source book. Harper & Row, New York, pp 463–471
Fenner F, Myers K (1978) Myxoma virsus and myxomatosis in retrospect: the first quarter century of a new disease. In: Kurstak E, Maramorosch K (eds) Viruses and environment. Academic Press, London, pp 539–570
Fenner F, Ratcliffe FN (1965) Myxomatosis. Cambridge University Press, Cambridge
File PC, McLeod JB (1977) The approach of solutions of nonlinear diffusion equations to traveling front solutions. Arch Rational Mech Anal 65: 335–361
Fisher RA (1922) On the dominance ratio, Proc R Soc Edinb 42: 321–431
Fisher RA (1928) The possible modification of the response of wild type to current mutations. Am Nat 62: 115–126
Fisher RA (1930) The genetical theory of natural selection. Oxford University press, Oxford
Fisher RA (1937) The wave of advance of advantageous genes. Ann Eugen 7: 355–369
Fisher RA (1941) Average excess and average effect of a gene substitution. Ann Eugen 11: 53–63
Fisher RA (1958) The genetical theory of natural selection, 2nd edn. Dover Publications Inc., New York
Flor HH (1955) Host-parasite interaction in flax rust-its genetic and other implications. Phytopathology 45: 680–685
Flor HH (1956) The complementary genic systems in flax and flax rust. Adv Genet 8: 29–54
Foin TC, Jain SK (1977) Ecosystems analysis and population biology: Lessons for the development of community biology. Bioscience 27: 532–538
Foster K, Jain SK, Smeltzer DG (1980) Response to 10 cycles of mass selection in an inbred population of grain sorghum. Crop Sci 20: 1–4
Fowler K, Partridge L (1986) Variation in male fertility explains an apparent effect of genotype diversity on success in larval competition in *Drosophila melanogaster*. Heredity 57: 31–36
Frazzetta TH (1975) Complex adaptation in evolving populations. Sinauer, Sunderland, Massachusetts
Futuyma DJ (1983) Evolutionary interactions among herbivorous insects and plants. In: Futuyma DJ, Slatkin M (eds) Coevolution. Sinauer, Sunderland, Massachusetts, pp 207–231
Futuyma DJ, Mayer GC (1980) Non-allopatric speciation in animals. Syst Zool 29: 254–271
Futuyma DJ, Slatkin M (1983) Coevolution, Sinauer, Sunderland, Massachusetts
Futuyma DJ, Leipertz SL, Mitter C (1981) Selective factors affecting clonal variation in the fall cankerworm *Alsophila pometaria* (Lepidoptera: Geometridae). Heredity 47: 151–172
Fyfe JL, Bailey NTJ (1951) Plant breeding studies in leguminous forage crops. 1. National cross-breeding in winter beans. J Agric Sci 41: 371–378
Gaines MS, Whittam TS (1980) Genetic changes in fluctuating vole populations: selective vs. non-selective forces. Genetics 96: 767–778
Gantmacher FR (1959) The theory of matrices. Chelsea, New York
Gause GF (1934) The struggle for existence. Williams & Wilkins, Baltimore. (Reprinted in 1964 by Hafner, New York)
Gause GF (1935) Verifications expérimentales de la théorie mathématique de la lutte pour la vie. Actualites scientifiques et industrielles 277. Hermann, Paris
Gause GF, Witt AA (1935) Behavior of mixed populations and the problem of natural selection. Am Nat 69: 596–609

Gazdag J, Canosa J (1974) Numerical solution of Fisher's equation. J Appl Probab 11: 445–457
Ghiselin MT (1974) The economy of nature and the evolution of sex. University Calif Press, Berkeley
Ghiselin MT (1983) Lloyd Morgan's canon in evolutionary context. Behav Brain Sci 6: 362–363
Giesel JT (1986) Genetic correlation structure of life history variables in outbred, wild *Drosophila melanogaster:* effects of photoperiod regimen. Am Nat 128: 593–603
Giesel JT, Murphy PA, Manlove MN (1982a) The influence of temperature on genetic interrelationships of life history traits in a population of *Drosophila melanogaster*: what tangled data sets we weave. Am Nat 119: 464–479
Giesel JT, Murphy PA, Manlove MN (1982b) An investigation of the effects of temperature on the genetic organizations of life history indices in three populations of *Drosophila melanogaster.* In: Dingle H, Hegmann JP (eds) Evolution and genetics of life histories. Springer, Berlin Heidelberg New York, pp 189–207
Gillespie JH (1977) Natural selection for variance in offspring number: A new evolutionary principle. Am Nat 111: 1010–1014
Gillespie JH (1979) A general model to account for enzyme variation in natural populations. V. The SAS-CFF model. Theor Popul Biol 14: 1–45
Gillespie JH (1984) The moclecular clock may be an episodic clock. Proc Natl Acad Sci USA 81: 8009–8013
Ginzburg LR (1983) Theory of natural selection and population growth. Benjamin & Cummings, Menlo Park
Ginzburg LR, Bingham PM, Yoo S (1984) On the theory of speciation induced by transposable elements. Genetics 107: 331–341
Gleeson SK (1987) Fitness, reproductive value, and Cole's result. Oikos 48: 116–119
Glesener RR, Tilman D (1978) Sexuality and the components of environmental uncertainty: clues from geographic parthenogenesis in terrestrial animals. Am Nat 112: 659–673
Gnaiger E (1983) Heat dissipation and energetic efficiency in animal anoxibiosis: economy contra power. J Exp Zool 228: 471–490
Goddard KA, Dawley RM, Dowling TE (1989) Origin and genetic relationships of diploid, triploid, and diploid-triploid mosaic biotypes in the *Phoxinus eos-neogaeus* unisexual complex. In: Dawley R, Bogart J (eds) Evolution and ecology of unisexual vertebrates. State University New York Press, Albany pp 268–280
Godfray HCJ (1986) Models for clutch size and sex ratio with sibling interaction. Theor Popul Bio 30: 215–231
Gojobori T, Li WH, Graur D (1982) Patterns of nucleotid substitution in pseudogenes and functional genes. J Mol Evol 18: 360–369
Golding GB, Strobeck C (1980) Linkage disequilibrium in a finite population that is partially selfing. Genetics 94: 777–789
Goldschmidt R (1940) The material basis of evolution. Yale University, New Haven
Gonzalez AM, Cabrera VM, Larruga JM, Gullon A (1982) Genetic distance in the sibling species of *Drosophila melanogaster, D. simulans* and *D. mauritiana.* Evolution 36: 517–522
Goodman D (1982) Optimal life histories, optimal notation, and the value of reproductive value. Am Nat 119: 803–823
Goodnight CJ (1985) The influence of environmental variaton on group and individual selection in a cress. Evolution 39: 545–558
Goodnight CJ (1987) On the effect of founder events on epistatic genetic variance. Evolution 41: 80–91
Gosler AG (1987a) Sexual dimorphism in the summer bill length of the great tit. Ardea 75: 91–98
Gosler AG (1987b) Pattern and process in the bill morphology of the great tit (*Parus major*). Ibis 129: 451–476
Gottlieb LD (1982) Does speciation facilitate the evolution of adaptation. In: Barigozzi c (ed) Mechanism of speciation. Liss, New York, pp 179–190
Gottlieb LD, Jain SK (eds) (1988) Plant evolutionary biology. Chapman & Hall, London
Gould SJ (1977) Ontogeny and phylogeny. Harvard University Press, Cambridge, Massachusetts, pp 501
Gould SJ (1980) Is a new and general theory of evolution emerging? Paleontology 6: 119–30

Gould SJ, Eldredge N (1977) Punctuated evolution: the tempo and mode of evolution reconsidered. Paleontology 3: 115–151
Gould SJ, Lewontin RC (1979) The spandrels of San Marco and the Panglossian paradigm: a critique of the adaptationist programme. Proc R Soc Lond B Biol Sci 205: 581–598
Gould SJ, Vrba ES (1982) Exaptation—a missing term in the science of form. Palaeobiology 8: 4–9
Graf JD, Polls M (1989) Evolutionary genetics of the *Rana esculenta* hybrid complex. In: Dawley R, Bogart J (eds) Evolution and ecology of unisexual vertebrates. State University New York Press, Albany
Grafen A (1982) How not to measure inclusive fitness. Nature 298: 425–426
Grafen A (1985) A geometric view of a relatedness. Oxford Surv Evol Biol 2: 28–89
Grant PR (1986) Ecology and evolution of Darwin's finches. Princeton University Press, Princeton
Grant V (1980) Gene flow and the homogeneity of species populations. Biol. Zentralbl 99: 157–169
Grant V (1981) Plant speciation, 2nd edn. Columbia University Press, New York
Gray R (1986) Faith and foraging. In: Kamil AC, Krebs JR, Pulliam HR (eds) Foraging behavior. Plenum, New York
Greaves JH, Redfern R, Ayres PB, Gill JE (1977) Warfarin resistance: a balanced polymorphism in the Norway rat. Genet Res 30: 257–263
Green RF (1984) Stopping rules for optimal foragers. Am Nat 123: 30–40
Green RF, Nunez AT (1986) Central-place foraging in a patchy environment. J. Theor Biol 123: 35–43
Greenbaum IF (1981) Genetic interactions between hybridising cytotypes of the tent-making bat (*Uroderma bilobatum*). Evolution 35: 306–321
Gregorius H-R, Namkoong G (1983) Conditions for protected polymorphism in subdivided plant populations. I. Uniform pollen dispersal. Theor Popul Biol 24: 252–267
Gregorius H-R, Ziehe M, Ross MD (1987) Selection caused by self-fertilization. I. Four measures of self-fertilization and their effects on fitness. Theor Popul Biol 31: 91–115
Grey DR (1980) Minimization of extinction probabilities in reproducing populations. Theor Popul Biol 18: 430–443
Griffin AR, Lindgreen D (1985) Effect of inbreeding on production of filled seed in *Pinus radiata*—experimental results and a model of gene action. Theor Appl Genet 71: 334–343
Griffiths RC (1980) Lines of descent in the diffusion approximation of neutral Wright-Fisher models. Theor Popul Biol 17: 37–50
Groenendael JM van (1985) Differences in life histories between ecotypes of *Plantago lanceolata*. In: White J (ed) Studies in plant demography. Academic Press, London, pp. 51–67
Groeters FR, Dingle H (1987) Genetics and maternal influences on life history plasticity in response to photoperiod by milkweed bugs (*Oncopeltus fasciatus*). Am Nat 129: 332–346
Groeters FR, Dingle H (1988) Genetic and maternal influences on life history plasticity in milkweed bugs (*Oncopeltus*): response to temperature. J Evol Biol 1: 317–333
Groot Mh De (1970) Optimal statistical decisions. MacGraw-Hill, New York
Gupta AN, Lewontin RC (1982) A study of reaction norms in natural populations of *Drosophila pseudoobscura*. Evolution 36: 934–948
Gustafsson A (1946) Apomixis in higher plants. I. The mechanism of apomixis. Acta Univ Lund Sect II Med Math Sci Resun Nat 42: 1–66
Hadeler KP, Liberman U (1975) Selection models with fertility differences. J Math Biol 2: 19–32
Haldane JBS (1931) A mathematical theory of natural selection. VI. Isolation. Trans Camb Philos Soc 26: 220–230
Haldane JBS (1932) The causes of evolution. Longmans, Green, London
Haldane JBS (1940) The conflict between selection and mutation of harmful recessive genes. Ann Eugen 10: 417–421
Haldane JBS (1948) The theory of a cline. J Genet 48: 277–284
Haldane JBS (1949a) The association of characters as a result of inbreeding and linkage. Ann Eugen 15: 15–23

Haldane JBS (1949b) Disease and evolution. Ric Sci Suppl 19: 68–76

Haldane JBS, Jayakar SD (1963) Polymorphism due to selection of varying direction. J Genet 58: 237–242

Hale LR, Singh RS (1987) Mitochondrial DNA variation and genetic structure in populations of *Drosophila melanogaster*. Mol Biol Evol 4 (6): 622–637

Hale LR, Singh RS (1989) Mitochondrial DNA variation and history of colonization of *Drosophila simulans*. Proc Natl Acad Sci USA (in press)

Halliday RB, Barton NH, Hewitt GM (1983) Electrophoretic analysis of a chromosomal hybrid zone in the grasshopper *Podisma pedestris*. Biol J Linn Soc 19: 51–62

Hamilton WD (1963) The evolution of altruistic behavior. Am 97: 354–356

Hamilton WD (1964a) The genetical evolution of social behavior. I. J Theor Biol 7: 1–16

Hamilton WD (1964b) The genetical evolution of social behavior. II. J Theor Biol 7: 17–52

Hamilton WD (1970) Selfish and spiteful behavior in an evolutionary model. Nature 228: 1218–1220

Hamilton WD, Zuk M (1982) Heritable true fitness and bright birds: a role for parasites. Science 218: 384–387.

Hamilton WD, Henderson P, Moran N (1981) Fluctuation of environmental and coevolved antagonist polymorphism as factors in the maintenance of sex. In: Alexander RD, Tinkle D (eds) Natural selection and social behavior. Chiron, New York, pp 363–381

Hamrick JL (1983) The distribution of genetic variation within and among natural plant populations. In: Schonewald-Cox C, Chambers SM (eds) Genetics and conservation. Addison Weseley

Hamrick JL, Linhart YB, Mitton JB (1979) Relationships between life history characteristics and electrophoretically detected genetic variation in plants. Ann Rev Ecol Syst 10: 173–200

Hancock JF, Bringhurst RS (1978). Interpopulational differentiation and adaptation in the perennial, diploid species *Fragaria vesca* L Am J Bot 65: 795–803

Harding JA, Barnes K (1977) Genetics of *Lupinus*. X Genetic variability, heterozygosity and outcrossing in colonial populations of *Lupinus succulentus*. Evolution 31: 247–255

Harpending H, Rogers A (1987) On Wright's mechanism for intergroup selection. J Theor Biol 127: 51–61

Harper JL (1977) Population biology of plants. Academic Press, London

Harper JL (1983) A Darwinian plant ecology. In: Bendall DS (ed) Evolution from molecules to man. Cambridge University Press, Cambridge

Harshman LG, Futuyma DJ (1985) The origin and distribution of clonal diversity in *Alsophila pometaria* (Lepidoptera: Geometridae). Evolution 39: 315–324

Hart D, Clark AG (1988) Principles of population genetics. Sinauer, Sunderland, Massachusetts

Hartl DL, Dykhuzen DE (1981) Potential for selection among nearly neutral allozymes of 6-phosphoglucomate dehydrogenase in *E. coli*. Proc Natl Acad Sci USA 78: 6344–6388

Harvey PH, Kavanagh MJ, Clutton-Brock TH (1978) Sexual dimorphism in primate teeth. J Zool 186: 475–485

Hassel MP (1978) The dynamics of arthropod predator-prey systems. Princeton University Press, Princeton

Hassel MP (1987) Detecting regulation in patchily distributed animal populations. J. Anim Ecol 56: 705–713

Hastings A (1984) Maintenance of high disequilibrium in the presence of partial selfing. Proc Natl Acad Sci 81: 4596–4598

Hastings A (1985) Stable equilibria at two loci in populations with large selfing rates. Genetics 109: 215–228

Hatchett JH, Gallun RL (1970) Genetics of the ability of the Hessian fly, *Mayetiola destructor*, to survive on wheats having different genes for resistance. Ann Entomol Soc Am 63: 1400–1407

Hauptli H, Jain SK (1984) Allozyme evidence on the evolutionary relationships in grain amaranths. Theor Appl Genet 68: 1–13

Hayman BI (1953) Mixed selfing and random mating when homozygotes are at a disadvantage. Heredity 7: 185–192

Hebert PDN (1974) Ecological differences among genotypes in a natural population of *Daphnia magna*. Heredity 33: 327–337

Hebert PDN (1981) Obligate asexuality in *Daphnia.* Am Nat 117: 784–789
Hebert PDN, Crease T (1983) Clonal diversity in populations of *Daphnia pulex* reproducing by obligate parthenogenesis. Heredity 51: 353–369
Hebert PDN, Beaton MJ, Schwartz SS (1989) Polyphyletic origins of asexuality in *Daphnia pulex.* I. Breeding system variation and levels of clonal diversity. Evolution 43 (in press)
Hedrick PW (1980a) The establishment of chromosomal variants. Evolution 35: 322–332
Hedrick PW (1980b) Hitchhiking: a comparison of linkage and partial selfing. Genetics 94: 791–808
Hedrick PW (1983) Genetics of populations. Jones & Bartlett, Boston, Massachusetts
Hedrick PW (1985a) Population biology. Jones & Bartlett, Boston, Massachusetts
Hedrick PW (1985b) Inbreeding and selection in natural populations. In: Gregorius H-R (ed) Population genetics in forestry. Lecture notes in biomathematics 60. Springer, Berlin Heidelberg New York, p 71
Hedrick PW (1986a) Average inbreeding or inbreeding equilibrium? Am J Hum Genet 38: 965–970
Hedrick PW (1986b) Genetic polymorphism in heterogeneous environments: A decade later. Annu Rev Ecol Syst 17: 535–566
Hedrick PW (1987a) Estimation of the rate of partial inbreeding. Heredity 58: 161–166
Hedrick PW (1987b) Population genetics of intragametophytic selfing. Evolution 4: 137–144
Hedrick PW (1987c) Genetic load and the mating system in homosporous ferns. Evolution 41: 1282–1289
Hedrick PW, Cockerham CC (1986) Partial inbreeding: Equilibrium heterozygosity and the heterozygosity paradox. Evolution 40: 856–861
Hedrick PW, Holden L (1979) Hitch-hiking: an alternative to coadaptation for the barley and slender wild oat examples. Heredity 43: 79–86
Hedrick PW, Ginevan ME, Ewing EP (1976) Genetic polymorphism in heterogeneous environments. Annu Rev Ecol Syst 7: 1–32
Hedrick PW, Jain S, Holden L (1978) Multilocus systems in evolution. Evol Biol 11: 101–182
Heithaus ER, Culver DC, Beattie AJ (1980) Models of some ant-plant mutualism. Am Nat 116: 347–361
Henrich VC, Denlinger DL (1982) A maternal effect that eliminates pupal diapause in progeny of the flesh fly, *Sarcophaga bullata.* J Insect Physiol 28: 881–884
Heslop-Harrison J (1964) Forty years of genecology. Adv Ecol Res 2: 159–247
Hewitt GM (1979) Animal cytogenetics. III. Orthoptera. Borntraeger Stuttgart
Hill WG (1975) Linkage disequilibrium among multiple neutral alleles produced by mutation in a finite population Theor Popul Biol 8: 117–126
Hill WG (1982) Rates of change in quantitative traits from fixation of new mutations. Proc Natl Acad Sci USA 79: 142–145
Hill WG, Robertson A (1968) Linkage disequilibrium in finite populations. Theor Appl Genet. 38: 226–231
Hilu KW (1983) The role of single-gene mutations in the evolution of flowering plants. Evol Biol 16: 97–122
Hinegardner R (1976) Darwin and the darwinian revolution. In: Ayala FJ (ed) Molecular evolution. Sinauer, Sunderland, Massachusetts
Hines WGS (1987) Evolutionary stable strategies: A review of the basic theory. Theor Popul Biol 31: 195–272
Hodges CM (1985) Bumblebee foraging: energetic consequences of using a threshold departure rule. Ecology 66: 188–197
Hoekstra RF, Bijlsma R, Dolman AJ (1985) Polymorphism from environmental heterogeneity: models are only robust if the heterozygote is close in fitness to the favoured homozygote in each environment. Genet Res 45: 299–314
Holden LR (1979) New properties of the two-locus partial selfing model with selection. Genetics 93: 217–236
Hollander J den, Pathak PK (1981) The genetics of the "biotypes" of the rice brown planthopper, *Nilaparvata lugens.* Entomol Exp Appl 29: 76–86

Holsinger KE (1986) Dispersal and plant systems: The evolution of self-fertilization in subdivided populations. Evolution 40: 405–413
Holsinger KL, Feldman MW (1985) Selection in complex genetic systems. VI. Equilibrium properties of two-locus selection models with partial selfing. Theor Popul Biol 28: 117–132
Holsinger KL, Feldman MW, Christiansen FB (1984) The evolution of self-fertilization in plants: a population genetics model. Am Nat 124: 446–453
Honeycutt RL, Wilkinson P (1989) Electrophoretic variation in the parthenogenetic grasshopper (*Warramaba virgo* and its sexual relatives. Evolution 43: 1027–1044
Horns HS (1978) Optimal tactics of reproduction and life-history. In: Krebs JR, Davies NB (eds) Behavioural ecology: an evolutionary approach. Sinauer, Sunderland Massachusetts, pp 411–429
Horn HS (1979) Adaptation from the perspective of optimality. In: Solbrig OT, Jain S, Johnson GB, Raven PH (eds) Topics in plant population biology. Columbia University Press, New York.
Horn HS, Rubenstein DI (1984) Behavioural adaptations and life history. In: Krebs JR, Davies NB (eds) Behavioural ecology: an evolutionary approach, 2nd edn. Sinauer, Sunderland, Massachusetts, pp 279–298
Houston AI, McNamara J (1982) A sequential approach to risk-taking Anim Behav 30: 1260–1261
Howlett RJ, Majerus MEN (1987) The understanding of industrial melanism in the peppered moth (*Biston betularia*) (Lepidoptera: Geometridae). Biol J Linn Soc 30: 31–44
Hudson RR (1985) The sampling distribution of linkage disequilibrium under an infinite allele model without selection. Genetics 109: 611–631
Hulscher JB (1985) Growth and abrasion of the oystercatcher bill in relation to dietary switches. Neth J Zool 35: 124–154
Hunt WG, Selander RK (1973) Biochemical genetics of hybridization in European house mice. Heredity 31: 11–33
Hutchinson GE (1957) Concluding remarks. Cold Spring Harbor Symp Quant Biol 22: 415–427
Hutchinson GE (1968) When are species necessary? In: Lewontin RC (ed) Population biology in evolution. Syracuse University Press, pp 177–186
Huxley J (1942, 1963) Evolution: The modern synthesis. Allen & Unwin, New York.
Intriligator M (1971) Mathematical optimization and economic theory. Prentice-Hall, New York
Istock CA (1978) Fitness variation in a natural population. In: Dingle H (ed) Evolution of insect migration and diapause. Springer, Berlin Heidelberg New York pp 171–190
Istock CA (1982) Some theoretical considerations concerning life history evolution. In: Dingle H, Hegmann JP (eds) Evolution and genetics of life histories. Springer, Berlin Heidelberg New York, pp 7–20
Istock CA (1984) Boundries to life history variation and evolution. In: Price, PW, Slobodchikoff CN, Gaud WS (eds) A new ecology: novel approaches to interactive systems. Wiley Interscience, New York, pp 143–168
Iwasa Y, Teramoto E (1980) A criterion of life history evolution based on density-dependent selection. J Theor Biol 84: 545–566
Iwasa Y, Higashi M, Yamamura N (1981) Prey distribution as a factor determining the choice of optimal foraging strategy. Am Nat 117: 710–723
Iwasa Y, Suzuki, Y, Matsuda H (1984) Theory of oviposition strategy of parasitoids: I. The effect of mortality and limited egg numbers. Theor Popul Biol 26: 205–227
Jackson JBC, Buss LW, Cook RE (1985) Population biology and evolution of clonal organisms. Yale University Press, New Haven
Jacobs J (1985) Konkurrenz und Einnischung. Hat Konkurrenz um Ressourcen eine evolutionsbiologische Bedeutung für die Artenmannigfaltigkeit der Tiere? Z Zool Syst Evolutions forsch 23: 243–258
Jacobs OLR (1974) Introduction to control theory. Clarendon Oxford
Jaenike J (1978) An hypothesis to account for the maintenance of sex within populations. Evol Theory 3: 191–194

Jaenike J, Parker ED Jr, Selander RK (1980) Clonal niche structure in the parthenogenetic earthworm *Octolasion tyrtaeum*. Am Nat 116: 196–205
Jaenike J, Ausubel S, Grimaldi DA (1982) On the evolution of clonal diversity in parthenogenetic earthworms. Pedobiologia 23: 304–310
Jain SK (1969) Comparative ecogenetics of two *Avena* species occurring in Central California. Evol Biol 3: 73–118
Jain SK (1975) Patterns of survival and microevolution in plant populations. In: Karlin S, Nevo E (eds) Population genetics and ecology. Academic Press, London, pp 49–89
Jain SK (1976) The evolution of inbreeding in plants. Annu Rev. Ecol Syst 7: 469–495
Jain SK (1979) Adaptive strategies: Polymorphism, plasticity, and homeostasis. In: Solbrig OT, Jain SK, Johnson G, Raven P (eds) Topics in plant population biology. Columbia University Press, New York, pp 160–187
Jain SK (1984) Breeding systems and the dynamics of plant populations. In: Chopra VL et al. (eds) Genetics: new frontiers. Proc XV Intern Congr Genet. IV. Oxford & IBH, New Delhi, p 291
Jain SK, Botsford L (eds) (1990) Applied population biology. Martinus-Nÿhoff, Dordrecht (in press)
Jain SK, Bradshaw AD (1966) Evolution in closely adjacent plant populations. I. The evidence and its theoretical analysis. Heredity 21: 407–441
Jain SK, Marshall DR (1969) Genetic changes in a barley population analyzed in terms of life cycle components of selection. Genetica 38: 355–374
Jain SK, Martins PS (1979) Ecological genetics of the colonizing ability of rose clover. Am J Bot 66: 361–366
Janetos AC (1980) Strategies of female choice: a theoretical analysis. Behav Ecol Sociobiol 7: 107–112
Janssen GM, Jong G de, Joosse ENG, Scharloo W (1988) A negative maternal effect in springtails. Evolution 42: 828–834
Janzen DH (1980) When is it coevolution? Evolution 34: 611–612
Janzen DH (1985) The natural history of mutualism. In: Boucher DH (ed) The biology of mutualism. Croom Helm, London, pp 40–49
Jayakar SD (1970) A mathematical model for interaction of gene frequencies in a parasite and its host. Theor Popul Biol 1: 140–164
Jayakar SD (1983) Mathematical models of coevolution of species. In: Chopra VL, Joshi BC, Sharma RP, Bansal HC (eds) Genetics: new frontiers. Oxford & IBH, New Delhi, 4: 273–282
Jayakar SD (1984a) A theoretical model for the coevolution of a host and its parasite. In: Levin SA, Hallan TG (eds) Mathematical ecology. Lectures Notes in Biomathematics. Springer, Berlin Heidelberg New York, pp 137–152
Jayakar SD (1984b) Coevolution at two trophic levels. In: Wörhmann K, Loeschcke V (eds) Population biology and evolution. Springer, Berlin Heidelberg New York, pp 247–253
Johannsen W (1909) Elements der exakten Erblichkeitsforschung. Fishcher, Jena
Jones DF (1916) Natural cross-pollination in the tomato. Science 43: 509–510
Jones JS, Leith BH, Rawlings P (1977) Polymorphism in *Cepaea*: a problem with too many solutions? Annu Rev Ecol Syst 8: 109–143
Jones JS, Bryant SH, Lewontin RC, Moore JA, Prout T (1981) Gene flow and the geographic distribution of a molecular polymorphism in *Drosophila pseudoobscura*. Genetics 98: 157–178
Jones LP, Frankham R, Barker JSF (1968) The effects of population size and selection intensity in selection for a quantitative character in *Drosophila*. II. Long-term response to selection. Genet Res Camb 12: 249–266
Kacelnik A (1984) Central-place foraging in starlings (*Sturnus vulgaris*). I. Patch residence time. J Anim Ecol 53: 283–300
Kacelnik A, Houston AI, Krebs JR (1981) Optimal foraging and territory defence in the great tit (*Parus major*). Behav Ecol Sociobiol 8: 35–40
Kalisz S (1986) Variable selection on the timing of germination in *Collinsia verna* (Scrophulariaceae). Evolution 40: 479–491
Karlin S (1969) Equilibrium behavior of population genetics models with non-random mating. Gordon & Breach, London

Karlin S (1982) Classification of selection-migration structures and conditions for a protected polymorphism. Evol Biol 14: 61–203
Karlin S, Feldman MW (1970) Linkage and selection. Two-locus symmetric viability model. Theor Popul Biol 1: 39–71
Karlin S, Lessard S (1986) Theoretical studies on sex ratio evolution. Princeton University Press Princeton
Karlin S, McGregor J (1972a) Polymorphisms for genetic and ecological systems with weak coupling. Theor Popul Biol 3: 210–238
Karlin S, McGregor J (1972b) Application of the method of small parameters to multi-niche population genetic models. Theor Popul Biol 3: 186–209
Karlin S, Nevo E (eds) (1986) Evolutionary processes and theory. Academic Press, London
Karron J, Linhart YP, Chaulk CA, Robertson CA (1988) Genetic structure of populations of geographically restricted and widespread species of *Astragalus*. Biol Conserv 39: 179–193
Keeler KH (1981) A model of selection for facultative, non-symbiotic mutualism. Am Nat 118: 488–498
Keeler KH (1985) Cost: benefit models of mutualism. In: Boucher DH (ed) The biology of mutualism. Croom Helm, London, pp 100–127
Keith TP (1983) Frequency distributions of esterase-5 alleles in two populations of *Drosophila pseudoobscura*. Genetics 105: 153–155
Keith TP, Brooks LD, Lewontin RC, Martinez-Cruzado JC, Rigby DL (1985) Nearly identical allelic distributions of xanthine dehydrogenase in two populations of *Drosophila pseudoobscura*. Mol Biol Evol 2: 206–216
Kelley SE, Antonovics J, Schmitt J (1988) A test of the short-term advantage of sexual reproduction. Nature 331: 714–716
Kesseli R, Jain SK (1984) New variation and biosystematic patterns detected by allozyme and morphological comparisons in *Limnanthes*, Section Reflexae. Plant Syst Evol 147: 133–165
Kesseli R, Jain SK (1985) Breeding systems and population structure in *Limnanthes*. Theor Appl Genet 71: 292–299
Kettlewell HBD (1955) Selection experiments on industrial melanism in the Lepidoptera. Heredity 10: 287–301
Kettlewell HBD (1973) The evolution of melanism. Clarendon, Oxford
Key KHL (1982) Species, parapatry, and the morabine grasshoppers. Syst Zool 30: 425–458
Kidwell MG, Kidwell JF, Sved JA (1977) Hybrid dysgenesis in *Drosophila melanogaster*: a syndrome of aberrant traits including mutation, sterility and male recombination. Genetics 86: 813–833
Kidwell MG, Novy JB (1979) Hybrid dysgenesis in *D. melanogaster:* sterility resulting from gonadal dysgenesis in the P-M System. Genetics 92: 1127–1140
Kiester AR, Nagylaki T, Shaffer B (1981) Population dynamics of species with gynogenetic sibling species. Theor Popul Biol 19: 358–369
Kimura M (1955) Solution of a process of random genetic drift with a continuous model. Proc Natl Sci USA 41: 144–150
Kimura M (1958) On the change of population fitness by natural selection. Heredity 12: 145–167
Kimura M (1964) Diffusion models in population genetics. J Appl Probab 1: 177–232
Kimura M (1968a) Evolutionary rate at the molecular level. Nature 217: 624–626
Kimura M (1968b) Genetic variability maintained in a finite population due to mutational production of neutral and nearly neutral isoalleles. Genet Res Camb 11: 247–269
Kimura M (1969) The rate of molecular evolution considered from the standpoint of population genetics. Proc Natl Acad Sci USA 63: 1181–1188
Kimura M (1983) The neutral theory of molecular evolution. Cambridge University Press, Cambridge
Kimura M, Crow JF (1964) The number of alleles that can be maintained in a finite population. Genetics 49: 725–738
Kimura M, Maruyama T (1971) Pattern of neutral polymorphism in a geographically structured population. Genet Res 18: 125–131
Kimura M, Ohta T (1971) Theoretical aspects of population genetics. Princeton University Press, Princeton
Kimura M, Weiss GH (1964) The stepping stone model of population structure and the decrease of genetic correlation with distance. Genetics 49: 561–576

King ABS (1974) Photoperiodic induction and inheritence of diapause in *Pionea forficalis* (Lepidoptera: Pyralidae). Entomol Exp Appl 17: 397–409
King CE (1972) Adaptation of rotifers to seasonal variation. Ecology 53: 408–418
King D, Roughgarden J (1982) Graded allocation between vegetative and reproductive growth for annual plants in growing seasons of random length. Theor Popul Biol 22: 1–16
King M (1981) Chromosomal change and speciation in lizards. In: Atchley WR, Woodruff D (eds) Evolution and speciation. Essays in honour of MJD White. Cambridge University Press, New York
King MC, Wilson AC (1975) Evolution at two levels in humans and chimpanzees. Science 188: 107–116
Kingman JFC (1961) A mathematical problem in population genetics. Proc Camb Philos Soc 57: 574–582
Kingman JFC (1982) The coalescent. Stoch Proc Appl 13: 235–248
Klausner E, Miller ER, Dingle H (1981) Genetics of Brachyptery in a lygaeid bug island population. J Hered 72: 288–289
Klekowski EJ (1979) The genetics and reproductive biology of ferns. In: Dyer AF (ed) The experimental biology of ferns. Academic Press, London, p 133
Kocher TD, Sage RD (1986) Further genetic analyses of a hybrid zone between Leopard frogs (*Rana pipiens* complex) in Central Texas. Evolution 40: 21–33
Kolmes SA (1986) Have hymenopteran societies evolved to be ergonomically efficient? JNY Entomol Soc 94: 447–457
Kolmogoroff A (1936) Sulla teoria di Volterra della lotta per l'esistenza. Giornale dell' Istituto Italiano Degli Attuari 7: 74–80
Kondrashov AS (1982) Selection against harmful mutations in large sexual and asexual populations. Genet Res Camb 40: 325–332
Kondrashov AS (1984) Deleterious mutations as an evolutionary factor. I. The advantage of recombination. Genet Res Camb 44: 199–217
Kornfield I, Smith DC, Gagnan PS, Taylor JN (1982) The cichlid fish of Cuetro Cieniges, Mexico: direct evidence of conspecificity among distinct trophic morphs. Evolution 36: 658–664
Kostitzin VA (1934) Symbiose, parasitisme et evolution (Etude mathématique), Hermann, Paris
Kostitzin VA (1936) Equations différentielles générales du problème de sélection naturelle. C R Acad Sci Paris 203: 157
Kostitzin VA (1937) Biologie mathématique. Colin, Paris
Kostitzin VA (1938) Sur les coefficients mendélians d'hérédité. CR Acad Sci Paris 206: 570–572
Krebs CJ, Myers JM (1974) Population cycles in small mammals. Adv Ecol Res 8: 267–399
Krebs JR (1970) Regulation of numbers in the Great Tit. J Zool (Lond) 162: 317–333
Krebs, JR, Kacelnik A, Taylor P (1978) Test of optimal sampling by foraging great tits. Nature 275: 27–31
Krebs JR, Stephens DW, Sutherland WJ (1983) Perspectives in optimal foraging In: Brush AH, Clark Jr GA (eds) Perspectives in ornithology. Cambridge University Press, New York, pp 165–216
Kreitman M (1983) Nucleotide polymorphisms at the alcohol dehydrogenase locus of *Drosophila melanogaster*. Nature 304: 412–417
Krieber M, Rose MR (1986) Molecular aspects of the species barrier. Annu. Rev Ecol Syst 17: 465–485
Lack D (1954) The natural regulation of animal numbers. Clarendon, Oxford
Lack D (1966) Population studies of birds. Clarendon, Oxford
Lakovaara S, Saura A, Falk CT (1972) Genetics distance and evolutionary relationship in the *Drosophila obscura* group. Evolution 26: 117–184
Lamarck JB de (1809) Philosophie zoologique, Paris. dt. Alfred Kröner Zoologische Philosophie, Leipzig 1909
Lande R (1976) Natural selection and random genetic drift in phenotypic evolution. Evolution 30: 314–334
Lande R (1979a) Rapid evolution of sexual dimorphism and character divergence in a cline. Evolution 36: 213–223
Lande R (1979b) Effective deme sizes during long-term evolution estimated from rates of chromosomal rearrangement. Evolution 33: 234–251
Lande R (1980) Genetic variation and phenotypic evolution during allopatric speciation. Am Nat 116: 463–479

Lande R (1982a) Elements of a quantitative genetic model of life history evolution. In: Dingle H, Hegmann JP (eds) Evolution and genetics of life histories. Springer Berlin Heidelberg New York, pp 21–29
Lande R (1982b) A quantitative genetic theory of life history evolution. Ecology 63: 607–615
Lande R (1984) The expected fixation rate of chromosomal inversions: Evolution 38: 743–752
Lande R (1985) The fixation of chromosomal rearrangements in a subdivided population with local extinction and colonization. Heredity 54: 323–332
Lande R (1988) Genetics and demography in biological conservation. Science 241: 1455–1460
Lande R, Barrowclough GF (1987) Effective population size, genetic variation, and their use in population management. In: Soule ME (ed) Viable populations for conservation. Cambridge University Press, New York, pp 87–123
Lande R, Schemske DW (1985) The evolution of self-fertilization and inbreeding depression in plants. I. Genetic models. Evolution 38: 24–40
Langley CH, Smith BD, Johnson FM (1978) Analysis of linkage disequilibria between allozyme loci in natural populations of *Drosophila melanogaster*. Genet Res 32: 215–229
Latter BDH (1973) The island model of population differentiation: a general solution. Genetics 73: 147–157
Laughlin R (1965) Capacity for increase: a useful population statistic. J Anim Ecol 34: 77–91
Laurie-Ahlberg CC, Stam LF (1987) Use of P-element mediated transformation to identify the molecular basis of naturally occurring variants affecting alcohol dehydrogenase expression in *Drosophila melanogaster*. Genetics 115: 129–140
Laurie-Ahlberg CC, Maroni G, Bewley GC, Lucchesi JC, Weir BC (1980) Quantitative genetic variation of enzyme activities in natural populations of *Drosophila melanogaster*. Proc Natl Acad Sci USA 77: 1073–1077
Law R (1979a) Optimal life-histories under age-specific predation. Am Nat 114: 399–417
Law R (1979b) Ecological determinants in the evolution of life histories. In: Anderson RM, Turner BD, Taylor LR (eds) Population dynamics, Blackwell, Oxford pp 81–103
Law R, Bradshaw AD, Putwain PD (1977) Life history variation in *Poa annua*. Evolution 31: 233–246
Ledig FT, Guries RP, Bonefeld BA (1983) The relation of growth to heterozygosity in pitch pine. Evolution 37: 1227–1238
Leipold M, Schmidtke J (1982) Gene expression in phylogenetically polyploid organisms. In: Dover GA, Flavell RB (eds) Genome evolution. Academic Press, London
Lenington S (1983) Social preferences for partners carrying 'good genes' in wild house mice. Anim. Behav 31: 325–333
Lenski RE, Levin BR (1985) Constraints on the coevolution of bacteria and virulent phage: a model, some experiments, and predictions for natural communities. Am Nat 125: 585–602
Léon JA (1976) Life histories as adaptive strategies. J Theor Biol 60: 301–335
Leonard KJ (1977) Selection pressures and plant pathogens. Ann NY Acad Sci 287: 207–222
Leonard KJ, Czochor RJ (1980) Theory of genetic interactions among populations of plants and their pathogens. Ann Rev Phytopathol 18: 237–258
Lerner IM (1954) Genetic homeostasis. Wiley, New York
Leslie JF, Vrijenhoek (1978) Genetic dissection of clonally inherited genomes of *Poeciliopsis*. I. Linkage analysis and preliminary assessment of deleterious gene loads. Genetics 90: 801–811
Leslie JF, Vrijenhoek RC (1980) Consideration of Muller's ratchet mechanism through studies of genetic linkage and genomic compatibilities in clonally reproducing *Poeciliopsis*. Evolution 34: 1105–1115
Levene H (1953) Genetic equilibrium when more than one niche is available. Am Nat 87: 331–333
Levin BR, Allison AC, Bremermann HJ et al. (1982) Evolution of parasites and hosts. In: Anderson RM, May RM (eds) Population biology of infectious diseases. Springer, Berlin Heidelberg New York, pp 213–243
Levin BR, Lenski RE (1983) Coevolution in bacteria and their viruses and plasmids. In: Futuyma DJ, Slatkin M (eds) Coevolution. Sinauer, Sunderland, Massachusetts, pp 99–127
Levin BR, Stewart FM, Chao L (1977) Resource-limited growth, competition, and predation: a model and experimental studies with bacteria and bacteriophage. Am Nat 111: 3–24
Levin DA (1975) Pest pressure and recombination systems in plants. Am Nat 109: 437–451
Levin DA (1979) The nature of plant species. Science 204: 381–384
Levin DA (1981) Dispersal versus gene flow in plants. Ann Mo Bot Gard 68: 233–253

Levin DA (1984) Immigration in plants: an exercise in the subjunctive, In: Dirzo R. Sarukhan (eds) Perspectives on plant population ecology: L Sinauer, Sunderland pp 242–260
Levin SA, Pimentel DS (1981) Selection of intermediate rates of increase in parasite-host systems. Am Nat 117: 308–315
Levin SA, Pimentel DS (1981) Selection of intermediate rates of increase in parasite-host systems. Am Nat 117: 308–315
Levin S, Cohen D, Hastings A (1988) Dispersal strategies in a heterogeneous environment. J Theor Biol (in press)
Levins R (1968) Evolution in changing environments. Princeton University Press, Princeton
Levy M, Levin DA (1975) Genic heterozygosity and variation in permanent translocation heterozygotes of the *Oenothera biennis* complex. Genetics 79: 493–512
Lewin B (1987) Genes, 3rd edn. Wiley, New York
Lewis II, Roberts MR (1956) The origin of *Clarkia lingulata*. Evolution 10: 126–138
Lewis WH (ed) (1980) Polyploidy: biological relevance. Plenum, New York
Lewontin RC (1962) Interdeme selection controlling a polymorphism in the house mouse. Am Nat 96: 65–78
Lewontin RC (1964) The interaction of selection and linkage. I. General considerations; heterotic models. Genetics 49: 49–67
Lewontin RC (1965) Selection for colonizing ability. In: Baker HG, Stebbins GL (eds) The genetics of colonizing species. Academic Press, London, pp 77–93
Lewontin RC (ed) (1968) Population biology and evolution. Syracuse University Press
Lewontin RC (1974) The genetic basis of evolutionary change. Columbia University Press, New York
Lewontin R (1978) Fitness, survival and optimality. In: Horn DH, Mitchell R, Stairs GR (eds) Analysis of ecological systems, Ohio State University Press, Ohio, pp 3–21
Lewontin (1985) Population genetics Annu Rev Genet 19: 81–102
Lewontin RC, Dunn LC (1960) The evolutionary dynamics of a polymorphism in the house mouse. Genetics 45: 705–722
Lewontin RC, Kojima K (1960) The evolutionary dynamics of complex polymorphism. Evolution 14: 458–472
Lewontin RC, Krakauer J (1973) Distribution of gene frequency as a test of the theory of the selective neutrality of polymorphisms. Genetics 74: 175–195
Lewontin RC, White MJD (1960) Interaction between inversion polymorphisms of two chromosome pairs in the grasshopper, *Moraba scurra*. Evolution 14: 116–129
Li WH (1967) Fundamental theorem of natural selection. Nature 214: 505–506
Li WH, Nei M (1974) Stable linkage disequilibrium without epistasis in subdivided populations. Theor Popul Biol 6: 173–183
Liberman U (1988) External stability and ESS: Criteria for initial increase of new mutant allele. J Math Biol 26: 477–485
Linhart YB (1974) Interpopulation differentiation in annual plants. I. *Veronica peregrina* L. raised under noncompetitive conditions. Evolution 28: 232–243
Lints, FA, Hoste C (1974) The Lansing effect revisited. I. Life Span Exp Gerontol 9: 51–69
Littler RA (1973) Linkage disequilibrium in two-locus finite random mating models without selection or mutation. Theor. Popul Biol 4: 259–275
Lloyd DG (1979) Some reproductive factors affecting the selection of self-fertilization in plants. Am Nat 113: 63–79
Lloyd DG (1980) Benefits and handicaps of sexual reproduction. Evol Biol 13: 69–111
Lloyd DG (1982) Selection of combined versus separate sexes in seed plants. Am Nat 120: 571–585
Loeschcke V (1984a) Species diversity and genetic polymorphism: Dual aspects of an exploitative competition model. In: Jayakar S, Zonta L (eds) Evolution and the genetics of populations: Atti Assoc Genet Ital XXIX (suppl): 53–66
Loeschcke V (1984b) The interplay between genetic composition, species number, and population sizes under exploitative competition. In: Wöhrmann K, Loeschcke V (eds) Population biology and evolution. Springer, Berlin Heidelberg New York, pp 235–246
Loeschcke V (1985) Coevolution and invasion in competitive guilds. Am Nat 126: 235–246
Loeschcke V (1987a) Niche structure and evolution in ecosystems. In: Schulze ED, Zwölfer H (eds) Ecological studies 61. Springer, Berlin Heidelberg New York, pp 320–332
Loeschcke V (1987b) Genetic constraints on adaptive evolution. Springer, Berlin Heidelberg New York, 188 pp

Loeschcke V, Christiansen FB (1984) Evolution and intraspecific exploitative competition. II. A two-locus model for additive gene effects. Theor Popul Biol 26: 228–264
Lokki J (1976) Genetic polymorphism and evolution in parthenogenetic animals. VIII. Heterozygosity in relation to polyploidy. Hereditas 83: 65–72
Lokki J, Suomalainen E, Saura A, Lankinen P (1975) Genetic polymorphism and evolution in parthenogenetic animals. II. Diploid and polyploid *Solenobia triquetrella* (Lepidoptera: Psvchidae). Genetics 79: 513–525
Lokki J, Saura A, Lankinen P, Suomalainen E (1976a) Genetic polymorphism and evolution in parthenogenetic animals. VI. Diploid and triploid *Polydrosus mollis* (Coleoptera: curculionidae). Hereditas 82: 209–216
Lokki J, Saura A, Lankinen P, Suomalainen E (1976b) Genetic polymorphism and evolution in parthenogenetic animals. V. Triploid *Adoxus obscurus* (Coleoptera: Chysomelidae). Genet Res Camb 28: 27–36
Lomnicki A (1988) Population ecology of individuals. Princeton University Press, Princeton, 224 pp
Long JC (1986) The allelic correlation structure of Gainj and Kalam speaking people. I The simulation and interpretation of Wright's F statistics. Genetics 112: 629–647
Lotka AJ (1925) Elements of physical biology. Baltimore: Williams & Wilkins (Reprinted 1956 as: Elements of mathematical biology: Dover Publications)
Lotka AJ (1932) The growth of mixed populations: two species competing for a common food supply. J Wash Acad Sci 22: 461–469
Loveless MD, Hamrick JL (1984) Ecological determinants of genetic structure in plant populations. Annu Rev Ecol Syst 15: 65–95
Lucas JR (1985) Time constraints and diet choice: different predictions from different constraints'. Am Nat 126: 680–705
Luckett DJ, Edwards KJR (1986) Esterase genes in parallel composite cross barley populations. Genetics 114: 289–302
Luckinbill LS, Clare MJ (1985) Selection for life span in *Drosophila melanogaster*. Heredity 55: 9–18
Luckinbill LS, Clare MJ (1986) A density threshold for the expression of longevity in *Drosophila melanogaster*. Heredity 56: 329–335
Lush JL (1945) *Animal breeding plans*. Collegiate Press, Ames, Iowa
Luskin M, Nagylaki T (1979) Numerical analysis of weak random drift in a cline. Genetics 92: 297–303
Lyman JC, Ellstrand NC (1984) Clonal diversity in *Taraxacum officinale* (Compositae), an apomict. Heredity 53: 1–10
Lynch M (1984) Destabilizing hybridization, general-purpose genotypes and geographical parthenogenesis. Q Rev Biol 59: 257–290
Lynch M (1985) Spontaneous mutations for life history characters in an obligate parthenogen. Evolution 39: 804–818
Lynch M, Gabriel W (1983) Phenotypic evolution and parthenogenesis. Am Nat 122: 745–764
MacArthur RH (1962) Some generalized theorems of natural selection, Proc Natl Acad Sci USA 48: 1893–1897
MacArthur RH (1970) Species packing and competitive equilibrium for many species. Theor Popul Biol 1: 1–11
MacArthur RH (1972) Geographical ecology. Harper & Row, New York
MacArthur RH, Levins R (1964) Competition, habitat selection and character displacement in a patchy environment. Proc Natl Acad Sci USA 51: 1207–1210
MacArthur RH, Levins R (1967) The limiting similarity, convergence and divergence of coexisting species. Am Nat 101. 377–385
MacArthur RH, Pianka E (1966) On optimal use of a patchy environment. Am Nat 100: 603–609
MacArthur RH, Wilson EO (1967) The theory of island biogeography. Princeton University Press, Princeton
Macevicz S, Oster GF (1976) Modeling social insect populations. II: Optimal reproductive strategies in annual eusocial insects colonies. Behav Ecol Sociobiol 1: 265–282
MacIntyre RJ (ed) (1985) Molecular evolutionary genetics, Plenum, New York
MacIntyre RJ, Collier GE (1986) Protein evolution in the genus *Drosophila*. In: Ashburner M, Carson HL, Thompson Jr JN (eds) The genetics and biology of *Drosophila*, vol 3. Academic Press, London, pp 39–146

Mackay TFC (1981) Genetic variation in varying environments. Genet Res 37: 79–95
MacKay PA, Wellington WG (1977) Maternal age as a source of variation in the ability of an aphid to produce dispersing forms. Res Popul Ecol (Kyot) 18: 195–209
Malecot G (1948) Les mathématiques de l'hérédité. Masson Paris
Mallet JLB (1985) Hybrid zones of *Heliconius* butterflies in Panama and the stability and movement of warning colour clines. Heredity 56: 191–202
Mani GS (1982) A theoretical study of morph frequency variation in the peppered moth over England and Wales. Biol J Linn Soc 17: 259–267
Marshall DR, Brown AHD (1974) Estimation of the level of apomixis in plant populations. Heredity 32: 321–333
Marshall DL, Beattie AJ, Bollenbacher WE (1979) Evidence for diglycerides as attractants in an ant-seed interaction. Chem Ecol 5: 335–344
Martin MJ, Perez Tome JM, Toro MA (1988) Competition and genotypic variability in *Drosophila melanogaster*. Heredity 69: 119–124
Maruyama T (1969) Genetic correlation in the stepping stone model with non-symmetrical migration rates. J Appl Probab 6: 463–477
Maruyama T (1970) Effective number of alleles in a subdivided population. Theor Popul Biol 1: 272–306
Maruyama T (1971) An invariant property of a structured population. Genet Res. 18: 81–84
Maruyama T (1972a) Rate of decrease of genetic variability in a two-dimensional continuous population of finite size. Genetics 70: 639–651
Maruyama T (1972b) Some invariant properties of a geographically structured finite population: distribution of neutral genes and of genes with small effect. Genet Res 20: 141–149
Maruyama T (1972c) Distribution of gene frequencies in a geographically structured finite population. I. Distribution of neutral genes and of genes with small effect. Ann Hum Genet 35: 411–423
Maruyama T (1977) Stochastic problems in population genetics. (Lecture notes biomathematics 17). Springer, Berlin Heidelberg New York
Maslin TP (1968) Taxonomic problems in parthenogenetic vertebrates. Syst Zool 17: 219–231
Matessi C, Jayakar SD (1976) Models of density-frequency dependent selection for exploitation of resources. I. In: Karlin S, Nevo E (eds) Population genetics and ecology, Academic Press, London, pp 707–721
Matessi C, Jayakar SD (1981) Coevolution of species in competition: A theoretical study. Proc Natl. Acad Sci 78: 1081–1084
Mather K (1953) The genetical structure of populations. Symp Soc Exp Biol 7: 66–95
Mather K (1955) Polymorphism as an outcome of disruptive selection. Evolution 9: 52–61
May RM (1973a) Stability and complexity in model ecosystems. Princeton University Press, Princeton
May RM (1973b) Stability in randomly fluctuating environments. Am Nat 107: 621
May RM (1976) Theoretical ecology: principles and applications. Saunders, Philadelphia
May RM (1981) Patterns in multi-species communities. In: May RM (ed) Theoretical ecology: principles and applications, 2nd edn. Blackwell, Oxford, pp 78–104
May RM (1984) An overview: real and apparent patterns in community structure. In: Ecological communities: conceptual issues and the evidence. Strong DR, Simberloff D, Abele LG, Thistle AB (eds) Princeton University Press, Princeton, pp 3–18
May RM (1985) Host-parasite associations. Their population biology and population genetics. In: Robinson D, Anderson RM (eds) Ecology and genetics of host-parasite interactions. Academic Press, London
May RM, Anderson RM (1983) Parasite-host coevolution. In: Futuyma DJ, Slatkin M (eds) Coevolution. Sinauer, Sunderland, Massachusetts, pp 186–206
May RM, Endler JA, McMurtrie RE (1975) Gene frequency clines in the presence of selection opposed by gene flow. Am Nat 109: 659–676
Maynard Smith J (1974) Group selection and kin selection. Nature 201: 1145–1147
Maynard Smith J (1966) Sympatric speciation. Am Nat 100: 637–650
Maynard Smith J (1968) Evolution in sexual and asexual populations. Am Nat 102: 469–473
Maynard Smith J (1970) Genetic polymorphism in a varied environment. Am Nat 104: 487–490.

Maynard Smith J (1971) The origin and maintenance of sex. In: Williams GC (ed) Group Selection. Aldine-Atherton, Chicago

Maynard Smith J (1976) Group selection. Q Rev Biol 51: 277–283

Maynard Smith J (1978a) The evolution of sex. Cambridge University Press, Cambridge

Maynard Smith J (1978b) Optimization theory in evolution. Annu Rev Ecol Syst 9: 31–56

Maynard Smith J (1981) Will a sexual population evolve to an ESS? Am Nat 117: 1015–1018

Maynard Smith J (1982) Evolution and the theory of games. Cambridge University Press, Cambridge

Maynard Smith J (1986) Contemplating life without sex. Nature 324: 300–301

Maynard Smith J (1987) How to model evolution. In: Dupré J (ed) The latest on the best. MIT Press, Cambridge, pp 119–132

Maynard Smith J, Haigh J (1974) The hitch-hiking effect of a favourable gene. Genet Res 23: 23–35

Maynard Smith J, Hoekstra R (1980) Polymorphism in a varied environment: how robust are the models. Genet Res 35: 45–57

Mayr E (1954) Change of genetic environment and evolution. In: Huxley J, Hardey AC, Ford EB (eds) Evolution as a process. Macmillan, New York, pp 157–180

Mayr E (1963) Animal species and evolution. Harvard, Cambridge, Massachusetts

Mayr E (1982a) The growth of biological thought: diversity, evolution and inheritance. Harvard, Cambridge, Massachusetts

Mayr E (1982b) Speciation and macroevolution. Evolution 36(6): 1119–1132

Mayr E (1982c) Process of speciation in animals. In: Barigozzi C (ed) Mechanisms of speciation. Liss, New York, pp 1–20

Mayr E (1983) How to carry out an adaptationist programm? Am Nat 21: 324–334

McCauley DE, Whittier DP, Reilly LM (1985) Inbreeding and the rate of self-fertilization in a grape fern, *Botrychium dissectum*. Am J Bot 72: 1978–1981

McCleery RH, Perrins CM (1985) Territory size, reproductive success and population dynamics in the Great Tit. *Parus major*. In: Sibly RM, Smith RH (eds) Behavioural ecology. BES Symp Blackwell, Oxford

McDonald JF, Ayala FJ (1974) Genetic response to environmental heterogeneity. Nature 250: 572–574

McFarland D, Houston AI (1981) Quantitative ethology: the state space approach. Pitman, Boston

McGraw JB, Antonovics J (1983) Experimental ecology of *Dryas octopetala* ecotypes. I. Ecotypic differentiation and life cycle stages of selection. J Ecol 71: 879–897

McKay FE (1971) Behavioral aspects of population dynamics in unisexual-bisexual *Poeciliopsis* (Pisces: Poeciliidae). Ecology 52: 778–790

McNair JN (1982) Optimal giving-up times and the marginal value theorem. Am Nat 119: 511–529

McNamara J (1982) Optimal patch use in a stochastic environment. Theor Popul Bio 21: 269–288

McNamara J, Houston AI (1980) The application of statistical decision theory to animal behaviour. J theor Biol 85: 673–690

McNamara J. Houston AI (1986) The common currency for behavioral decisions. Am Nat 127: 358–378

McNeill CI, Jain SK (1983) Genetic differentiation studies and phylogenetic inference in the plant genus *Limnanthes* (Section Inflexae). Theor Appl Genet 66: 257–269

McWilliams JR, Schroder HE, Marshall DR, Oram RN (1971). Genetic structure of Australian phalaris under domestication. Aust J Agric Res 22: 895–908

Meagher TR (1986) Analysis of paternity within a natural population of *Chamaelirium lateum*. I. Identification of most likely male parents. Am Nat 128: 199–215

Michod RE, Levin BR (1988) The evolution of sex. Sinauer, Sunderland, Massachusetts

Miescher F (1869) Über die chemische Zusammensetzung der Eiterzellen. Medizinisch-chemische Untersuchungen aus dem Laboratorium für angewandte Chemie zu Tübingen. Heft 4: 441–460

Milkman R (1973) Electrophoretic variation in *E. coli* from natural sources. Science 182: 1024–1026

Mitchell-Olds T, Rutledge R (1986) Quantitative genetics in natural plant populations: A review of the theory. Am Natur 127: 379–402
Mitter C, Brooks DR (1983) Phylogenetic aspects of coevolution. In: Futuyma DJ, Slathin M (eds) Coevolution. Sinauer, Sunderland, Massachusetts, pp 65–98
Mitter C, Futuyma DJ, Schneider JC (1979) Genetic variation and host plant relations in a parthenogenetic moth. Evolution 33: 777–790
Mitton JB, Grant MC (1984) Associations among protein heterozygosity, growth rate developmental homeostasis. Annu Rev Ecol Syst 15: 479–499
Mode CJ (1958) A mathematical model for the coevolution of obligate parasites and their hosts. Evolution 12: 158–165
Mode CJ (1961) A generalized model of a host-pathogen system. Biometrics 17: 386–404
Moore WS (1976) Components of fitness in the unisexual fish *Poeciliopsis monachaoccidentalis*. Evolution 30: 564–578
Moore WS (1984) Evolutionary ecology of unisexual fishes. In: Turner BJ (ed) Evolutionary genetics of fishes. Plenum, New York, pp 329–398
Moore WS, McKay FE (1971) Coexistence in unisexual-bisexual species complexes of *Poeciliopsis* (Pisces: Poeciliidae). Ecology 52: 791–799
Moran C (1979) The structure of the hybrid zone in *Caledia captiva*. Heredity 42: 13–42
Moran GF, Brown AHD (1980) Temporal heterogeneity of outcrossing rates in Alpine ash (*Eucalyptus delegatensis* R. T. Bak.). Theor Appl Genet 57: 191–105
Moritz C, Brown WM, Densmore LD et al. (1989) Genetic diversity and the dynamics of hybrid pathenogenesis in *Cnemidophorus* (Teiidae) and *Heteronotia* (Gekonidae). In: Dawley R, Bogart J (eds) Evolution and ecology of unisexual vertebrates. State University New York Press, Albany pp. 87–112
Motro V, Thomson G (1982) On heterozygosity and the effective size of populations subject to size changes. Evolution 36: 1059–1066
Mousseau TA, Roff DA (1987) Natural selection and the heritability of fitness components. Heredity 59: 181–197
Mukai T, Chigusa ST, Mettler LE, Crow JF (1972) Mutation rate and dominance of genes affecting viability in *Drosophila melanogaster*. Genetics 72: 335–355
Muller HJ (1932) Some genetic aspects of sex. Am Nat 66: 118–138
Muller HJ (1964) The relation of mutation to mutational advance. Mutat Res 1: 2–9
Muona O (1982) A multilocus study of an experimental barley population. Hereditas 96: 247–254
Murphy GI (1968) Pattern in life history and the environment. Am Nat 102: 390–404
Murray MG, Peters DL, Thompson WF (1981) Ancient repeated sequences in the pea and mung bean genomes and implications for genome evolution. J Mol Evol 17: 31–42
Nadeau JH, Collins RL (1983) Does associative overdominance account for the extensive polymorphism of H-2 linked loci. Genetics 105: 241–244
Nadeau JH, Collins RL, Klein J (1982) Organization and evolution of the mammalian genome. I. Polymorphism of H-2 linked loci. Genetics 102: 583–598
Nagylaki T (1975) Conditions for the existence of clines. Genetics 80: 595–615
Nagylaki T (1976a) A model for the evolution of self-fertilization and vegetative reproduction. J Theor Biol 58: 55–58
Nagylaki T (1976b) Clines with variable migration. Genetics 83: 867–886
Nagylaki T (1978a) A diffusion model for geographically structured populations. J Math Biol 6: 375–382
Nagylaki T (1978b) Random genetic drift in a cline. Proc Natl Acad Sci USA 75: 423–426
Nagylaki T (1983) The robustness of neutral models of geographical variation. Theor Popul Biol 24: 268–294
Nagylaki T (1985) Homozygosity, effective number of alleles and interdeme differentiation in subdivided populations. Proc Natl Acad Sci USA 82: 8611–8613
Nagylaki T (1989) The maintenance of genetic variability in two-locus models of stabilizing selection. Genetics 122: 235–248
Nagylaki T, Luicer B (1980) Numerical analysis of random drift in a cline. Genetics 94: 497–517

Nagylaki T, Moody M (1980) Diffusion model for genotype-dependent migration. Proc Natl Acad Sci USA 77: 4842–4846
Namkoong G, Gregorius H-R (1985) Conditions for protected polymorphism in subdivided plant populations II. Seed versus pollen migration. Am Nat 125: 521–534
Naveira H, Fontdevila A (1986) The evolutionary history of *Drosophila buzzatti*. XII. The genetic basis of sterility in hybrids between *D. buzzatti* and its sibling *D. serido* from Argentina. Genetics 114: 841–857
Nei M (1973) Analysis of gene diversity in subdivided populations. Proc Natl Acad Sci USA 70: 3321–3323
Nei M (1975) Molecular population genetics and evolution. North Holland, Amsterdam
Nei M (1987) Molecular evolutionary genetics. Columbia, New York
Nei M, Chesser RK (1983) Estimation of fixation indices and gene diversities. Ann Hum Genet 47: 253–259
Nei M, Li WH (1973) Linkage disequilibrium in subdivided populations. Genetics 75: 213–219
Nei M, Li WH (1976) The transient distribution of alleles under mutation pressure. Genet Res 28: 205–214
Nei M, Murata M (1966) Effective population size when fertility is inherited. Genet Res 8: 257–260
Nei M, Maruyama T, Chakraborty R (1975) The bottleneck effect and genetic variation in populations. Evolution 29: 1–10
Nevo E (1978) Genetic variation in natural populations: patterns and theory. Theor Popul Biol 13: 121–177
Nevo, E. (1983) Population genetics and ecology: the interface. In: Bendall D (ed), Evolution from molecules to men. Cambridge University Press, pp 283–321
Nevo E, Beiles A, Ben-Shlamo R (1984) Evolutionary significance of genic diversity: ecological, demographic and life-history correlates. Lect Notes Biomath 53: 13–213
Nevo E, Zohary D, Brown AHD, Haber M (1979) Genetic diversity and environmental associations of wild barley, *Hordeum spontaneum*, in Israel. Evolution 33: 815–833
Nilsson-Ehle H (1909) Kreuzungsuntersuchungen an Hafer und Weizen. Lunds Univ Aarskr N F Afd Ser 2, 5: 1–122
Noordwijk AJ van (1984) Problems in the analysis of dispersal and a critique on its 'heritability' in the great tit. J Anim Ecol 53: 533–544
Noorwijk AJ van (1987) On the implications of genetic variation for ecological research. Ardea 75: 13–19 (21)
Noordwijk AJ van (1988) Sib competition as an element of genotype-environment interaction for body size in the great tit. In: Jong G de (ed) Population genetics and evolution. Springer Berlin Heidelberg New York, pp 124–137
Noordwijk AJ van (1988) Two stage selection for body size in the great tit (Parus major). Proc IXXth Int Orn Congr, Ottawa 1986, pp 1408–1415
Noordwijk AJ van, Balen JH van (1988) The great tit. In: Clutton Brock T (ed) Variation in lifetime reproductive success. University Chicago Press, Chicago, pp 119–135
Noordwijk AJ van, Gebhardt M (1987) Reflections on the genetics of quantitative traits with continuous environmental variation. In: Loeschcke V (ed) Genetic constraints on adaptive evolution. Springer, Berlin Heidelberg New York, pp 73–90
Noorwijk AJ van, Jong G de (1986) Acquisition and allocation of resources: their influence of variation in life history tactics. Am Nat 128: 137–142
Noordwijk AJ van, Klerks PLM (1984) Heritability of bill dimensions in the great tit. Verh K Ned Akad Wet Afd Natuurkd 2e reeks, 81: 7–12
Noordwijk AJ van, Balen JH van, Scharloo W (1981) Genetic and environmental variation in clutch size of the great tit (*Parus major*). Neth J Zool 31: 342–372
Noordwijk AJ van, Balen JH van, Scharloo W (1988) Heritability of body size in a natural population of the great tit (*Parus major*) and its relation to age and environmental conditions during growth. Genet Res Camb 53: 149–162
Norton HTJ (1928) Natural selection and Mendelian variation. Proc Roy Soc Lond (Ser 2) 28: 1–45
Nunney L (1985) Group selection, altruism, and structured-deme models. Am Nat 126: 212–230

Nur N (1984) The consequence of brood size for breeding blue tits. I. Adult survival, weight change and the cost of reproduction. J Anim Ecol 53: 497–517
Oakeshott JG, Gibson JB, Anderson PR, Knibb WR, Anderson DG, Chambers GK (1982) Alcohol dehydrogenase and glycerol-3-phosphate dehydrogenase in *Drosophila melanogaster* in different continents. Evolution 36: 86–96
Ochman H, Selander RK (1984) Evidence for clonal population structure in *Escherichia coli.* Proc Natl Acad Sci USA 81: 198–201
Ochman HB, Stille B, Niklasson R, Salender R, Templeton AR (1980) Evolution of clonal diversity in the parthenogenetic fly *Lonchoptera dubia.* Evolution 34: 539–547
O'Donald P (1980) Genetic models of sexual selection. Cambridge University Press, Cambridge
O'Donald P (1983) The Arctic skua. Cambridge University Press, Cambridge
Ohnishi S, Kawanishi M, Watanabe TK (1983) Biochemical phylogenies of *Drosophila*: protein differences detected by two-dimensional electrophoresis. Genetica 61: 55–63
Ohno S (1970) Evolution by gene duplication, Springer, Berlin Heidelberg New York
Ohno S (1984) Birth of unique enzyme from an alternative reading frame of the preexisted, internally repititions coding sequence. Proc Natl Acad Sci USA 81: 2421–2425
Ohta T (1976) Role of very slightly deleterious mutations in molecular evolution and polymorphism. Theor Popul Biol 10: 254–275
Ohto T (1982) Linkage disequilibrium due to random genetic drift in finite subdivided populations. Proc Natl Acad Sci USA 79: 1940–1944
Ohta T, Cockerham CC (1974) Detrimental genes with partial selfing on a neutral locus. Genet Res 23: 191–200
Ohto T, Kimura M (1969) Linkage disequilibrium at steady state determined by random genetic drift and recurrent mutation. Genetics 63: 229
Ohta T, Kimura M (1970) Development of associative overdominance through linkage disequilibrium in finite populations. Genet Res 16: 165–177
Ohta T, Kimura M (1971) Linkage disequilibrium between two segregating nucleotide sites under the steady flux of mutations in a finite population. Genetics 68: 571–580
Oliver CG (1979) Genetic differentiation and hybrid viability within and between some Lepidopteran species. Am Nat 114: 681–694
Oliver JH Jr, Herrin CS (1976) Differential variation of parthenogenetic and bisexual *Haemapysalis longicornis* (Acari: Ixodidae). J Parsitol 62: 475–484
Olivieri A, Jain SK (1977) Variation in *Helianthus exilis-H. bolanderi* complex: a reexamination. Madrono 24: 177–189
Orr AH (1987) Genetics of male and female sterility in bybrids of *D. pseudoobscura* and *D. persimilis.* Genetics 116: 555–563
Oster GF, Wilson EO (1978) Caste and ecology in the social insects. Princeton University Press, Princeton
Palmer JO, Dingle H (1986) Direct and correlated responses to selection among life-history traits in milkweed bugs (*Oncopeltus fasciatus*). Evolution 40: 767–777
Park T (1962) Beetles, competition and populations. Science 138: 1369–1375
Parker ED (1979) Ecological implications of clonal diversity in parthenogenetic morphospecies. Am Zool 19: 753–762
Parker ED, Selander RK (1976) The organization of genetic diversity in the parthenogenetic lizard *Cnemidophorus tesselatus.* Genetics 84: 791–805
Parker ED, Selander RK, Hudson RO, Lester LJ (1977) Genetic diversity in colonizing parthenogenetic cockroaches. Evolution 31: 836-842
Parker GA, Courtney SP (1984) Models of clutch size in insect oviposition. Theor Popul Biol 26: 27–48
Parker GA, Hammerstein P (1985) Game theory and animal behaviour. In: Greenwood PJ, Harvey PH, Slatkin M (eds) Evolution. Cambridge University Press, Cambridge, pp 73–94
Parsons PA (1961) The initial progress of new genes with viability differences between sexes and with sex linkage. Heredity 16: 103–107
Parsons PA (1983) The evolutionary biology of colonizing species. Cambridge University Press, Cambridge
Partridge L (1986) Sexual activity and life span. In: Collatz KG, Sohal RS (eds) Insect aging, Springer, Berlin Heidelberg New York, pp 45–54

Partridge L, Harvey PH (1985) Costs of reproduction. Nature 316: 20

Partridge L, Harvey PH (1988) The ecological context of life history evolution. Science 241: 1449–1455

Patterson C (ed) (1986) Molecules and morphology in evolution: conflict or compromise? Cambridge University Press Cambridge

Pearl R, Read L (1920) On the rate of growth of the population of the United States since 1790 and its mathematical representation. Proc Natl Acad Sci USA 6: 275–288

Perez-Tome JM, Toro M (1982) Competition of similar and non-similar genotypes. Nature 299: 153–154

Perrins CM (1965) Population fluctuations and clutch size in the great tit *Parus major*. J Anim Ecol 44: 601–647

Perrins CM (1979) British tits. Collins, London

Petit C (1954) L'isolement sexual chez *Drosophila melanogaster*. Elude du mutant *White* et de son allélomorphe *sauvage*. Bull Biol Fr Belg 88: 435–443

Pianka ER (1983) Evolutionary ecology, 2nd edn, Harper & Row, New York

Pierce GJ, Ollason JG (1987) Eight reasons why opimal foraging theory is a complete waste of time. Oikos 49: 111–118

Pietiäinen II, Saorola P, Väisänen RA (1986) Parental investment in clutch size and egg size in the ural owl (*Strix uralensis*). Ornis Scand 17: 309–325 (36)

Pimentel D (1968) Population and genetic feedback. Science 159: 1432–1436

Pimentel DS, Levin SA, Olson D (1978) Coevolution and the stability of exploiter-victim systems. Am Nat 112: 119–125

Piper JG, Charlesworth B, Charlesworth D (1984) A high rate of self-fertilization and increased seed fertility of homostyle primroses. Nature 310: 50–51

Pollak E (1978) With selection for fecundity the mean fitness does not necessarily increase. Genetics 90: 383–389

Poulsen ET (1979) A model for population regulation with density- and frequency-dependent selection. J Math Biol 8: 325–343

Powell JR (1971) Genetic polymorphisms in varied environments. Science 174: 1035–1056

Powell JR (1975) Protein variation in natural populations of animals. In: Dobzhansky T, Hecht MK, Steere WC (eds) Evolutionary biology, vol 8. Plenum, New York, pp 79–119

Powell JR, Wistrand H (1978) The effect of heterogeneous environment and a competitor on genetic variation in *Drosophila*. Am Nat 112: 935–947

Price GR (1970) Selection and covariance. Nature 227: 520–521

Prout T (1965) The estimation of fitnesses from genotypic frequencies. Evolution 19: 546–551

Prout T (1980) Some relationships between density independent selection and density dependent growth. Evol Biol 13: 1–96

Pyke GH (1984) Optimal foraging theory: a critical review. Annu Rev Ecol Syst 15: 523–575

Pyke GH, Pulliam HR, Charnov EL (1977) Optimal foraging: a selective review of theory and tests Q Rev Biol 52: 137–154

Quinn JA (1987) Complex patterns of genetic differentiation and phenotypic plasticity versus an outmoded ecotype terminology. In: Urbanska (ed) Differentiation patterns in higher plants. Academic Press, London

Raff RA, Kaufman TC (1983) Embryos, genes and evolution. MacMillan, New York

Rai K, Jain SK (1982) Population biology of *Avena*. IX. Gene flow and neighborhood size in relation to microgeographic variation in *A. barbata*. Oecologia (Berl) 53: 399–405

Raiffa H (1968) Decision analysis. Addison-Wisley, Reading MA

Ralls K, Harvey PH, Lyles AM (1986) Inbreeding in natural populations of birds and mammals. In: Soule ME (ed) Conservation biology, Sinauer, Sunderland, Massachusetts, pp 35

Reddingius J, Boer PJ Den (1970) Simulation experiments illustrating stabilization of animal number by spreading of risk. Oecologia 5: 240–248

Reichman OJ, Aitchison S (1981) Mammal trails on mountain slopes: Optimal paths in relation to slope angle and body weight. Am Nat 117: 416–420

Rendel JM (1967) Canalization and gene control. Logos, London

Rescigno A, Richardson IW (1967) The struggle for life. I. Two species. Bull Math Biophysics 29: 377–388

Reynolds J, Weir BS, Cockerham CC (1983) Estimation of the coancestry coefficient: basis for a short-term genetic distance. Genetics 105: 767–779
Reznick D (1981) "Grandfather effects": the genetics of interpopulation differences in offspring size in the mosquito fish. Evolution 35: 941–933
Reznick D (1985) Costs of reproduction: an evaluation of the empirical evidence. Oikos 44: 257–267
Reznick DN, Perry E, Travis J (1986) Measuring the cost of reproduction: a comment on papers by Bell. Evolution 40: 1338–1344
Rice WR (1983) Parent-offspring pathogen transmission: a selective agent promoting sexual reproduction. Am Nat 121: 1317–1320
Richards AJ (1973) The origin of *Taraxacum* agamospecies. Bot J Linn Soc 66: 189–211
Richards AJ (1986) Plant breeding systems. Longman, London
Riesberg LH, Soltis DE, Palmer JD (1988) A molecular reexamination of introgression between *Helianthus annuus* and *H. bolanderi*. Evolution 42: 205–214
Ritland K (1984) The effective proportion of self-fertilization with consanguineous matings in inbred populations Genetics 106: 139–152
Ritland K (1986) Joint maximum likelihood estimation of genetic and mating structure using open-pollinated progenies. Biometrics 42: 25–43
Ritland K, Ganders FR (1985) Variation in the mating system of *Bidens mengiesii* (Asteraceae) in relation to population structure. Heredity 55: 235–244
Ritland K, Jain SK (1984a) A comparative study of floral and electrophoretic variation with life history characters in *Limnanthes alba*. Oecologia (Berl) 23: 243–251
Ritland K, Jain SK (1984b) The comparative life histories of two annual *Limnanthes* species in a temporally variable environment. Am Nat 124: 656–679
Roberts JJ, Gallun RL (1984) Chromosome location of the H_5 gene for resistance to the Hessian fly in wheat. J Hered 75: 147–148
Robertson A (1966) A mathematical model of the culling process in dairy cattle. Anim Prod 8: 95–108
Robertson A (1975) Gene frequency distribution as a test of selective neutrality. Genetics 11: 775–785
Roelofs W, Glover T, Tang X-H (1987) Sex pheromone production and perception in European Corn borer moth is determined by both autosomal and sex-linked genes. Proc Natl Acad Sci USA 84: 7585–7589
Roff DA (1984) The evolution of life history parameters in teleosts. Can J Fish Aquat Sci 41: 989–1000
Roff DA (1986) Evolution of wing polymorphism and its impact on life cycle adaptation in insects. In: Taylor F, Karban R (eds) The evolution of insect life cycles. Springer, Berlin Heidelberg New York, pp 204–221
Roff DA, Mousseau TA (1987) Quantitative genetics and fitness: lessons from *Drosophila*. Heredity 58: 103–118
Rogers A (1987) A model of kin-structured migration. Evolution 41: 417–426
Rose MR (1983) Theories of life-history evolution. Am Zool 23: 15–23
Rose MR, Charlesworth B (1981) Genetics of life history evolution in *Dorosophila melanogaster*. I. Sib analysis of adult females. Genetics 97: 173–186
Rose MR, Doolittle WF (1983) Molecular biological mechanisms of speciation. Science 220: 157–162
Rosenzweig ML (1978) Competitive speciation. Biol J Linn Soc 10: 275–289
Ross J (1982) Myxomatosis: the natural evolution of the disease. In: Animal disease in relation to animal conservation. Symp Zool Soc Lond 26: 26–27
Ross SM (1983) Introduction to stochastic dynamic programming. Academic Press, London
Rossi L, Fano ED (1979) Role of fungi in the trophic niche of the congeneric detritivorous *Asellus aquaticus* and *Proasellus coxalis*. Oikos 32: 380–385
Roughgarden J (1971) Density-dependent natural selection. Evolution 52: 453–468
Roughgarden J (1972) Evolution of niche width. Am Nat 106: 683–718
Roughgarden J (1975) Evolution of marine symbiosis—a simple cost-benefit model. Ecology 56: 1201–1208

Roughgarden J (1976) Resource partitioning among competing species: a coevolutionary approach. Theor Popul Biol 9: 388–424

Roughgarden J (1979) Theory of population genetics and evolutionary ecology: an introduction. Macmillan, New York

Roughgarden J (1983a) The theory of coevolution. In: Futuyma DJ, Slatkin M (eds) Coevolution. Sinauer, Sunderland, Massachusetts, pp 33–64

Roughgarden J (1983b) Coevolution between competitors. In: Futuyama DJ, Slatkin M (eds) Coevolution. Sinauer, Sunderland, Massachusetts, pp 383–403

Roughgarden J, Gaines S, Possingham II (1988) Recruitment dynamics in complex life cycles. Science 241: 1460–1466

Rouhani S, Barton NH (1987) Speciation and the "shifting balance" in a continuous population. Theor Popul Biol 31: 465–492

Roy J, Mooney HA (1982) Physiological adaptations and plasticity to water stress of coastal and desert populations of *Heliotropium curassaricum* L. Oecologia (Berl) 52: 370–375

Sachs JM (1967) A stable equilibrium with minimum average fitness. Genetics 56: 705–708

Sammeta K, Levins R (1970) Genetics and ecology. Ann Rev Genet 4: 469–488

Saura A, Lokki J, Lankinen P, Suomalainen E (1976a) Genetic polymorphism and evolution in parthenogenetic animals. III. Tetraploid *Otiorrhynchus scaber* (Coleoptera: Curculionidae). Hereditas 82: 79–100

Saura A, Lokki, J, Lankinen P, Suomalainen E (1976b) Genetic polymorphism and evolution in parthenogenetic animals. IV. Triploid *Otiorrhynchus salicis* Strom (Coleoptera: Curculionidae). Ent Omol Scand 7: 1–6

Schaal BA (1984) Life-history variation, natural selection, and maternal effects in plant populations. In: Dirzo R, Sarukhán J (eds) Perspectives in plant population ecology. Sinauer, Sunderland, Massachusetts pp 188–206

Schaffer WM (1974a) Optimal reproductive effort in fluctuating environments. Am Nat 108: 783–790

Schaffer WM (1974b) Selection for optimal life histories: the effects of age structure. Ecology 55: 291–303

Schaffer WM (1981) On reproductive value and fitness. Ecology 62: 1683–1685

Schaffer WM (1983) The application of control theory and the general life history problem. Am Nat 121: 418–431

Schaffer WM, Rosenzweig ML (1977) Selection for optimal life histories. II. Multiple equilibrium and the evolution of alternative reproductive strategies. Ecology 58: 60–72

Schaffer WM, Rosenzweig ML (1978) Homage to the Red Queen. I. Coevolution of predators and their victims. Theor Popul Biol 14: 135–157

Schaffer WM, Inouye RS, Whittam TS (1982) Energy allocation by an annual plant when the effects of seasonality on growth and reproduction are decoupled. Am Nat 120: 787–815

Scharloo W (1966) Pattern formation and canalization. Genen Phaenen 11: 1–15

Scharloo W (1987) Constraints in selection response. In: Loeschcke V (ed) Constraints on adaptive evolution. Springer, Berlin Heidelberg New York, pp 125–149

Scharloo W (1988) Selection on morphological patterns. In: Jong G de (ed) Population genetics and evolution. Springer, Berlin Heidelberg New York, pp 230–250

Scheiner SM, Goodnight CJ (1984) The comparison of phenotypic plasticity and genetic variation in populations of the grass *Danthonia spicata*. Evolution 38: 845–855

Schemske DW (1984) Population structure and local selection in *Impatiens pallida* (Balsaminaceae), a selfing annual. Evolution 38: 817–832

Schemske DW, Lande R (1985) The evolution of self-fertilization and inbreeding depression in plants. II. Empirical observations. Evolution 39: 41–52

Schenck RA, Vrijenhoek RC (1986) Spatial and temporal factors affecting coexistence among sexual and clonal of *Poeciliopsis*. Evolution 40: 1060–1070

Schenck RA, Vrijenhoek RC (1989) Coexistence among sexual and asexual forms of *Poeciliopsis*: foraging behavior and microhabitat selection. In: Dawley R, Bogart J (eds) Evolution and ecology of unisexual vertebrates. State University New York Press, Albany pp. 39–48

Schmalhausen II (1949) Factors of evolution. Reprinted by University of Chicago Press, Chicago, 1986
Schmid-Hempel P (1989) Reproductive competition and the evolution of work load in social insects. Am Nat 134 (in press)
Schmid-Hempel P, Kacelink A, Houston AI (1985) Honeybees maximize efficiency by not filling their crop. Behav Ecol Sociobiol 17: 61–66
Schmidt KP, Levin DA (1985) The comparative demography of reciprocally sown populations of *Phlox drommondii* (Hook). Evolution 39: 396–404
Schmitt J, Antonovics J (1986) Experimental studies of the evolutionary significance of sexual reproduction. IV. Effect of neighbor relatedness and aphid infestation on seedling performance. Evolution 40: 830–836
Schnee BK, Waller DM (1986) Reproductive behavior of *Amphicarpaea bracteata* (Leguminosae), an amphicarpic annual. Am J Bot 73: 376–386
Schoen DJ (1982) The breeding system of *Gilia achilleifolia:* variation in floral characteristics and outcrossing rate. Evolution 36: 352–360
Schoen DJ, Clegg MT (1984) Estimation of mating system parameters when outcrossing events are correlated. Proc Natl Acad Sci 81: 5258–5262
Schoen DJ, Clegg MT (1986) Montle Carlo studies of plant mating system estimation models: the one-pollen parent and mixed mating models. Genetics 112: 927–945
Schoener TW, Roughgarden J, Fenchel T (1986) The body-size-prey-size hypothesis: a defence. Ecology 67: 260–261
Schultz RJ (1969) Hybridization, unisexuality and polyploidy in the teleost *Poeciliopsis* (Poeciliidae) and other vertebrates. Am Nat 103: 605–619
Schultz RJ (1971) Special adaptive problems associated with unisexual fishes. Am Zool 11: 351–360
Schultz RJ (1973) Unisexual fish: laboratory synthesis of a "species". Science 179: 180–181
Schwanitz F (1974) Genetick und Evolutionsforschung bei Pflanzen. In: Heberer G (Hrg) Die Evolution der Organismen Bd II/1. Fischer, Stuttagart, pp 1–189
Scudo FM, Ziegler JR (1976) Vladimir Aleksandrovich Kostitzin and theoretical ecology. Theor Popul Biol 10: 395–412
Scudo FM, Ziegler JR (1978) The golden age of theoretical ecology: 1923–1940 (Lecture Notes in Biomathematics 22). Springer, Berlin Heidelberg New York
Sebens KP, Thorne BL (1985) Coexistence of clones, clonal diversity and the effects of disturbance In: Jackson JBC, Buss LW, Cook RE (eds) Population biology and evolution of clonal organisms. Yale University Press, New Haven, pp 357–398
Seiler J (1961) Untersuchungen über die Entstehung der Parthenogenese bei *Solenobia triquetrella* FR (Lepidoptera, Psychidae). III Die geographische Verarbeitung der drei Rassen von *Solenobia triquetrella* (bisexuell, diploid, und tetraploid parthenogenetisch) in der Schweiz und in den angrenzenden Länder und die Beziehung zur Eiszeit. Bem Z Vererb 92: 261–316
Selander RK (1976) Genic Variation in natural populations. In: Ayala FJ (ed) Molecular evolution. Sinauer, Sunderland, Massachusetts, pp 21–45
Selander RK, Kaufman DW (1973) Genetic variablity and strategies of adaptation in animals. Proc Natl Acad Sci USA 70: 1875–1877
Selander RK, Levin BR (1980) Genetic diversity and structure in *Escherichia coli* populations. Science 210: 545–547
Selgrade JF, Ziehe M (1987) Convergence to equilibrium in a genetic model with differential viability between the sexes. J Math Biol 25: 477–490
Shapiro AM (1976) Seasonal polyphenism. Evol Biol 9: 259–333
Shapiro JA (1983) Variation as a genetic engineering process. In: Bendall DS (ed) Evolution from molecules to men. Cambridge University Press, pp 253–271, Cambridge
Shaw DD, Coates DJ, Arnold ML, Wilkinson P (1985) Temporal variation in the chromosomal structure of a hybrid zone and its relationship to karyotype repatterning. Heredity 55: 293–307
Sheppard PM (1958) Natural selection and heredity. Hutchinson, London
Shields W (1982) Philopatry, inbreedig, and the evolution of sex. State University New York Press, Albany, New York

Siegismund HR, Christiansen FB (1985) Selection component analysis of natural polymorphisms using population samples including mother-offspring combinations. III. Theor Popul Biol 27: 268–297
Silander J (1985) Microevolution in clonal plants. In: Jackson BC, Buss LW, Cook LE (eds) Population biology and evolution of clonal organisms. Yale University Press, New Haven, pp 107–151
Simmons MJ, Crow JF (1977) Mutations affecting fitness in *Drosophila* populations. Annu Rev Genet 11: 49–78
Simpson GG (1953) The major features of evolution. Columbia, New York
Singh DP (1986) Breeding for resistance to diseases and insect pests. Crop Predation Monographs. Springer, Berlin Heidelberg New York
Singh RS (1983) Genetic differentiation for allozyme and fitness characters between mainland and Bogota populations of *Drosophila pseudoobscura*. Can J Genet Cytol 25: 590–604
Singh RS (1988a) The genetic studies of species differences and their relevance to the problem of species formation. In: Loxadale H (ed) Electrophoretic studies of agricultural pests. Oxford University Press, Oxford (in press)
Singh RS (1988b) A comprehensive study of genic variation in natural populations of *Drosophila melanogaster*. IV. Molecular functional constraints and their effects on genic diversity (in preparation)
Singh RS (1989) A macroevolutionary perspective of molecular variation and species divergence in *Drosophila* (in preparation)
Singh RS, Rhomberg LR (1987a) A comprehensive study of genic variation in natural populations of *Drosophila melanogaster*. I. Estimates of geneflow from rare alleles. Genetics 115: 313–322
Singh RS, Rhomberg LR (1987b) A comprehensive study of genic variation in natural populations of *Drosophila melanogaster*. II. Estimates of heterozygosity and patterns of geographic differentiation. Genetics 117: 255–271
Skellam JG (1951) Random dispdersal in theoretical populations. Biometrika 38: 196–218.
Skellam JG (1973) The formulation and interpretation of mathematical models of diffusionary processes in population biology. In: Bartlett MS, Hiorns RW (eds) The mathematical theory of the dynamics of biological populations. Academic Press, London, pp 63–85
Skinner SW (1985) Clutch size as an optimal foraging problem for insects. Behav Ecol Sociobiol 17: 231–238
Slatkin M (1973) Gene flow and selection in a cline. Genetics 75: 733–756
Slatkin M (1975) Gene flow and selection in a two-locus system. Genetics 81: 209–222
Slatkin M (1978) Spatial patterns in the distribution of polygenic characters. J Theor Biol 70: 213–228
Slatkin M (1979) Frequency- and density-dependent selection on a quantitative character. Genetics 93: 755–771
Slatkin M (1980a) The distribution of mutant alleles in a subdivided population. Genetics 95: 503–524
Slatkin M (1980b) Ecological character displacement. Ecology 61: 163–177
Slatkin M (1981a) Estimating levels of gene flow in natural populations. Genetics 99: 323–335
Slatkin M (1981b) Population heritability. Evolution 35: 859–871
Slatkin M (1988) Rare alleles as indicators of gene flow. Evolution (in press)
Slatkin M (1985) Gene flow in natural populations. Ann Rev Ecol Syst 16: 393–430
Slatkin M, Maruyama T (1975) Genetic drift in a cline. Genetics 81: 209–222
Slatkin M, Maynard Smith J (1979) Models of coevolution. Q Rev Biol 54: 233–263
Slatkin M, Takahata N (1985) The average frequency of private alleles in a partially isolated population. Theor Popul Biol 28: 314–331
Smith CC (1975) The coevolution of plants and seed predators. In: Gilbert LE, Raven PH (eds) Coevolution of animals and plants. University of Texas Press, Austin, pp 53–77
Smith JNM (1981) Does high fecundity reduce survival in song sparrows? Evolution 35: 1142–1148
Smithies O (1964) Chromosomal rearrangements and protein structure. Cold Spring Habor Symp Quant Biol 29: 309–319

Smyth CA, Hamrick JL (1984) Variation in estimates of outcrossing in musk thistle populations. J Hered 75: 303–307
Snaydon RW, Davies MS (1982) Rapid divergence of plant populations in response to recent changes in soil conditions. Evolution 36: 289–297
Sobel MJ, Arnold J, Sobel M (1986) Statistics of natural populations. III. Sequential sampling plans for the estimation of gene frequencies. Biometrics 42: 45–65
Solbreck C (1986) Wing and flight muscle polymorphism in a lygaeid bug, *Horvathiolus gibbicollis:* determinants and life history consequences. Ecol Entomol 11: 435–444
Solbrig OT, Simpson BB (1974) Components of regulation of a population of dandelions in Michigan, J Ecol 62: 473–486
Solbrig OT, Simpson BB (1977) A garden experiment on competition between biotypes of the common dandelion (*Taraxacum officinale*). J Ecol 65: 427–430
Solignac M, Monnerot M (1986) Race formation, speciation, and introgression within *Drosophila simulans, D. mauritiana*, and *D. sechellia* inferred from mitochondrial DNA analysis. Evolution 40: 531–39
Soltis DE, Soltis PS (1987) Polyploidy and breeding systems in homosporous pteridophyta: A reevaluation. Am Nat 130: 219–232
Sorensen FC (1982) The roles of polyembryony and embryo viability in the genetic system of conifers. Evolution 36: 725–733
Southwood TRE (1976) Bionomic strategies and population parameters. In: May R (ed) Theoretical Ecology: Principles and applications. Saunders, Philadelphia, pp 26–48
Souza HML DE, Cunha AB DA, Santos EP Dos (1970) Adaptive polymorphism of behavior evolved in laboratory populations of *Drosophila willistoni.* Am Nat 104: 175–189
Spinella DG, Vrijenhoek RC (1982) Genetic dissection of clonally inherited genomes of *Poeciliopsis.* II. Investigation of a silent carboxylesterase allele. Genetics 100: 279–286
Stanley SM (1975) Clades versus clones in evolution: why we have sex. Science 190: 382–383
Stanley SM (1979) Macroevolution: pattern and process. Freeman, San Francisco
Stanley SM (1982) Speciation and the fossil record. In: Barigozzi C (ed) Mechanisms of speciation. Liss, New York, pp 41–49
Stearns SC (1976) Life-history tactics: a review of the ideas. Q Rev Biol 51: 3–47
Stearns SC (1977) The evolution of life history traits: a critique of the theory and a review of the data. Annu Rev Ecol Syst 8: 145–171
Stearns SC (1980) A new view of life history evolution. Oikos 35: 266–281
Stearns SC, Crandall RE (1984) Plasticity for age and size at sexual maturity: a life-history response to unavoidable stress. In: Potts G, Wootton R (eds) Fish reproduction. Academic Press, London pp 13–33
Stearns SC, Koella JC (1986) The evolution of phenotypic plasticity in life-history traits: predictions of reaction norms for age and size at maturity. Evolution 40: 893–913
Stearns SC, Schmid-Hempel P (1987) Evolutionary insights should not be wasted. Oikos 49: 118–125
Stebbins GL (1950) Variation and evolution in plants. Columbia University Press, New York
Stebbins GL (1958) Longevity, habitat, and release of genetic variability in the higher plants. Cold Spring Harbor Symp Quant Biol 23: 365–378
Stebbins GL (1982a) Perspectives in evolutionary theory. Evolution 36: 1109–18
Stebbins GL (1982b) Plant speciation. In: Barigozzi C (ed) Mechanisms of speciation. Liss, New York, pp 21–39
Stebbins GL (1988) Essays in comparative evolution: The need for evolutionary comparisons. In: Gottieb LD, Jain SK (eds) Plant evolutionary biology. Chapman & Hall, London, pp 3–20
Stebbins GL, Daly K (1961). Changes in variation pattern of hybrid population of *Helianthus* over an eight-year period. Evolution 15: 60–71
Stenseth NC, Maynard Smith J (1984) Coevolution in ecosystems: Red Queen evolution or stasis. Evolution 38: 870–880
Stenseth NC, Kirkendall LR, Moran N (1985) On the evolution of pseudogamy. Evolution 39: 294–307

Stephens DW (1981) The logic of risk-sensitive foraging preferences. Anim Behav 29: 628–629
Stephens DW (1987) On economically tracking a variable environment. Theor Popul Biol 32: 15–25
Stephens DW, Charnov EL (1982) Optimal foraging: some simple stochastic models. Behav Ecol Sociobiol 10: 251–263
Stephens DW, Krebs JR (1986) Foraging theory. Princeton University Press, Princeton
Stephens JC, Nei M (1985) Phylogenic analysis of polymorphic DNA sequences at the Adh locus in *Drosophila melanogaster* and its sibling species. J Mol Evol 22: 289–300
Stephenson AG, Winsor JA (1986) *Lotus corniculatus* regulates offspring quality through selective fruit abortion. Evolution 40: 453–458
DeSteven D (1980) Clutch size, breeding success, and parental survival in the tree swallow (*Iridoprocne bicolor*). Evolution 34: 278–291.
Stokes AN (1976) On two types of moving front in quasilinear diffusion. Math Biosc 31: 307–315
Strickberger MW (1985) Genetics, 3rd edn. Macmillan, New York
Strobeck C (1974) Sufficient conditions for polymorphism with N niches and M mating groups. Am Nat 108: 152–156
Strobeck C (1979) Partial selfing and linkage: The effect of a heterotic locus on a neutral locus. Genetics 92: 305–315
Sultan SE (1987) Evolutionary implications of phenotypic plasticity in plants. Evol Biol 21: 127–178
Suomalainen E (1947) Parthenogenese und Polyploidie bei Russelkäfern (Curculionidae). Hereditas 33: 425–456
Suomalainen E, Saura A, Lokki J (1987) Cytology and evolution in parthenogenesis. CRC, Boca Raton, Florida
Suzuki DT, Griffiths AJF, Miller JH, Lewontin RC (1986) An introduction to genetic analysis, 3rd edn. Freeman, New York, 612 pp
Sved JA, Reed TE, Bodmer WF (1967) The number of balanced polymorphisms that can be maintained by natural selection. Genetics 55: 469–481
Swennen C, Bruijn LLM de, Duiven P, Leopold MF, Marteijn ECL (1983) Differences in bill form of the oystercatcher *Haematopus ostralegus*; a dynamic adaptation to specific foraging techniques. Neth J Sea Res 17: 57–83
Szymura JM, Barton NH (1986) Genetic analysis of a hybrid zone between the fire-bellied toads *Bombina bombina* and *B. variegata*, near Cracow in Southern Poland. Evolution 40: 1141–1159
Tabachnik WJ, Munstermann LE, Powell JR (1979) Genetic distinctness of sympatric forms of *Aedes aegypti* in East Africa. Evolution 33: 287–295
Tajima F (1983) Evolutionary relationships of DNA sequences in finite populations. Genetics 105: 437–460
Takahata N, Maruyama T (1981) A mathematical model of extranuclear genes and the genetic variability maintained in a finite population. Genet Res 37: 291–302
Takahata N, Nei M (1984) F_{ST} and G_{ST} statistics in the finite island model. Genetics 107: 501–504
Takahata N, Nei M (1985) Gene genealogy and variance of interpopulation nucleotide differences. Genetics 110: 325–344
Takahata N, Palumbi SR (1985) Extranuclear differentiation and gene flow in the finite island model. Genetics 109: 441–457
Taper ML, Case TJ (1985) Quantitative genetic models for the coevolution of character displacement. Ecology 66: 355–371
Taylor CE (1976) Genetic variation in heterogeneous environments. Genetics 83: 887–894
Taylor CE, Pereda AD, Ferrari JA (1987) On the correlation between mating success and offspring quality in *Drosophila melanogaster*. Am Nat 129: 721–729
Taylor F (1980) Optimal switching to diapause in relation to onset of winter. Theor Popul Biol 18: 125–133
Taylor HM, Gourby RS, Lawrence CE, Kaplan RS (1974) Natural selection of life-history attributes: an analytical approach. Theor Popul Biol 5: 104–122

Taylor PD (1979) An analytical model for a short-term advantage for sex. J Theor Biol 81: 407–421
Templeton AR (1980) The theory of speciation via the founder principle. Genetics 94: 1011–1038
Templeton AR (1981) Mechanisms of speciation—a population genetics approach. Annu Rev Ecol Syst 12: 23–48
Templeton AR (1982) The prophecies of parthenogenesis. In: Dingle H, Hegmann JP (eds) Evolution and genetics of life histories. Proc life Sci, Springer, Berlin Heidelberg New York, 250 pp
Templeton AR, Rothman Ed (1981) Evolution in fine-grained environments. II. Habitat selection as a homeostatic mechanism. Theor Popul Biol 7: 326–340
Thibault RE (1974) Genetics of cannibalism in a viviparous fish and its relationship to population density. Nature 251: 138–140
Thoday JM, Gibson JB (1970) The probability of isolation by disruptive selection. Am Nat 104: 219–230
Thomson G (1977) The effect of a selected locus on linked neutral loci. Genetics 85: 753–788
Thompson P, Sites JW (1986) Comparison of population structure in chromosomally polytypic and monotypic species of *Sceloporus* in relation to chromosomally-mediated speciation. Evolution 40: 303–314
Throckmorton LH (1977) *Drosophila* systematics and biochemical evolution. Annu Rev Ecol Syst 8: 235–54
Tienderen PH von, Noordwijk AJ van (1988) Dispersal, kinship and inbreeding in an island population of the great tit. J Evol Biol 1: 117–137
Tinbergen JM (1981) Foraging decisions in Starlings (*Sturnus vulgaris*) L. Ardea 68: 1–67
Tomnicki A (1988) Population ecology of individuals. Princeton University Press, Princeton, 224 pp
Townsend CR (1987) Bioenergetics: linking ecology, biochemistry, and evolutionary theory, TREE 2: 3–4
Treisman M (1983) Errors in the theory of the structure deme. J Theor Biol 102: 339–346
Tunner HG, Nopp H (1979) Heterosis in the common European water frog. Naturwissenschaffen 66: 268–269
Tuomi J, Hakala T, Haukioja E (1983) Alternative concepts of reproductive effort, costs of reproduction, and selection in life-history evolution. Am Zool 23: 25–34
Turkington R (1985) Variation and differentiation in populations of *Trifolium repens* in permanent pastures. In: White J (ed) Studies in plant demography, Academic Press, London, pp 69–81
Turner BJ, Brett BLH, Rasch EM, Balsano JS (1980) Evolutionary genetics of a gynogenetic fish, *Poecilia formosa*, the amazon molly. Evolution 34: 246–258
Turner ME, Stephens JC, Anderson WW (1982) Homozygosity and patch structure in plant populations as a result of nearest-neighbor pollination. Proc Natl Acad Sci USA 79: 203–207
Usberti JA, Jain SK (1978) Variation in *Panicum maximum:* a comparison of sexual and asexual populations Bot Gaz 139: 112–116
Uyenoyama MK (1979) Evolution of altruism under group selection in large and small populations in fluctuating environments. Theor Popul Biol 15: 58–85
Unenoyama MK (1984) Inbreeding and the evolution of altruism under kin selection: effects on relatedness and group structure. Evolution 38: 778–795
Uyenoyama MK (1986) Inbreeding and the cost of meiosis: The evolution of selfing in populations practicing biparental inbreeding. Evolution 40: 388–404
Uyenoyama MK, Feldman MW (1980) Theories of kin and group selection: a population genetics persepective. Theor Popul Biol 17: 380–414
Uyenoyama MK, Feldman MW (1981) On relatedness and adaptive topography in kin selection. Theor Popul Biol 19: 87–123
Uyenoyama MK, Feldman MW (1982) Population genetic theory of kin selection. II. The multiplicative model. Am Nat 120: 614–627

Uzzell TM, Berger L (1975) Electrophoretic phenotypes of *Rana lessonae, Rana ridibunda*, and their hybridogenetic associate *Rana esculenta*. Proc Acad Nat Sci Phila 127: 13–24
Uzzell TM, Darevsky IS (1975) Biochemical evidence for the hybrid origin of the parthenogenetic species of the *Lacerta saxicola* complex (Sauria: Laceridae), with a discussion of some ecological and evolutionary implications. Copeia 1975: 204–222
Valen L van (1973) A new evolutionary law. Evol Theory 1: 1–30
Valen L van (1976) Ecological species, multispecies, and oaks. Taxon 25: 233–239
Vandel A (1928) La parthenogenese geographique contribution a l'etude biologique et cytologique de la parthenogenese naturelle. Bull Biol Fr Belg 62: 164–281
Vanderberghe F, Dod B, Bellis M, Bonhomme F (1986) Absence of Y-chromosome introgression across the hybrid zone between *Mus musculus domesticus* and *M.m. musculus*. Genet Res 48: 191–197
Venable DL (1984) Using intraspecific variation to study the ecological significance and evolution of plant life-histories. In: Dirzo R, Sarukhán J (eds) Perspectives in plant population ecology. Sinauer, Sunderland, Massachusetts pp 166–187
Verhulst JH (1838) Notice sur la loi que population suit dans son accroissement. Corr Math Phys 10: 113–121
Vervey J (1930) Coral reef studies. I. The symbiosis between damsel fishes and sea anemones in Batavia Bay. Treubia 12: 305–366
Vetukhiv M (1956) Fecundity of hybrids between geographic populations of *Drosophila pseudoobscura*. Evolution 10: 139–146
Via S (1984) The quantitative genetics of polyphagy in an insect herbivore. II. Genetic correlations in larval performance within and across host plants. Evolution 38: 896–905
Via S (1987) Genetic constraints on the evolution of phenotypic plasticity. In: Loeschcke V (ed) Genetic constraints on adaptive evolution. Springer, Berlin Heidelberg New York, pp. 47–71
Via S, Lande R (1985) Genotype-environment interaction and the evolution of phenotypic plasticity. Evolution 39: 505–522
Vincent TL, Pulliam HR (1980) Evolution of life history strategies for an asexual annual plant model. Theor Popul Bio 17: 215–231
Vogel F, Kopun M (1977) Higher frequencies of transitions among point mutations. J Mol Evol 9: 159–180
Volterra V (1927) Variazioni e fluttuazioni del numero d'individui in specie animali coviventi. R. Comitato Talassografico Italiano, Memoria 131: 1–142 (English translation in FM Scudo and JR Ziegler: the golden age of theoretical ecology: 1923–1940. Lecture Notes in Biomathematics 22, Springer, Berlin Heidelberg New York, 1979).
Vrijenhoek RC (1978) Coexistence of clones in a heterogeneous environment. Science 199: 549–552
Vrijenhoek RC (1979) Factors affecting clonal diversity and coexistence. Am Zool 19: 787–797
Vrijenhoek RC (1984a) The evolution of clonal diversity in *Poeciliopsis*. In: Turner BJ (ed) Evolutionary genetics of fishes. Plenum, New York, pp 399–429
Vrijenhoek RC (1984b) Ecological differentiation among clones: the frozen niche variation model. In: Wöhrmann K, Loeschcke V (eds) Population biology and evolution. Springer, Berlin Heidelberg New York, pp 217–231
Vrijenhoek RC (1985) Animal population genetics and disturbance: the effects of local extinctions and recolonizations on heterozygosity and fitness. In: Pickett STA, While P (eds) The ecology of natural disturbance and patch dynamics. Academic Press, London, pp 265–285
Vrijenhoek RC (1989) Genetic and ecological constraints on the origins and establishment of unisexual vertebrates. In: Dawley R, Bogart J (eds) Evolution and ecology of unisexual vertebrates. State University New York Press, Albany
Vrijenhoek RC, Marteinsdottir G, Schenck RE (1987) Genotypic and phenotypic aspects of niche diversification in fishes In: Mathews W, Heins D (eds) Community and evolutionary ecology of North American stream fishes. University Oklahoma Press, Norman, Oklahoma, pp 245–250
Waage JK, Godfray CHJ (1985) Reproductive strategies and population ecology of insect

parasitoids. In: Sibly R, Smith RH (eds) Behavioural ecology: the ecological consequences of adaptive behaviour. Blackwells, Oxford
Waddington CH (1957) The strategies of genes. Allen & Unwin, London
Waddington CH (1960) Experiments on canalizing selection. Genet Res 1: 140–150
Wade MJ (1978) A critical review of the models of group selection. Q Rev Biol 53: 101–114
Wade MJ (1980) Kin selection: its components. Science 210: 665–667
Wade MJ (1982) Group selection: migration and the differentiation of small populations. Evolution 36: 949–961
Wade MJ (1985) Soft selection, hard selection, kin selection, and group selection. Am Nat 125: 61–73
Wade MJ, McCauley DE (1980) Group selection: the phenotypic and genotypic differentiation of small populations. Evolution 34: 799–812
Walbot V (1985) On the life strategies of plants and animals. Trends in Genetics 25: 165–169
Walker J, Stamps J (1986) A test of optimal caste ratio theory using the ant *Camponotus (Colobopsis) Impressus*. Ecology 67: 1052–1062
Wallace B (1975) Hard and soft selection revisited. Evolution 29: 465–473
Waller DM (1980) Environmental determinants of outcrossing in *Impatiens capensis* (Balsaminaceae). Evolution 34: 747–761
Waller DM (1986) Is there disruptive selection for self-fertilization? Am Nat 128: 421–426
Waller DM, Green D (1981) Implications of sex for the analysis of life histories. Am Nat 117: 810–813
Walsh JB (1982) Rates of accumulation of reproductive isolation by chromosomes rearrangements. Am Nat 120: 510–532
Walsh JB (1984) Hard lessons for soft selection. Am. Nat. 124: 518–526
Watt MB (1986) Power and efficiency as indexes of fitness in metabolic organization. Am Nat 127: 629–653
Watterson GA (1974) The sampling theory of selectively neutral alleles. Adv Appl Probab 6: 463–488
Watterson GA (1978) The homozygosity test of neutrality. Genetics 88: 405–417
Watterson GA (1984a) Lines of descent and the coalescent. Theor Popul Biol 26: 77–92
Watterson GA (1984b) Allele frequencies after a bottleneck. Theor Popul Biol 26: 387–407
Weir BS, Cockerham CC (1973) Mixed self and random mating at two loci. Genet Res 21: 247–262
Weir BS, Cockerham CC (1984) Estimating F statistics for the analysis of population structure. Evolution 38: 1358–1370
Weir BS, Allard RW, Kahler AL (1972) Analysis of complex allozyme polymorphisms in a barley population. Genetics 72: 505–523
Wells H (1979) Self-fertilization: advantageous or deleterious. Evolution 33: 252–255
Wetherington JD, Kotora KE, Vrijenhoek RC (1987) A test of the spontaneous heterosis hypothesis for unisexual vertebrates. Evolution 41: 721–731
Wetherington JD, Weeks SC, Kotora KE, Vrijenhoek RC (1989) Genotypic and environmental components of variation in growth and reproduction of fish hemiclones (*Poeciliopsis*.: Poeciliidae). Evolution 43. (in press)
White MJD (1973) Animal cytology and evolution, 3rd edn. Cambridge University Press, Cambridge
White MJD (1978) Modes of speciation. Freeman, San Francisco
Whittam TS, Ochman H, Selander RK (1983a) Geographic components of linkage disequilibrium in natural populations of *Escherichia coli*. Mol Biol Evol 1: 67–83
Whittam TS, Ochman H, Selander RK (1983b) Multilocus genetic structure in natural populations of *Escherichia coli*. Proc Natl Acad Sci USA 80: 1751–1755
Whittam TS, Clark AG, Stoneking M, Cann RL, Wilson AC (1986) Allelic variation in human mitochondrial genes based on patterns of restriction site polymorphism. Proc. Natl. Acad Sci USA 83: 9611–9615
Wijsman E, Cavalli-Sforza LL (1984) Migration and genetic population structure with special reference to man. Annu. Rev. Ecol. Syst 15: 279–302
Williams BK, Nichols JD (1984) Optimal timing in biological processes. Am Nat 123: 1–19

Williams GC (1957) Pleiotropy, natural selection, and the evolution of senescence. Evolution 11: 398–411

Williams GC (1966) Adaptation and natural selection. Princeton University Press, Princeton

Williams GC (1975) Sex and evolution. Princeton University Press, Princeton

Williams GC (1985) A defense of reductionism in evolutionary biology. In: Dawkins R, Ridley M (eds) Oxford Survey in Evolutionary Biology, vol 2. Oxford University Press, Oxford

Williams GC, Mitton JB (1973) Why reproduce sexually? J. Theor Biol 39: 545–554

Williams KS, Gilbert LE (1981) Insects as selective agents on plant vegetative morphology: egg mimicry reduces egg laying by butterflies. Science 212: 467–469

Wills CJ (1978) Rank-order selection is capable of maintaining genetic polymorphisms. Genetics 89: 403–417

Willson MF, Burley N (1984) Mate choice in plants: Tactics, mechanisms, and consequences. Princeton University Press, Princeton

Wilson AC, Carlson SS, White TJ (1977) Biochemical evolution. Annu Rev Biochem 46: 573–639

Wilson AC, Maxson LR, Sarich VM (1974) Two types of molecular evolution. Evidence from studies of interspecific hybridization. Proc Natl Acad Sci USA 71: 2843–2827

Wilson DS (1975) A theory of group selection. Proc Natl Acad Sci USA 72: 143–146

Wilson DS (1977) Structured demes and the evolution of group-advantageous traits. Am Nat 11: 157–185

Wilson DS (1982) The natural selection of populations and communities. Series in evolutionary biology, Institute of Ecology. Davis, California

Wilson DS (1983) The group selection controversy: history and current status. Annu Rev Ecol Syst 14: 159–188

Wilson DS (1988) The diversification of single gene pools by density- and frequency-dependent selection. In: Otte D (ed) Speciation and adaptation. Philadelphia Academy of Sciences, Philadelphia

Wilson DS, Turelli M (1986) Stable underdominance and the evolutionary invasion of empty niches. Am Nat 127: 835–850

Wilson EO (1975) Sociobiology: the new synthesis. Harvard University Press, Cambridge, Massachusetts

Wöhrmann K, Tomiuk J (1988) Life cycle strategies and genotypic variability in populations of aphids. J Genet 67: 43–52

Wolf TJ, Schmid-Hempel P, Ellington CP, Stevenson RD (1989) Physiological correlates of foraging efforts in honeybees: oxygen consumption and nectar load. Funct Ecol 3: (in press)

Wolin CL (1985) The population dynamics of mutualistic systems. In: Boucher DH (ed) The biology of mutualism. Croom Helm, London, pp 248–269

Woltereck R (1909) Weitere experimentelle Untersuchungen über Artveränderung, speziell über das Wesen quantitativer Artenunterschiede bei Daphniden. Verh Dtsch Zool Ges 1909: 110–172

Workman PL (1969) The analysis of simple genetic polymorphism. Hum Biol 40: 260–279

Wright JW, Lowe CH (1968) Weeds, polyploids, parthenogenesis and the geographical and ecological distribution of all-female species of *Cnemidophorus*. Copeia 1968: 128–138

Wright S (1929) Fisher's theory of dominance. Am Nat 63: 274–279

Wright S (1931) Evolution in Mendelian populations. Genetics 16: 97–159

Wright S (1932) The roles of mutation, inbreeding, crossbreeding and selection in evolution. Proc Sixth Int Congr Genet 1: 356–366

Wright S (1935) Evolution in populations in approximate equilibrium. J Genet 30: 257–266

Wright S (1937) The distribution of gene frequencies in populations. Proc Natl Acad Sci USA 23: 307–320

Wright S (1941) On the probability of fixation of reciprocal translocations. Am Nat 74: 513–522

Wright S (1943) Isolation by distance. Genetics 28: 114–138

Wright S (1945a) The differential equation of the distribution of gene frequencies. Proc Natl Acad Sci USA 31: 382–389

Wright S (1945b) Tempo and mode in evolution: a critical review. Ecology 26: 415–419

Wright S (1951) The genetical structure of populations. Ann Eugen 15: 323–354

Wright S (1969) Evolution and genetics of populations vol II. University of Chicago Press, Chicago
Wright S (1978) Evolution and the genetics of populations. IV. Variability within and among populations. University of Chicago Press, Chicago
Wright S (1980) Genic and organismic selection. Evolution 34: 825–843
Wright S (1982a) The shifting balance theory and macroevolution. Annu Rev Genet 16: 001–19
Wright S (1982b) Character change, speciation and the higher taxa. Evolution 36(3): 427–443
Wright S, Dobzhansky T (1943) Genetics of natural populations. X. Dispersion rates in *Drosophila pseudoobscura*. Genetics 28: 304–340
Wright S, Dobzhansky T (1947) Genetics of natural populations. XV. Rate of diffusion of a mutant gene through a population of *Drosophila pseudoobscura*. Genetics 32: 303–324
Wright S, McPhee HC (1925) Mendelian analysis of the pure breeds of livestock. III. The short-horns. J Hered 16: 205–215
Wright S, Dobzhansky T, Hovanitz W (1942) Genetics of natural populations. VI. The allelism of lethals in the third chromosome of *Drosophila pseudoobscura*. Genetics 27: 363–394
Wyatt R (1988) Phylogenetic aspects of the evolution of self pollination. In: Gottlieb LD, Jain SK (eds) Plant evolutionary biology. Chapman & Hall, London, pp 109–132
Wu KK, Jain SK (1978) Genetic and plastic responses of geographic differentiation of *Bromus rubens* populations. Canad J Bot 56: 873–879
Wynne-Edwards VC (1962) Animal dispersion in relation to social behaviour. Oliver & Boyd, Edinburgh
Yang SY, Wheeler LL, Bock IR (1972) Isozyme variations and phylogenetic relationships in the *Drosophila bipectinata* species complex. In: Wheeler MR (ed) Studies in genetics. University Texas Publ 7213, pp 213–27
Yodzis P (1981) Concerning the sense in which maximizing fitness is equivalent to maximizing reproductive value. Ecology 62: 1681–1682
Yokoyama S, Schaal B (1983) A note on multiple-niche polymorphisms in plant populations. Am Nat 125: 158–163
Yoo BH (1980) Long term selection for a quantitative character in large replicate populations of *Drosophila melanogaster*. Ganet Res Camb 35: 1–17, 19–31
Young JPW (1979) Enzyme polymorphism and cyclical parthenogenesis in *Daphnia magna*. II. Heterosis following sexual reproduction. Genetics 92: 971–982
Young JPW (1981) Sib competition can favor sex in two ways. J Theor Biol 88: 755–756
Ziegler JR (1976) Evolution of the migration response: emigration by *Tribolium* and the influence of age. Evolution 30: 579–592
Zimmer EA, Martin SL, Beverly SM, Kan YW, Wilson AC (1980) Rapid duplication and loss of genes for the alpha chains of hemoglobin. Proc Natl Acad Sci USA 77: 1058–2162
Zouros E (1973) Genic differentiation associated with the early stages of speciation in the *mulleri* subgroup of *Drosophila*. Evolution 27: 601–21
Zouros E (1981) The chromosomal basis of sexual isolation in two sibling species of *Drosophila*: *D. arizonensis* and *D. majavensis*. Genetics 97: 703–18

Subject Index

Q

R

S

T

U

V

W

Z